NETHERLANDS YEARBOOK FOR HISTORY OF ART

NEDERLANDS KUNSTHISTORISCH JAARBOEK

NETHERLANDS YEARBOOK FOR HISTORY OF ART 2011 VOLUME 61

NEDERLANDS KUNSTHISTORISCH JAARBOEK 2011 DEEL 61

ART AND SCIENCE
IN THE EARLY MODERN NETHERLANDS

KUNST EN WETENSCHAP
IN DE VROEGMODERNE NEDERLANDEN

Editors / *Redactie:*
Eric Jorink
Bart Ramakers

W BOOKS

Founded in 1947, the *Netherlands Yearbook for History of Art* / Nederlands Kunsthistorisch Jaarboek, or *NKJ*, is a peer-reviewed journal, which has established an international reputation for publishing outstanding articles that reflect the variety and diversity of approaches to the study of Netherlandish art and culture. The *NKJ* aims to foster traditional art historical scholarship and to open up the field to innovative cross-disciplinary developments. The *NKJ* is ranked in the European Reference Index for the Humanities (ERIH) as an International 1 (INT1) journal and is listed in the Arts & Humanities Citation Index (Thomson Reuters).

Publication of / *Uitgave van*:
The Foundation for Dutch Art Historical Publications /
Stichting Nederlandse Kunsthistorische Publicaties

Editorial office / *Redactiesecretariaat*:
WBOOKS
Carijn Oomkens
P.O. Box / *Postbus* 1129, 8001 BC Zwolle, The Netherlands

Volume / *Deel* 62 will be devoted to / *zal gewijd zijn aan*:
Materials and materiality in Netherlandish art

Editors: H. Perry Chapman, Ann Sophie Lehmann, Frits Scholten

Published by / *Uitgave*: WBOOKS
Design / *Ontwerp*: Annelies Mikmak
Copy editor / *corrector*: Ruth Gaskill
Printed by / *Druk*: Epospress

ISBN 978 90 400 7808 8
NUR 654

Contents / *Inhoud*

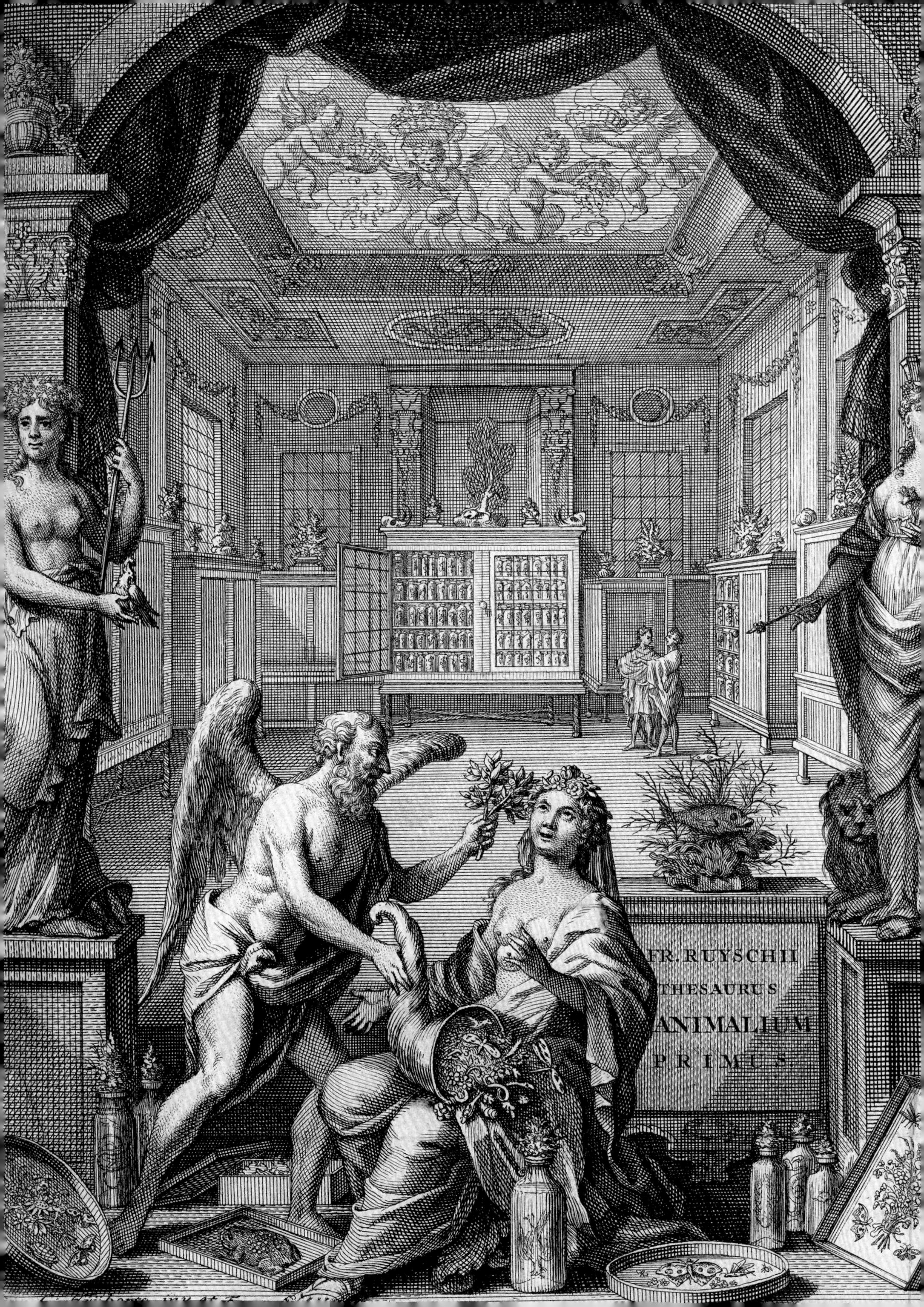
FR. RUYSCHII
THESAURUS
ANIMALIUM
PRIMUS.

Undivided territory
'Art' and 'science' in the early modern Netherlands

Eric Jorink and Bart Ramakers

On 2 February 1677, the famous Amsterdam professor of anatomy Frederik Ruysch (1638-1731) executed a complicated operation during which the patient, Lysbeth Jans, nearly died. Subsequently, Lysbeth was brought to the municipal hospital, where the chief physician, Bonaventura Dortmont, observed her wounds. Dortmont, a man of disputed reputation, accused Ruysch of medical malpractice. The result was a war of pamphlets, in which virtually the entire Amsterdam medical world became involved. Ruysch was by that time already famous for his collection of anatomical preparations, in which were displayed artful compositions of human bodies (or parts of them) and all kinds of *naturalia* and *artificialia* (figs. 1, 2).[1] In an anonymous pamphlet, written by a group of hostile colleagues, Ruysch's skill and integrity were called into question. The accusations were without doubt the result of professional jealousy, but they also touched upon important epistemological questions. A physician, his opponents insisted, should read the works of Hippocrates, Galen and all other classical authors. Instead, Ruysch, 'a person of little study' ('een Persoon van weynig studie'), used his *hands* 'in order to dissect parts of dead animals as well as human bodies'.[2] Even worse was the fact that the professor devoted much of his time to the art of painting ('syn overige tijd noch besteed aen de Schilderkonst').[3] Obviously, a *medicinae doctor* should just use his mind, and not dirty his hands.

However, in a pamphlet written in defence of Ruysch, another anonymous writer made a twofold claim. Referring to Karel van Mander (1548-1606), the writer maintained that the art of painting was a noble occupation for kings and princes.[4] Moreover, 'I say that neither an *Anatomicus* without the skill of painting, nor a painter without [the knowledge of] *Anatomy*, will be able to climb the ladder of perfection'.[5] The painter must extensively study the outward appearance, form, structure and functioning of the human body as a whole, as well as of its respective parts. And, vice versa, the anatomist should be trained in the art of drawing and painting:

> Which *Anatomist* will be able to approve of, or improve the engraved images of this or that part of the human body (...) if he has never practised the art of drawing and painting (...) And which dissector will

1
Frontispieces, slightly idealised, depicting Frederik Ruysch's cabinet at the Amsterdam Bloemgracht, from F. Ruysch, *Thesaurus anatomicus primus* (1701)
The Hague, Koninklijke Bibliotheek, 556 D 30:2.

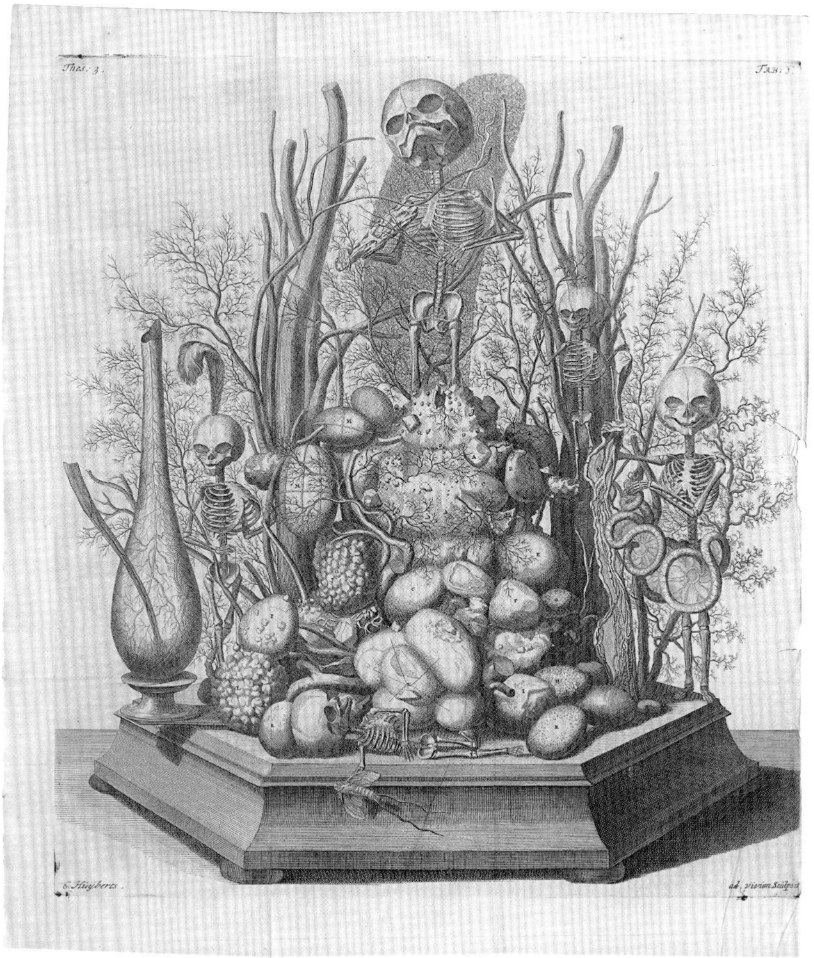

2

Engraving, from F. Ruysch, *Thesaurus anatomicus tertius* (1703)
The Hague, Koninklijke Bibliotheek, 556 D 30:4.

> be able to have engraved into copper or silver the newly found ducts, (...) muscles and other [concealed organs] if he does not master the art of arithmetic and painting.[6]

Thus, the skill of drawing was an essential condition for anatomical research. The quotation even suggests an analogy between cutting the body and cutting the print. If the anatomist himself was not able to give a visual representation of his observations and discoveries, he had to rely on other skilled artists. If a representation was not lifelike, the result would be as ridiculous as 'a painter who has to add a legend to his painting, A is a dog, B is an hare'.[7] Ruysch, the pamphleteer maintained, was to be praised for his ability to draw and paint. These were noble skills in themselves, but also the necessary preconditions for the study of nature.

The point made by Ruysch's anonymous defendant on the interdependence of the artist and the anatomist, on drawing and knowing, on the hand and the mind, is a commonplace in contemporary treatises on art theory. For example, a case like Ruysch's is described almost verbatim in Willem Goeree's *Inleyding tot de practijck der algemeene schilderkonst*, of which the first edition was published seven years prior to this dispute.[8]

The case of the artist and the anatomist is just one of the many examples of the fluid boundaries between 'art' and 'science' in the early modern era, which is the overall theme of this volume of the *Netherlands Yearbook for History of Art.* Before introducing the respective contributions to this volume and pointing at some potential new themes for research, we will first give a short historiographical account of the history of art and science. Since most readers of the *NKJ* will be more familiar with the former than the latter, we put slightly more emphasis on recent developments within the history of science.

For a long time, art and science were considered as two distinct expressions of human culture. Yet, the concepts are to a certain extent constructs, and the distinction between the two is much the result of the rise of the modern academic disciplines in the nineteenth and twentieth centuries. Highly important is the fact that in the early modern era the contemporary Latin terms *ars* and *scientia* were complementary rather than opposites, covering a spectrum of meanings such as skill, experience and knowledge. We find this semantic fact reflected in many ways. For example, just a quick glimpse at treatises on art theory published in the Netherlands confirms the fluid boundaries between *konst* and *wetenschap*. The meanings of the words are often interchangeable. Of equal importance is that in treatises like Goeree's *Inleyding*, artists are constantly encouraged to study nature, both with their own eyes and hands as well as guided by books and experts. In an interesting passage, Goeree refers to a contemporary Amsterdam collection of antiquities and *naturalia*.[9] Moreover, Goeree repeatedly underlines the importance of *wetenschap* (here understood in the broad sense as both knowledge and science) for the painter. As Van de Roemer shows in his contribution on Goeree in this volume, the painter should not only practise the arts of drawing, painting, perspective, astronomy, architecture and colour theory; he should also know the Bible by heart, know classical history and poetry, the *artes liberales* and contemporary developments in the natural sciences.[10] The real artist ('een groot en algemeen meester') should have all the necessary knowledge and science ('alle de noodige kennissen en wetenschappen').[11]

The examples of Ruysch and Goeree make it abundantly clear how *konst* and *wetenschap* were interwoven in the early modern Netherlands. However, as a result of the emergence, establishment and demarcation of the disciplines of the history of art and the history of science, the fluid boundaries between *ars* and *scientia* have often not been recognised. Art historians have been putting much of their attention to the style and iconography of artworks. For a long time, their interest in science was limited to a few themes, such as the discovery and use of linear perspective in painting, the *camera obscura* and the 'naturalistic' images in the herbals by Otto Brunfels (*Herbarum vivae eicones*, 1530) and Leonhart Fuchs (*De historia stirpium*, 1552).

Much has changed over the last decades. Art historians have increasingly paid attention to the historical, intellectual, social and material contexts in which works of art were produced and viewed.[12] They have become very interested in the way works of art were produced and marketed, which also involved a great deal of knowledge and practical skills. Different authors have pointed out that artists must have had considerable learning or actually moved in learned circles. Scholars trained as art historians, such as Martin Kemp, David Freedberg and Horst Bredekamp, have become much interested in the modes of representation and have started highly important and ongoing research on the role of images in the works of, for example, Leonardo da Vinci (1452-1519), Galileo Galilei (1564-1642) and Maria Sibylle Merian (1647-1717).[13] Thomas DaCosta Kaufmann extensively discussed the work and strategies of visual representation of Joris Hoefnagel (1541-1601) and some of his contemporaries in their intellectual context.[14] This turn towards scientific culture was prompted by the interest in early modern collections of *artificialia* and *naturalia*. A conference held in Oxford in 1983, 'The origins of museums', led to both an epoch-making book edited by the art historians Oliver Impey and Arthur MacGregor and a focus of research on the history of collecting that still

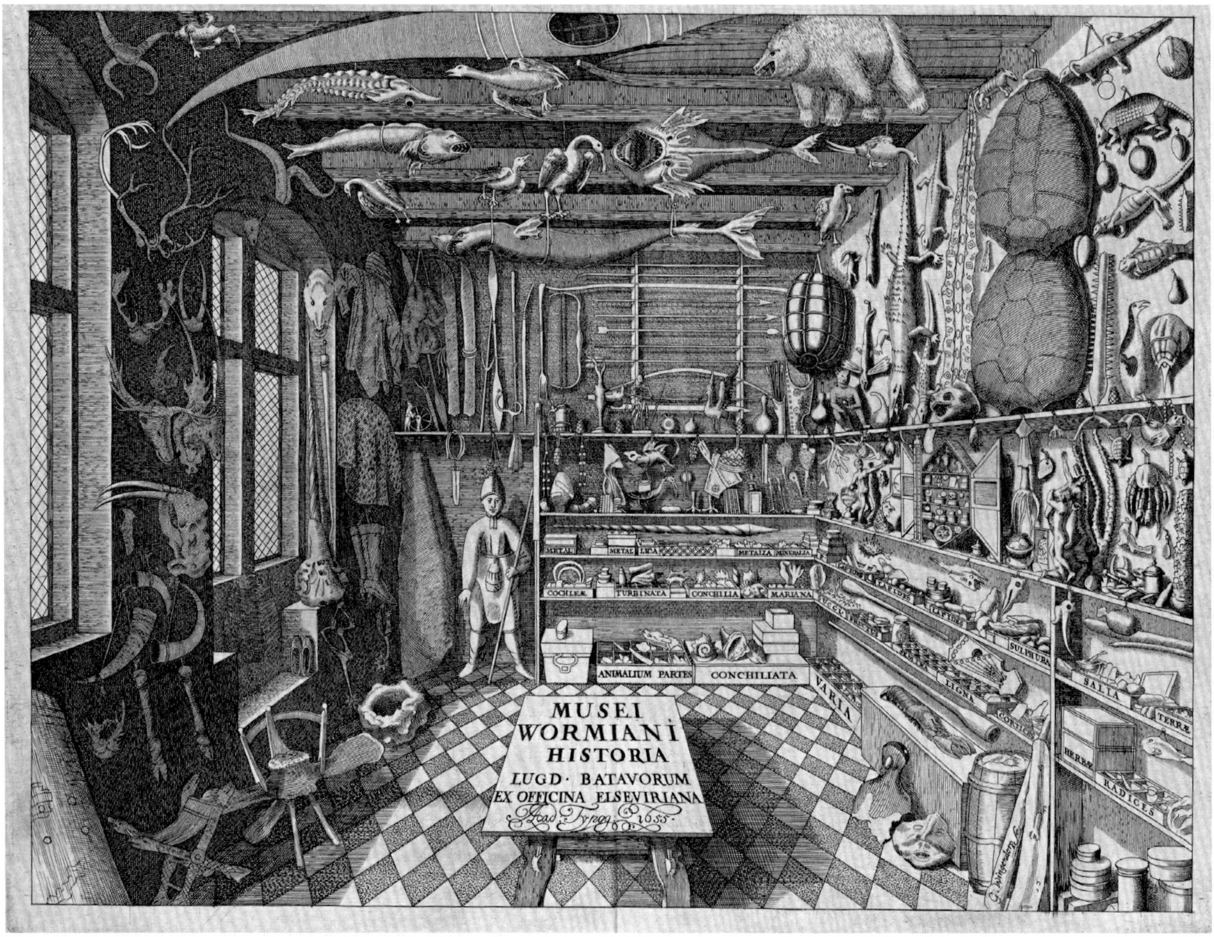

3
Engraving, from O. Worm, *Museum Wormianum* (1655)
The Hague, Koninklijke Bibliotheek, 40 E 10.

has not reached its peak.[15] It has become clear that the early modern cabinets of curiosities were paradigmatic for what had aptly been called 'the age of the marvelous' (figs. 3, 4).[16] The collections show an interplay between words and things, nature and art, art and technique. We can only hope to understand their physical appearance, epistemological background, metaphysical connotations and possible meanings if we challenge disciplinary boundaries.

To a certain extent, the historiography of the history of science shows the same pattern as that of history of art. Until quite recently, the study of the history of science was very much an internal affair – focusing on the rise of astronomy, mathematics and physics – and basically interested in only one question: who was first?[17] However, historians of science have increasingly focused on broader issues. Here, the turn towards cultural history has led to the study of botany, alchemy and astrology in the early modern period. Historians of science are no longer exclusively focused on formulas; they take a strong interest in unpublished manuscripts, letters and images, as well as the social context of scientific culture and the rhetorical strategies used by its

4
Frans Francken II, *Art cabinet*, c. 1650
(160 cm x 108 cm x 49.5 cm)
Amsterdam, Rijksmuseum, BK-NM-4789.

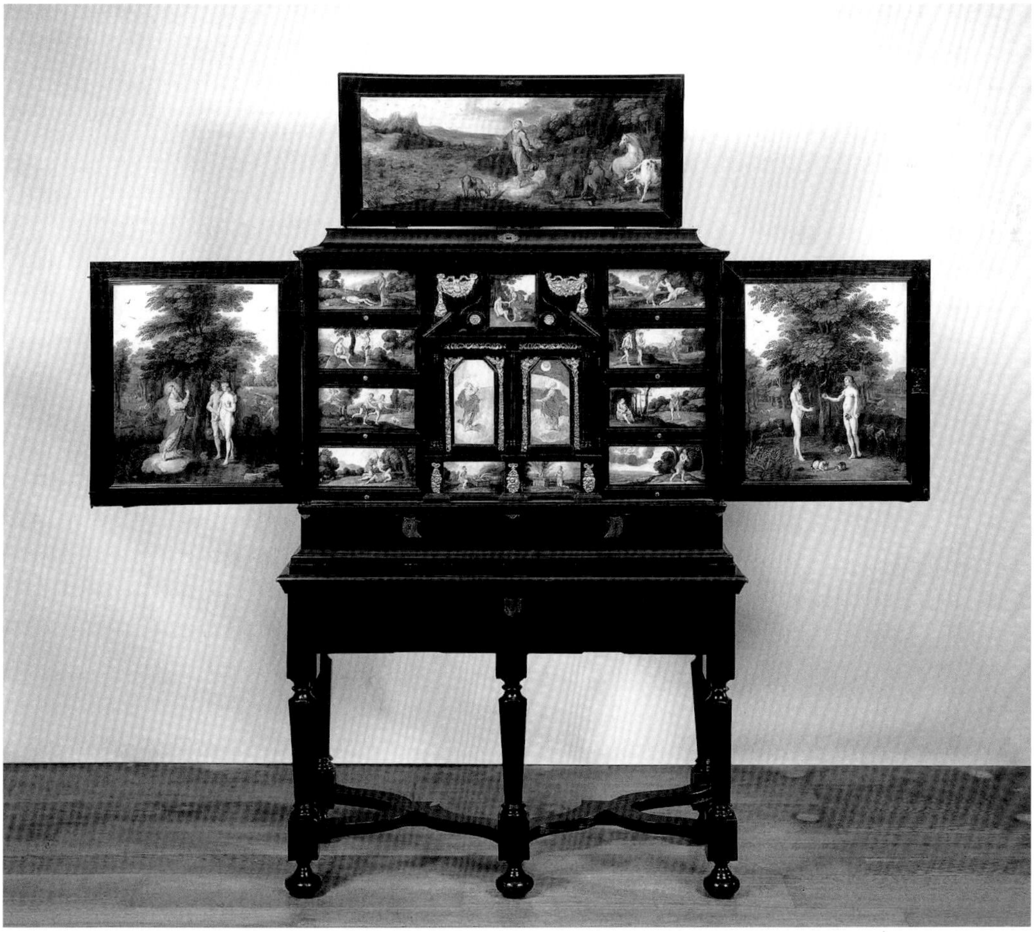

practitioners.[18] It has become clear that science is not a timeless, fixed entity, but an umbrella term for a contingent range of theories and practices. It refers to a broad spectre of activities, stretching from abstract geometry in the tradition of Euclid to the practical skills of surgeons, and from the humanistic study of Dioscorides to the art of engineering.[19] The early modern Netherlands played an extremely important role in this process. Seen from an early modern point of view, science could include not only natural history, natural philosophy, medicine, astronomy and mathematics but also astrology, alchemy, teratology as well as millenarian speculations. Recently, much attention has been devoted to botanical gardens, anatomical theatres, instruments, book illustrations and other themes that can be qualified as the material culture of science.[20]

This shift in focus is also reflected in the institutional basis of the history of science, which is gradually moving from faculties of science to the humanities. Part of this broader perspective is an increasing interest in the history of art and what has been called visual culture. Here are just a few examples: Wolfgang Harms and William Ashworth Jr. have focused on the relation between emblematics and early modern natural history, convincingly demonstrating that the latter was not a neutral descriptive activity but an essentially humanistic form of *scientia*, in which the textual references, symbolic value and hidden qualities of *naturalia* were as important as their outward appearance.[21] Others, such as Paula Findlen, have underlined the great epistemological importance of the cabinets of curiosities.[22]

Besides this broader range of subjects, historians of science have become very sensitive to the question of what makes 'science'. This is not clear at all. The process between the publication of a theory or fact and the acceptance of it is far from self-evident and is much guided by rhetoric, hidden agendas or sheer coincidence. In the wake of James Secord's article 'Knowledge in transit' (2004), a lot of research has been initiated to the theme of the circulation of *knowledge* (not essentially science): the ways in which it was made, diffused, received and adapted.[23] Pamela H. Smith has brought to our attention the importance of the knowledge and skills of artisans such as goldsmiths, potters and painters, arguing that the so-called Scientific Revolution was much less a theoretical affair than we tend to think.[24] Others have made similar claims with regard to what Smith has called 'artisanal knowledge', underlining the importance of knowledge of the classics by Renaissance engineers, of the properties of materials by artists, or the skill of drawing or model making by scientists. This inclusion of attention to objects and materials undoes the traditional dichotomies between the scholar and the craftsman, between knowledge and handicraft, between the mind and the hand.[25] Debora Harkness has brilliantly described this fruitful interaction between artisans and natural philosophers on a microlevel in *The jewel house. Elizabethan London and the Scientific Revolution*.[26]

A similar account could be written on Antwerp, Leiden or Amsterdam. As is well known, both the arts and the sciences were flourishing in the Netherlands in the early modern period. Besides the Brueghel family and Rubens, Rembrandt and Vermeer, the Netherlands also fostered men like Andreas Vesalius (1514-1564), Gerardus Mercator (1512-1594), Lambertus Dodonaeus (?-1607), Carolus Clusius (1526-1609), Simon Stevin (1548-1620), Christiaan Huygens (1629-1695), Johannes Swammerdam (1637-1680), Antoni van Leeuwenhoek (1632-1723) and a host of lesser-known figures who all contributed to what is usually called the Scientific Revolution (fig. 5).[27] In a recent volume, Sven Dupré and Christoph Lüthy write:

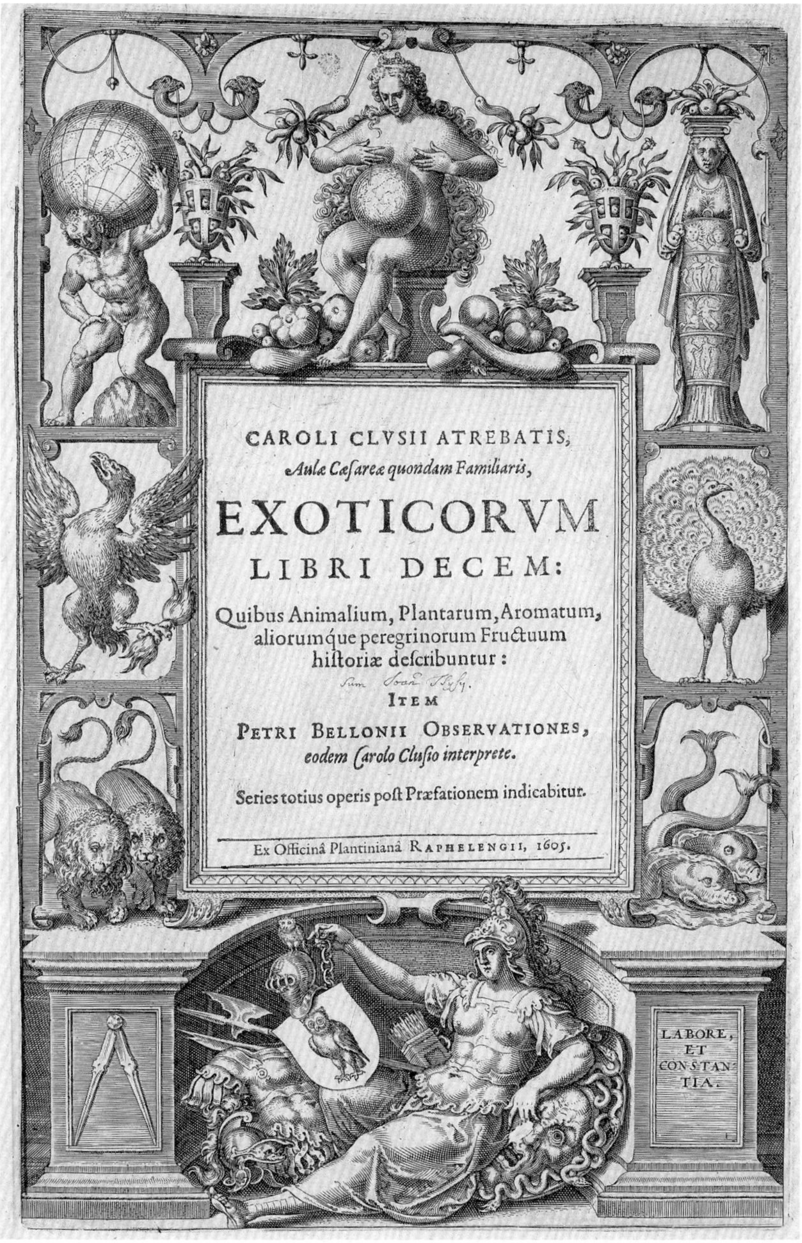

5

Frontispieces of C. Clusius, *Exoticorum libri decem* (1605)

Universiteitsbibliotheek Leiden, Collectie Bibliotheca Thysiana, THYSIA 2202.

6

The compound microscope came into use around 1620; this decorated specimen was manufactured by Lucas Cramer, c. 1740 (see also fig. 19, page 175).
Leiden, Museum Boerhaave, inv. no. V07083.

> (...) the 'circulation of knowledge' was perhaps nowhere more intense as in the early modern Low Countries, and this had to do as much with the circulation of scholars which was, in the *Carrefour de la République des Lettres*, particularly lively, as with the extraordinary nodal points that cities like Antwerp and Amsterdam represented in the international exchange of goods, news, and skill.[28]

Within less than two centuries, the image of a man-centred, finite cosmos, consisting of the four Aristotelian elements and ruled by hidden properties and divine power, was replaced by the concept of a heliocentric, infinite universe, obeying the laws of nature. During the sixteenth century, all kinds of previously unknown creatures, objects and materials were brought in from the East and West Indies. The invention of the telescope in 1608 and the introduction of the compound microscope around 1620 opened the way to intriguing new discoveries, such as the moons around Jupiter and the rings of Saturn, the fascinating anatomy of insects and the previously unknown microorganisms (fig. 6). The Netherlands played a crucial role in this process, in terms of both single observations of plants and animals and in the development of a new conceptual framework. René Descartes (1596-1650), the founder of rationalism and the mechanistic worldview, spent most of his life in the Dutch Republic, and it was there that his revolutionary New Philosophy was first published and debated. As it will be shown in this volume, Descartes's work had also a profound influence on reflections on *ars* and *scientia*.

The importance of the Frenchman on Dutch culture can hardly be overestimated. Some points are of relevance here: first, Descartes rejected bookish knowledge – that is, the writings of the ancients and their humanistic followers in philosophical and scientific matters. Also, the senses could be deceitful; only the human *ratio* could provide clear and distinct certainties. Moreover, in contrast to scholastic Aristotelianism, in which the explanatory factors were the different elements and qualities and their diverse aims and potentials, Descartes regarded the universe as a vast mechanism in which the only factors that truly counted were matter and motion. Planets, clouds, people and animals consisted purely of moving particles. Stones fell to the earth because particles pressed on them. Light consisted of emitted subtle matter moving into the eyes of the viewer. As Dijksterhuis notes in this volume, Descartes discarded the classical distinction between real and apparent colours. Besides developing a mathematical description of the rainbow to calculate the angles at which the colours are seen, he also gave an account of how light is modified by the successive refractions in a raindrop to create the various colours of the rainbow. Colour is the result of the interaction between light and matter, and not an intrinsic quality of any given material used by artists (lapis lazuli, sulphur, madder). The impact of Descartes's theory of matter on the meaning ascribed to materials is a subject deserving more research.

Be this as it may, the mechanistic model was instrumental in the acceptance of challenging new scientific theories in the Dutch Republic. It was in the context of Cartesianism that Nicolaus Copernicus's theory of

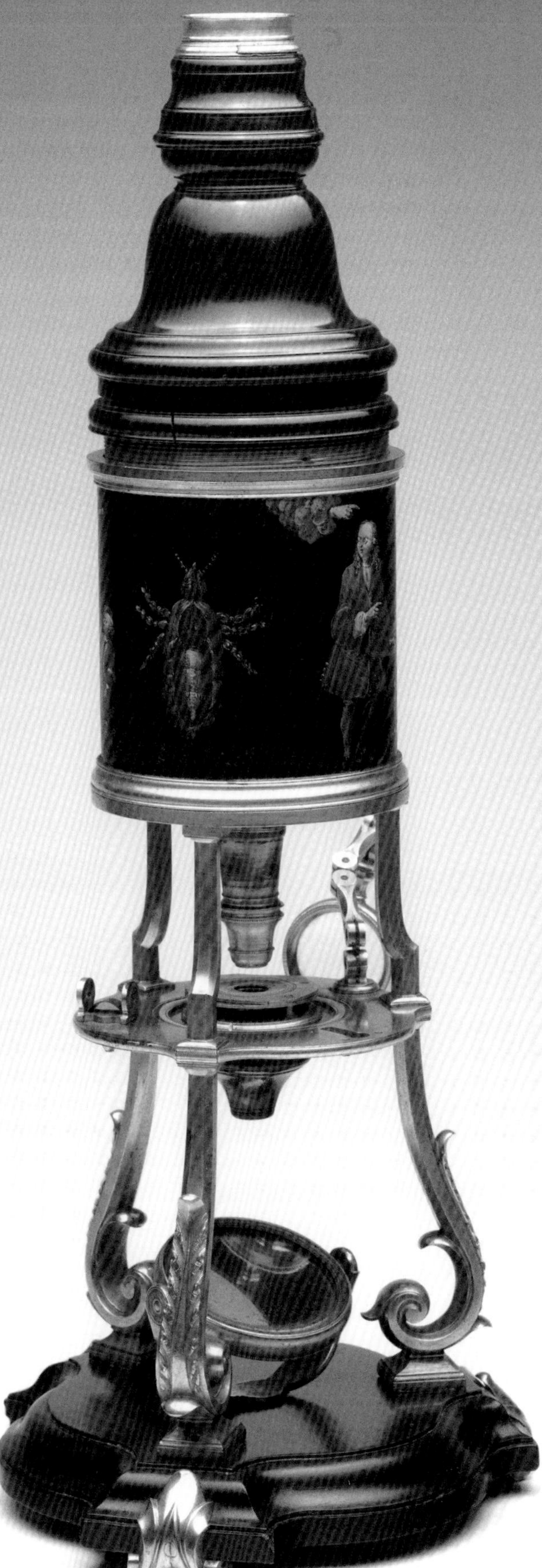

the twofold motion of the earth became accepted (*De revolutionibus orbium coelestium*, 1543), as well as William Harvey's theory of the circulation of the blood (*De motu cordis*, 1628). The idea that animals were just ingenious coalescences of material, and their bodies hydraulic machines, provided an agenda as well as an excuse for very elaborate and artful ('seer konstige') experiments on living dogs and frogs, as Swammerdam, for example, would perform (fig. 7). From Descartes's point of view nature was orderly and uniform, governed by fixed laws.[29] Chance did not exist: nature was stable, not contingent. There was no ontological distinction between the spheres under and above the moon,

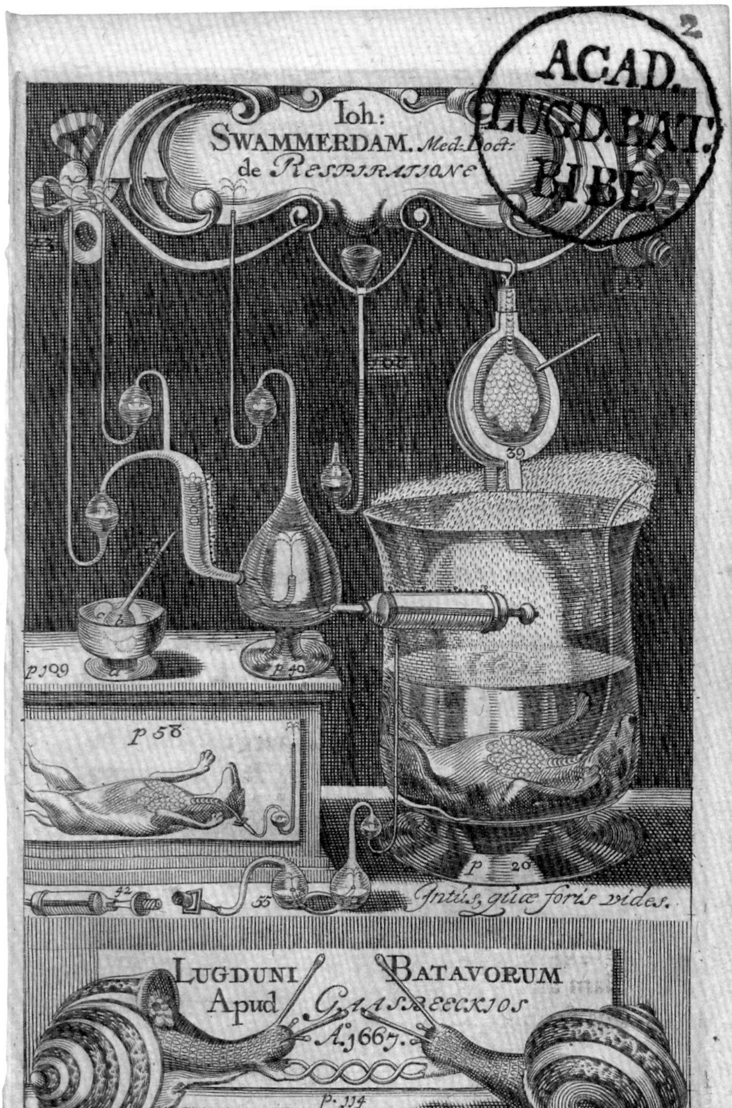

7
Frontispieces, showing many 'artful' experiments on respiration, from J. Swammerdam, *Tractatus de respiratione* (1667)
Universiteitsbibliotheek Leiden, 615 F 15: 2.

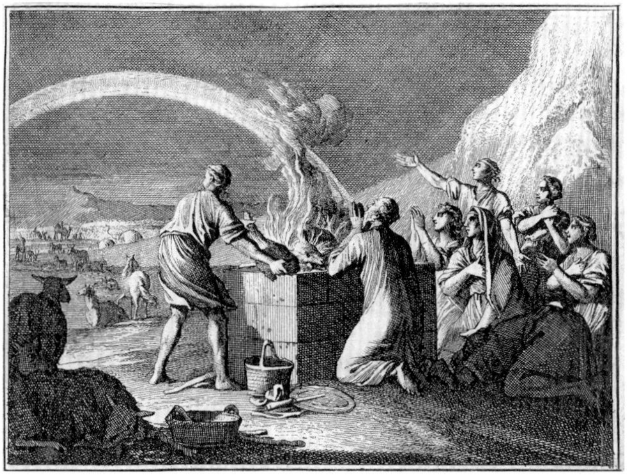

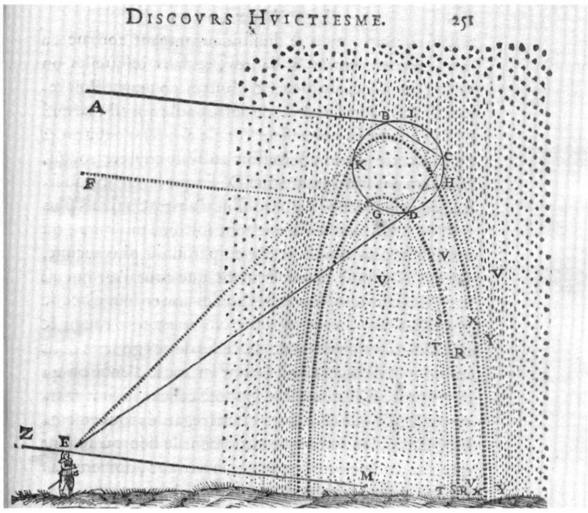

8

Engraving, illustrating Genesis 9:13, from J. Luyken, *Schriftuurlyke geschiedenissen* (1712)
The Hague, Koninklijke Bibliotheek, 1754 C 12.

9

Rainbow, as depicted in R. Descartes's *Dioptrique* (1637). The appearance is stripped of its Biblical connotations and reduced to a purely physical phenomenon The Hague, Koninklijke Bibliotheek.

nor between big and small, rare and common, male and female. The powerful traditional ideas on the strict hierarchy of things, 'the great chain of being', which formed the basis for Platonic as well as Aristotelian concepts of the world, were no longer relevant here.

According to Descartes, everything was merely matter-in-motion; the cosmos was essentially a meaningless mechanism. For example, the rainbow was not a sign of the bond between God and humanity (Genesis 9:13-16) but rather an optical phenomenon that could be explained and described mathematically (see figs. 8, 9). Peter Harrison has characterised the development initiated by Descartes as follows:

> Natural objects have been stripped of their intrinsic meanings, and even their qualities and essences have gone. In the physics of Descartes and Newton, simple natural objects are denuded of all but basic quantitative properties. In this new language of nature, syntax has triumphed over semantics. Henceforth, the science of nature will deal with the mathematical or classificatory rules which govern the relations between natural objects.[30]

In the view of many of Descartes's Dutch adherents, God's creation was no longer an emblematic web of symbolic references and religious meanings, but a divine piece of engineering, referring to and proving the existence of God, the almighty architect. To what extent this idea influenced artistic representations of nature – for example, in still life or landscape paintings – is still a matter of discussion.

The few historians of science working in the Netherlands have traditionally focused on the rise of mathematics, astronomy and physics and stressed the importance of Simon Stevin, Descartes and others in what E.J. Dijksterhuis (1892-1965) called 'The mechanization of the world picture'.[31] For a long time, the history of science in the Netherlands was a narrative that started in the Southern Netherlands with Gemma

Frisius (1508-1555) and his son Cornelius Gemma (1535-1578), then via Stevin and the Dutch engineering school at Leiden University (established in 1600), via Isaac Beeckman (1588-1637) to Descartes, culminating in the mathematical and descriptive work of Christiaan Huygens.[32] According to a tradition that goes back at least to Johan Huizinga (1872-1945), Dutch scientific culture of the seventeenth century was considered rational, pragmatic and non- (if not anti-) metaphysical, and the image of Dutch science seemed to fit neatly in this pattern. According to this view, Netherlandish science was practical: it was primarily concerned with observing, surveying, mapping and finding longitude at sea. It was this anachronistic concept of science that lay behind Svetlana Alpers's *The art of describing*, a book that for many art historians might have been their introduction to the theme under discussion here.[33]

However, over the last two decades the perspective has broadened extensively. It has become clear that scientific culture in the Southern and Northern Netherlands was extremely diverse and rich. Although during the early modern period the centre of activity shifted from the South to the North, from Leuven and Antwerp to Leiden and Amsterdam, the Low Countries at large kept attracting students, professors, princes and *curieux* and *virtuosi* from abroad. Numerous travel accounts illustrate how these foreigners were fascinated not only by the paintings and other works of art produced in the Netherlands, but also by the botanical gardens, the cabinets of curiosities and the anatomical dissections. It has become increasingly clear that creating works of art and studying nature were considered as belonging to the same set of cultural practices, laden with religious meanings.[34] Since nature was considered the creation of God, the study of nature, or the representation of it in art, had strongly metaphysical undertones. It could be argued that the acts of depicting and describing nature were by no means 'neutral' or 'objective', but to a great extent were implicit or explicit tributes to the Creator. A good example of the latter is provided by Johannes Goedaert (1617-1668), who was mostly known for his still life paintings but was also engaged in the business of alchemy and extensively published on the generation and religious meaning of insects. Goedaert paid little attention to the *vanitas* motif, but all the more to praising God, the almighty architect and creator of even the humblest and smallest of creatures – a point that adds a new dimension to the continuing discussion on the interpretation of paintings portraying nature, such as shells, insects and bouquets of flowers, or landscapes.[35] Anatomists such as Johannes Swammerdam and Frederik Ruysch became famous for their artful preparations of humans and animals and were skilled draftsmen themselves. Other collectors, like Levinus Vincent (1658-1727) and Albertus Seba (1665-1736), displayed their *naturalia* in elaborate, aesthetic, 'artful' compositions. They hired artists – often anonymous – to illustrate the catalogues of their collections, thus disseminating knowledge of previously unknown species throughout Europe in an accessible, accurate and appealing way (figs. 10, 11; see also fig. 22, page 268).

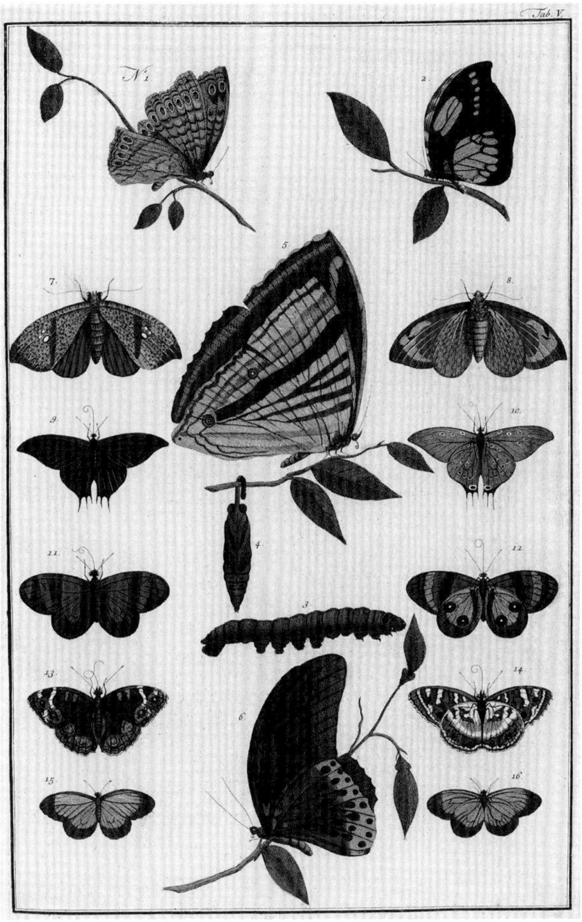

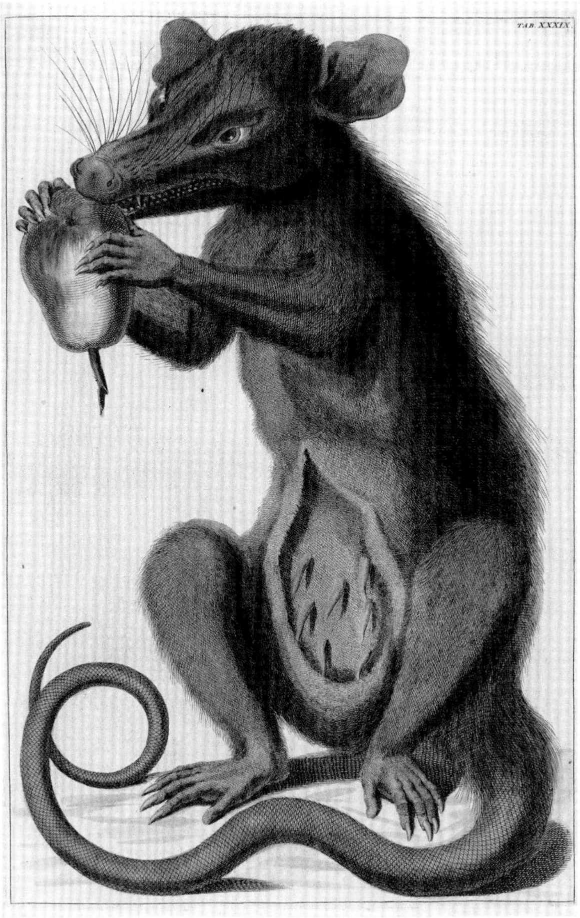

10

Artful ensemble of exotic butterflies, depicted by an anonymous artist in A. Seba, *Thesaurus* (1734-1765)
The Hague, Koninklijke Bibliotheek, 394 B 26.

11

Southern common opossum, depicted by an anonymous artist in A. Seba, *Thesaurus* (1734-1765)
The Hague, Koninklijke Bibliotheek, 394 B 26.

Although the relation between art and science in the Netherlands has not to date been studied in a programmatic or systematic way and lacks a sound institutional basis, many valuable, interesting contributions to this topic have seen the light over the last decades. These range from studies of the scholarly and (natural) philosophical interests of Pieter Paul Rubens, the work of Vesalius and his followers, the Dutch anatomical theatres and the genre of 'anatomische lessen' to the work of Ruysch and contemporaries and successors like Bernhard Siegfried Albinus (1697-1770).[36] The culture of gardening and botany, inaugurated in the Southern Netherlands by Dodonaeus, Matthias Lobelius (1538-1616) and Clusius, has been the subject of many penetrating studies, stressing social context as well as the role of art and science, with regard both to experiments on cultivation as to the design of gardens and to the pictorial representation of plants.[37] The intersecting activities of artists and scholars with connections to Leiden University around 1600, such as Clusius, Theodorus Clutius (1546-1598) and Jacques de Gheyn II (c. 1565-1629), has been the subject of many publications.[38] Highly relevant in this context is, of course, the way in which Flemish and Dutch scholars and artists perceived the non-European world. The encounter with indigenous peoples — their physical appearance as well as their cultures and histories, customs and religions, the local flora and fauna – is a subject of the utmost significance.[39] Closely linked is the important culture of collecting *naturalia* and *artificialia,* the study of which received a major impulse with the research project and exhibition 'De wereld binnen handbereik. Nederlandse kunst- en rariteitenverzamelingen 1585-1735' (1992).[40] This event inspired many other research projects.[41] It also revived attention towards the culture of collecting in the Southern Netherlands, most notably Antwerp, a theme that only recently has become the subject of research.[42] A comparative approach, analysing the similarities as well as the differences between the cultures of collecting in the South and the North, is a desideratum at the moment. The same could be noted about early modern theories of optics – more notably vision, light, sight and colour – a theme addressed by Sven Dupré and Fokko Jan Dijksterhuis in the present volume, and also the subject of a recent dissertation by Katrien Vanagt on Vopiscus Fortunatus Plempius (1601-1671).[43]

In the tradition of the *Netherlands Yearbook for History of Art*, the editors have tried to incorporate new trends and signal potential new fields of research. Earlier volumes on *The body and embodiment* (2008) and *The passions* (2010) already touched upon the intersections between art and science. Here, we have explicitly tried to explore this fascinating territory in a more explicit and programmatic way, taking early modern practices as well as theories as our point of departure. Earlier versions of the papers published here were presented at a workshop held at the Huygens Institute for Netherlands History, The Hague, in May 2009. In the present volume the essays are published in more or less chronological order, although it will be clear that some important subjects and themes recur in many of the contributions. It was the aim of the editors to make the whole more than the sum of its parts and to publish a volume as coherent as possible. In this introduction, we will single out four of the main themes addressed in this *Yearbook*: the problem of visual perception and colour theory; the way in which newly discovered life-forms were represented; the impact of the philosophy of René Descartes; and the religious connotations of the study of nature.

In the opening essay, Sven Dupré reassesses the historiography of perspective and optics, taking as his point of departure the work of Panofsky and Gombrich, who tended to distinguish between the history of art after development of the central perspective in Italy, and the reflection of light on various surfaces, as painted by Northern artists like Jan van Eyck (c. 1390-1441). This distinction, Dupré argues, is merely a twentieth-century construction, which had seriously blurred our vision of the ways Renaissance and early modern artists dealt with optics – both in Italy and in the Netherlands, in theory as well as in practice. *Optica* involved more than *perspectiva*. The dominant focus of art historians on the origins and influence of the invention of linear perspective has ignored the knowledge of early modern artists and scholars regarding the effects of light and the properties of material. Dupré substantiates his claim with two case studies. Taking a closer look at Stevin's *Vande devrsichtighe* (1605), Dupré argues that this is not a text devoted exclusively to perspective, as is usually assumed, but to optics in general. In the second part, he demonstrates that the field of optical inquiry of artists ranged to *reflexy-const*, optical and meteorological effects such as the rainbow, and the properties of paint.

A century later, these themes were still the focus of artistic and scholarly attention, as Fokko Jan Dijksterhuis argues in his contribution on the optical experiments of Lambert ten Kate (1674-1731) and Hendrik van Limborch (1681-1759). These two men – a wealthy polymath and a painter – were engaged in an extremely interesting project of colour mixings, in order to describe in a rational and systematic way the making of gradations of bright and dark colours. Their main focus was on the *Lijfverf*, the much sought-after colour of paint used to create a lifelike representation of human flesh. Although this project became known to a larger audience only through Hessel Miedema's edition of the correspondence of Ten Kate and Van Limborch (2006), it was extremely fascinating from a contemporary epistemological point of view. Although a rational, 'scientific' approach lay behind the experiments on the nature and properties of paint and light, the main focus was not on philosophical abstraction but on sense perception and practical knowledge – the reason why this enterprise thus far largely went unnoticed by historians of art and historians of science. Implicitly, Dijksterhuis endorses the argument made by Smith on 'artisanal knowledge'.

Experiments and the properties of paint are also the subject of Karin Leonhard's essay. Taking the early modern fascination of some painters with the depiction of snakes, lizards, locusts, fungi and toads as her point of departure, Leonhard elaborates on the complex relationship between painting, toxicology and pharmacology. The painter and student of nature Otto Marseus van Schrieck (c. 1620-1678) plays an important role in Leonhard's article. Marseus, active in Rome between 1648 and c. 1660 and working in Amsterdam until his death in 1678, was highly fascinated by the toxicological experiments and the rituals surrounding the production of medicine taking place in Italy, most notably at the Medici court in Florence. A keen observer of nature, he became the best-known exponent of the so-called *sottobosco* genre. Van Schrieck pasted real butterfly wings in the wet surface of his paintings (thereby provocatively challenging theories of representation), but, as Leonhard claims, above all he was intrigued by the authenticity of materials: painting rock with rock pigments, earth with earth, and herbs with herbal dyes.

That an empirical, experimental attitude towards nature was one of the crucial factors in the so-called Scientific Revolution is beyond dispute. However, research in the last decades has made it abundantly clear that sensory experience and the act of observing were not unproblematic. A basic epistemological axiom until the time of Descartes was that of a lost pristine knowledge – the *sapientia veterum* as humanist scholars called it, or *de Wijsentijt* ('the age of the sages') as Stevin put it so eloquently – to be restored only by careful research. The subject of scholarly attention – be it of natural philosophers, physicians, cartographers, or artists and engineers

– was the lost Golden Age of the Greeks and their Roman successors, who had left so many writings, and so few pictures. The scholarly turn towards empiricism in the sixteenth century of Vesalius, Conrad Gesner (1516-1565), Tycho Brahe (1546-1601) and others must not be seen as a revolt against the Ancients, but as an attempt to supplement classical knowledge with physical, observational and visual evidence. In what sometimes is called the Age of Discovery, it was not immediately clear what was new. Many of the finest spirits actually argued that no new things could be discovered – but only *re*discovered. For example, the construction of the compound microscope around 1620 led to a discussion about whether the Ancients had not already known such an instrument. The description of a red-haired anthropoid ape found in Sumatra led some to conclude that this must be one of the monsters living in the outer parts of the world, as described by Pliny.[44]

The growing awareness of the existence of as yet unexplored continents, unknown species of animals and unseen anatomical details on the one hand, and the strong intellectual and pictorial tradition on the other, are dealt with in the essays by Margócsy, Rikken and Smith, and Jorink. Dániel Margócsy analyses how Netherlandish artists struggled with the visual representation of previously unknown animals. Using the case of the camel as an example, Margócsy demonstrates the various techniques artists like Maarten de Vos (1532-1603) and Adriaen Collaert (1560-1618) relied on, varying from hearsay reports to the strategy of metonymic composition: the re-use of images of parts of animals already known, assembled into a composite image. A basic presumption, then, was the conviction that nature was immutable and stable. For example, terrestrial life was potentially mirrored with life underwater or in the sphere above the moon – a premise that even led to counterfeit representations of fabulous creatures, not to speak of speculation of extraterrestrial life (a theme also mentioned in the contributions by Weststeijn, Spaans and Ramakers).

Marrigje Rikken and Paul Smith describe how subsequent versions of Jan Brueghel's *Allegory of air* display an increasing knowledge of exotic birds. While Brueghel initially relied much on the depictions by Gessner and others, later versions show lifelike representations based on observations, presumably done by the artist himself. The versions document the rapid influx of exotic birds in European collections, as well as the growing awareness that really new species were being witnessed. Brueghel self-consciously started a visual dialogue with his fellow countryman Clusius, by improving on the images of the latter's *Exoticorum libri decem* (1605; see fig. 5). In the same spirit he was the first artist to depict one of the icons of the Scientific Revolution, the telescope, on the later versions of the *Allegory of air*. On the other hand, Brueghel's range of volatile creatures is framed by traditional, scholastic and Aristotelian views of nature, imbued with symbolic meanings, hidden qualities and overt sympathies and antipathies.

Eric Jorink explores the way in which the microscopist Johannes Swammerdam explored another unchartered territory: the anatomy of insects and other creatures belonging to the lowest states of being in the

Aristotelian hierarchy. Swammerdam, moving in the same Amsterdam circle of *curiosi* as Otto Marseus (with whom he often collaborated), was a talented anatomist as well as a draftsman. He had outspoken ideas on the epistemological value of drawings *ad vivum* and anatomical preparations, and he was successful in establishing a new mode in representing the outward appearance as well as minute internal anatomy of the smallest creature, culminating in his dissection of the louse in 1678. Nature, Swammerdam maintained, was a work of art created by God, and it was man's duty to study and represent God's creatures as faithfully as possible. Although Swammerdam rejected the symbolic and emblematic meanings traditionally attributed to the animal kingdom, his lifelike representations of it were not objective, religiously neutral descriptions – much to the contrary. Representing nature 'naer het leeven' was a tribute to God, the divine Artist. Swammerdam, like so many others in the Netherlands, was profoundly influenced by the New Philosophy of Descartes. It was the Cartesian idea of the uniformity and stability of nature that was the driving force behind Swammerdam's investigations into the 'humblest creatures', which, according to classical theories, had their origin in spontaneous generation – that is, from decaying organic material. Referring to Descartes, Swammerdam dismissed the theory of spontaneous generation as blasphemous: it would leave the door open to chance or, in other words, would imply that God's divine creation was contingent.

In his contribution to this volume, Gijsbert van de Roemer investigates the possible influence of the Cartesian frame of mind on art theory, thereby also touching upon the issues of chance and fixed rules. Van de Roemer notes that recently much has been written on Samuel van Hoogstraten (1627-1678), who has been credited for translating the New Philosophy into art theory and has even been connected to Descartes's most radical adherent, Benedictus de Spinoza (1632-1677). Van de Roemer compares Van Hoogstraten's *Inleyding tot de hooge schoole der schilderkonst* (1678) with the less studied treatise by Willem Goeree, *Inleyding tot de practijck der algemeene schilderkonst* (1670), and makes some intriguing observations and remarks. He argues that Van Hoogstraten's *Inleyding* is less 'modern' than is usually assumed and, regarding form and content, is still embedded in the scholastic, emblematic humanist tradition. In a rather associative way, Van Hoogstraten presents an almost unrestricted accumulation of data. For example, in the part of his work that deals with colours, Van Hoogstraten not only gives practical information on pigments, but also extensively writes on the sympathies and antipathies between colours and their correspondences with the planets. Goeree, on the other hand, seems much more influenced by the New Philosophy. Van de Roemer demonstrates Goeree's preoccupation with rules and order and his tendency towards regulation and intellectualisation of the arts. Usually, these tendencies are seen as characteristic of Dutch classicism. Few attempts have been made so far to link this development with the natural sciences. Van de Roemer suggests that Goeree, although far from being a Spinozist, was profoundly influenced by Cartesianism: 'in Goeree's thinking there is no sign of equating God with nature. On the contrary, he conceived nature as a precise and mathematical artwork crafted and ruled by the Supreme Artist'. One is reminded that this is exactly the position held by Swammerdam and many others. Goeree's love for regulation and order and his aversion to chance led him to interesting observations and suggestions. He noted, for example, that the use of colours and the mixing of paint was still a matter of guesswork or trial and error. Instead, he pleaded for a regularised, quantitative system for mixing colours – the project Ten Kate and Van Limborch would embark on some decades later.

Although Descartes had a profound influence in the Dutch Republic, the philosopher was not uncontested. To many, his philosophy as expounded in *Discours de la méthode* (1637) and its three annexes were enchanting, not least because of the eye-catching illustrations by the mathematician Frans van Schooten Jr. (1615-1660), which seemed to make the universe clear and comprehensible (fig. 9). Shortly after the publication of the *Discours*, orthodox theologians led by the Utrecht professor Gisbertus Voetius declared war on the New Philosophy, resulting in a battle that would continue for decades. The Voetians took offence to Descartes's materialist view of nature, which, they maintained, would reduce God to a retired engineer and was in plain contradiction with the Bible. Moreover, Cartesian methodological doubt and the famous dictum *cogito ergo sum* would inevitably lead to scepticism and atheism. The publication of Spinoza's *Tractatus theologico-politicus* (1670) – in which, among other things, the reality of the Biblical miracles was denied, since this would imply that nature was contingent – made Voetius's worst nightmares come true.

The furious reactions to Descartes and Spinoza have been studied in great detail over the past decades, exclusively, however, on the basis of books, pamphlets, letters and official or ecclesiastical documents. Rienk Vermij adds a new dimension to this theme, by examining the frontispieces of some books issued in the wake of the Cartesian-Spinozist debate. At a time when books did not have dust jackets, the title page and frontispiece were the few places where the content could be summarised and visualised. A frontispiece, with its allegories and images, could be used to give potential readers some foretaste of the contents of the work. Moreover, it could be used as a non-, half- or even meta-verbal medium to enter into debate. Frontispieces have only recently been discovered to be such a rewarding resource by historians of science and ideas, historians of art, and historians of the book alike. In his contribution, Vermij analyses a number of Dutch frontispieces, ranging from the 1690 Amsterdam edition of Descartes's *Principia philosophiae, of beginselen der wijsbegeerte* to Bernard Nieuwentijt's *Het regt gebruik der wereltbeschouwingen* ('The right way of contemplating the works of the Creator', 1715), considered the most powerful answer to the threat of Spinoza. It is often not clear who commissioned a frontispiece, or who executed it. Vermij pays much attention to the role of Jan Luyken, the artist and engraver who had strong sympathies for spiritualistic ideas and executed many frontispieces for works within the broad spectrum between the natural sciences and theology, thereby underlining the fluid boundaries between art and science in this period (fig. 12). The Nieuwentijt frontispiece was executed by Jan Goeree (1670-1731), a son of Willem Goeree. The examples presented by Vermij make it clear that the artists borrowed much from the Western pictorial and allegorical traditions but adapted them to local demands. The imagery of light, so prominent in the frontispieces analysed by Vermij, had many connotations. It could refer either to the divine light, according to the orthodox Calvinist the only way to true knowledge and revelation, or to the inner light, a concept much at the heart of Spinozist and spiritualist movements. Goeree used the imagery of light deliberately, Vermij concludes.

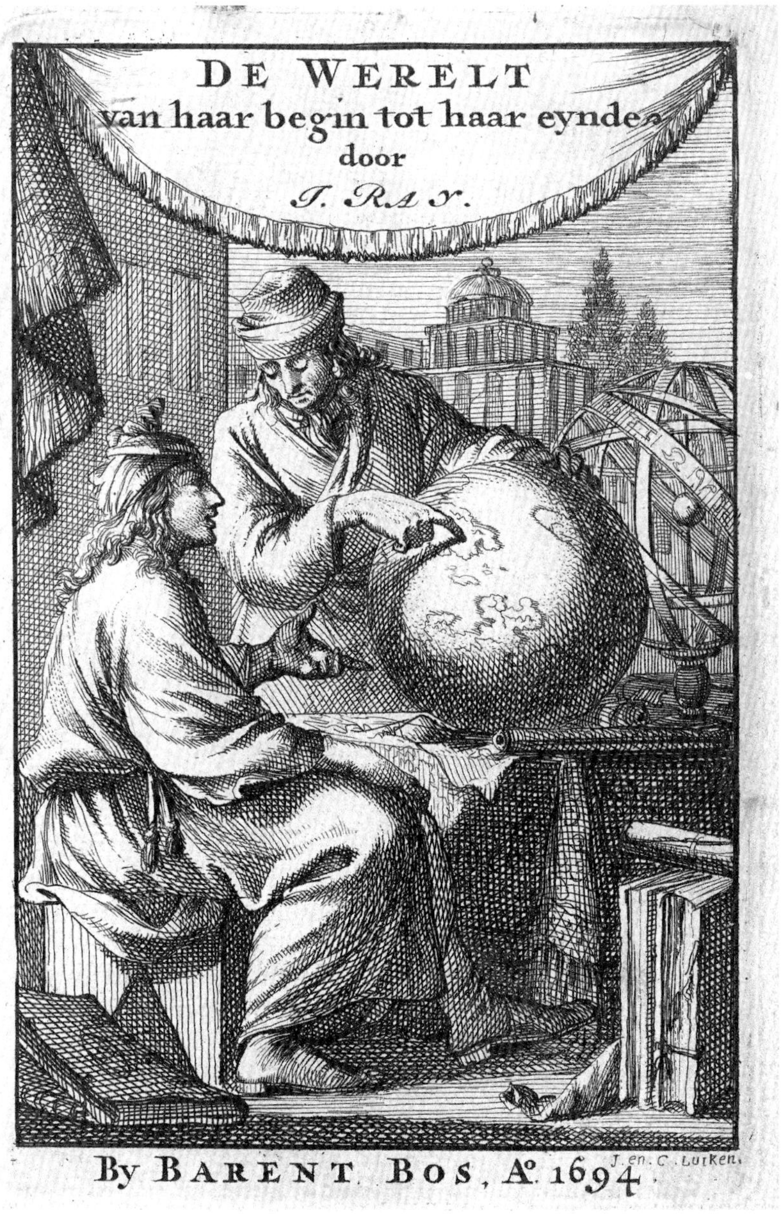

12

J. and C. Luyken, frontispieces of J. Ray, *De werelt van haar begin tot haar eynde* (1694); see the contribution by Rienk Vermij in this volume.
The Hague, Koninklijke Bibliotheek, 2105 A 70.

Images and words play a central role in Thijs Weststeijn's contribution as well. In a broad panorama, Weststeijn brings attention to yet another neglected theme at the junction of *ars* and *scientia*: pictography. As Weststeijn notes, early modern scholars were fascinated by hieroglyphs, Chinese characters and Meso-American writing for several, overlapping reasons. By some scholars these signs were considered as *pictures* and, as such, more powerful than words. This fostered utopian ideals of a universal language and a pure communication of knowledge, as, for example, expressed by Jan Amos Comenius (1592-1670) and Gottfried Wilhelm Leibniz (1646-1716). Even more important, the graphic nature of these hieroglyphs (which, it should be remembered, were deciphered only in the nineteenth century) had deeply religious connotations. They were considered as the silent remains of the *prisca scientia*, the time before God gave man the Hebrew script. In hieroglyphs, *res* and *verba* were identical. The closer one got to Moses and Adam, the closer one got to the Word of God. Before the Fall, the Creation seemed to have been a book in which words and objects were identical. In those days there was no Bible, only the 'speechless teachings of heaven and earth', as Calvin believed. After the Fall, the Original Sin and Babel, however, the tie between divine

13
Engraving, illustrating Genesis 11, from J. Luyken, *Schriftuurlyke geschiedenissen* (1712)
The Hague, Koninklijke Bibliotheek, 1754 C 12.

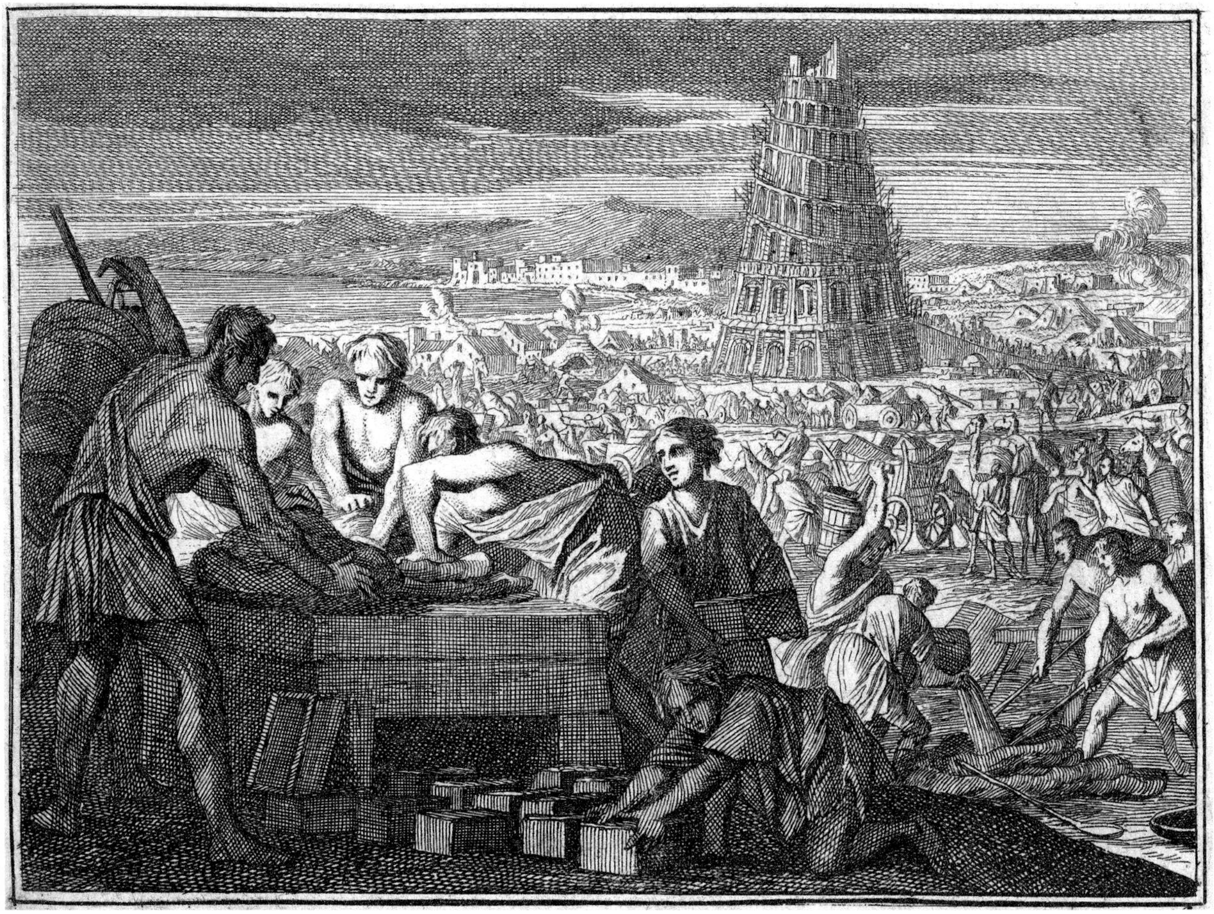

knowledge and human knowledge was severed. Language became confused and the peoples were scattered (fig. 13). Thus, the fascination for hieroglyphs was essentially a quest for the lost pristine knowledge. The pictorial character of hieroglyphs also fitted nicely into the art theoretical idea that drawing and writing were essentially the same activities. In his posthumous treatise on painting, *De graphice* (1650), the learned humanist Gerardus Johannes Vossius (1577-1649) noted that the Greek verb γράφειν meant 'painting' as well as 'writing'. His son, the idiosyncratic Isaac Vossius, expressed similar ideas about Chinese characters.

Weststeijn discusses Egyptian mummies and seventeenth-century theories of art, encyclopaedic collections and early modern conceptions of America. He introduces us also to the wonderful world of Athanasius Kircher (1602-1680), 'the last man who knew everything', who had many contacts in the Dutch Republic.[45] His books on mnemotechnics, the subterranean world, Noah's Ark, symbolic logic, contemporary China and, of course, hieroglyphs were printed in Amsterdam, lavishly illustrated by Gérard de Lairesse (1640-1711) and Romeyn de Hooghe (1645-1708) amongst others, and read by Huygens and Spinoza.

These are also some of the themes and people in Joke Spaans's contribution on Romeyn de Hooghe. Although De Hooghe is a well-known artist, mainly for his many engravings if not for his libertine sympathies, no research has been done so far on his posthumous book *Hieroglyphica of merkbeelden der oude volkeren* (1735). In more than 400 pages, De Hooghe presents in words and, above all, in images his views on religion. Firmly based on contemporary philosophical debates, recent scientific discoveries and accounts from non-European countries, De Hooghe dismisses the Christian Church as essentially a deceitful human construction. That is not to say that De Hooghe is irreligious or an atheist. Like so many of his contemporaries, De Hooghe rejected the visible churches and endorsed an enlightened spirituality. Spaans does not hesitate to situate De Hooghe in the camp of Spinozists. Moreover, she presents some interesting connections between the artist and scientists like Swammerdam and Huygens that had not been noticed before.

In the final essay, by Bart Ramakers, we turn from the turmoil of religious strife and philosophical debate to the peace and quiet of the Dutch countryside. In *Het land, in brieven* (*The countryside, in letters*) from 1788, an epistolary novel by Elisabeth Maria Post (1755-1812), the female characters whose letters make up the book, apart from a keen observational eye, bring two additional means to their approach of nature: their imaginations and their emotions. In their letters, the main character, Emilia, and her bosom friend Eufrozyne describe their intense confrontation with nature during long solitary and joint walks through the countryside. Their imaginations gradually lead them from empirical description to mental impressions and, finally, to emotional outpourings. Their thoughts get carried away to different places, both inside and outside of observable reality, leading to, in some instances, ecstatic raptures of the heart and, ultimately, even to experiences of the sublime. Their intimate friendship intensifies their observations, imaginations and feelings. They start describing external *Nature* – with a capital letter – in order to ultimately arrive at experiences, feelings and thoughts that touch on their own, internal *nature* – written in lower case. It is in this fusion of object and subject, Ramakers argues, that scientifically inquisitive minds like Post's – and the characters featured by her – try to reach the source and essence of Creation, which, in line with the physico-theological tradition that dominated eighteenth-century Netherlandish science, was metaphysical in character, leading up to God as creator of all life. What distinguishes Post's approach from mainstream physico-theology, however, is that she juxtaposes mere classification and admiration of nature with this 'sympathetic', sentimental approach, which from the middle of the eighteenth century started to inform scientific empiricism and preluded romantic science of the late eighteenth and early nineteenth centuries. Parallels and possible inspiration for Post's approach can be found in (the theory of) landscape art – both verbal and visual – from abroad, especially from the German lands. Literature, like painting, offered an aesthetic way to reach the deeper meaning of Creation. Thus, with the help of his subjective powers, the artist was able to reach a level of knowledge about nature that was just as high – or maybe even higher – than that of the scientist operating in a purely empirical manner.

The present volume may serve to indicate the richness as well as the potential rewards of future research into art and science in the early modern Netherlands. Many other subjects still remain to be explored, such as the use of schemes and diagrams by philosophers and mathematicians; the concept of *ad vivum*, which seems to be more problematic than is usually assumed; the visualisation of Biblical models of knowledge (the Ark, the Tower, the Temple); the role of images in the work of Descartes; the way scholars commissioned – often invisible – artists; the iconography of maps; or the meaning of materials, to mention just a few. To some of these we hope to return in future volumes of the *NKJ*.

Notes

1 On Ruysch, see Kooijmans 2004; Van de Roemer 2008; and Margócsy 2011.

2 Anonymous 1677a, 4: '(...) zich heeft begeven tot het manuael, om de deelen der dieren en doode menschen van een te scheyden'.

3 Ibid.

4 Anonymous 1677b, 13.

5 Ibid.: 'Ick seg dat een Anatomicus sonder schilderkonst, en een schilder sonder de *Anatomy*, geen van beyden de trap der volmaaktheid sullen beklimmen'.

6 Ibid., 56: 'wat *Anatomicus*, sal de gemaeckte en gesnede plaeten van een, of 't ander deel des menselijcke lichaems, volgens haer naturelijcke loop der *fibren, fibrillen, situatie, insertie*, en articulatie konnen voor goet keuren, en verbeteren, indien hy de teken en schilderkonst noyt geoefend heeft, of wat *ontleder* (...) sal sijn nieuw gevonden *duchtus, meavis, musculen* en ander verholentheden, konnen nae hun eys doen in 't kooper snijden, of silver snijden, ingeval hem de reken en schilderkunst gebreeckt'.

7 Ibid.: '(...) een Schilder, die by sijn schildery met duidelijcke letteren bij moet voegen, dit is een hont, en dat is de haes'.

8 Goeree 1670, 73-74 passim.

9 Ibid., 55. The reference is to the Reynst collection. Gerard Reynst (1599-1658) operated in the Amsterdam community of collectors, which also included Rembrandt, Jan Jacobsz Swammerdam (1603-1678; the father of Johannes Swammerdam) and, somewhat later, Ruysch. See Logan 1979; Bergvelt & Kistemaker 1992; and Van den Boogert 1999.

10 Goeree 1670, 55.

11 Ibid., 80.

12 We mention here only Silver 2006.

13 Kemp 1990; Kemp 2006; Kemp 2007; Freedberg & De Vries 1991; Freedberg 2002; Bredekamp 2000; Bredekamp 2007.

14 Kaufmann 1993.

15 Impey & MacGregor 1985. In its wake, the *Journal of the History of Collections* was established in 1989. The research

agenda inspired by the Oxford conference has stimulated and initiated publications such as Bergvelt & Kistemaker 1992; Grote 1994; Bredekamp 1995; Findlen 1996; Van de Roemer 2005; and Jorink 2010.

16 Kenseth 1991; cf. Daston and Park 1998.

17 For a historiographical overview of the period under discussion here, see Lindberg & Westman 1990, and Cohen 1995. For an overview of recent trends, see Smith 2009.

18 Paradigmatic works are Shapin & Schaffer 1985, and Biagioli 1993.

19 Recent examples are Smith & Findlen 2002, and Roberts *et al.* 2007.

20 A comprehensive work, reflecting the state of the art, is Park & Daston 2006.

21 Harms 1985; Ashworth 1990.

22 Findlen 1994.

23 Secord 2004.

24 Smith 2004.

25 Kusukawa & Maclean 2006; Roberts *et al.* 2007; Klein & Spary 2010; Dupré & Lüthy 2011.

26 Harkness 2007.

27 Yet to be written is an overview of the role of the Netherlands ('Netherlands' in the sense used by the *NKJ*, i.e., roughly present-day Flanders and the Netherlands). Valuable surveys are Halleux *et al.* 1998-2001; Van Berkel 1983; Van Berkel *et al.* 1999; Van Berkel 2010; Cook 1992; Cook 2007; and Jorink 2010.

28 Dupré and Lüthy 2011, 1-2.

29 Vermij 1999; Jorink 2010, 85-88.

30 Harrison 1998, 263-264.

31 Dijksterhuis 1969. The book was originally published in 1951 as *De mechanisering van het wereldbeeld*. See also Van Berkel 1983.

32 See, for example, Van Berkel 1983, and Van Berkel *et al.* 1999, 3-67. To a certain extent, this interpretation of Dutch scientific culture as being practical and non-metaphysical is also put forward in Cook 2007.

33 Alpers 1983; cf. De Mare 1987; Franits 1997.

34 Van de Roemer 1998; Van de Roemer 2004; Van de Roemer 2005; Huisman 2009; Jorink 2010.

35 On Goedeart, see Bol 1982, and Jorink 2010, 201-214. Cf. Taylor 1995. For a recent debate on the interpretation of landscape paintings, see Bakker 2004a, Stumpel 2004, and Bakker 2004b.

36 See, for example, Morford 1991; Punt 1983; Kooijmans 2004; Swan 2008; Huisman 2009; Zuidervaart 2009; Margócsy 2009; and Margócsy 2011.

37 De Koning *et al.* 2008; Egmond *et al.* 2007; Goldgar 2007; Van Gelder 2011.

38 De Jong 2000; Swan 2002; Swan 2005; Egmond 2005.

39 We mention here only Van den Boogaart 2003; Parker Brienen 2006; Cook 2007; Schmidt 2007; Reitsma 2008; Mason 2009; Hunt *et al.* 2010.

40 Bergvelt & Kistemaker 1992.

41 Van de Roemer 2004; Margócsy 2009; Driessen-Van het Reve 2006.

42 See, for example, Van Suchtelen & Van Beneden 2009, and Dupré 2010.

43 Vanagt 2010.

44 On this point, more in general, see Grafton 1992, and Jorink 2010, 58-77.

45 Cf. Findlen 2004.

Bibliography

Anonymous 1677a
Anonymous, *Den aftocht ter eeren van de krakkeelende doctoren en chirurgijns van Amsterdam, over het verschil tusschen den Hr. Doctor Bonaventura van Dortmont, en den Vroet-doctor Fredericus Ruysch*, Amsterdam 1677.

Anonymous 1677b
Anonymous, *Luyelack, of een t'samen-spraeck tusschen wit en swart. Anders genaemt antwoort aen den auteur van Den aftocht ter eeren van de krakkeelende doctoren en chirurgijns van Amsterdam*, Amsterdam 1677.

Alpers 1983
S. Alpers, *The art of describing. Dutch art in the seventeenth century*, Chicago 1983.

Ashworth 1985
W. Ashworth Jr., 'The persistent beast. Recurring images in early zoological illustrations', in: J. Ackerman (ed.), *The natural sciences and the arts. Aspects of interaction from the Renaissance to the 20th century*, Uppsala 1985, 46-66.

Ashworth 1990
W. Ashworth Jr., 'Natural history and the emblematic worldview', in: D.C. Lindberg & R.S. Westman (eds.), *Reappraisals of the Scientific Revolution*, Cambridge 1990, 303-323.

Bakker 2004a
B. Bakker, *Landschap en wereldbeeld van Van Eyck tot Rembrandt*, Bussum 2004.

Bakker 2004b
B. Bakker, 'Some notes on method. Regarding Jeroen Stumpel's review of my *Landschap en wereldbeeld van Van Eyck tot Rembrandt*', *Simiolus* 31 (2004), 260-262.

Bergvelt & Kistemaker 1992
E. Bergvelt & R. Kistemaker (eds.), *De wereld binnen handbereik. Nederlandse kunst- en rariteitenverzamelingen 1585-1735*, Zwolle 1992.

Biagioli 1993
M. Biagioli, *Galileo, courtier. The practice of science in the culture of absolutism*, Chicago 1993.

Bol 1982
L.J. Bol, *'Goede onbekenden'. Hedendaagse verkenning en waardering van verscholen, voorbijgezien en onbekend talent*, Utrecht 1982.

Bredekamp 1995
H. Bredekamp, *The lure of antiquity and the cult of the machine. The Kunskammer and the evolution of nature, art and technology*, Princeton 1995.

Bredekamp 2000
H. Bredekamp, 'Gazing hands and blind spots. Galileo as draftsman', *Science in Context* 13 (2000), 423-463.

Bredekamp 2007
H. Bredekamp, *Galilei der Künsteler. Der Mond. Die Sonne. Die Hand*, Berlin 2007.

Cohen 1994
H.F. Cohen, *The Scientific Revolution. A historiographical inquiry*, Chicago 1994.

Cook 1992
H.J. Cook, 'The New Philosophy in the Low Countries', in: R. Porter and M. Teich (eds.), *The Scientific Revolution in national context*, Cambridge 1992, 115-149.

Cook 2007
H.J. Cook, *Matters of exchange. Commerce, medicine, and science in the Dutch Golden Age*, New Haven 2007.

Daston & Park 1998
L. Daston & K. Park, *Wonders and the order of nature 1150-1750*, New York 1998.

De Jong 2000
E. de Jong, *Nature and art. Dutch garden and landscape architecture, 1650-1740*, Philadelphia 2000.

De Koning *et al.* 2008
J. de Koning e.a. (eds), *Drawn after nature. The complete botanical watercolours of the 16th Century: Libri Picturati*, Zeist 2008.

De Mare 1987
H. de Mare, 'De verbeelding onder vuur. Het realisme-debat der Nederlandse kunsthistorici', *Theoretische geschiedenis* 24 (1987) 113-137.

Dijksterhuis 1969
E.J. Dijksterhuis, *The mechanization of the world picture*, Oxford 1969.

Driessen-Van het Reve 2006
J. Driessen-Van het Reve, *De* Kunstkamera *van Peter de Grote. De Hollandse inbreng, gereconstrueerd uit brieven van Albert Seba en Johann Daniel Schumacher uit de jaren 1711-1752*, Hilversum 2006.

Dupré 2010
S. Dupré, 'Trading luxury glass, picturing collections and consuming objects of knowledge in early seventeenth-century Antwerp', *Intellectual History Review* 20 (2010) 53-78.

Dupré & Lüthy 2011
S. Dupré & C. Lüthy (eds.), *Silent messengers. The world of goods and the circulation of knowledge in the early modern Netherlands*, Berlin 2011.

Egmond 2005
F. Egmond, 'Clusius, Cluyt, Saint Omer. The origins of the sixteenth-century botanical and zoological watercolours in *Libri Picturati* A. 16-30', *Nuncius. Journal of the History of Science* 20 (2005) 11-67.

Egmond *et al.* 2007
F. Egmond, P. Hoftijzer & R. Visser (eds.), *Carolus Clusius. Towards a cultural history of a Renaissance naturalist*, Amsterdam 2007.

Findlen 1994
P. Findlen, *Possessing nature. Museums, collecting, and scientific culture in early modern Italy*, Berkeley 1994.

Findlen 2004
P. Findlen (ed.), *Athanasius Kircher. The last man who knew everything*, New York 2004.

Franits 1997
W. Franits (ed.), *Looking at seventeenth-century Dutch art. Realism reconsidered*, Cambridge 1997.

Freedberg 2002
D. Freedberg, *They eye of the lynx. Galileo, his friends, and the beginnings of natural history*, Chicago 2002.

Freedberg & De Vries 1991
D. Freedberg & J. de Vries (eds.), *Art in history, history in art. Studies in seventeenth-century Dutch culture*, Chicago 1991.

Goeree 1670
W. Goeree, *Inleyding tot de practijck der algemeene schilderkonst*, Middelburg 1670.

Goldgar 2007
A. Goldgar, *Tulipmania. Money, honor, and knowledge in the Dutch Golden Age*, Chicago 2007.

Grafton 1992
A. Grafton, *New worlds, ancient texts. The power of tradition and the shock of discovery*, Cambridge 1992.

Grote 1994
A. Grote (ed.), *Macrocosmos in Microcosmo, die Welt in der Stube. Zur Geschichte des Sammelns, 1450 bis 1800*, Opladen 1994.

Halleux *et al.* 1998-2001
R. Halleux, C. Opsomer, Carmelia & J. Vandersmissen (eds) (1998), *De geschiedenis van de wetenschappen in België van de Oudheid tot1815*, Brussels 1998-2001.

Harkness 2007
D. Harkness, *The jewel house. Elizabethan London and the Scientific Revolution*, New Haven 2007.

Harms 1985
W. Harms, 'On natural history and emblematics in the 16th century', in: J. Ackerman (ed.), *The natural sciences and the arts. Aspects of interaction from the Renaissance to the 20th century*, Uppsala 1985, 67-83.

Harrison 1998
P. Harrison, *The Bible, Protestantism and the rise of natural science*, Cambridge 1998.

Huisman 2009
T. Huisman, *The finger of God. Anatomical practice in 17th-century Leiden*, Leiden 2009.

Hunt *et.al.* 2010

L. Hunt, M. Jacob & W. Mijnhardt (eds.), *Bernard Picard and the first global vision of religion*, Los Angeles 2010.

Impey & MacGregor 1985
O. Impey & A. MacGregor (eds.), *The origins of museums. The cabinet of curiosities in sixteenth- and seventeenth-century Europe*, Oxford 1985.

Jorink 2010
E. Jorink, *Reading the Book of Nature in the Dutch Golden Age, 1575-1715*, Leiden & Boston 2010.

Kaufmann 1993
T.D. Kaufmann, *The mastery of nature. Aspects of art, science and humanism in the Renaissance*, Princeton 1993.

Kemp 1990
M. Kemp, *The science of art. Optical themes in Western art from Brunelleschi to Seurat*, New Haven 1990.

Kemp 2006
M. Kemp, *Leonardo da Vinci. Experience, experiment, and design*, Princeton & Oxford 2006.

Kemp 2007
M. Kemp, *The human animal in Western art and science*, Chicago 2007.

Kenseth 1991
J. Kenseth (ed.), *The age of the marvelous*, Hanover 1991.

Klein & Spary 2009
U. Klein & E.C. Spary (eds), *Materials and expertise in early modern Europe: between market and laboratory*, Chicago 2010.

Kooijmans 2004
L. Kooijmans, *De doodskunstenaar. De anatomische lessen van Frederik Ruysch*, Amsterdam 2004.

Kusukawa & Maclean 2006
S. Kusukawa & I. Maclean (eds.), *Transmitting knowledge. Words, images, and instruments in early modern Europe*, Oxford 2006.

Lindberg & Westman 1990
D.C. Lindberg & R.S. Westman (eds.), *Reappraisals of the Scientific Revolution*, Cambridge 1990.

Logan 1979
A.-M.S. Logan, *The 'cabinet' of the brothers Gerard and Jan Reynst*, Amsterdam 1979.

Margócsy 2009
D. Margócsy, 'Advertising cadavers in the Republic of Letters. Anatomical publications in the early modern Netherlands', *British Journal for the History of Science* 42 (2009) 187-210.

Margócsy 2011
D. Margócsy, 'A museum of wonders or a cemetery of corpses? The commercial exchange of anatomical collections in the early modern Netherlands', in: S. Dupré & C. Lüthy (eds.), *Silent messengers. The circulation of material objects of knowledge in the early modern Low Countries*, Berlin 2011, 185-216.

Mason 2009
P. Mason, *Before disenchantment. Images of exotic animals and plants in the early modern world*, London 2009.

Morford 1991
M. Morford, *Stoics and neostoics. Rubens and the circle of Lipsius*, Princeton 1991.

Park & Daston 2006
K. Park & L. Daston (eds.), *The Cambridge history of science. Early modern science*, vol. 3, Cambridge 2006.

Parker Brienen 2006
R. Parker Brienen, *Visions of savage paradise. Albert Eckhout, court painter in colonial Dutch Brazil*, Amsterdam 2006.

Punt 1983
H. Punt, *Bernard Siegfried Albinus (1697-1770). On 'human nature'. Anatomical and physiological ideas in eighteenth-century Leiden*, Amsterdam 1983.

Reitsma 2008
E. Reitsma, *Maria Sibylla Merian & daughters. Women of art and science*, Zwolle 2008.

Roberts *et al.* 2007
L. Roberts, S. Schaffer & P. Dear (eds), *The mindful hand. Inquiry and invention from the late Renaissance to early industrialisation*, Amsterdam, 2007.

Schmidt 2007
B. Schmidt, '"Imperfect chaos". Tropical medicine and exotic natural history c.1700', in: O.P. Grell & A. Cunningham (eds.), *Medicine and religion in Enlightenment Europe*, Aldershot 2007, 145-172.

Secord 2004
J. Secord, 'Knowledge in transit', *Isis* 95 (2004) 654-672.

Shapin & Schaffer 1985
S. Shapin & S. Schaffer, *Leviathan and the air-pump. Hobbes, Boyle and the experimental life*, Princeton 1985.

Silver 2006
L. Silver, 'Arts and minds. Scholarship on early modern art history (Northern Europe)', *Renaissance Quarterly* 59 (2006) 351-373.

Smith 2004
P.M. Smith, *The body of the artisan. Art and experience in the Scientific Revolution*, Chicago 2004.

Smith 2006
P.M. Smith, 'Art, science, and visual culture in early modern Europe', *Isis* 97 (2006) 83-100.

Smith 2009
P.M. Smith, 'Science on the move. Recent trends in the history of early modern science', *Renaissance Quarterly* 62 (2009) 345-375.

Smith & Findlen 2002
P.M. Smith & P. Findlen (eds.), *Merchants and marvels. Commerce, science, and art in early modern Europe*, New York & London 2002.

Stumpel 2004
J. Stumpel, review of Bakker 2004a, *Simiolus* 31 (2004), 115-124.

Swan 2002
C. Swan, 'From blowfish to still life paintings. Classification and its images, circa 1600', in: Smith & Findlen 2002, 109-137.

Swan 2005
C. Swan, *Art, science, and witchcraft in early modern Holland. Jacques de Gheyn II (1565-1629)*, Cambridge 2005.

Swan 2008
C. Swan, 'Making sense of medical collections in early modern Holland. The uses of wonder', in: P.H. Smith & B. Schmidt (eds.), *Making knowledge in early modern Europe. Practices, objects and texts, 1400-1800*, Chicago 2008, 199-213.

Taylor 1995
P. Taylor, *Dutch flower painting 1600-1720*, New Haven & London 1995.

Vanagt 2010
K. Vanagt, *The emancipation of the eye. V.F. Plempius'* Ophthalmographia *and early-modern medical conceptions of sight*, Enschede 2010.

Van Berkel 1983
K. van Berkel, *Isaac Beeckman (1588-1637) en de mechanisering van het wereldbeeld*, Amsterdam 1983.

Van Berkel 2010
K. van Berkel, 'The Dutch Republic. Laboratory of the Scientific Revolution', in: K. van Berkel & L. de Goei (eds.), *The international relevance of Dutch history* (= special issue *Bijdragen en medelingen betreffende de geschiedenis der Nederlanden* 125 (2010) 3-367).

Van Berkel *et.al.* 1999
K. van Berkel, A. van Helden & L. Palm (eds.), *A history of science in the Netherlands. Survey, theme and reference*, Leiden 1999.

Van den Boogaart 2003
E. van den Boogaart, *Civil and corrupt Asia. Image and text in the* Itinerario *and the* Icones *of Jan Huygen van Linschoten*, Chicago 2003.

Van den Boogert 1999
B. van den Boogert (ed.), *Rembrandts schatkamer*, Zwolle & Amsterdam 1999.

Van de Roemer 1998
G.M. van de Roemer, 'God en het rariteitenkabinet. Het religieuze motief van Noord-Nederlandse rariteitenverzamelaars einde zeventiende en begin achttiende eeuw', *Theoretische geschiedenis* 25 (1998), 242-255.

Van de Roemer 2004
G.M. van de Roemer, 'Neat nature. The relation between art and nature in a Dutch cabinet of curiosities from the early eighteenth century', *History of Science* 42 (2004), 47-84.

Van de Roemer 2005
G.M. van de Roemer, *De geschikte natuur. Theorieën over natuur en kunst in de verzameling van zeldzaamheden van Simon Schijnvoet (1652-1727)*, Amsterdam 2005 (unpub. diss. University of Amsterdam).

Van de Roemer 2008
G.M. van de Roemer, 'Het lichaam als borduursel. Kunst en kennis in het anatomisch kabinet van Frederick Ruysch', in: A.-S. Lehmann & H. Roodenburg (eds.), *Body and embodiment in Netherlandish art*, Zwolle 2008 (= *Nederlands Kunsthistorisch Jaarboek/Netherlands Yearbook for History of Art* 58 [2008]), 216-240.

Van Gelder 2011
E. van Gelder, *Tussen hof en keizerskroon. Carolus Clusius en de ontwikkeling van de botanie aan de Midden-Europese hoven (1573-1593)*, Leiden 2011.

Van Suchtelen & Van Beneden 2009
A. van Suchtelen & B. van Beneden, *Kamers vol kunst in zeventiende-eeuws Antwerpen*, Antwerpen & Zwolle 2009.

Vermij 1999
R. Vermij, 'Een nieuw concept. De wetten der natuur', in: F. Egmond, E. Jorink & R. Vermij (eds.), *Kometen, monsters en muilezels. Het veranderende natuurbeeld en de natuurwetenschap in de zeventiende eeuw*, Haarlem 1999, 105-119.

Zuidervaart 2009
H.J. Zuidervaart, 'Het in 1658 opgerichte theatrum anatomicum te Middelburg. Een medisch-wetenschappelijk en cultureel convergentiepunt in een vroege stedelijke context', in: *Archief. Mededelingen van het Koninklijk Zeeuwsch Genootschap der Wetenschappen*, Middelburg 2009, 73-140.

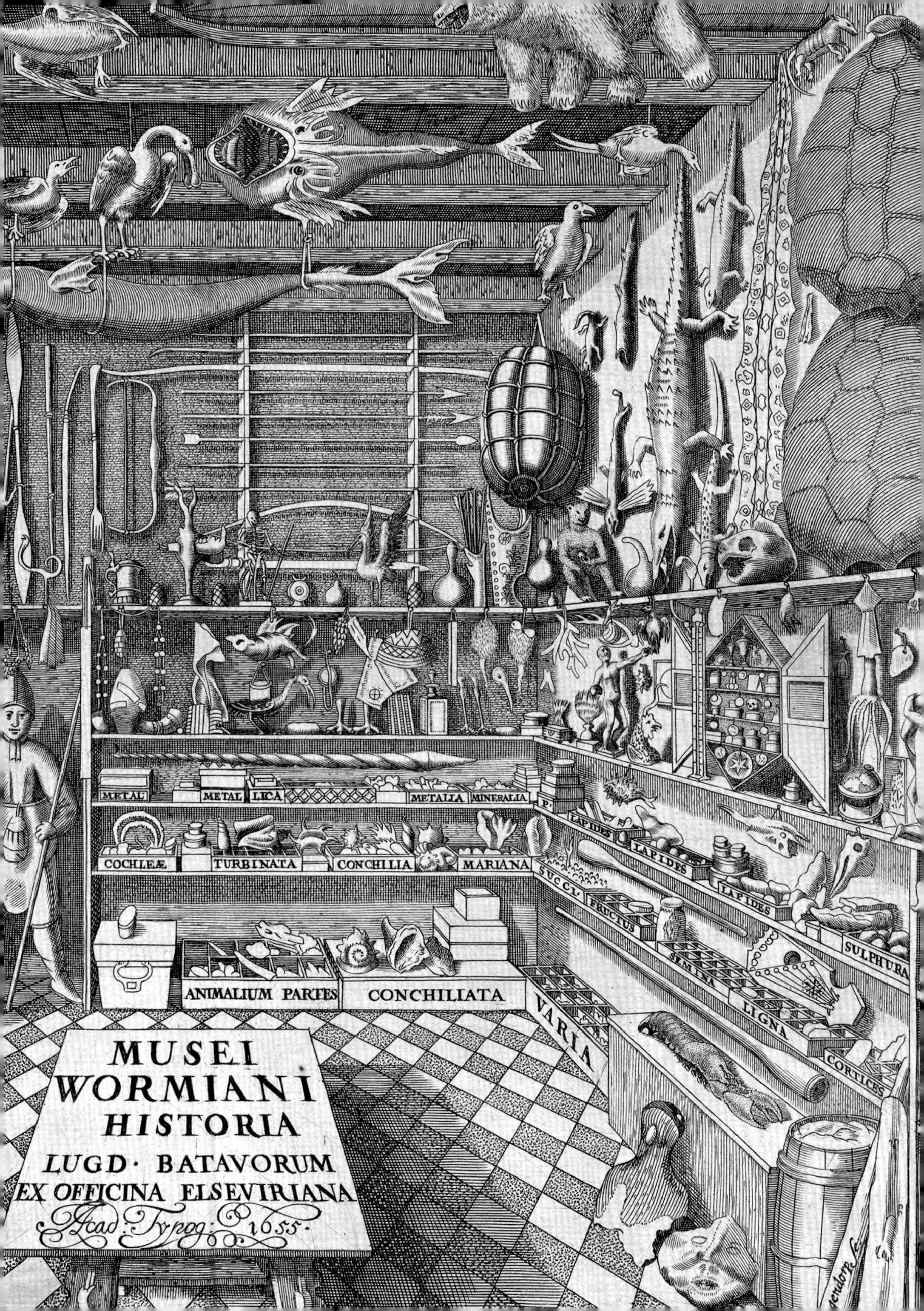
METAL
METAL LICA
METALLA
MINERALIA
COCHLEÆ
TURBINATA
CONCHILIA
MARIANA
LAPIDES
LAPIDES
LAPIDES
SUCCI
FRUCTUS
SEMINA
LIGNA
CORTICES
SULPHURA
ANIMALIUM PARTES
CONCHILIATA
VARIA
MUSEI
WORMIANI
HISTORIA
LUGD· BATAVORUM
EX OFFICINA ELSEVIRIANA
Acad: Typog: 1655·

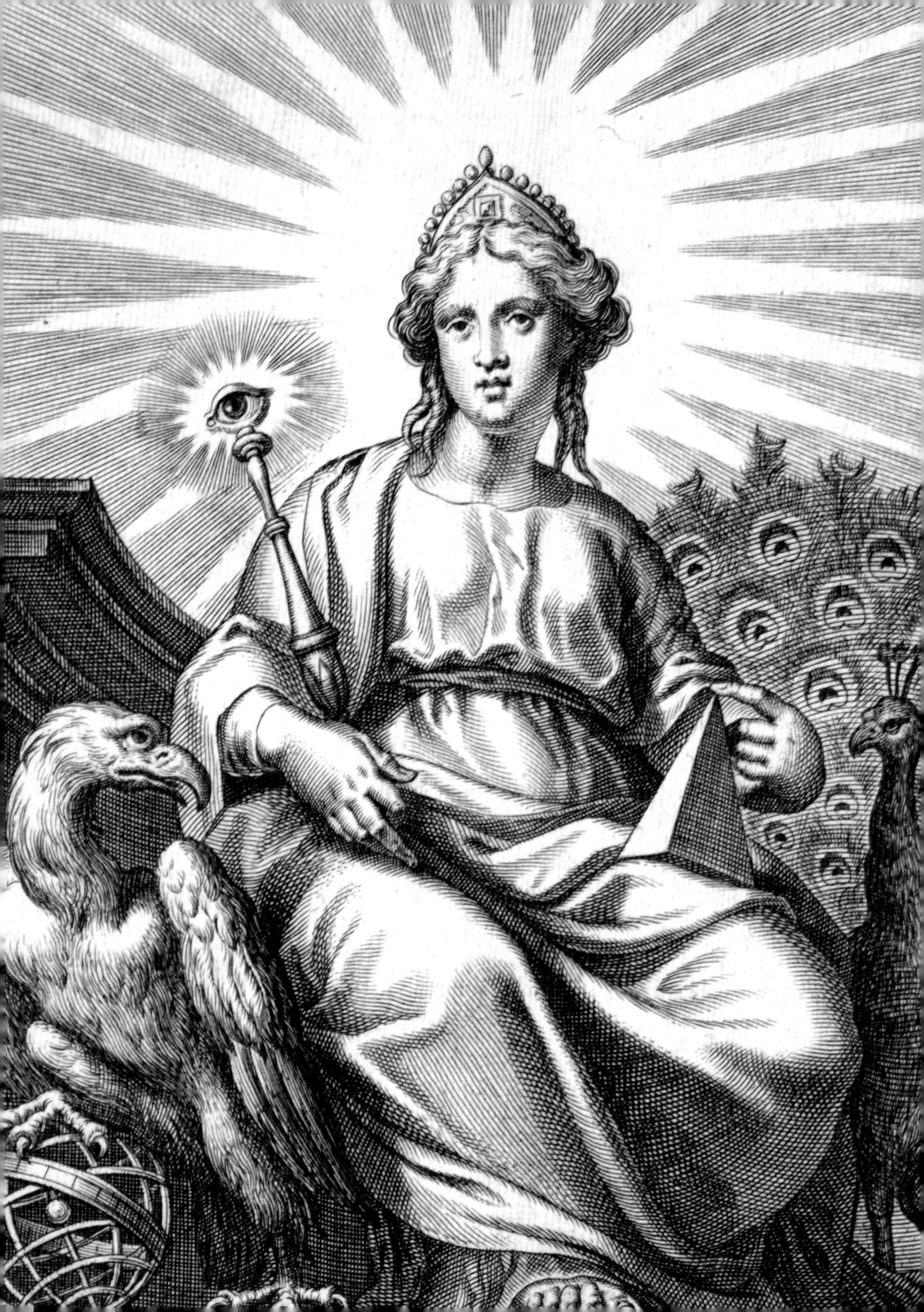

The historiography of perspective and *reflexy-const* in Netherlandish art

Sven Dupré

Introduction

In *The heritage of Apelles* (1976), Ernst Gombrich famously drew a distinction between art North and South of the Alps along optical lines. Painters in fifteenth-century Florence, such as Domenico Veneziano (c.1410-c.1461), used different optical features than did their contemporaries in Bruges, the likes of Jan van Eyck (1390-1441), to create the illusion of space. Gombrich commented:

> We all associate Florentine art with the development of central perspective, and thus with the mathematical method of revealing form in ambient light. The other aspect of optical theory, the reaction of light to various surfaces, was first explored in modern times by painters North of the Alps. It was there that the mastery of lustre, sparkle and glitter was first achieved, permitting the artist to convey the peculiar character of materials. Indeed, for a time, during the first decades of the fifteenth century, the two schools of painting appeared thus to have divided the kingdom of appearances between them.[1]

While Italian painters of the fifteenth century used perspective, Netherlandish artists studied and painted the reflection of light from the surfaces of different textures and materials to create the illusion of space. For Gombrich, the point of contact between North and South was Leonardo da Vinci (1452-1519), 'the greatest explorer of natural appearances', who must have been 'a keen student of Northern painting'.[2]

Here I am interested less in Gombrich's geography of optics, but more in how his juxtaposition of perspective and painterly light reflects and complicates the historiography of perspective. For reasons that fall outside the scope of this paper, linear perspective, since the seminal work of Erwin Panofsky, has become a *locus classicus* of the study of artistic practice and science; and the history of perspective, which Panofsky in *Die Perspektive als symbolische Form* (1927) disconnected from optics, has become a field of research of its own, independent from the history of science and the history of art. Panofsky projected the categories of *perspectiva naturalis* and *perspectiva artificialis* back into the period of the Renaissance. While *perspectiva naturalis* or *communis* referred to the general category of the science of optics – including questions of

Detail fig. 4
Franciscus Aguilonius, *Opticorum libri sex*, 1613
Frontispiece. Ghent, University Library, Math. 101.

psychology, physiology, anatomy, physics, and mathematics – *perspectiva artificialis* was the more limited domain of the geometrical *technique* (not the science) of drawing in perspective. We know now that the definition of two types of *perspectiva* underscores a desire of finding a rupture between the Middle Ages and the Renaissance but fails to find support in the sources.[3] But due to this historical separation between perspective and optics, which at the same time was constitutive of a field of study of art and science, the study of painterly light – associated by Gombrich with artists North of the Alps, as we have seen – has received much less attention. Perhaps unsurprisingly, the most important exception is Gombrich's 'keen student of Northern painting', Leonardo da Vinci.[4]

In this paper I am interested in the historiographical disconnection of optics and perspective since Panofsky, to which Gombrich's geography of painterly optics is a reaction and (as I hope to show) only a partial remedy. My paper consists of two parts. In the first part I will argue that the historiographical disconnection of optics and perspective has entailed serious distortions of the nature of the enterprises, consisting of treatises of mathematicians, canonized in the historiography of perspective. The presentation of *perspectiva naturalis* and *perspectiva artificialis* as two different and largely independent enterprises also had serious distortive consequences for the presentation of the role of perspective in the history of optics. The standard narrative on the history of optics – that of David Lindberg in his *Theories of vision* (1976) – constructed a so-called perspectivist canon, following the reception of the work of Ibn al-Haytham (or Alhacen, 965-c. 1040) in the Latin West. Lindberg discussed painters' perspective as an impoverished 'application' of perspectivist theory, without a development of its own and without much consequence for the route taken by the discipline of perspectivist optics since the fifteenth century. He concluded a section on Filippo Brunelleschi (1377-1446), Leon Battista Alberti (1404-1472), and Lorenzo Ghiberti (1378-1455) as follows:

> The inventors of linear perspective (...) possessed at least an elementary knowledge of medieval visual theory, and they employed the visual cone or pyramid of the perspectivists to justify the geometrical operations by which the three-dimensional field was projected onto the two-dimensional panel. (...) But it must be emphasized that only the barest framework, the mathematical skeleton, of the perspectivist theory of vision was required.[5]

The separation between optics and perspective has obscured the fact that treatises, in the historiography of perspective considered to be strictly on perspective, were in fact on optics. I will illustrate this point with Simon Stevin's *Vande deursichtighe* (1605), a book that not only played a significant role in the building of the now dominant historiography of perspective but also is considered, in the words of Kirsti Andersen, 'the origin of a Dutch academic approach to perspective' (and not optics).[6] It is worth pointing out though that other Dutch treatises on perspective, such as that of Hans Vredeman de Vries, discussed below, reached a much larger audience in the seventeenth century.

My focus in the second part of this paper will be on another genre of texts – art theory treatises. This discussion will support my contention that historians' separation of the histories of perspective and optics has often, in the literature on art and science, reduced artists' optical knowledge to knowledge of perspective, leaving aside the effects of light and material. However, to make this point, we do not need to embrace Gombrich's geography of optics, in which (as we have seen) he associated light and texture with paintings North of the Alps. My contention about the historiographical distortion of the separation of *perspectiva naturalis* and *perspectiva artificialis* is equally valid for Italy. In the second part of the paper I will therefore first turn to the art theory work of Giovanni Paolo Lomazzo (1538-1592) to show how his definitions of perspective and artists' knowledge of optics did not leave the study of the effects of light and texture aside. Next, I will turn to Karel Van Mander's *Schilder-boeck* (1604). The choice of connecting the treatises of Lomazzo and Van Mander in this context is most appropriate. Lomazzo was very familiar with Netherlandish art (mostly derived from his reading of Domenicus Lampsonius's *Pictorum aliquot celebrium Germaniae Inferioris Effigies* [1572]) and thought highly of it.[7] Moreover, because of his attention to light and reflection, his work was particularly suited for adoption, and was indeed highly influential, in Dutch art theory. Van Mander made the study of light and its reflection from different surfaces (*reflexy* and *reflexy-const*) foundational to the shaping of the Netherlandish canon of art of which Van Eyck was the origin and beginning. This paper is no direct response to Gombrich's call. We do not intend to study how, for example, Van Eyck painted. But we do hope that the study of art theory treatises might provide a middle ground between painterly practices in Netherlandish art and the study of light in contemporary optics, which should be fruitful for the future study of interactions between artists' optical knowledge and painterly practice.[8]

Stevin's *Vande deursichtighe*. Optics and perspective in the mathematical treatises

Kirsti Andersen's recent history of perspective is a scholarly tour de force, in that it provides the reader with a narrative from the fifteenth to the nineteenth century. However, as already announced in the title of her book, *The geometry of an art*, she considers perspective theory as geometry and a self-sustained discipline, explicitly supporting a separation between *perspectiva naturalis* and *perspectiva artificialis* in her introduction. Simon Stevin (1548-1620) is a crucial figure in her story. She writes:

> The real progress in the entire history of mathematical theory of perspective was made by a few mathematicians, among whom the most outstanding are Guidobaldo, Stevin, 's Gravesande, Taylor and Johann Heinrich Lambert. One of my interests has been to find connections between these scholars, and I hope to have convinced the reader that Stevin played an important role in a continuous development from Guidobaldo to Taylor.[9]

Andersen then firmly places Stevin's treatise within a history of mathematization of linear perspective – as did Rocco Sinisgalli before her or, more recently, Jean-Pierre Le Goff – which increasingly moves the study of perspective away from painterly practice and in the direction of projective geometry.[10] The point of the work of these historians of perspective is to determine, on the one hand, the influence of the work on perspective of the Urbino mathematicians, and especially Guidobaldo del Monte (1545-1607) on Stevin's *Vande deursichtighe*, and on the other, to show how Stevin's work on perspective was received by Girard Desargues (1591-1661).

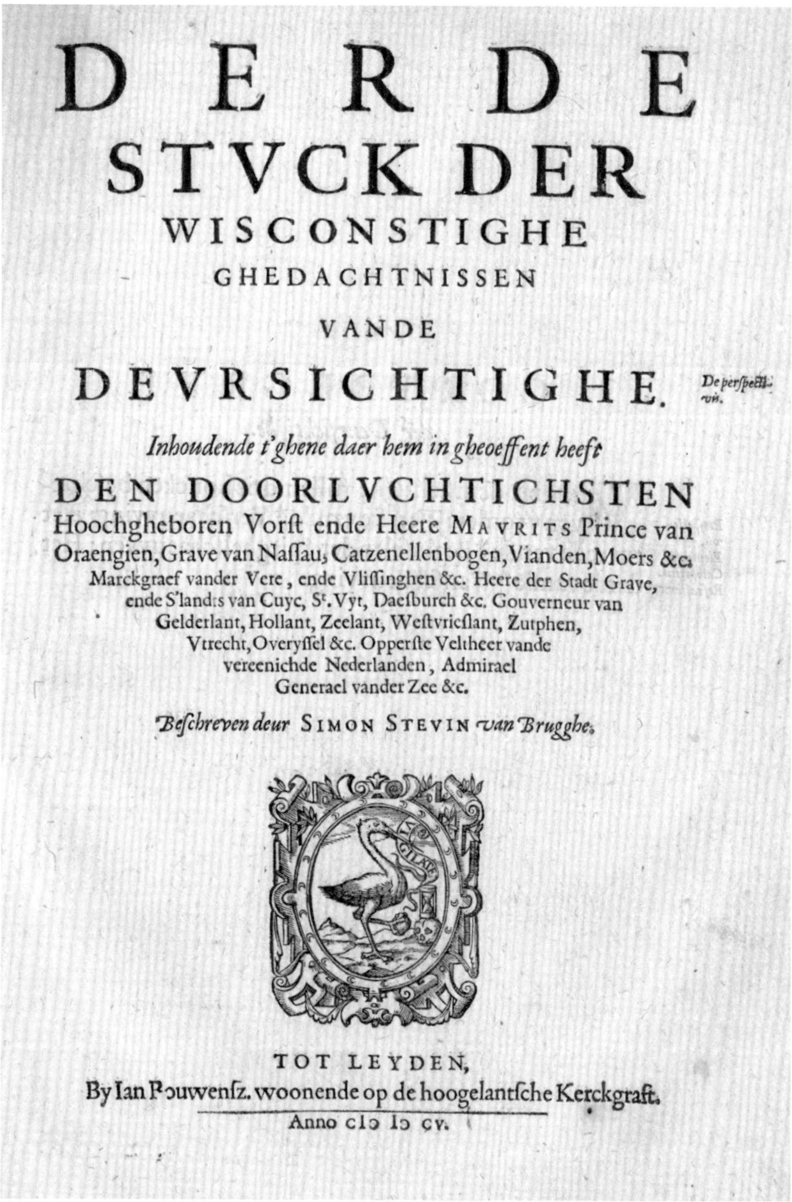
DERDE
STVCK DER
WISCONSTIGHE
GHEDACHTNISSEN
VANDE
DEVRSICHTIGHE.

Inhoudende t'ghene daer hem in gheoeffent heeft
DEN DOORLVCHTICHSTEN
Hoochgheboren Vorst ende Heere MAVRITS Prince van Oraengien, Grave van Nassau, Catzenellenbogen, Vianden, Moers &c. Marckgraef vander Vere, ende Vlissinghen &c. Heere der Stadt Grave, ende S'lands van Cuyc, S^t. Vyt, Daesburch &c. Gouverneur van Gelderlant, Hollant, Zeelant, Westvrieslant, Zutphen, Vtrecht, Overyssel &c. Opperste Veltheer vande vereenichde Nederlanden, Admirael Generael vander Zee &c.

Beschreven deur SIMON STEVIN *van Brugghe.*

TOT LEYDEN,
By Ian Pouwensz. woonende op de hoogelantsche Kerckgraft.
Anno clɔ Iɔ cv.

1
Simon Stevin, *Vande deursichtighe*, 1605
Deursichtighe is the translation of *perspectiva.*
Ghent, University Library, Acc. 56517.

However, Dominique Raynaud has convincingly shown that in the fifteenth century the distinction between *perspectiva artificialis* and *perspectiva naturalis* was constructed only a posteriori to mark a rupture between the Middle Ages and the Renaissance.[11] All aspects of *perspectiva* – the anatomical, physiological, physical, geometrical, psychological – were in fact inseparable, and authors, such as Ghiberti, Alberti, and Piero della Francesca (c. 1420-1492), considered themselves as working on *perspectiva* as much as their thirteenth-century predecessors, such as Roger Bacon (c. 1214- c. 1294) and John Pecham (c. 1230-1292). In fact, Raynaud shows that Ghiberti, Alberti and Piero della Francesca, among other fifteenth-century authors on optics, borrowed extensively from a specific group of thirteenth-century optical texts, those authored by the Franciscans residing in Oxford: Robert Grosseteste (c. 1170-1253), Bacon, and Pecham.

Raynaud's *Hypothèse d'Oxford* undercuts the foundations of the historiography of perspective as found in Andersen's account. But Raynaud discusses developments only up to around 1500: what about a century later? Had perspective not developed into a discipline of its own, independent of optics, when Stevin's *Vande deursichtighe* was published as part of the *Wisconstige ghedachtenissen* in the first decade of the seventeenth century? Here I will show that the historiography of perspective has also seriously misrepresented Stevin's project. We should not forget that Stevin's book on perspective (which he called in Dutch *verschaeuwing*, indicating that this was a translation of *scenographia*) was one of only two published books belonging to the *Vande deursichtighe*, and that more books were planned but never published (see figs. 1 and 2). In the second book of *Vande deursichtighe* Stevin discussed *spiegelschaeuwen*, or catoptrics: the study of reflection in plane, convex, and concave mirrors. It is telling that, though occasionally mentioned in

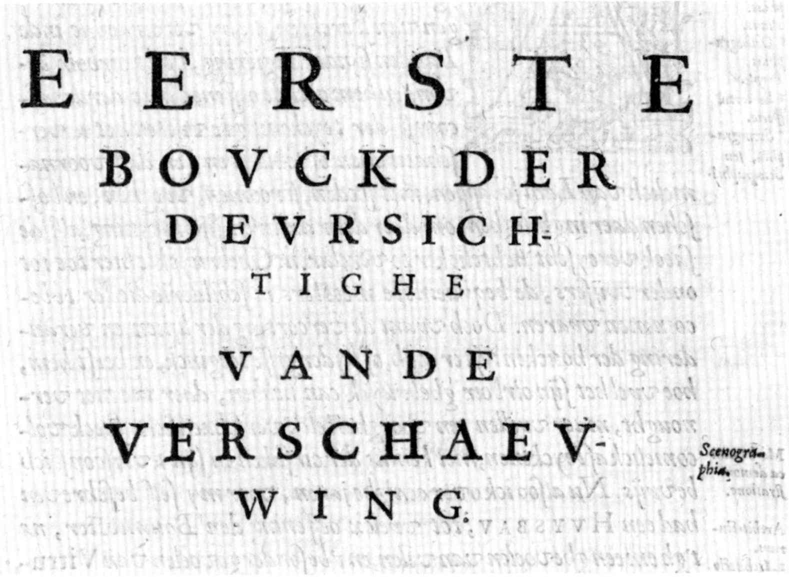
EERSTE
BOVCK DER
DEVRSICH-
TIGHE
VANDE
VERSCHAEV-
WING.

Scenographia.

2

Simon Stevin, *Vande deursichtighe*, 1605

Verschaeuwing is the translation of *scenographia*.

Ghent, University Library, Acc. 56517.

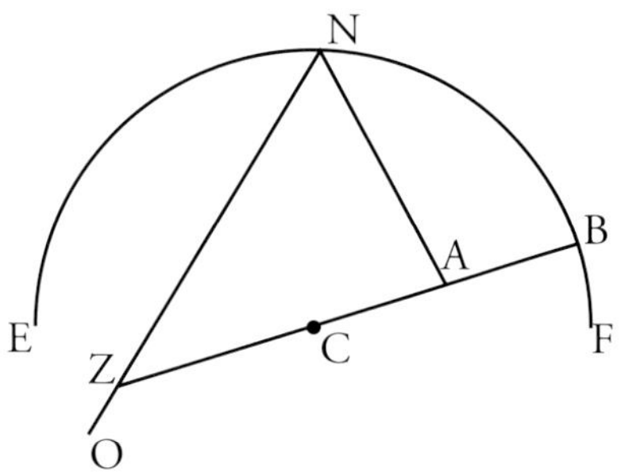

3
The cathetus rule for curved mirrors. If the mirror is curved, the cathetus is the line drawn from the object to the centre of curvature of the mirror. In the concave mirror EF, the locus of the image Z of the object A seen by the eye O is at the intersection of the cathetus AB and the reflected ray NO.

passing, this second book has never been analyzed, nor was it included in Dirk Struik's edition and translation of the collected works of Stevin, which did include the first book of *Vande deursichtighe*.[12]

What then was the content of this obscured second book on catoptrics? Stevin focused on image location in plane, convex, and concave mirrors. The appendix to this book is a critique of the so-called cathetus rule – a geometrical construction that was used in the period to determine the location of an image in a mirror (fig. 3).[13] In a concave mirror, the image was said to be at the intersection of the line of incidence and the cathetus – the line connecting the object-point and the centre of curvature of the concave mirror. Stevin took issue with this. He denied that the image was located at this point when perceived with one eye – a condition that the cathetus rule took for granted. He referred to the appropriate passages in the works of Euclid, Alhacen, and Witelo. These opticians had been misled, according to Stevin, because the image is indeed perceived in a concave mirror at the point predicted by the cathetus rule when the observer looks in the concave mirror with two eyes – a condition contrary to the one-eye assumption on which the cathetus rule is based. Stevin's critique of the cathetus rule, and his introduction of the effects of binocular vision and how this is connected to the issue of image location in mirrors, finds a point of comparison in the third chapter of Johannes Kepler's *Paralipomena ad Vitellionem*, published in 1604.[14] Perhaps Stevin knew the work of Kepler, although Kepler's approach after his critique of the cathetus rule was different from Stevin's. Kepler embraced binocular vision and developed an alternative rule for the location of images perceived with both eyes, while Stevin offered an alternative rule for the location of images perceived with one eye only. Whatever the lines of influence, the important point in this context is that Stevin's project in *Vande deursichtighe* was optical, not 'perspectival'.

The remarkably aborted nature of Stevin's optical project – only two books were published – reminds us of another optical work published by the Jesuit mathematician François de Aguilón (1567-1617) only a few years later in the Low Countries, in 1613 (fig. 4).[15] Aguilón's *Opticorum libri sex* is similarly the result of a larger publication project on optics that was never fully realized. Aguilón planned three books, but only the book on direct vision was published, not those on catoptrics and dioptrics (the study of refraction). I have argued elsewhere that Aguilón's optical project (and its aborted result) was partly shaped by his activity and interests in architecture.[16] Aguilón was the architect of the Carolus Borromeus Church in Antwerp and, more importantly for his ideas on the scope and disciplinary identity of optics, familiar with Vitruvianism. A similar influence is at work in Stevin's image of optics. For now, I point to only two aspects of this influence. First, Vitruvius expressed the architect's need to be able to draw an *ichnography* (or a plan), an *orthography* (or an elevation), and a *scenography* (or a shaded perspective rendering of a building). *Scenography* was precisely the term that Stevin used as a title for his book on perspective. Second, for Vitruvius, the architect needed to have knowledge of a whole range of disciplines, which was reflected in the wide scope of Vitruvian architecture.

4
Franciscus Aguilonius, *Opticorum libri sex*, 1613
Frontispiece. Ghent, University Library, Math. 101.

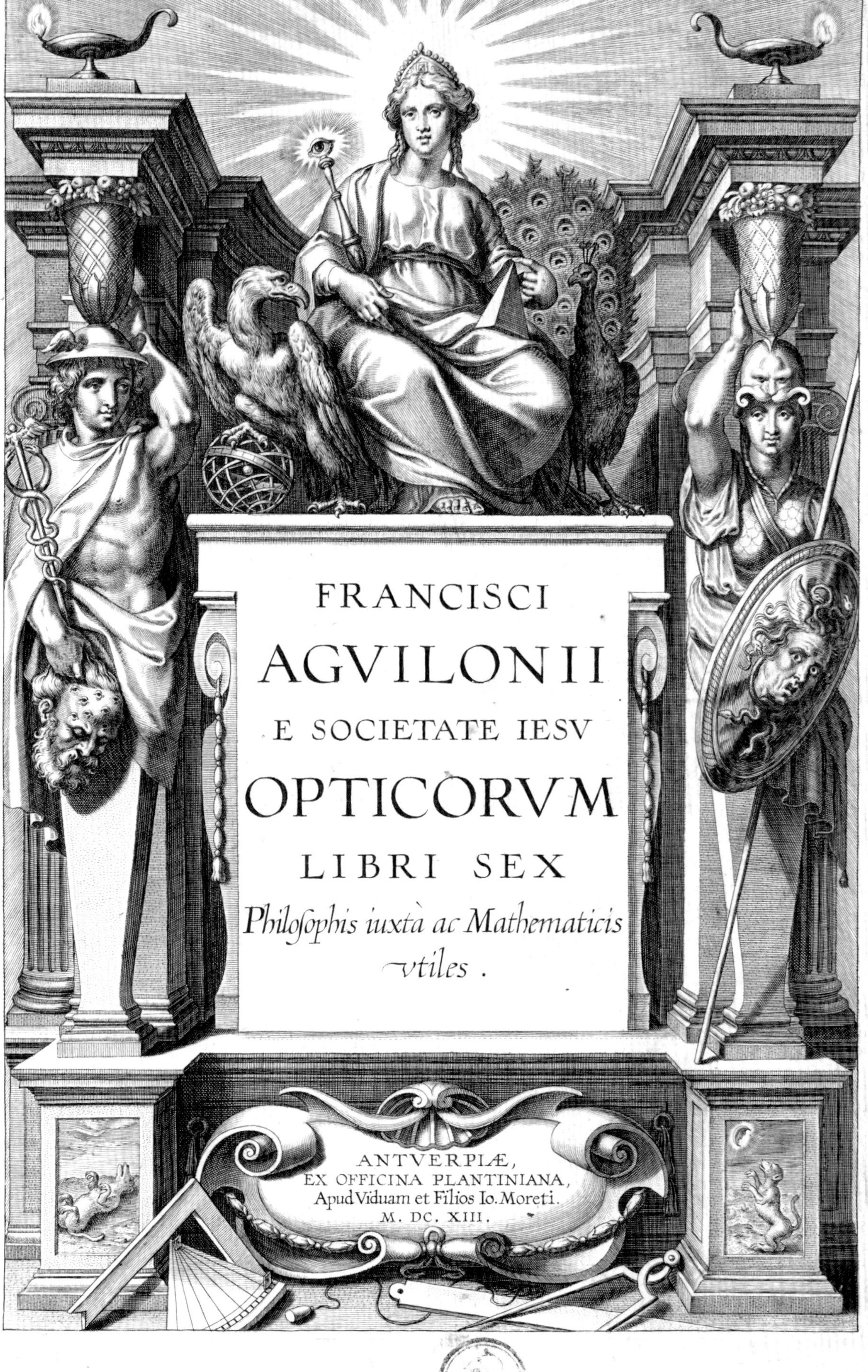
FRANCISCI
AGVILONII
E SOCIETATE IESV
OPTICORVM
LIBRI SEX
Philosophis iuxtà ac Mathematicis
vtiles.
ANTVERPIÆ,
EX OFFICINA PLANTINIANA,
Apud Viduam et Filios Io. Moreti.
M. DC. XIII.

Vitruvius's book on architecture contained sections on optics, astronomy, clocks, sundials, and machines, among other things. Optics included the study of projections used in sundials and astrolabes, as it did for Stevin. At the beginning of the first book of *Vande deursichtighe*, he laid out his optical enterprise and the publication project that he would only partly realize:

> Optics as genus has several species, such as Catoptrics, Dioptrics, Planispheres, Sun-dials, *verschaeuwing* [foreshortening], and several others, which have something in common in pertaining to optics, but since their effects serve different ends and consequently they have to be performed in different ways, each species is discussed separately as a special art and described in due order.[17]

Stevin migrated to the North during a period of war between the revolting Dutch provinces in the North and the Spanish army. Events would evolve into the independence of seven northern provinces and their separation from the southern provinces, which would remain under Spanish control. An important factor in these events was the Stadholder Maurits (1567-1625), who gained wide renown for his army leadership and whose successes in war are said to have relied a great deal on the army reforms that were carried out under his leadership. Stevin and Maurits probably met when they both studied at the University of Leiden in 1582-1584.[18] Stevin became Maurits's personal mathematics teacher. Some of Stevin's many works were written at the request of Maurits. The clearest evidence of their close cooperation is offered by the *Wisconstige ghedachtenissen*, published by Stevin in 1605-1608, and of which *Vande deursichtighe* is part. This book contains the mathematics lessons originally given by Stevin to Maurits. Maurits's court in The Hague has recently been portrayed as an important centre of patronage of the arts and sciences, and Maurits as especially fond of engineers and inventions. These preferences were also reflected in the kind of mathematics offered.[19]

Originally, Stevin seems to have been part of Maurits's personal household, but in 1593, Maurits persuaded the States to give Stevin a paid position as quartermaster in the States' army, and he accompanied Maurits during his military campaigns. Stevin's involvement with military affairs is reflected in his publications. In 1594 he published a treatise on fortifications, *De sterctenbouwing*, closely connected to international – that is, Italian – trends in fortification design, which had been introduced in the Netherlands in the second half of the sixteenth century by Italian engineers in the service of the Spanish.[20] Two later works on military affairs more closely reflect Stevin's own experiences in the Dutch war: *Nieuwe Maniere van Sterctebou door Spilsluysen*, and *Castrametatio*, on the establishment and ideal layout of a military camp, both published in 1617. Stevin was also involved in the development of the programme of the Duytsche Mathematique, a school associated with the University of Leiden and established by Maurits in 1600, where students were taught in the vernacular in order to supply the new state, the Dutch Republic, with engineers and surveyors.[21]

In *Vande deursichtighe*, Stevin naturally portrayed Maurits as an eager student of perspective. Stevin suggested that he himself became involved with the subject following up on a request of Maurits. The context of this request was Maurits's supervision of fortifications:

> As his *Princely Grace* frequently exercised himself in drawing ground and vertical plans for fortification, which he erected in the lands under his government, he found it useful to exercise himself as well in the third species of drawing, to wit perspective or painting, such mainly of landscapes, with cities, rivers, roads, and woods situated therein, thus to explain more easily to others his views, as required by the matter.[22]

This same connection between Vitruvius and the use of drawings as a means of communication between different parties involved in military fortification, including Maurits himself, returns in the continuation of this same passage, in which Stevin also refers to a work of his on architecture, *Huysbou*, never published during his lifetime and recently reconstructed by the Dutch historian of architecture, Charles van den Heuvel.

> Now since I had some years ago myself described an *Architecture*, to the practice of which, in the common opinion of many and the special opinion of Vitruvius in the 2nd chapter of his 1st book, knowledge of perspective is conducive, I perused and examined, more fully than before, several writers who deal with this subject and made a description thereof in my own words. And after his *Princely Grace* had looked it through and helped to correct the imperfections that are commonly found in first attempts, had also fundamentally understood the common rule of finding the perspective of any given figure, and to his satisfaction practised it, I included this description among his *Mathematical Memoirs* (...)[23]

To learn perspective, Maurits had first turned to the painters, but he had been disappointed, according to Stevin:

> But because the foreshortening of the lines and the change of the angles was obtained by sight or by guessing – which have their use – he [Maurits] was not satisfied with this, but wished to design exactly the perspective of any given figure, with knowledge of the causes and its mathematical proof.[24]

Were there any grounds for this dissatisfaction?

An important book on perspective in the Netherlands was *Perspective*, published in 1604, by Hans Vredeman de Vries (1527-c. 1607) and his son Paul. This book contains a series of drawings, so-called *perspectiven*, representations of architecture, with a minimal amount of text. Vredeman revealed the construction method of his drawings, but only in the second book of *Perspective* (fig. 5).[25] The construction he used is similar to the distance-point method. Vredeman draws a horizon line, on which he

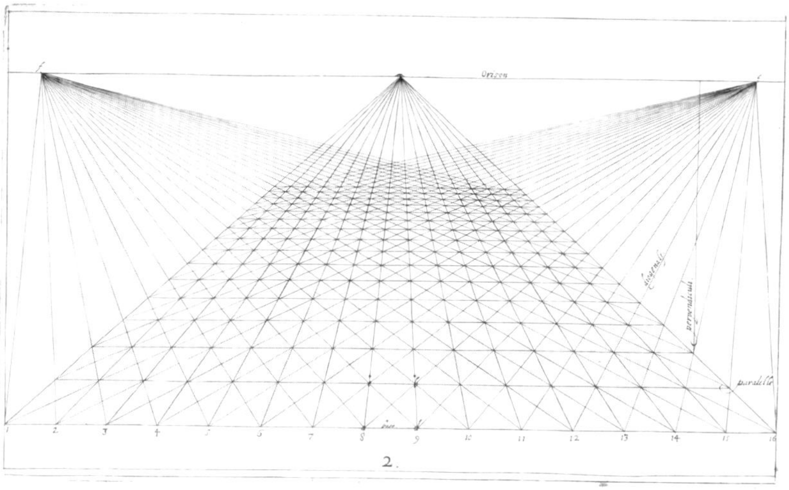

5
Hans Vredeman de Vries, *Perspective,* 1604-1605
Brussels, Royal Library of Belgium, WBS CL 13961b.

situates the 'eye point' (a central vanishing point), parallel to and 5.5 feet above the base line, the assumed height of a human figure. He divides the base line in equal intervals and connects the 'eye point' with the interval points set out on the base line. To draw the perspectival squares, he makes the shortening of a square equal to the amount it becomes less broad. Next he draws the diagonals of the perspectival square, and where they intersect the horizon line, he locates two 'horizon points'. The intersections of the lines connecting to the 'horizon points' and the lines connecting to the 'eye point' are then used to construct a perspectival checkerboard floor. The rules for the foreshortening of the square lack any optical or geometrical rationale. Moreover, Vredeman realizes that lines not perpendicular to the image plane converge to their own 'accidental' vanishing points, but the precise location of these vanishing points by Vredeman is arbitrary, as we can see in figure 6.

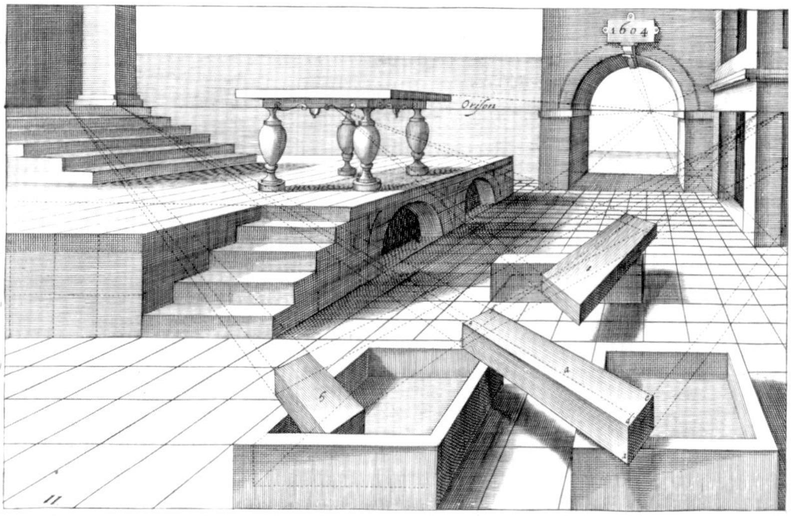

6
Hans Vredeman de Vries, *Perspective,* 1604-1605
Brussels, Royal Library of Belgium, WBS CL 13961b

There were then indeed grounds for dissatisfaction with the painters for those, like Maurits, who 'wished to design exactly the perspective of any given figure', as Stevin wrote. The first book of Stevin's *Vande deursichtighe* would have provided satisfaction by offering a general theorem of the vanishing point to Maurits and his other readers (fig. 7).[26] Stevin shows that parallel lines seen through a picture plane – Stevin speaks of a 'glass' – which is not parallel to said parallel lines, converge in one vanishing point, and he locates this vanishing point at the intersection of the line through the eye, parallel to the parallel lines, and the picture plane. This theorem allows Stevin perspective constructions like the following for the case of a point in the ground plane (see fig. 8). A is the object-point; B the intersection line of the ground and picture planes; DE the observer, with D the foot in the ground plane and E the eye, perpendicular to the ground plane. First, he draws a line in the ground plane, which intersects the line B in F. Second, he draws two lines from object-point A: one line parallel to DF, intersecting the line B in F; a second line connecting object-point A with D, intersecting the line B in I. Third, he draws a line FG, perpendicular to BF with FG equal to DE. Fourth, he connects G to H and calls the resulting line *saemlijn*, because G is the vanishing point (*saempunt*) of all lines parallel to DF. Fifth, he draws a line perpendicular to BF proceeding from I, which connects HG in K. In this way Stevin constructs K, which is the image in the picture plane of the object-point A.

Those interested in placing Stevin in a history of perspective pointing forwards to projective geometry, while denying that Stevin's enterprise in *Vande deursichtighe* is optical, typically refer to this theorem to support their narrative.[27] They have similarly pointed to Stevin's first postulate in support of Stevin's place in a history of increasing mathematical abstraction away from the connections between optics and perspective. This first postulate of *Vande deursichtighe* states that the object-point, the image, and the eye are all on the same line.[28] For Stevin this was a postulate – in Dutch, a *begheerte* – thus, something desirable rather than proven. Desirable, according to Stevin, only because of experiments on binocular vision, which go back to Ptolemy. Stevin notes that if one presses the eye, the location of the image is different from the one seen when the eye is not pressed. He even says that the angle between those two images is 33 degrees. There are then two images, but only one can be on the line between the eye and the object-point. Similarly, Stevin notes that under normal conditions of binocular vision it is impossible to ascertain that the object-point, the image, and the eye are on one straight line. Stevin's only purpose with this postulate seems then to reduce the eye to a mathematical point and to a centre of projection out of a desire for mathematical abstraction. But this interpretation misses the fact that, in the often ignored second book of *Vande deursichtighe*, Stevin similarly points to experiments with binocular vision to reveal (as we have seen) binocular vision as the source of the error of the (prevalent, at that time in optics) geometrical construction of image location in mirrors. Stevin's postulate is then not only an expression of a desire for mathematical abstraction, but part of a more broadly defined *optical* program.

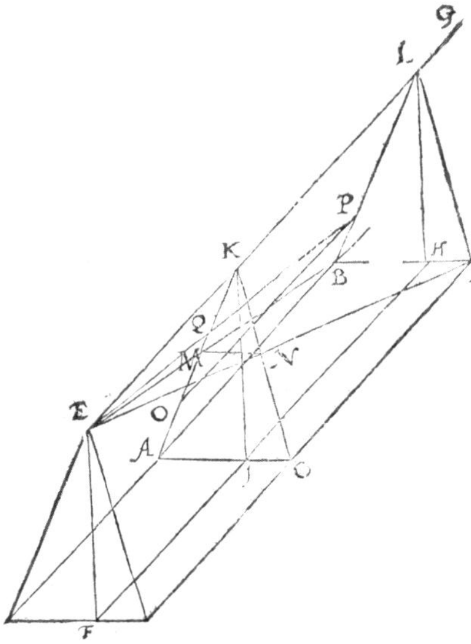

7

Simon Stevin, *Vande deursichtighe*, 1605

Stevin's general theorem of the vanishing point.

Ghent, University Library, Acc. 56517.

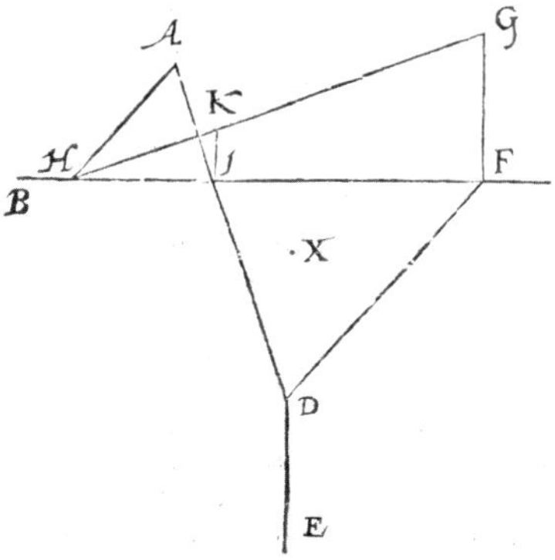

8

Simon Stevin, *Vande deursichtighe*, 1605

Projection of a point in the ground plane with the plane of projection perpendicular to the ground plane.

Ghent, University Library, Acc. 56517.

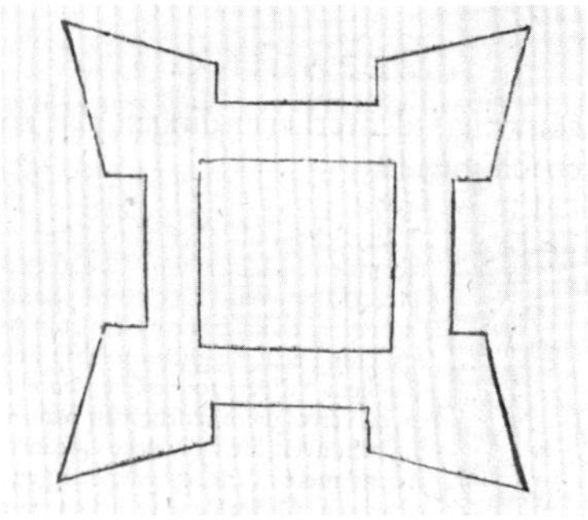

9

Simon Stevin, *Vande deursichtighe*, 1605
Fortification – ground plan. Ghent, University Library, Acc. 56517.

As for Aguilón, Stevin's involvement with architecture and military fortifications also shaped the scenography in the first book of *Vande deursichtighe*. It was closely connected to the practice of engineering. Some of its seemingly highly mathematized theorems, which appear to have no pictorial relevance at all, are, in fact, on closer examination, closely related to the use of drawings in the design of military fortifications. For Stevin, drawings were important tools, both in education and in the decision processes around the design of military fortifications. In the preface to *Vande deursichtighe*, Prince Maurits is portrayed as a very active student. Stevin writes that Maurits practised drawing, as this could help him to understand the mathematical proofs. But he also wanted to learn how to draw in perspective – after exercises in drawing ground plans and elevations – to help him communicate his views when discussing a military fortification with all those involved in its planning. Significantly, Maurits is not portrayed as an ignorant patron who could be easily impressed with a presentation drawing in perspective, but as a knowledgeable military leader who took an active part in complex military engineering projects. It comes then as no surprise that Stevin's *Vande deursichtighe* contains a ground plan, elevation, and a perspective drawing of a military fortress (see figs. 9, 10, and 11). These drawings signal that *Vande deursichtighe* is a book on perspective drawing as it could serve the design practice of engineering, and of military fortification in particular. As such, the book nicely connects to Stevin's earlier publications on military fortifications, which, as Charles van den Heuvel has shown, were books on the *design* (including the drawing, but not the building) of fortifications.[29]

Stevin's early work on military fortification, the *Sterctenbouwing*, is foremost a book on the design of military fortifications. In the dedication to Hendrick van Brienen, Stevin writes:

> I do not wish you to reckon me among the simple 'sham-fighters', I have sent you this actual example, which, though still they are only (as is said) castles in the air, or, even more properly expressed, bastions of paper, yet comes much nearer to the matter than theoretical ideas of magnitudes separated from matter. For since drawings and descriptions have to precede the practical work, it seems that it might to some extent be called a part of the practical work.[30]

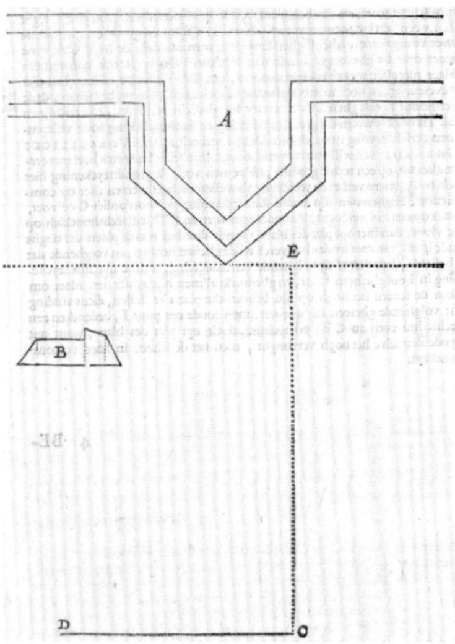

10

Simon Stevin, *Vande deursichtighe*, 1605
Fortification – elevation. Ghent, University Library, Acc. 56517.

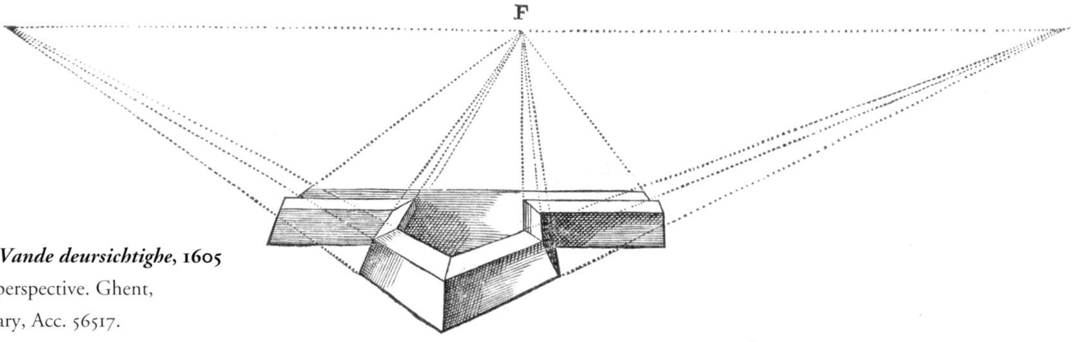

11

Simon Stevin, *Vande deursichtighe*, 1605
Fortification – perspective. Ghent, University Library, Acc. 56517.

The book contains – following the Vitruvian definition of drawings – ground plans, elevations, and one so-called bodily drawing of a military fortress (fig. 12). Stevin stressed the role of drawings and models in the design of fortifications:

> All fortresses must be designed (or at least ought to be so) before the building is undertaken, lest through errors anything should be ill-handled, which afterwards could be but hardly remedied, because of the great cost such works involve. Therefore is it a custom that we should first make drawings of various figures, according to the situation and the place and nature of the work, that all those whom it concerns, through such designs may understand each other, and finally arrive at the best plan. These figures are to be made in two ways, first plain by a ground plan upon paper and afterwards bodily with potter's earth, wax, wood or other matter.[31]

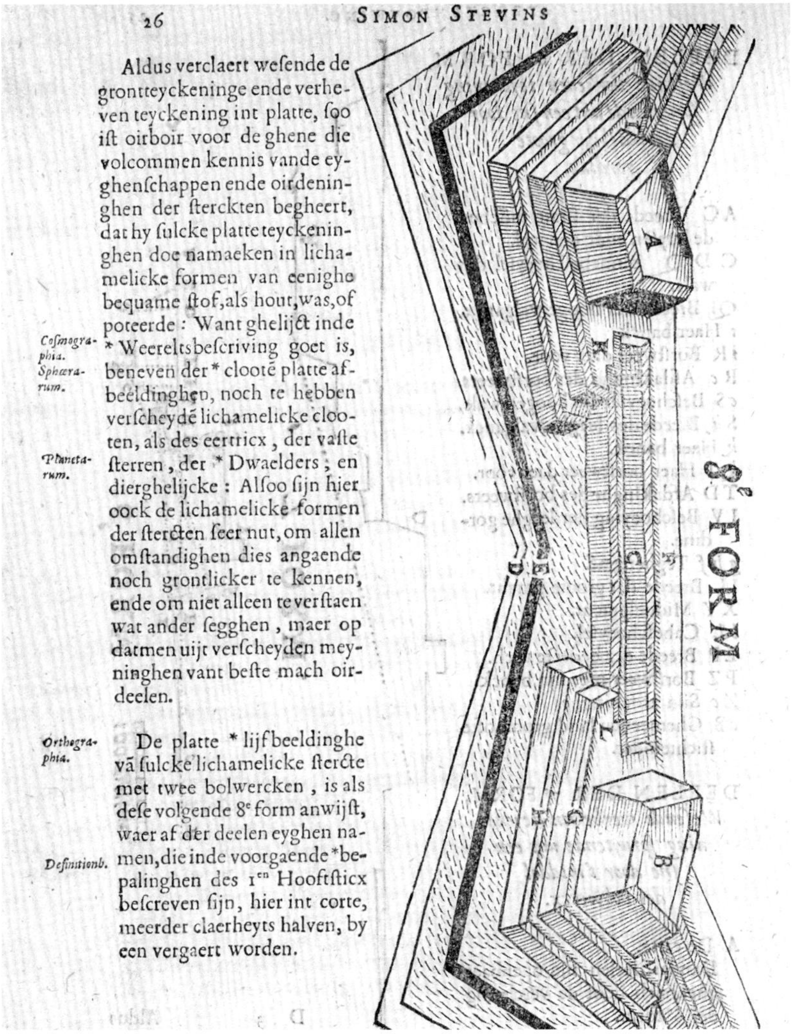

26 SIMON STEVINS

Aldus verclaert wesende de grontteyckeninge ende verheven teyckening int platte, soo ist oirboir voor de ghene die volcommen kennis vande eyghenschappen ende oirdeninghen der sterckten begheert, dat hy sulcke platte teyckeninghen doe namaeken in lichamelicke formen van eenighe bequame stof, als hout, was, of poteerde: Want ghelijct inde * Weereltsbescriving goet is, beneven der * clooté platte afbeeldinghen, noch te hebben verscheydé lichamelicke clooten, als des eertricx, der vaste sterren, der * Dwaelders; en dierghelijcke: Alsoo sijn hier oock de lichamelicke formen der stercten seer nut, om allen omstandighen dies angaende noch grontlicker te kennen, ende om niet alleen te verstaen wat ander segghen, maer op datmen uijt verscheyden meyninghen vant beste mach oirdeelen.

De platte * lijfbeeldinghe vã sulcke lichamelicke stercte met twee bolwercken, is als dese volgende 8^e form anwijst, waer af der deelen eyghen namen, die inde voorgaende *bepalinghen des 1^en Hooftsticx bescreven sijn, hier int corte, meerder claerheyts halven, by een vergaert worden.

Cosmographia. *Sphærarum.* *Planetarum.* *Orthographia.* *Definitionb.*

12

Simon Stevin, *De sterctenbouwing*, 1594

Bodily drawing. Antwerp, Museum Plantin-Moretus/Prentenkabinet, Antwerpen – UNESCO Werelderfgoed, MPM A 1007 (1).

13
Simon Stevin, *Vande deursichtighe*, 1605
Stevin's *glas*. Ghent, University Library, Acc. 56517.

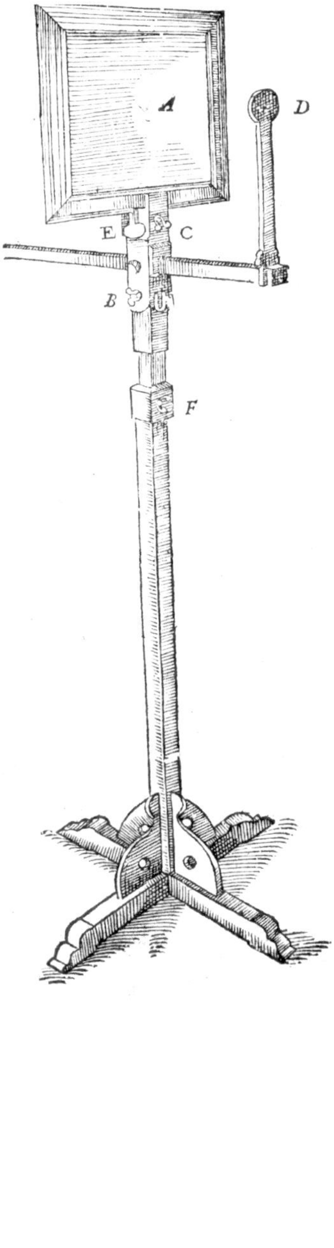

Stevin recommended the use of models of different aspects of a fortress, and of different scales so that they could be used for experimentation and demonstration of the principles of flank defence, that is, 'for finding out, with a stretched thread, all flanking proceeding from the striking angles, as from the parapets upon the walls, and cavaliers, unto all places to be flanked'.[32] Also, Stevin's instructions for the Duytsche Mathematique, the engineering school in Leiden, stressed the role of models in explaining the main terms to the students and of drawings in military fortification:

> (...) having sufficient experience in this, will be capable of undertaking the fortification or building of defences, for which wooden or earthen models of ramparts and bastions shall be prepared, and having learnt the authentic names, the drawing of maps or ground plans of cities will be easy, which will soon get them working for the cities. They will also draw on paper the perimeter of forts or towns with four bastions or more from which they will be given the measurements to mark out the fortifications in the field with markers.[33]

Stevin's other student, of quite a different social status than the soon-to-be professionals at the Duytsche Mathematique, Prince Maurits, similarly appreciated the didactic use of a three-dimensional model. In *Vande deursichtighe*, Stevin described a perspective instrument that he called the glass and that was inspired by similar devices by Albrecht Dürer (compare figs. 13 and 14). It consisted of a plate of crystal mirror glass on a tripod

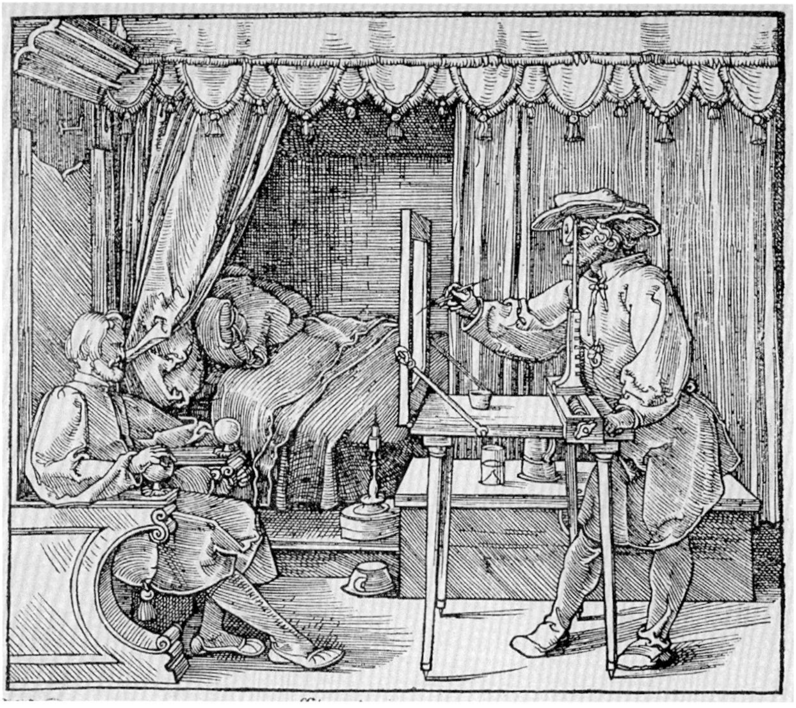

14
Albrecht Dürer, *Underweysung der Messung*, 1538
Instrument for drawing in perspective.

(which could be tilted at wish) and a sighting device to position the eye, of which the distance to the glass could be varied, when drawing an object on the glass. This instrument – a truly three-dimensional model of the picture plane intersecting the visual pyramid, the principle on which perspective drawing was based – was, according to Stevin, not only a handy tool for drawing in perspective but also an educational tool, as it helped Maurits, and other users, to understand the proofs of perspective.

> We wanted to describe this form of the glass (by means of which his Princely Grace drew perspective images both of men and of other things in such a way that it seems it may be said in truth that postures of men cannot possibly be drawn so perfectly at sight, without a glass) (...) it promotes a thorough knowledge of perspective.[34]

Moreover, Maurits is again presented as an active student. Stevin claims that the use of the glass helped Maurits 'to correct several imperfections that were present in my [Stevin's] first conception of this perspective'.[35] The most important imperfection that Stevin singled out had to do with the so-called inverse problem of perspective – that is, the problem of finding the eye, given a perspective drawing.[36] It is considered one of Stevin's more important original contributions to perspective but also a contribution that clearly shows that Stevin's *Vande deursichtighe* was driven by a desire for mathematical abstraction, because the problem is of no relevance for pictorial practice. True as this is for painters *as painters*, this is not the case for painters involved with military fortification or for anyone else, like Maurits, involved in the decision processes around military fortification, in which case, as we have seen, communication between the different parties took place by means of drawings.[37] In such a context of military engineering it was in fact most useful to be able to find the eye when confronted with a perspective drawing of a military fortification, since this was necessary to understand the design of the fortress presented. It then comes as no surprise that it was precisely on this point that, according to Stevin, Maurits discovered by means of the glass an imperfection in Stevin's conception of the inverse perspective problem:

> Further it happened that in the finding of the eye we had described several propositions in which the object figure had been put as given beside its given image, as it had stood in the projection, by which means the finding of the eye was easier. But his Princely Grace, understanding the matter more thoroughly, said that there was an imperfection, because in practice we do not meet with this, for in paintings such object figures are not drawn beside the images in their places. And since we saw that this was reasonable, we changed those propositions and replaced them by others (...)[38]

We see then that the inverse problem of perspective, seemingly remote from practice, was instead a problem intrinsic to engineering practice when drawings moved between different parties – including that of the commander of the Dutch troops, Prince Maurits – in projecting a design of a military fortification that would work under all foreseeable conditions of attack.

In summary, we should characterize Stevin's *Vande deursichtighe* as a Vitruvian optical project. This does not entail that Stevin uncritically took on board Vitruvius's views. Vitruvius argues that

> sight does not always produce true effects; indeed, the mind is quite frequently deceived by visual judgments. (…) Therefore, if things that are true appear false, and many things are taken to be other than they are by our eyes, I think there should be no doubt that it is proper to make additions and subtractions according to the natures and requirements of sites.[39]

Stevin disagreed with Vitruvius by rejecting such optical corrections.[40] The usefulness of perspective drawings in the representation of architecture was a debated issue in the period. The reception of Vitruvius in the work of Daniele Barbaro, who preferred the *profilo* above the drawing in perspective, provided an important voice in this debate. But Barbaro's reception had little effect on the drawing practices in architecture in the Netherlands.[41]

Although the characterization of Stevin's *Vande deursichtighe* as a Vitruvian optical project might thus be open to misunderstanding, I still think that it is valid in two ways. First, it stresses that *Vande deursichtighe* is an optical project. The historiography of perspective has obscured the fact that *Vande deursichtighe* is a book on optics: Stevin's concerns in catoptrics were similar to Kepler's, and his conception of the scope of optics is Vitruvian. Second, it also stresses that it is misleading to reduce Stevin's perspective to a step in a process of mathematization towards projective geometry. In contrast, Stevin's perspective was closely connected to design practices in military engineering. In this context, it is important to stress, though, that Stevin was against optical corrections to perspective drawings and Vitruvius's own advice.

Reflexy-const and artists' knowledge of optics

Art theory treatises of the same period confirm the view that optics was about more than just perspective. However, contemporary terminology might easily have led the reader astray. The *Trattato dell'arte della pittura, scoltura, et architettura* (1585) of Lomazzo, who is otherwise best known for his role in the transmission of the work of Leonardo, underlines this point. Lomazzo's fifth chapter is devoted to 'perspective'. With reference to the ancient Greek mathematician Geminus, Lomazzo divided 'perspective' into three different kinds: *perspectiva* or *optica*; *sciographica* (devoted to the drawing of shadows); and *specularia* (that is, the study of mirrors). Moreover, *optica* is subdivided into two kinds: *physiologica*, which is about 'the universall principles, causes, and elements of visible things'; and *grammica*, or 'the art of delineation'. Lomazzo considered the latter 'the most necessary part of painting'. In fact, the fifth chapter of his *Trattato* is mostly devoted to this art of delineation, which he confessed, he 'will not handle like a Mathematician, but speake of them according to the usuall practice of the painters, and mine owne observations out of pictures of all sortes of men.'[42]

However, for Lomazzo, artists' optical knowledge was not limited to this art of delineation. The fourth chapter was completely devoted to light. Lomazzo emphasized that the translation of the effects of light to a painter's canvas or panel was an epistemic issue. It was a matter of knowledge, which, according to Lomazzo, not all painters equally shared:

> Into which error most of painters of our time running, loose much of the worth of their works, making them seeme, (as indeede they are) rather painted, then counterfeited; and done rather to satisfie the eie of the rude and ignorant, then to content the conceit of the iudicious. Which custome doth so increase daily, that I feare me the true knowledge of this art will fresh begin to decay.[43]

The chapter is devoted to a discussion of the different effects of light on not only different types of texture and surfaces but also different materials. There is a section devoted to each of the four Aristotelian elements: water, air, fire, earth. So begins the section on water:

> Whereas the matters of precious stones are transparent more or lesse, (as those which are commonly called gemmes) they must needs receive the light more sharpely; which easily passeth through them, carrying their virtue along with it: as may bee seene in the Sunne, which casting his beames upon the stone Iris, causeth the rain-bowe to appeare therein: so that the light passing through these stones carrieth with it their true and perfect colour; like as the colour of the wine or water in a vessel or glasse, is cast upon the table whereon it standeth.[44]

Lomazzo's *Trattato* is important in two respects. First, like Leonardo, he grounded his discussion of how light in art is recorded by painters in the light and reflections in nature. Second, he discussed the effects of light not just as changes of texture but as material transformations. Both these aspects are also present in Van Mander's *Schilder-boeck*.[45] Van Mander discussed *reflexy-const* in chapter seven of the 'Groundwork' of his *Schilder-boeck*, a text chronologically almost coincidental with Stevin's *Vande deursichtighe*. Van Mander's description of painterly recording of reflected light – and its associated vocabulary of mirroring (*spiegeling*), reflection (*reflectie*), polish (*glans*), re-reflection (*weerschijn*), and reverberation (*reverberatie*) – would, however, have failed to find a source in Stevin's chapter on catoptrics or in the works of any other contemporary mathematicians or opticians who divided the science of optics in the three parts of direct vision, catoptrics, and dioptrics.[46] In fact, the effects that Van Mander described in words and that the artists whom he discussed recorded in paint went far beyond anything that was available in contemporary writings on optics.

Van Mander's discussion of the art of depicting reflections is also grounded in nature. The chapter on *reflexy-const* opens with the statement that the Sun is the source of all reflected light, without which the celestial bodies would be invisible, and moves on to descriptions of the reddening of the skies at sunrise and sunset, the mirroring effects of water, halos and the illusions of double suns, and, above all, rainbows – all natural phenomena that were discussed in meteorology, a vibrant tradition of textual studies commenting on Aristotle's *Meteorologica*.[47] Van Mander includes classical references to Pliny alongside his own observations, such as the rainbows he witnessed on his trip through Italy at the waterfall near Terni (between Rome and Venice) or in the fountains of Villa d'Este in Tivoli, a sight that, as Simon Werrett has argued, also inspired Descartes's discussion of the rainbow in *Les météores*.[48] Van Mander refers to Iris, 'who is the rainbow', the appropriate companion of the goddess of weather, Juno.[49] In the Ovidian myth, Juno sprinkled the two hundred eyes of Argus on the peacock's tail, which we have already seen in Aguilón's frontispiece, designed by Rubens, who repeated the motive in his painting *Juno and Argus* (1611), and in which he appropriately also shows a seven-coloured rainbow in the skies (fig. 15).[50]

Van Mander describes these colours of the rainbow, referring to the peacock:

> Closest to us is purple, then *incarnatich* or light carmine, to paint it well; thereafter orange-like, or rich red; then masticot yellow, then soft green, then pure blue, like the neck feathers of peacocks, and finally purple again. This messenger of Juno is used to show off such a multi-coloured mantle.[51]

The hope that a study of the rainbow could clarify the rules for colour harmonies and mixtures was already expressed by Leonardo:

> Treat of the rainbow in the last book on Painting, but first write the book on colours produced by the mixture of other colours, so as to be able to prove by those painters' colours how the colours of the rainbow are produced.[52]

'The rainbow allows painters to see which colours like to stand next to each other', noted Van Mander. 'For example, blue with purple and purple with red, and with red yellow that looks like orange. But light yellow is *bevriend* with green, and green likes the company of blue. Ash blue and yellow are thus mixed to make green'.[53] For Leonardo, painters' colours would be able to produce an explanation of the rainbow, whereas Van Mander maintained that art was founded in nature.

Van Mander's biography of the Van Eycks establishes Jan (and Hubert) as the founders of a Netherlandish canon of painting based on *reflexy-const*.[54] According to Van Mander, Van Eyck was able to approximate the transparency, saturation, and radiance of the rainbow's hues. Van Eyck's masterful ability to depict reflections and refractions of light – on which the rainbow was also based – has long been noted, even if Rudolf Preimesberger's assertion that Van Eyck would have been familiar with the optical work of Pecham and his contemporaries is doubtful;[55] Van Eyck goes well beyond anything found in those writings. Van

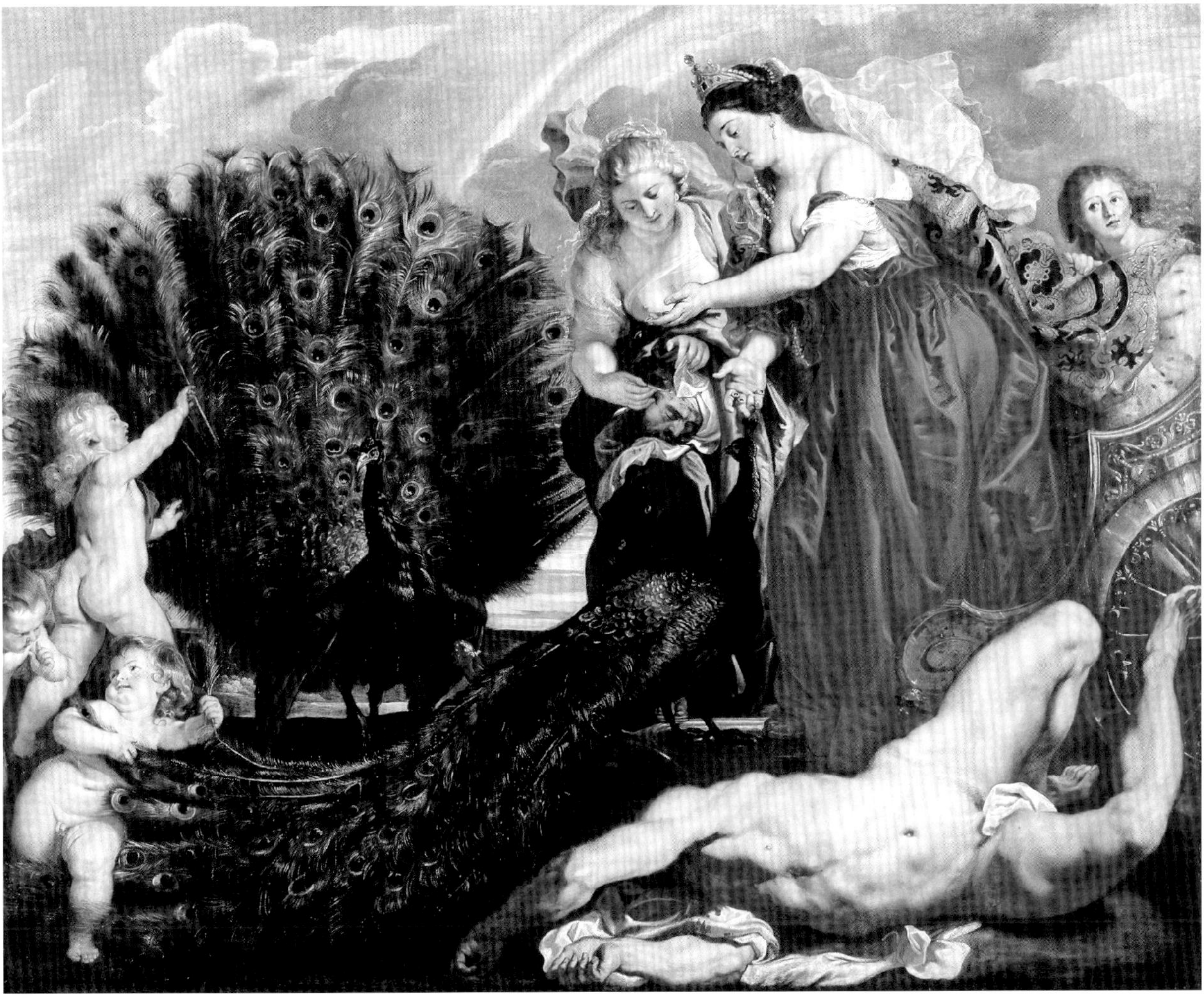

15
Peter Paul Rubens, *Juno and Argus*, 1611
Cologne, Wallraf-Richartz Museum & Fondation Corboud: WRM 1040, © Rheinisches Bildarchiv Köln, rba_c002830.

Mander emphasized that Van Eyck's mastery consisted of rendering all surfaces as if mirrors; Van Eyck likened painting to mirroring; Van Mander's Van Eyck established that representation both imitates and ensues from the process of reflection. Van Eyck's visual work seems to corroborate Van Mander's point. As much is evident from Walter Melion's description of Van Eyck's *Arnolfini wedding*, in which he argues for the relatedness of mirroring and picturing:

> the pigments are translucent, allowing light to penetrate to the white ground, where it is refracted back through the intervening layers of color. The panel's overall sheen, like the sheen of the various surfaces described in the image, results from the action of colored light, entering and exiting the layers of semitransparent glaze. (…) His representational means and ends conspire to an unprecedented extent: the interaction of light with Jan's media intensifies the representation of lustrous surfaces interacting with light.[56]

According to Van Mander, Van Eyck's *reflexy-const* was based on his invention of oil-based pigments. Van Eyck is said to have experimented with varnishes and binding agents, 'and found after much investigation that pigments mixed with such oils became malleable and dried hard, and having dried became impermeable, and that the oil made colours livelier, and that they themselves became lustrous with varnishing'.[57] The sheen of oil colours gave the painted surfaces the impression of instantaneity, as if the products of reflection instead of a painter's brush, and they allowed the portrayal of precious objects, such as transparent crystal, jewels, and golden ornaments, which possessed the lucidity, refulgence, and gem-like hue of oil colours. Van Mander's emphasis on oil-based pigments as foundational to Van Eyck's and Netherlandish art was based on an image of Van Eyck as alchemist. Again in Melion's words, 'Jan vows the kinship of paint and light, subsuming both into an image that shares the translucency of glass, the descriptive precision of the mirror, and the luster of enamel'.[58]

In his chapter on *reflexy-const*, Van Mander himself established a connection between alchemy and light. His discussion of the rainbow is followed by an elaboration of painting night scenes, especially scenes of fires by night, noting especially the reflections in water:

> The dangerous fire of disastrous conflagrations seizes with fright the human heart when raising with its sparkles a fierce sputtering. The blacker and thicker the dark veil of the night is, the brighter its powerful flames light; and they give the houses, temples and other buildings a re-reflection of the same colour; and they also give the water a horrible look.[59]

Van Mander names Gillis Coignet (c. 1538-1599) as a painter of night fires, but we could equally well think of Mostaert, Blondeel, or Brueghel, whom Vasari mentioned in this connection. Van Mander thus established a connection between painting night fires, spectacular optical effects, vivid colours, and the 'destructive force of flickering flames'.[60] The work of Bosch, which Van Mander associated with the lively depiction of fires in his *vita* of this painter, was no doubt inspirational. Van Mander might also have been familiar with Brueghel's poetic hells.[61]

However, one theme stood out for Van Mander, allowing the discernment of the true master of reflection: Vulcan's forge.

> Those who succeed to paint the rage of Vulcan, such an atrocious disaster, have become most expert in this art, because depending upon the nutrient, that is the material with which he feeds his violent flames, which difficult to tame, flare up towards heaven, they receive their colour, either bowing to red, to purple, to blue, or to green.[62]

In contemporary Netherlandish art, the only depiction of Vulcan's forge seems to be Maarten Van Heemskerck's *Venus and Cupid in Vulcan's forge* (1536) (fig. 16).[63] But the appearance of Vulcan in this context is not surprising. Vulcan was the governor of fire, but following Paracelsus

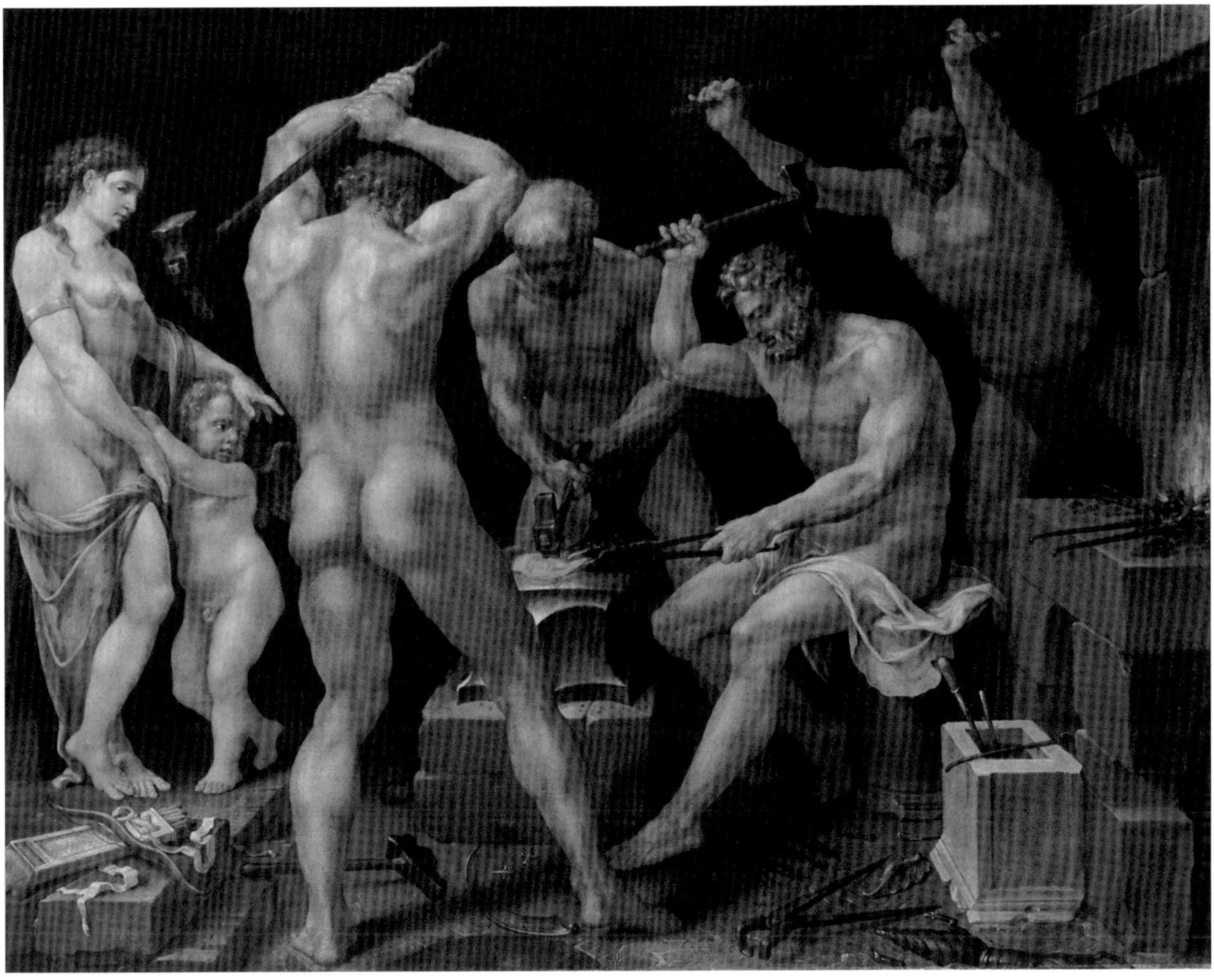

16
Maarten Van Heemskerck, *Venus and Cupid in Vulcan's Forge*, 1536
Prague, National Gallery.

(1493-1541), he also became the patron of alchemy. In the commentary to the *Metamorphoses* of Ovid, included in the *Schilder-Boeck*, Van Mander named Vulcan the inventor of the alchemical arts.[64] Thus, Van Mander's reference to Vulcan supports connections between alchemy, light, and paint, which, Van Mander argued, were foundational for Van Eyck's and Netherlandish art. Van Mander claimed that Van Eyck's depiction of reflections depended upon intimate knowledge and sustained investigation into the reflective properties of different materials.[65] In summary, Van Mander brought out the painter's knowledge of reflections, which he likened to the alchemical transformation of materials through light and fire.

Conclusion

The emergence of a self-contained discipline of history of perspective in the twentieth century has allowed the development of a substantial body of knowledge on the geometry of perspective. However, as I have attempted to show, the associated and dominant historiography of perspective has also produced unfortunate distortions. Perhaps it is best to characterize these distortions in terms of a reduction of the fields of investigation. In this paper I have illustrated two aspects of this reduction. First, the historiography of perspective – and its treatment of Stevin's *Vande deursichtighe* is constitutive of this approach – has reduced the scenes of inquiry developed in treatises on optics to the geometry of perspective teleologically pointing forward to the invention of projective geometry. I have argued that Stevin's *Vande deursichtighe* was an optical project. Moreover, I have also suggested that this project be considered 'Vitruvian', not in the sense that Stevin would have agreed with Vitruvius on issues of architecture (which he did not), but in his embracing of the Vitruvian scope of optics as well as in the grounding of his *Vande deursichtighe* in design practices of military engineers. Second, the dominant historiography has also reduced artists' optical knowledge to perspective. Art theory treatises, such as Lomazzo's and Van Mander's, support a much broader basis of artists' optical knowledge. Lomazzo's and Van Mander's representations of artists' optical knowledge share two characteristics: both grounded their discussion of the art of depicting light and reflections in nature; and both discussed the effects of light not just as changes of texture but as material transformations. Whereas for the first aspect, art theorists turned to sources in optics and meteorology, for the second, they were often more innovative. But such innovations become evident only once the field of inquiry of artists is no longer reduced to the geometry of perspective.

Notes

1 Gombrich 1976, 20.
2 Ibid., 33.
3 For *perspectiva naturalis* versus *artificialis*, see Raynaud 1998, 16-23.
4 Indeed, the most important exception to Gombrich's call for studies of means other than perspective to create the illusion of space is perhaps Ernst van de Wetering's famous study of how Rembrandt applied paint to his surfaces to create the illusion of space and to imitate texture. See Van de Wetering 1997. For the connection of Rembrandt's painterly practices to the art theory of his apprentice Samuel van Hoogstraten, see Weststeijn 2008.
5 Lindberg 1976, 154. Lindberg's downgrading of perspective is also only possible if one favours a monolithic canon of optics above the recognition of different images of optics that circulated since the fifteenth century. See, for example, Dupré & Korey 2009, and Dupré 2008.
6 Andersen 1990.
7 Van Son 1993.
8 In this way we hope to move beyond the identification of *Lichtgefüge*, conceptual orders, or structures of light, which would equally be at work in art, science, and philosophy. See Bohlmann *et al.* 2008.
9 Andersen 1990, 46.
10 Sinisgalli 1978; Le Goff 2004; Le Goff 2001.
11 Raynaud 1998, 16-23.
12 Struik 1958.
13 Stevin 1605, 105-108.
14 Kepler 1939, 84; translation in Kepler 2000, 100. For Kepler and the cathetus rule, see Chen-Morris & Unguru 2001, Shapiro 1990, 122-124 and Simon 1976, 464-477.
15 Aguilonius 1613.
16 Dupré 2008.
17 'De deursichtige als gheslacht hebben verscheyden akcomsten, ghelijck sijn Spiegelschaeuwen, Breekschaeuwen, Platclooten, Sonwijsers, Verschaeuwing, en meer ander die met malcander inde deursiening eenighe ghemeenschap hebben, doch alsoo heur daden tot verscheyden manier van wercking behouven, soo wort elcke afcomst als besonder const met onderscheuyt ghenoemt, en oirdentlick beschreven'. Stevin 1605, 7. Translation in Struik 1958, 807.
18 For this biographical context to Stevin and Maurits, see Dijksterhuis 1970, and several contributions to Cat. Brussels 2004.
19 Cat. Amsterdam 2000. See also Vermij 2010.
20 Van den Heuvel 1991, 139-148.
21 Van den Heuvel 2004.
22 'Alsoo sijn vorstelicke ghenade hem dickwils oeffende in te trecken grontteyckeninghen, en stantteyckeningen van sterckten, die hy veroirdende inde landen sijnder regiering, heeft oirboir bevonden hem oock te oeffenen inde derde afcomst der teyckening te weten het verschaeuwen of schilderen, en dat voornamelick van lantschappen, met steeden, stroomen, weghen, en bosschen daar ingheleghen, om daer deur anderen sijn meyning, alst de saeck vereyscht, lichtelicker te verclaren'. Stevin 1605, 4. Translation in Struik 1958, 801.
23 'Nu also ick over eenighe jaren, voor mij self beschreven had een HVYSBAV, tot

welcx oeffening den Boumeester, na t'ghemeen ghevoelen van velen, en t'besonder gevoelen van Vitruvius int 2 hooftsick sijns 1 boucxn, kennis der verschaeuwing voorderlick is, soo heb ick breeder dan daer te vooren, deursien en ondersocht verscheyden schrijvers van dese of handelende, en na mijn stijl van dies een beschrijving ghemaeckt: Welcke nadiense sijn VORSTELICKE GHENADE oversien hadde, en helpen verbeteren de onvolcomentheden die ghemeenlick in eerste vonden sijn, oock grondelick verstaen de ghemeene reghel om alle voorghestelde verschaeulicke saeck te verschaeuwen, en dat hy tot sijn vernoughen dadelick verschaeude: Soo heb ick dese beschrijving onder sijn WISCONSTIGHE GHEDACHTENISSEN vervought, (...)'. Stevin 1605, 4. Translation in Struik 1958, 801. For Stevin's treatise on architecture, see Van den Heuvel 2005.

24 'Doch want de vercorting der linien, en verandering der houckenuyter oogh, of byder gisse toeginck, en heeft hem, hoe wel het sijn oirboir ghebruyck can hebben, daer me niet vernought, maer willen een voorghestelde verschaeulicke saeck volcomelick afteyckenen, mer kennis der oirsaken en sijn wisconstich bewijs'. Stevin 1605, 4. Translation in Struik 1958, 801.

25 Vredeman de Vries 1979, 29-30. See Dubourg Glatigny 2002.

26 Stevin 1605, 21-22.

27 Andersen 2007, 237-276. For the connection between perspective and Desargues's projective geometry, see Field 1988 and Field 1997.

28 Stevin 1605, 14-15.

29 Van den Heuvel 1991, 140-141. See also Van den Heuvel 2006.

30 '(...) ende dat ick niet en begheer datse my onder de eenvoudighe spieghelvechters rekent, soo heb ick haer dit werckelic voorbeelt toegeschickt: Het welcke, hoewel dattet noch al maer (ghelijckmen seght) burghen inde locht, ofte noch eyghentlicker, papiere bolwercken en sijn. Doch commet de sake veel naerder, dan spieghelsche ghedachten der grootheden ghescheyden van stof: want anghesien teyckeninghen ende redenen der omstandighen, voor het dadelick werck moeten gaen, soo schynet datment eenichsins deel des dadelicx wercx soude meughen noemen'. Stevin 1594, 3-4. Translation in Schukking 1965, 46-47.

31 'Wantmen de sterckten al veroirdent moet hebben, oft immers behoirt te hebben, eermen ant bouwen comt, om deur misgripinghe niet qualick te maken t'gene men daer na s'waerlick soude connen veranderen, deur de groote costen die daer an hanghen, soo ist inde ghebruijck, datmen van te vooren verscheyden formen teeckent, na gheleghentheyt der plaets, ende den eysch van allen omstandighen, op dat de personen die daer af te spreken hebben, deur soodanighe formen malcander wel verstaen meughen, en de eintlick het beste daer af besluyten. Dese formen worden ghemaeckt op tweederley wijse, eerst int platte op papier, daer na lichamelick van poteerde, was, hout, of ander stof'. Stevin 1594, 7. Translation in Schukking 1965, 65.

32 'om te ondersoucken met een ghespannen draet, alle strijckinghen commende soo uyt de strijckhoucken, als van borstweeren der wallen en catten, na alle strijckelicke plaetsen'. Stevin 1594, 27. Translation in Schukking 1965, 103.

33 'Hier in genoech ervaren wesende sullen bequaem sijn om totte fortificatie of sterckbouwinge te comen, waertoe bereyt sullen worden houtten ofte eerden botsen van schantsen ende bolwercken ende daer mede geleert hebbende die eygen naemen, soe sal het trecken van planten ofte grontteyckeningen van steden heurlyeden licht vallen, t welck men hun aen steden oeck datelyck zal doen te werck stellen. Sy sullen oeck teeckenen op papyer den omtreck van schantsen of steden met vyer, vijff ende meer bolwercken, waer aff men alsdan henl. de maten sal geven ende sullen daer nae sulcke stercten opt velt teeckenen met baecken'. Molhuysen 1913, 390*.

34 'Dese ghedaente vant glas (waer me sijn VORSTELICKE GENADE schaeuwen teyckende so van menschen als anders, sulcx dat schijnt Mette waerheyt te meughen gheseyt worden, dat standen van menschen niet meugelick en sijn uyter oogh sonder glas, soo volcomelick gheteyckent te worden) hebben wy willen beschrijven, ... dattet tot grondelicke kennis der verschaeuwing voorderlick is'. Stevin 1605, 89. Translation in Struik 1958, 961.

35 'gheholpen heeft om te mercken en verbeteren ettelicke onvolcomentheden die in mijn eerste begrijp deser verschaeuwing waren'. Stevin 1605, 89. Translation in Struik 1958, 961.

36 Stevin 1605, 60-81.

37 This is not to say that Stevin's *Vande Deursichtighe* would have been useless for painterly practice. The Dutch painter of church interiors, Pieter Saenredam, for example, is known not only for having mastered the theory of perspective to an unsurpassed level of mathematical intricacy, which allowed him to draw the same church interior from a bewildering variety of view points, but also to have looked for specific aesthetic effects offered by wide-angle views. In particular, he looked for depicting the marginal distortions of columns offered by a correct application of perspective. In this he followed Stevin's advice not to introduce optical corrections for a 'pleasant' perspective. See Kemp 1986; Dupré 2001, 117-130. For the differences of representation of architecture between painters, architects, and mathematicians (including Stevin) in the context of the Dutch Republic, see Van den Heuvel 1994.

38 'Voort ist ghebeurt dat wy inde vinding des ooghs ettelicke voorstellen beschreven hadden, waer in de verschaeulicke form als gegheven by haer ghegheven schaeu, ghestelt was ghelijcke int verschaeuwen ghestaen hadde, deur t'welck het vinden des ooghs lichter viel: Doch sijn VORSTELICKE GHENADE de saeck grondelicker insiende, seyde hier in onvolcommentheyt te wesen, omdat ons sulcx inde daet niet en ontmoet, wantmen inde schilderijen soodanighe verschaeulicke formen by de schaeuwen t'haerder plaets niet en teyckent'. Stevin 1605, 89. Translation in Struik 1958, 961.

39 Vitruvius 1999, 78.

40 Stevin 1605, 83.

41 Van den Heuvel 1994.

42 'non come matematico, mà liberamente procederò, & parlerò secondo la pratica tenuta da pittori, & come hollo anch'io osservato, & fatto vedere nelle figure'. Lomazzo 1585, 255. Translation in Lomazzo 1598, 190.

43 'nelqual errore incorrendo quasi tutti I pittori di questa età togliono à le opera quella forza che si gli ricercarebbe, & fanno che paiono, come sono più pinte che finte, & fatte più per diletto, de' gossi che per pasto, & nutrimeno de gli intelligenti; la cui usanza và di giorno, in giorno avazandosi tanto, & pigliando tanta forza, ch'io dubito che di nuovo la vera cognition di quest' arte non si

smarrisca'. Lomazzo 1585, 238. Translation in Lomazzo 1598, 168.

44 'Essendo la materia delle pietre preciose, & sine trasparenti, o più o meno; come quelle che volgarmente sogliamo chiamar gemme, è necessario che ricevano il mume più acuto, in maniera che ha forza passare per entro loro l'arco celeste; si che in ogni mod oil lume porta seco, passando per le pietre il medemo colore ch'elle tiene in quella guise chef à il color del vino, o de l'acqua in un vaso di vetro sopra il piano dove posa'. Lomazzo 1585, 229. Translation in Lomazzo 1598, 157.

45 For Van Mander's reception of Leonardo's work, see Heck 2009.

46 Van Mander 1973, vol. 1, 182-203.

47 For discussions of the rainbow in this tradition, see Boyer 1987.

48 Werrett 2001.

49 Van Mander 1973, 189.

50 Kemp 1990, 276.

51 'Het dichtst bij ons is hij purper, dan 'incarnaet' of lichtkarmijnrood, om het goed te schilderen; daarna oranje-achtig, of rijk rood; dan masticotgeel, dan zachtgroen, dan zuiver blauw, als de halsveren van de pauwen, en tenslotte weer purper. Met zo'n veelkleurige mantel is deze bode van Juno gewend te pronken'. Van Mander 1973, 189.

52 'Fa l'arco cieleste nell' ultimo libro della picture ma fa prima il libro delli colori nati dalla mistio nelli altri colori, acciò che tu possa provare mediante essi colori de' pictori la gieneratio de' colori de l'arco'. Leonardo Windsor 19076r. See Da Vinci 1977, vol. 1, 229.

53 'aan de regenboog kunnen schilders waarnemen welke kleuren graag bij elkaar staan'; 'bijvoorbeeld blauw bij purper en purper bij rood, en bij rood geel dat er oranje-achtig uitziet. Maar lichtgeel is bevriend met groen, en groen heeft graag omgang met blauw. Asblauw en geel mengt men dan ook tot groen.' Van Mander 1973, 190.

54 Van Mander 1604, 199-203.

55 Preimesberger 1991.

56 Melion 1991, 83.

57 'bevont hy met veel ondersoeckens, dat de verwe gemengelt met sulcke Olyen haer seer wel liet temperen, en wel hardt drooghde, en drooge wesende, het water wel verdraghen mocht, dat d'Oly oock de verwen veel levender maeckten, en van selfs een blinckentheyt deden hebben, sonder datmense verniste'. Van Mander 1604, 199v. Translation in Melion 1991, 79.

58 Melion 1991, 83.

59 'Het gevaarlijke vuur van rampzalige branden, een schrik voor het mensenhart, als het opsteekt met zijn vonken een fel geknetter. Hoe zwarter en dikker de donkere sluier van de nacht is, des te heller lichten zijn krachtige vlammen op; en ze delen aan huizen, tempels of andere gebouwen een weerschijn mee van dezelfde kleur; en ook in het water [geven ze] een verschrikkelijke aanblik'. Van Mander 1973, vol. 1, 193.

60 This connection is the subject of a most inspiring discussion in Göttler 2010, 362-366. However, this is not to say that Van Mander embraced an alchemical theory of colour (which he did not). For alchemical theories of colour, see Leonhard 2011.

61 Göttler 2008.

62 'Diegenen hebben heel wat vat op de kunst, die het woeden van Vulcanus, zo'n gruwelijke ramp, goed met verf uitbeelden; want afhankelijk van de voedingsstof, dat is de materie waarmee hij zijn heftige vlammen voedt, die moeilijk te temmen, hemelwaarts oplaaien, [afhankelijk daarvan] krijgen ze hun kleur, hetzij nijgend naar rood, naar purper, naar blauw of naar groen'. Van Mander, 1973, vol. 1, 193.

63 Veldman 1977, 19-42. It was probably motivated by Italian examples. See Göttler 2010, 380. Brueghel's first depiction of Vulcan's forge came two years too late to be of influence on Van Mander.

64 Göttler 2010, 382.

65 Recent research corroborates Van Manders's contention that Van Eyck possessed knowledge of the material sources (natural, mineral, alchemical) of pigments, a body of knowledge shared with pharmacists. See Seidel 2007.

Bibliography

Aguilonius 1613
F. Aguilonius, *Opticorum libri sex philosophis ac mathematicis utiles*, Antwerp 1613.

Cat. Amsterdam 2000
Cat. Amsterdam, Rijksmuseum, *Maurits Prins van Oranje* (K. Zandvliet *et al.*, eds.), Amsterdam/Zwolle 2000.

Andersen 1990
K. Andersen, 'Stevin's theory of perspective. The origin of a Dutch academic approach to perspective', *Tractrix. Yearbook for the history of science, medicine, technology and mathematics* 2 (1990), 25-62.

Andersen 2007
K. Andersen, *The geometry of an art. The history of the mathematical theory of perspective from Alberti to Monge*, New York 2007.

Bohlmann *et al.* 2008
C. Bohlmann, T. Fink & P. Weiss (eds.), *Lichtgefüge des 17. Jahrhunderts. Rembrandt und Vermeer. Leibniz und Spinoza*, Munich 2008.

Boyer 1987
C.B. Boyer, *The rainbow. From myth to mathematics*, Princeton 1987.

Cat. Brussels 2004
Cat. Brussels, Bibliothèque Royale de Belgique, *Simon Stevin 1548-1620. L'émergence de la nouvelle science*, Turnhout 2004.

Chen-Morris & Unguru 2001
R.D. Chen-Morris & S. Unguru, 'Kepler's critique of the medieval perspectivist tradition', in: G. Simon & S. Débarbat (eds.), *Optics and astronomy. Proceedings of the XXth International Congress of History of Science (Liège, 20-26 July 1997)*, Turnhout 2001, 83-92.

Da Vinci 1977
L. da Vinci, *The literary works* (J.-P. Richter & C. Pedretti, eds.), Oxford 1977.

Dijksterhuis 1970
E.J. Dijksterhuis, *Simon Stevin. Science in the Netherlands around 1600*, The Hague 1970.

Dubourg Glatigny 2002
P. Dubourg Glatigny, 'Hans Vredeman de Vries und die Perspektive',

in: H. Borggrefe, V. Lüpkes, P. Huvenne & B. van Beneden (eds.), *Hans Vredeman de Vries und die Renaissance im Norden*, Munich 2002, 127-132.

Dupré 2001
S. Dupré, *De Optica van Galileo Galilei. Interactie tussen kunst en wetenschap*, Brussels 2001.

Dupré 2008
S. Dupré, 'Aguilón, Vitruvianism and his *Opticorum libri sex*', in: Piet Lombaerde (ed.), *Innovation and experience in early Baroque in the Southern Netherlands. The case of the Jesuit Church in Antwerp*, Turnhout 2008, 53-66.

Dupré & Korey 2009
S. Dupré & M. Korey, 'Inside the Kunstkammer. The circulation of optical knowledge and instruments at the Dresden Court', *Studies in History and Philosophy of Science* 40 (2009), 405-420.

Field 1988
J.V. Field, 'Perspective and the mathematicians. Alberti to Desargues', in: C. Hay (ed.), *Mathematics from manuscript to print: 1300-1600*, Oxford 1988, 71-99.

Field 1997
J.V. Field, *The invention of infinity. Mathematics and art in the Renaissance*, Oxford 1997.

Gombrich 1976
E.H. Gombrich, *The heritage of Apelles. Studies in the art of the Renaissance*, Oxford 1976.

Göttler 2008
C. Göttler, 'Fire, smoke and vapour. Jan Brueghel's "Poetic Hells". "Ghespoock" in early modern European art', in: C. Göttler & W. Neuber (eds.), *Spirits unseen. The representation of subtle bodies in early modern European culture*, Leiden & Boston 2008, 19-46.

Göttler 2010
C. Göttler, *Last things. Art and the religious imagination in the Age of Reform*, Turnhout 2010.

Heck 2009
M.-C. Heck, 'The reception of Leonardo da Vinci's *Trattato della Pittura*, or *Traité de la Painture*, in seventeenth-century Northern Europe', in: C. Farago (ed.), *Re-reading Leonardo. The treatise on painting across Europe, 1550-1900*, Farnham 2009, 377-414.

Kemp 1986
M. Kemp, 'Simon Stevin and Pieter Saenredam. A study of mathematics and vision in Dutch science and art', *Art Bulletin* 68 (1986), 237-252.

Kemp 1990
M. Kemp, *The science of art. Optical themes in Western art from Brunelleschi to Seurat*, New Haven / London 1990.

Kepler 1939
J. Kepler, *Gesammelte Werke*, vol. 2, Munich 1939.

Kepler 2000
J. Kepler, *Optics. Paralipomena to Witelo & optical part of astronomy* (W.H. Donahue, transl.), Santa Fe 2000.

Le Goff 2001
J.-P. Le Goff, *La scènographie de Simon Stevin de Bruges, edition critique compare des éd. fr. De 1605 (Jean Tuning) et de 1634 (Albert Girard), precede de nombreux elements de la fortune critique et d'une introduction par J.-P. Le Goff*, Caen 2001.

Le Goff 2004
J.-P. Le Goff, 'Le traité d'ombragement de Simon Stevin et sa place dans l'histoire de perspective', in: *Simon Stevin 1548-1620. L'émergence de la nouvelle science*, Turnhout 2004, 83-102.

Leonhard 2011
K. Leonhard, 'White earth, or, how to cultivate color in the field of painting. Still life and Baroque color theory', in: Alina Payne (ed.), *The object as event. Vision and its instruments in the early modern period*, New Haven 2011 (in preparation).

Lindberg 1976
D.C. Lindberg, *Theories of vision. From al-Kindi to Kepler*, Chicago & London 1976.

Lomazzo 1585
G.P. Lomazzo, *Trattato dell'arte della pittura, scoltura, et architettura*, Milan 1585.

Lomazzo 1598
G.P. Lomazzo, *A tracte containing the artes of curious paintinge caruinge building* (R. Haydocke, trans.), Oxford 1598.

Melion 1991
W.S. Melion, *Shaping the Netherlandish canon. Karel van Mander's Schilder-boeck*, Chicago & London 1991.

Molhuysen 1913
P.C. Molhuysen, *Bronnen tot de geschiedenis der Leidsche universiteit*, vol. 1, The Hague 1913.

Preimesberger 1991
R. Preimesberger, 'Zu Jan Van Eycks Diptychon der Sammlung Thyssen-Bornemisza', *Zeitschrift fur Kunstgeschichte* 54 (1991), 459-489.

Raynaud 1998
D. Raynaud, *L'hypothèse d'Oxford. Essai sur les origins de la perspective*, Paris 1998.

Schukking 1965
W.H. Schukking (ed.), *The principal works of Simon Stevin*, Amsterdam 1965, vol. 4, 29-231.

Seidel 2007
L. Seidel, 'Visual representation as instructional text. Jan Van Eyck and *The Ghent Altarpiece*', in: P.H. Smith & B. Schmidt (eds.), *Making knowledge in early modern Europe. Practices, objects, texts, 1400-1800*, Chicago & London 2007, 45-67.

Shapiro 1990
A.E. Shapiro, 'The optical lectures and the foundations of the theory

of optical imagery', in: Mordechai Feingold (ed.), *Before Newton. The life and times of Isaac Barrow*, Cambridge 1990, 105-178.

Simon 1976
G. Simon, *Structures de pensée et objets du savoir chez Kepler*, Paris 1976 (unpub. diss. Université de Paris IV).

Sinisgalli 1978
R. Sinisgalli, *Il contributo di Simon Stevin allo sviluppo della prospettiva artificiale*, Rome 1978.

Stevin 1594
S. Stevin, *De sterctenbovwing*, Leiden 1594.

Stevin 1605
S. Stevin, *Derde stvck der wisconstige ghedachtenissen Vande devrsichtighe, inhoudende t'ghene daer hem in gheoeffent heeft ... Maurits Prince van Oraengien ...*, Leiden 1605.

Struik 1958
D.J. Struik (ed.), *The principal works of Simon Stevin*, Amsterdam 1958, vol. 2B, 783-965.

Van den Heuvel 1991
C. van den Heuvel, *Papiere bolwercken. De introductie van de Italiaanse stede- en vestingbouw in de Nederlanden (1540-1609) en het gebruik van tekeningen*, Alphen aan den Rijn 1991.

Van den Heuvel 1994
C. van den Heuvel, '"T samenspreeckinge betreffende de Architecture ende Schilderkonst". Schilders, Architecten en Wiskundigen over de uitbeelding van architectuur', *Incontri. Rivista europea di Studi italiani* 1 (1994), 69-85.

Van den Heuvel 2004
C. van den Heuvel, 'Le traité incomplète de l'art militaire et l'instruction pour une école des ingénieurs de Simon Stevin', in: Cat. Brussels 2004, 103-113.

Van den Heuvel 2005
C. van den Heuvel, *De Huysbou. A reconstruction of an unfinished treatise on architecture, town planning and civil engineering by Simon Stevin*, Amsterdam 2005.

Van den Heuvel 2006
C. van den Heuvel, 'De vesting als mathematisch en cultureel kennissysteem. Het onderwijs in de vesting-bouw aan hovelingen, kooplui en ambachtslieden in de Hollandse Republiek (17de eeuw)', *Scientiarum Historia* 32 (2006), 99-117.

Van de Wetering 1997
E. van de Wetering, *Rembrandt. The Painter at Work*, Amsterdam: Amsterdam 1997.

Van Mander 1604
K. Van Mander, *Het schilder-boeck waerin voor eerst de leerlustighe iueght den grondt der edel vry schilderconst in verscheyden deelen wort voorghedraghen*, Haarlem 1604.

Van Mander 1973
K. van Mander, *Den grondt der edel vry schilder-const* (H. Miedema, ed.), Utrecht 1973.

Van Son 1993
R. van Son, 'Lomazzo, Lampsonius en de noordelijke kunst', *Nederlands Kunsthistorisch Jaarboek* 44 (1993), 185-196.

Veldman 1977
I.M. Veldman, *Maarten van Heemskerck and Dutch humanism in the sixteenth century*, Amsterdam 1977.

Vermij 2010
R. Vermij, 'The telescope at the court of Stadtholder Maurits', in: A. van Helden, S. Dupré, R. van Gent & H. Zuidervaart (eds.), *The origins of the telescope*, Amsterdam 2010, 73-92.

Vitruvius 1999
M. Vitruvius, *Ten books on architecture* (I.D. Rowland, T.N. Howe & M.J. Dewar, eds.), Cambridge 1999.

Vredeman de Vries 1979
H. Vredeman de Vries, *Perspective* (P. Karstkarel, ed.), Mijdrecht 1979.

Werrett 2001
S. Werrett, 'Wonders never cease. Descartes's "Météores" and the rainbow fountain', *British Journal for the History of Science* 34 (2001), 129-147.

Weststeijn 2008
T. Weststeijn, *The Visible world. Samuel van Hoogstraten's art theory and the legitimation of painting in the Dutch Golden Age*, Amsterdam 2008.

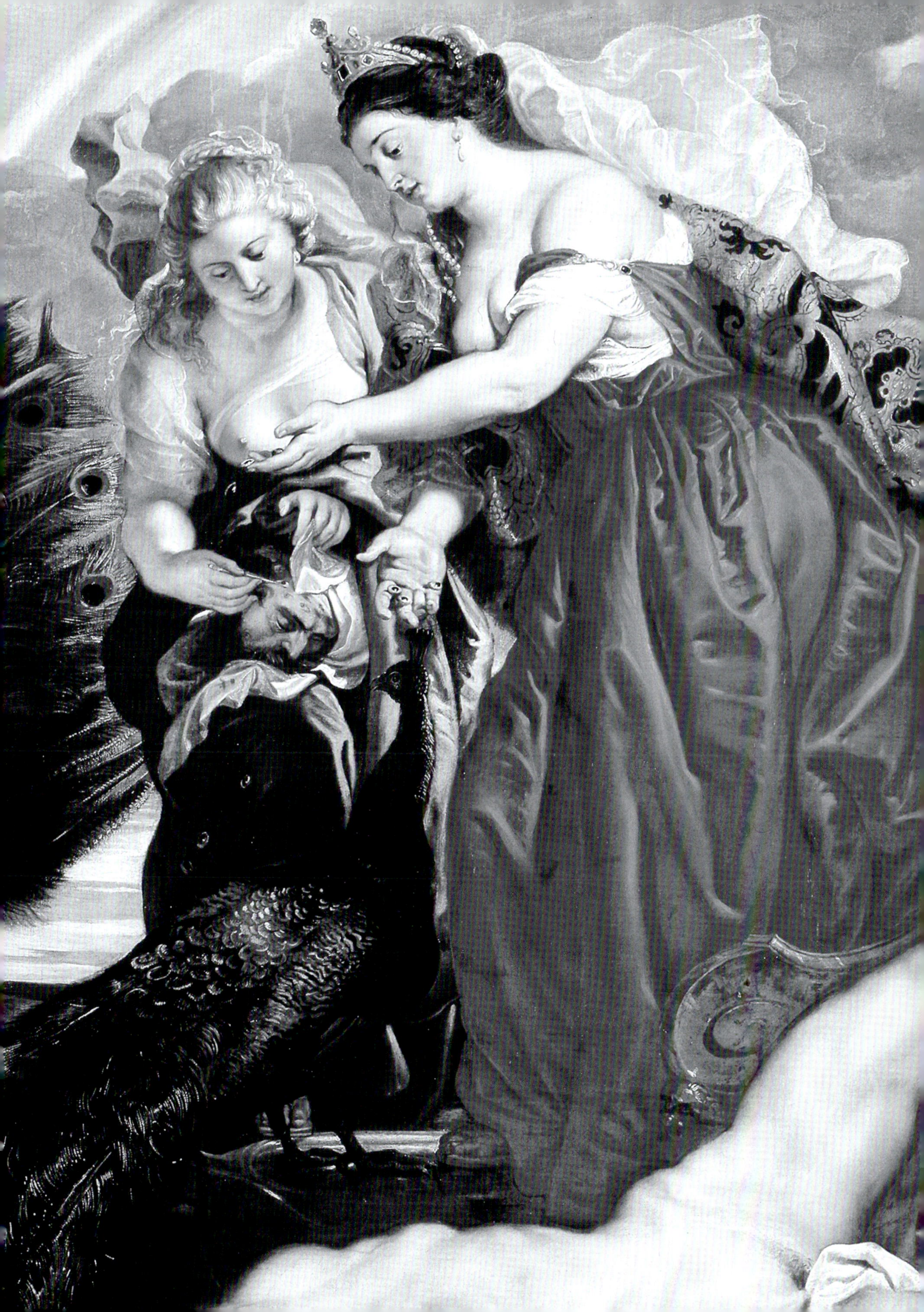

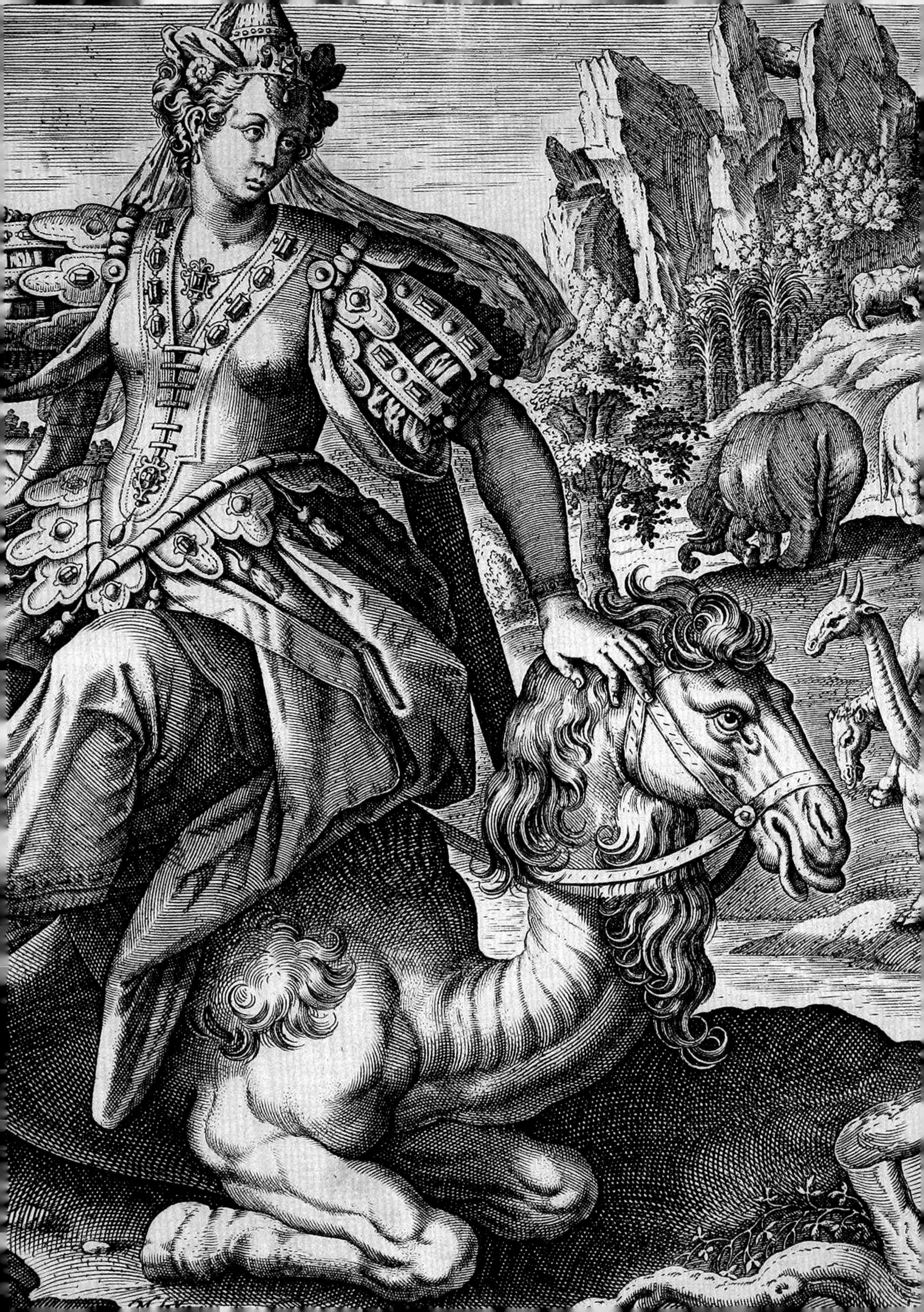

The camel's head
Representing unseen animals in sixteenth-century Europe*

Dániel Margócsy

La mer a tout ainsi que l'Element voisin,
Sa Rose, son Melon, son Oeïllet, son Raisin.[1]

Start with a print designed by Maarten de Vos (1532-1603) and engraved by Adriaen Collaert (1560-1618). *Asia*, part of a series on the *Four continents* from the end of the sixteenth century, features an allegorical representation (fig. 1).[2] Donning a revealing dress and a pointed hat with a veil, the female personification of Asia sits on a camel with an incense burner in her hands. In the background, a battle scene with Turkish warriors can be observed, and several exotic animals. In the foreground, a few tulips call attention to the Asian plant's growing popularity in the Low Countries. For a European audience, a woman, a battle scene, a plant and exotic animals could stand in for a whole continent.

Given the prominence accorded to plants and animals, one could be tempted to cite *Asia* as a manifestation of the rise of naturalism in late sixteenth-century art. In transposing De Vos's sketch onto the copperplate, Collaert's attention to detail was exemplary: the animals' musculature and the tulip's venations are portrayed in high resolution, even surpassing the original drawing. *Asia* echoes the Dürer Renaissance of the Rudolphine court, where the arts and the sciences collaborated in the exploration and description of nature. De Vos and Collaert were based in Antwerp, and neither was a stranger to the world of natural history and other scientific disciplines. De Vos portrayed a variety of exotic animals, devoted several series to the astrological powers of the planets, satirized alchemy, allegorized the seven liberal arts and designed cartouches for his friend the cartographer Abraham Ortelius (1527-1598).[3] Collaert also maintained good relations with Ortelius and would go on to publish several print series, titled *Icones*, on birds, fish, flowers and quadrupeds.[4] In these engravings, he would liberally borrow from the Swiss naturalist Conrad Gesner's *Historiae animalium* or from the Flemish naturalist Rembert Dodoens's *Cruydeboeck*.[5] Originally designed as model books, the *Icones* themselves would also penetrate the circles of natural history.[6] They were consulted by the Leiden naturalist Carolus Clusius (1526-1609) and copied by Anselmus de Boodt (c. 1550-1632), a Flemish physician and collector in Hapsburg Prague.[7]

Despite these artists' credentials for scientific naturalism, one still encounters some jarring elements when scrutinizing the details of *Asia*.

Detail fig. 1
Adriaen Collaert (after Maarten de Vos), *Asia* (from the *Four continents*), c. 1588-1589.

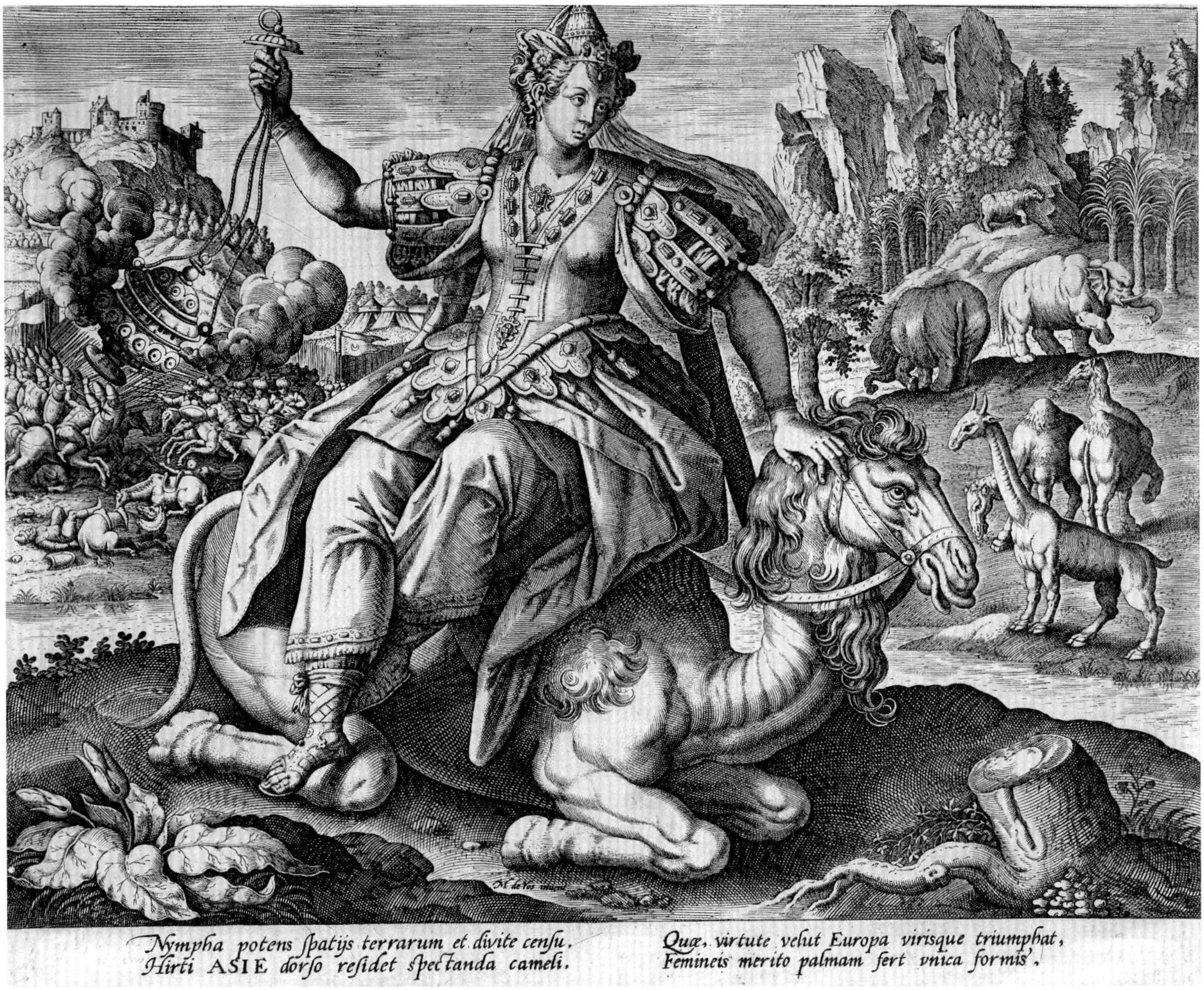

1

Adriaen Collaert (after Maarten de Vos), ***Asia (from the Four continents),*** **c. 1588-1589**
engraving, Cambridge, Harvard Art Museum. (Copyright President and Fellows of Harvard College. Harvard Art Museum, Fogg Art Museum, Gift of Belinda L. Randall from the collection of John Witt Randall, R13561).

2

Adriaen Collaert (after Jan van der Straet), ***The triumph of Caesar,*** **c. 1612**
engraving, 327 x 220 mm, London, The British Museum (© Trustees of the British Museum).

The tulip is shown with pinnately netted leaves, and not with the actual parallel venation. The giraffe sports goat-like horns, and not blunter, hairy and smaller protuberances.[8] Prominently positioned and elegantly executed, the camel's head is not the head of a camel. Its features closely follow the head of a horse. The curved lips, the flattened nostrils, the elongated nose and the rounded eyes are taken from the artist's massive knowledge of equine anatomy. While De Vos and Collaert were probably not familiar with live camels, they must have been intimately acquainted with horses.[9] As the Dutch art theorist Karel van Mander (1548-1606) emphasized, they were the noblest animal, and painters needed to study them, extensively.[10] A comparison of the camel with the horses in Collaert's engraving of *The triumph of Caesar* reveals close similarities in the treatment of the two species (fig. 2).[11] The mane, the arrangement of the muscles above the eyes, and even the caparison are practically identical. One can tell apart the heads only because of the horses' pointed ears and prominent teeth. In order to create a plausible image of a little-

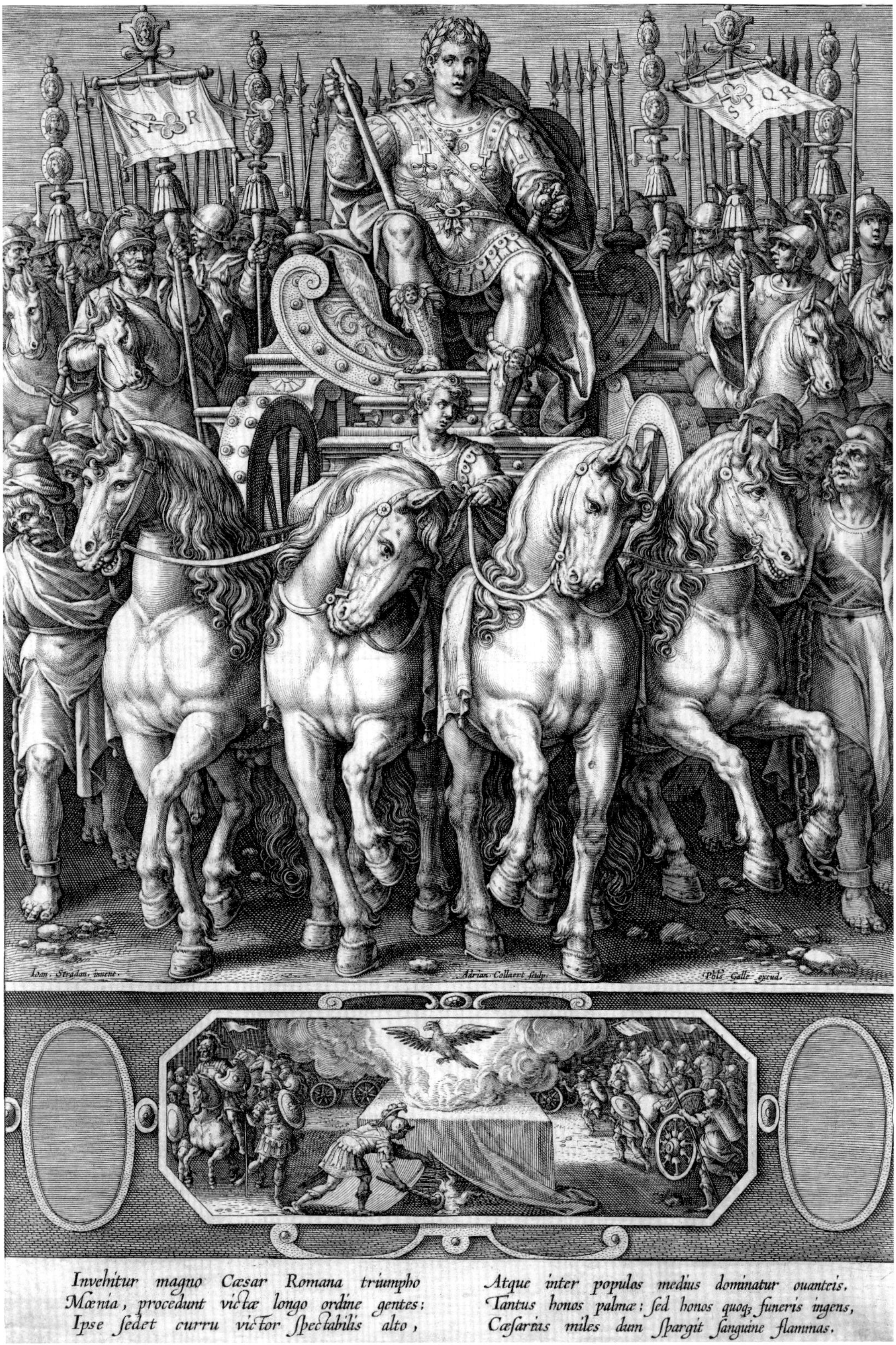
SPQR
SPQR
Ioan. Stradan. inuent.
Adrian. Collaert sculp.
Phls Galle excud.
Invehitur magno Cæsar Romana triumpho
Mœnia, procedunt victæ longo ordine gentes:
Ipse sedet curru victor spectabilis alto,
Atque inter populos medius dominatur ouanteis.
Tantus honos palmæ: sed honos quoq; funeris ingens,
Cæsareas miles dum spargit sanguine flammas.

known animal, De Vos and Collaert used a strategy that Peter Mason dubbed metonymic composition.[12] According to Mason, exotic animals were frequently imagined and represented in the early modern period by juxtaposing the body parts of diverse known animals. Since De Vos and Collaert were not closely familiar with camels, they used this strategy to visualize the unseen beast. A horse's head was appended to a humped trunk. This illustration is an educated guess at the camel's form, and not the result of firsthand observation.

This article asks how contemporary audiences reacted to the camel in the engraving of *Asia*, and other animals created in this manner. What was the epistemological status of such composite animals in early modern visual culture, and especially in the emerging discipline of natural history? Would natural historians and erudite scholars class this camel as a grotesque fantasy? Would unseen, exotic creatures, depicted with the help of metonymy, reinforce the traditional boundaries between the arts and sciences? In recent years, historians bridged the gap between these two disciplines by pointing out that Renaissance artists increasingly turned towards firsthand observation.[13] Some of this historiography has suggested that empirical science and naturalism were born at the same time, and both relied on the careful observation of life. As images of unseen animals were not taken from the life, could they still have a claim for scientific interest?

Viewed from the perspective of art history, De Vos and Collaert's camel could be perceived to belong to the tradition of the grotesque. Blending the parts of various known animals into a new creature, after all, had already been employed by Leonardo da Vinci (1452-1519) for the representation of fantastic animals.[14] Only this way could the artistic imagination take shape in a form with a sufficient reality effect. According to the Italian art theorist Giorgio Vasari (1511-1574), Leonardo's *Medusa* was produced according to this method. The Italian artist collected and observed 'crawling reptiles, green lizards, crickets, snakes, butterflies, locusts, bats, and other strange species of this kind, and by adapting various parts of this multitude, he created a most horrible and frightening monster with poisonous breath that set the air on fire'.[15]

Vasari's description might have been one of the sources of inspiration for Flemish artists at the close of the sixteenth century. A *Head of Medusa* in the Uffizi, by an unknown Flemish artist, fit the description closely enough to fool eighteenth-century authors into believing it was the original Leonardo.[16] A more famous version by Frans Snyders (1579-1657) and Peter Paul Rubens (1577-1640), an acquaintance of Collaert, illustrates how the artists produced a mythological creature from the close observation of nature and from the consultation of Pliny the Elder and contemporary works of natural history.[17] The Medusa's hair, turning into a variety of snakes, is evidence of the artists' close familiarity with and careful attention to the various shapes, folds and textures of snakeskin. In all three variants of *Medusa*, a mythical creature was created by the juxtaposition and joining of the carefully observed snakes and the human head.

Yet *Asia* is different from the paintings of Leonardo, his unknown Flemish imitator, and Rubens and Snyders. The camel does not strike viewers as a horrifying monster, and it is not purported to be a mythological creature. It is an informed guess at a little-known animal's live form. In the inscription of the print, Collaert explicitly claims that he represents a camel. Unlike the *Medusas*, this image maintains a claim to truth, and, as this paper argues, contemporary natural historians carefully considered such claims. Not only were composite animals an artistic strategy, but they populated the world of natural historians, as well.

A natural history of the future

For the natural historians of early modern Europe, the sixteenth century was oriented towards the future. With the discovery of America and the exploration of Africa and Asia, unknown plants and animals arrived in Europe in large numbers.[18] Naturalists eagerly described these new species and waxed lyrical about what still remained to be discovered. Further expeditions would bring to Europe even more unknown creatures, and natural history would never fail to expand. Turning 75 in 1600, the aging Carolus Clusius prefaced his own *Exoticorum libri decem* (1605) with the caveat that it was too early to write a definitive history of the exotic. As the Dutch were just beginning to send ships to 'India, Ethiopia and to America', they would surely discover a whole new flora and fauna. Only then would someone 'be able to describe the history of all those things better and more perfectly'.[19] Foreign lands were not the only source of surprising discoveries. The marine life of the Mediterranean, the Atlantic, the Indian and, eventually, the Pacific Oceans held much potential for exploration. In the 1550s the French naturalist Pierre Belon (1517-1564), who would later be translated by Clusius, argued that the discovery of America could be rivaled by the investigation of the sea, 'where the forms of things one sees are so varied and stupefying that their inventory and their contemplation can never be complete'.[20] Natural history would always remain a projective enterprise.

The number of known species indeed grew exponentially in the sixteenth century. While only a few hundred plant species were known by naturalists around 1500, the list had grown to over 6 000 by 1623.[21] Exotic animals similarly expanded the horizons of zoology.[22] Yet novelty was not restricted to only exotic places. Strange and previously unseen monsters emerged within Europe at an alarming rate. Treated as an integral part of descriptive natural history, these preternatural creatures called into doubt the standard, Aristotelian concept of the species itself.[23] In Spain, a fish was found in the 1540s with ships tattooed on its side, and news of it soon reached the Netherlands.[24] Conjoined twins, hermaphrodites and even conjoined hermaphroditic twins were reported across European countries.[25] While naturalists, theologians and astrologers widely debated the origins of these prodigies, many of them agreed that their numbers were growing. Girolamo Cardano (1501-1576) wrote that 'I was born in an epoch in which I was privileged to see many marvels', and the Louvain professor Cornelius Gemma (1535-1578) could only agree.[26] Writing in the second half of the sixteenth century, Gemma argued vehemently that time was out of joint. In a period of religious strife, with portents like the supernova of 1572 and the plague of 1574, could anyone be surprised that monstrous births appeared everywhere?[27]

In the years around 1600, two novel inventions complemented these sources of exotic creatures. The early seventeenth-century Netherlands saw the invention of the microscope. The Dutch polymath Constantijn Huygens (1596-1687) claimed that this instrument opened up a 'new theater of nature, another world', shedding light on the anatomy of insects and bees, and he encouraged the painter Jacques de Gheyn II (c. 1565-1629) to turn his burin to engraving it.[28] Within a few decades, Johannes Swammerdam (1637-1680) and Antoni van Leeuwenhoek (1632-1723) became the first geographers of these uncharted territories, where a drop of water was discovered to teem with previously unknown life forms.[29] And if this was not enough, the Dutch invention of the telescope opened up the possibility of extraterrestrial life. For those who accepted a Copernican, heliocentric universe, the Aristotelian theory about the incorruptibility of the planetary spheres became highly problematic.[30] As a result, scholars began to speculate whether other planets would hold life. In Johannes Kepler's *Somnium*, a scientific fantasy conceived around 1609 at the Rudolphine court, the renowned astronomer furnished the Moon with daemonic creatures, whose legs 'far surpass those of our camels', and then went on to speculate about their anatomy and behavior.[31] Several decades later, the astronomer Christiaan Huygens (1629-1695), son of Constantijn, also believed that other planets would be populated. While Christiaan Huygens openly acknowledged that he could only hypothesize, he argued that 'conjectures [were] not useless, because not certain'. He suggested that extraterrestrial plants and animals were 'not to be imagin'd too unlike ours'. In theory, nature could produce life in any shape. But in practice, the basic structures of life were probably the same everywhere. As Huygens argued,

> Who doubts but that God, if he had pleased, might have made the Animals in *America* and other distant Countries nothing like ours? (...) yet we see he has not done it. (...) Their Animals have Feet and Wings like ours, and like ours have Heart, Lungs, Guts, and the Parts serving to generation.[32]

Huygens's focus on the extraterrestrial was a seventeenth-century development. When talking about the exotic, however, sixteenth-century natural historians often agreed with his main claims. Although they preferred facts, they did not always shun conjectures and potentially false information. They believed in nature's variability, but they also believed that this variability was usually bound by some rules. Exotic and monstrous creatures tended not to drastically differ from their European counterparts.

This claim is not particularly new. Historians have argued ad nauseam that when artists, travelers and naturalists described and depicted exotic animals, plants and humans, they constantly compared them to their European counterparts.[33] These comparisons could take various forms. Some decided to endow the people of the new world with the features of ancient and contemporary models. The German artist Hans Burgkmair (1473-1531) grafted the features of Dürer's Adam onto the natives of Allago, while natural philosophers and explorers argued endlessly whether native Americans were related to Indians, Ethiopians, Mongols or Norwegians.[34] Other natural historians and theologians interpreted American plants within a Biblical framework, equating tomatoes with the forbidden fruit and the peoples of America with the lost tenth tribe of Israel.[35] Insofar as the marine world was concerned, the French erudite poet Guillaume du Bartas (1544-1590), quoted in the epigraph, was one among many to argue that most terrestrial plants and animals had their counterparts underwater. Writing in the 1550s, the first golden age of zoology, Pierre Belon and the Montpellier physician and professor Guillaume Rondelet (1507-1566) were both convinced that the oceans held marine rams, elephants and hyenas, as well as sea monks and sea bishops, two monsters that closely resembled Catholic clergymen.[36]

Yet not all exotic and monstrous creatures could be pictured by drawing straight parallels with European forms of life. Both natural historians and artists described most of these other exotica with the help of metonymy, by mixing and matching the body parts of known animals. Belon himself, a prolific author of travel narratives and encyclopedic works, used this technology to describe a variety of seashells, as well as numerous other animals. In *La nature et diversité des poissons*, he argued that the purpura shell 'justly resembled a large terrestrial snail, had it not been for the spikes on it'. The marine animal 'carried its tentacle like a snail, and moved in the same manner'.[37] Other shells were constructed by comparison to the purpura. The murex, for instance, looked very similar, except that the shell 'turned in a spiral, but less so than the *purpura*', 'its spikes were not so long as the *purpura*', and 'its flesh was harder'.[38] Through one intermediary, the murex itself was described as an analogy of the common, terrestrial snail, wearing a fancy, spiked shell.

Interestingly, the Hapsburg Kunstkammer has a bronze life cast–like sculpture in its holdings that closely follows Belon's description of the murex (fig. 3).[39] Life cast sculptures, in vogue throughout the sixteenth century, were three-dimensional models of small animals, cast directly from the carcass. The resulting sculpture offered a detailed, naturalistic image of the animal, and it appeared to eliminate the artist's hand from the process of representation. As a result, it could have a particularly powerful claim to truth; there was literally no possibility for the artistic imagination to intervene.[40] The bronze murex played on the conventions of this genre. This counterfeit animal resembled life casts, but it was produced by the artist's manual intervention, with a marine shell attached to the back of a common, terrestrial snail. The unknown sculptor's artwork attempted to imitate life casts, yet was not faithful to life. The sculptor conjectured that marine shells were inhabited by a species similar to their terrestrial counterparts and created a spurious, illusionistic artwork. The sculpture appears even livelier and more convincing by the prominent display of the tentacles, an impossible

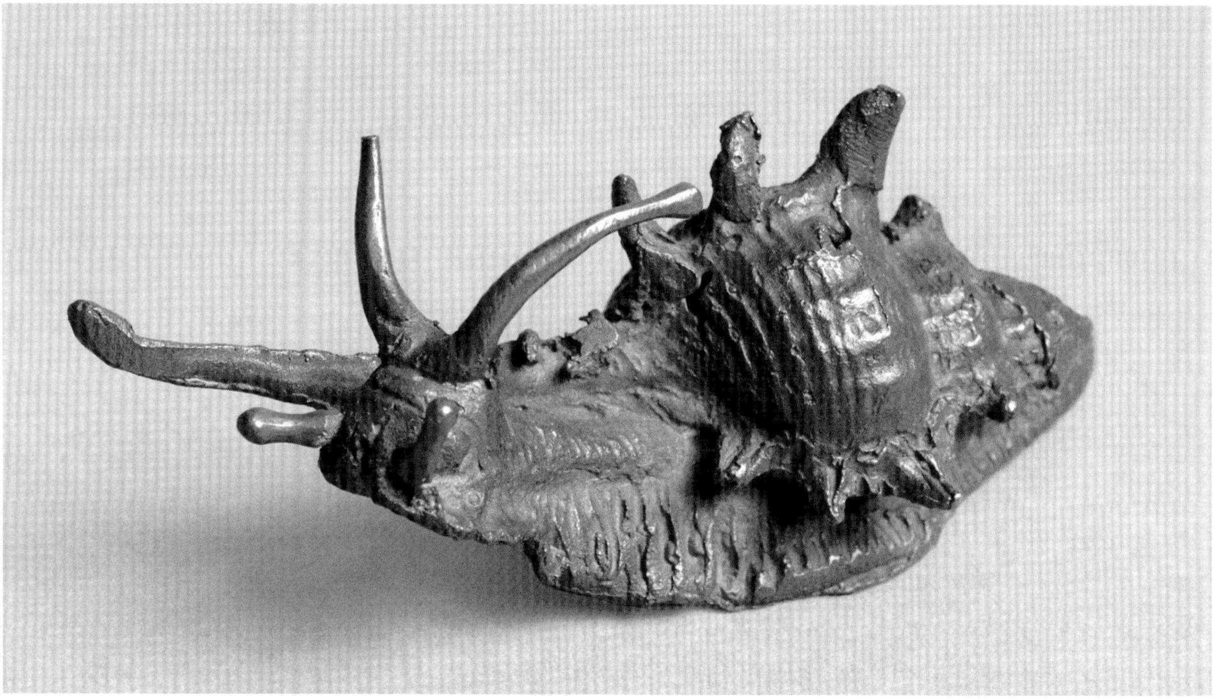

3
Artist unknown, *Murex*, 16th century
bronze, Vienna, Kunsthistorisches Museum, inv. no. KK 5940.

configuration in life casts because the tentacle withdrawal reflex persists even after death. The end result was thus a naturalistic imitation without an actual referent. The parts of known species were appropriated to carefully reconstruct the as yet unobserved anatomy of an exotic marine creature.

Given its provenance from the Hapsburg Kunstkammer, one might interpret the bronze murex as a grotesque, three-dimensional equivalent of Giuseppe Arcimboldo's renowned composite portraits. Yet, read together with Pierre Belon's natural historical account, this sculpture does not appear to be fantastic. The exaggerated tentacles of the bronze rhyme well with Belon's statement on the purpura, and the spikes are of moderate length. Intentionally or unintentionally, the unknown artisan followed Belon's recipe relatively closely and produced a trustworthy representation of nature. In using metonymic composition, his imagination only followed those rules of nature that Belon's history had already made explicit.

Nature's rules for producing exotica

As we have seen, the various shapes of exotic plants, animals and humans were discussed by poets, artists, philosophers and natural historians alike. Not all of these authors sought a reason for nature's variability. It became the domain of medical professionals, natural historians and natural philosophers to discuss these questions in their publications on the exotic world, on prodigies and monsters, and on astrology. For this set of authors, it was a surprising fact that animals on all continents looked similar to each other. Given the omnipotence of God, one could well expect to encounter unique life forms in America and elsewhere, shaped like no European creature. Although exotic nature offered an infinite variety of species, this variability was guided by some rules. The regularity of nature, the limited variability of species across the continents and nature's own reliance on metonymic composition needed explanation. Throughout the sixteenth century, scholars offered a host of reasons for the limits imposed on variability. Most of these explanations combined natural causes with divine intentions.[41]

Minor geographical variations within species were usually explained through the theory of climates.[42] The physiology of each animal was influenced by the four humors. As temperature and humidity varied by latitude, so did the balance of the humors fluctuate. Consequently, the constitution, behavior and shape of species were somewhat different in Europe, Asia, Africa and America. Had the weather been the same everywhere, Indian and African elephants, Bactrian and African camels, and lions in India, Syria and Libya would have looked the same.

Theories of climate could not account for more exotic marvels and monsters. Natural historians offered controversial arguments as to how composite creatures came into being.[43] Throughout the Renaissance, a host of authors supported the Aristotelian theory that the overabundance, or lack, of male semen in conception could result in a fetus with duplicated, or diminished, body parts: humans with two heads, goats with six feet. Others believed that such marvels were the result of divine or astral influences. For Cornelius Gemma, to give but one example, monstrous births were just one portent, signifying divine wrath on the wretched vices of humankind.[44] Discussing his own birth, Cardano suggested that, had he been born under another constellation, he might well have become a half-human, half-animal monster. The animal-like signs of the zodiac could bring forth animal-like monsters.[45]

Zoophilia was another frequent explanation as to why hybrid monsters could come into being. Although Aristotle cautioned against it, both teratological authors, like the French surgeon Ambroise Paré (c. 1510-1590), and natural historians, like Edward Topsell (c. 1572-1625) in England and John Jonstonus (1603-1675) in Poland, accepted that such creatures could be born from the wedlock of two different species.[46] Nature itself encouraged such an explanation through the well-known example of the mule, born to a horse and a donkey. Yet Topsell and others were aware of several other, similar examples. An ape and a weasel could produce a creature known as *sagoin*, a hyena and a lioness could conceive a *gulon*, and the tall, spotted giraffe was obviously born from the union of a camel and a panther.[47]

Camels were especially promiscuous. Topsell did not stop at the giraffe, and listed other curious offspring, as well. In these discussions, he came close to providing an explanation for De Vos and Collaert's visual representation of the camel, at least from the naturalist's perspective. Topsell did not completely agree with the Flemish artists, however. While the engraver depicted the camel with the head of a horse, the English naturalist claimed that the exotic animal's head and neck were 'different in proportion from all others'. Yet the camel of *Asia* was not far from the truth. Topsell did describe other, related species that bore striking resemblance to horse-like camels. The naturalist had heard of reports about a 'beast called Nabim, which in his neck resembleth a Horse, and in his head a camell'. While the origins of the *nabim* were unclear, mating across species played a role in the creation of the *allocamelus*, or llama: 'a beast which hath the heade, necke, and eares of a Mule, but the body of a Camell, wherefore it is probable, that it is conceiued by a Camell and a Mule'.[48] Clearly, De Vos and Collaert were not so far off naturalists' expectations. Even if his engraving did not correctly depict the camel in itself, such an animal might well have come into existence through the intercourse of camels and horses.[49] If a mule, usually considered barren, was able to mate with the camel, would Topsell have believed that horses wouldn't?

Naturalist scholars might well have had other reasons for not treating the camel in the engraving of *Asia* as a fiction. Images had a particular power. If a woman cast a glance on a painting or a sculpture during conception, the visual image could be impressed on the fetus. As John Jonstonus wrote, recounting the oft-repeated myth from Heliodorus, the Ethiopian queen Persina, 'seeing the Image of a white child when she lay with a man, had a child with a white face'.[50] Collaert's contemporary, the Flemish physician Jan Baptist van Helmont (1579-1644) argued similarly.[51] He reported that, if a pregnant woman is frightened by ducks, her imagination will transform the embryo into a duck. One could speculate, whether, even if De Vos and Collaert's camel did not represent an animal in the real world, flashing it in front of a female camel might have brought about a horse-like offspring. The artist's imagination, sparking the mother's imagination, could potentially lead to the creation of reality.

Uncertain animals

If the divine will, mating across species, and the artistic imagination all produced exotic animals in a similar manner, naturalists had fewer reasons to reject artists' representations of exotica as purely fictitious. If both nature and artists relied on the powers of metonymy in creating exotic animals, naturalists could justly call the latter's images an educated guess, and not outlandish fantasy. As we have seen in the case of the murex, both the unknown sculptor and Pierre Belon constructed the marine animal by positioning a spiked house on top of a terrestrial snail. This common strategy, I would suggest, allowed for the increasing circulation of images of exotic nature between artists and authors of natural history.[52]

Naturalists and the makers of visual representations had a complex relationship throughout the early modern period. Most illustrated volumes of natural history required the collaboration of author, draftsmen and woodcutters, which was oftentimes fraught with difficulties. Naturalists clearly appreciated expert artists and frequently expressed their admiration for the skill of artists like Albrecht Dürer.[53] Yet they were also ready to express their dissatisfaction with artists whose images proved to be inaccurate upon closer scrutiny. To avoid such situations, the naturalist would ideally provide the specimen, supervise the work of both the draftsman and the woodcutter in person and ensure that their images remained faithful to nature.[54] In less fortunate circumstances, the naturalist had no access to an actual specimen, and the makers of his illustrations were at a distance. The fame of the artist, the testimony of trustworthy witnesses and philological comparison with other sources were some of the means to vouchsafe for the credibility of such images.[55] Yet not all images of exotica came with warrants for credibility, and some had little to recommend them to naturalists. Despite their doubtful origins, I would argue that such images could also enter works of natural history. If they were produced with the help of metonymic composition, they were at least plausible, if not completely trustworthy representations. And works of natural history did not necessarily have to contain only undisputed facts. The plausible was oftentimes good enough.

While naturalists emphasized firsthand observation and the careful examination of evidence, they were aware that they could not fill their encyclopedias with only tight and neatly examined evidence. In the expanding world of the sixteenth century, the discipline of natural history also included the cataloguing of the possible, which would be verified or rejected by future discoveries. It was both a descriptive and a projective enterprise. Pierre Belon, for example, did not claim that his *La nature et diversité des poissons* contained only images from the life. The title page to his work offered the more qualified statement that they were 'au plus pres du naturel', or, in the slightly different Latin version, 'ad vivam effigiem, quoad ejus fieri potuit'.[56] Within the volume, some entries focused on highly controversial species. When describing the hippocampus, Belon could only say that the animal was a miniature combination of a horse and a caterpillar, a slightly grotesque case of composite animals. The accompanying illustration was not bulletproof, either. It was 'the fabulous portrait of the ancient horse of Neptune, as the old marbles and ancient medals taught us'.[57] While this illustration came from the arts and was deemed fabulous, it could still stand in for a real object.

Even Conrad Gesner (1516-1566), the Swiss doyen of sixteenth-century natural history, was ready to incorporate illustrations from artists into his masterly *Historiae animalium*.[58] This monumental, multivolume project aimed at cataloguing all the information that the Ancients and more recent authors had ever written on any animal species. When his sources contained contradictory information on an animal, Gesner often tried to adjudicate between the competing claims, but this was not always possible. In such cases, Gesner listed the differing authors' opinions and left it for future readers to make a judgment. When it came to providing illustrations for his entries, Gesner was similarly open to including images whose credibility was uncertain.[59] Artists' representations, possibly tainted by excessive imagination, were certainly game. For his entry on the unicorn, an animal whose existence was

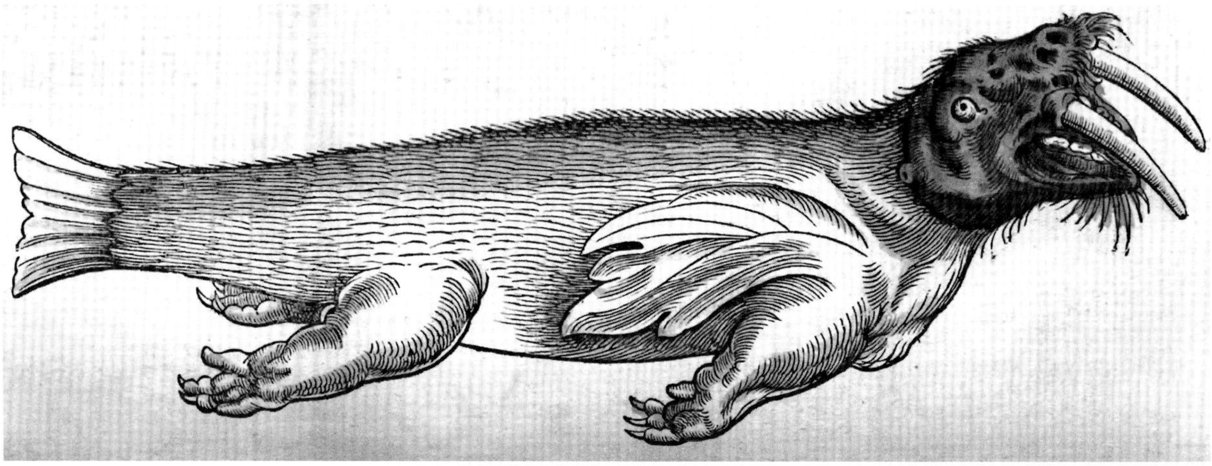

4
Artist unknown (after Albrecht Dürer), *Walrus*, 1558
woodcut with hand-coloring, Conrad Gesner. *Historiae animalium* (Zürich 1551-1558), vol. 4, 249 (Courtesy of the New York Academy of Medicine Library).

widely debated, Gesner explicitly relied on a source from the realm of the arts. As he wrote, 'this image is as it is nowadays generally depicted by painters, of which I know nothing for sure'.[60] Painters might not have been trustworthy, but the *Historiae animalium* could not completely ignore their evidence.

Metonymy and the analogy between marine and terrestrial animals played an important role in making it difficult to decide whether certain images were the figment of the imagination or a playful creation of nature. For his entry on the walrus, a little-known animal from the North, Gesner selected a woodcut that showed a fairly accurate head appended to a body with feet and claws (fig. 4). The naturalist expressed his reservations about the image and wrote that 'I heard that the head was made after the skull of a real head, and the rest of the body was added from conjecture or from a report'.[61] Gesner's hunch was probably correct. The head appears to be copied after Albrecht Dürer's *Head of a walrus*, drawn from the life, while the rest of the body was probably inspired by Hans Baldung Grien's fantastic pastiche of the Dürer walrus in Emperor Maximilian I's *Gebetbuch* (fig. 5).[62] Like Gesner's woodcut, Baldung Grien added a body, feet and claws to the Dürerian head.

Yet Gesner did not completely discredit the addition of feet. The engraver of the woodcut might have exaggerated it, but it was 'possible in the skeletons of fish, especially the larger ones, that, to a large extent, the fins were artfully fashioned according to the shape of feet and claws'.[63] The artist only accentuated how nature itself worked. Fins and feet, performing the same function in water and on earth, could well take similar shapes. The teratologist Ambroise Paré had even fewer doubts about the walrus's feet. In his *Des monstres et prodiges*, the French author reproduced Gesner's woodcut but left out the Swiss author's reservations. The image was a good representation. Paré called the walrus a marine elephant and argued that it had 'two teeth similar to an elephant'.[64] Like Du Bartas and Belon, the French author worked in a framework of correspondences between terrestrial and marine life. The claws and feet of the walrus fit well in this system and gave no grounds for suspicion.

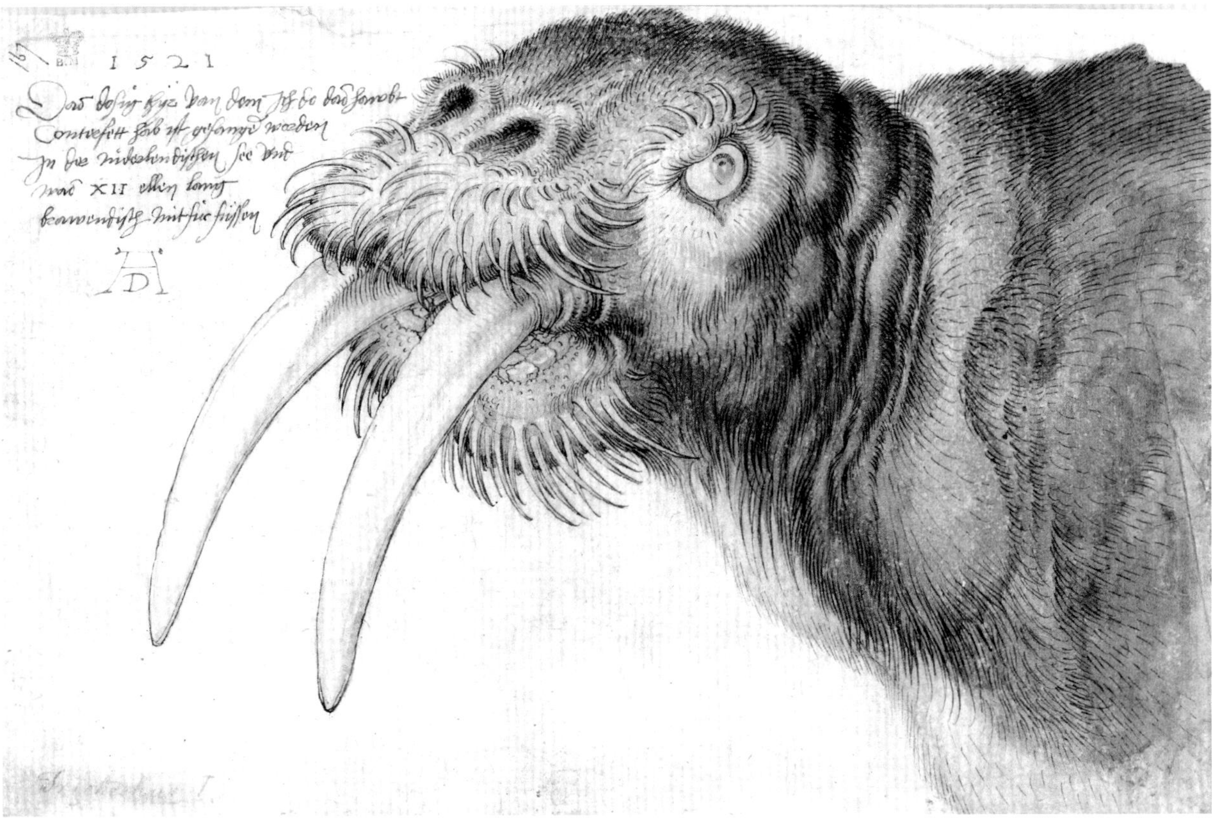

5
Albrecht Dürer, *Head of a walrus*, 1521
pen and brown ink, watercolor,
211 x 312 mm, London, The British Museum
(© Trustees of the British Museum).

Fins did not need to look like feet. They could also be mistaken for ears. In 1577, when a sperm whale was beached near Antwerp, a large number of broadsheets were published soon after the event.[65] Although purportedly based on firsthand observation, these broadsheets transposed the lateral fin next to the eyes and endowed it with an ear-like shape. While most broadsheets maintained that this body part was a fin, albeit with an auricular shape, some later copies would explicitly call it an ear.[66] This representation soon entered natural history, as well. Always a fan of composite creatures, Paré incorporated it in his *Des monstres et prodiges.* The Scheveningen fish merchant Adriaen Coenen (1514-1587) also made a copy of it for his manuscript *Whale book.* Within a few decades, the image was well entrenched in the zoological imagination. When another whale was beached in 1598, this time near Scheveningen, representations of this stranding bore a striking resemblance to those of the earlier broadsheets. The renowned artist Hendrick Goltzius (1558-1617) made a drawing of the whale, which would be turned into an engraving by his nephew Jacob Matham (1571-1631) (fig. 6).[67] Although Goltzius probably went to see the stranded whale in person, he decided to adopt the ear-like fin for his own drawing. His personal experience was colored by the tradition. In Goltzius's drawing, and Matham's print, the whale is pictured in the same position as in the earlier broadsheets. The enlarged penis is prominently displayed, and a few people are shown on top of the

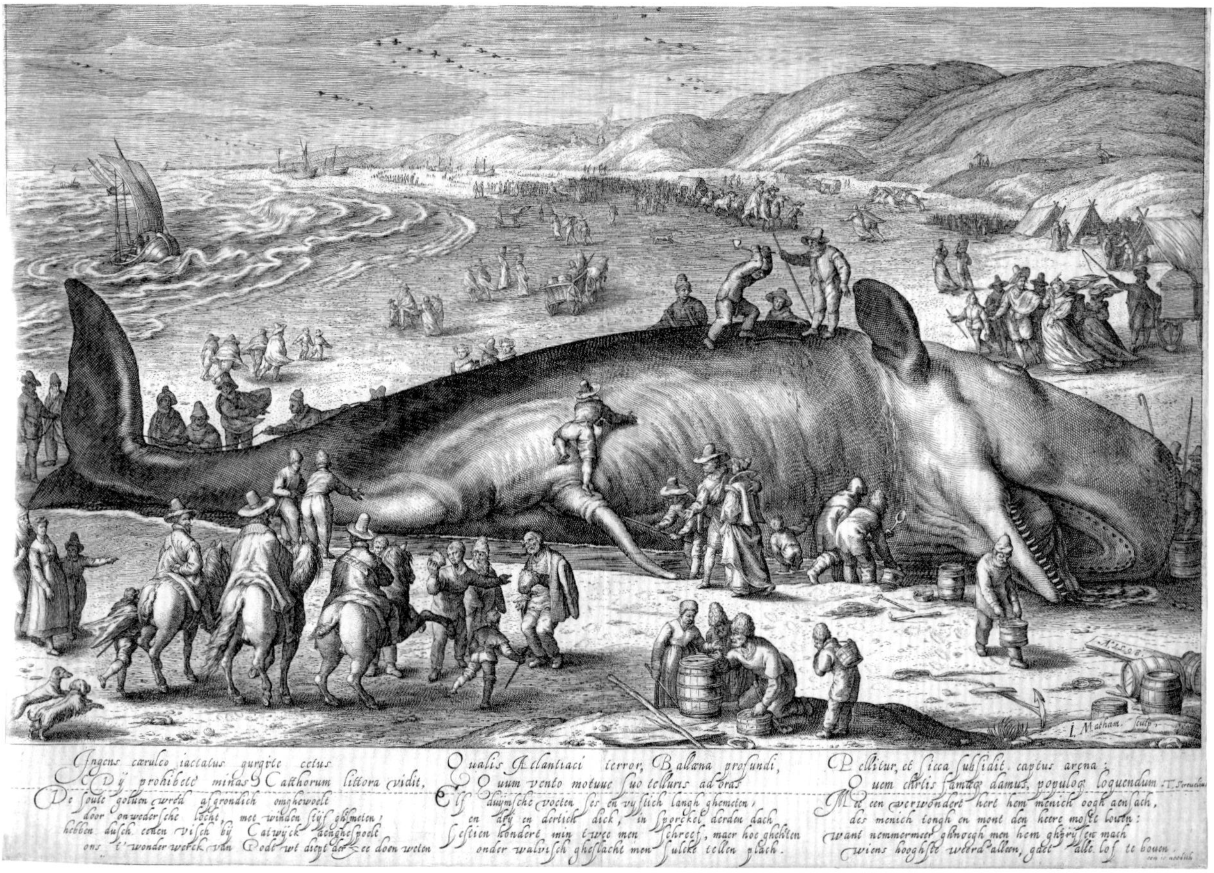

6

Jacob Matham (after Hendrick Goltzius), *The beached sperm whale near Berkhey*, 1598

engraving, 317 x 428 mm, London, The British Museum (© Trustees of the British Museum).

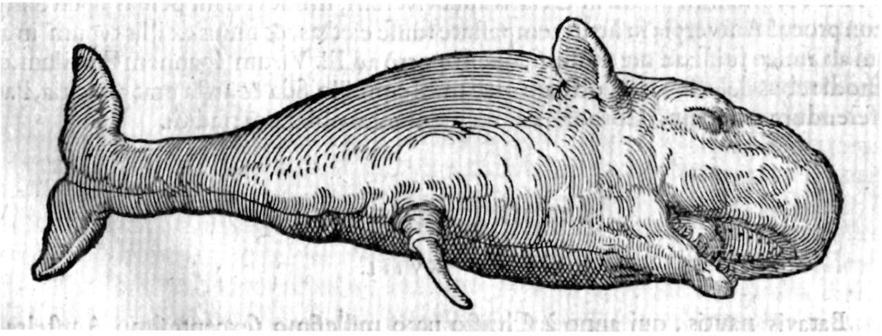

7

Artist unknown (after Jacob Matham), *Beached whale*, 1605

woodcut, Carolus Clusius, *Exoticorum libri decem* (Leiden 1605), 131 (Courtesy of the New York Academy of Medicine Library).

whale. As in De Vos and Collaert's *Asia*, the engraving's high resolution offers the appearance of naturalism, but the whale in the picture does not directly correspond to an observed specimen. Metonymic composition trumped firsthand observation.

Like the feet of the walrus, the whale's ear also entered natural history. Within a few years it showed up on the pages of the Carolus Clusius's *Exoticorum libri decem* (fig. 7).[68] Clusius's choice might appear surprising, as the Flemish-Dutch naturalist usually paid careful attention to verifying his sources. Much of the *Exoticorum libri decem* was devoted to items in the naturalist's own collection, and these entries show exotic specimens in the state they reached Clusius's cabinet.[69] Various tree barks, roots and nuts are pictured as fragments, while the illustrations of the whole plants are lacking from the book. Yet, as Sachiko Kusukawa and Peter Mason have shown, financial reasons, untrustworthy woodcutters and the lack of reliable information forced Clusius to include some imperfect images in his books.[70] In the *Rariorum plantarum historia*, the woodcutter's shoddy work ensured that different types of anemone were wrongly shown with identical roots and leaves.[71] While the naturalist complained about this extensively in his correspondence, the printed work contained no warnings about this inaccuracy.[72] Similarly, the illustration of the *ficus indica*, the very first image in the *Exoticorum libri decem*, was drawn on the basis of an oral report and not from observation of the actual tree.[73] The artist's negligence and the limits of available evidence resulted in not completely trustworthy illustrations.

Clusius's decision to include Matham's whale in his book might have been influenced by his realization of the limits of firsthand observation and the logic of metonymy. He could have selected another print of another contemporary beaching. In Jan Saenredam's engraving, for example, the whale appeared with a regularly shaped fin.[74] Yet it was not easy to decide which representation was correct. Clusius did not go and see the carcass of the animal in person, but he had heard that it was impossible to observe it carefully, anyway. Before finally expiring, the whale tossed and turned until large parts of it became buried underneath the sand. As a result, even those present were not able to examine all parts of the animal, and no one could agree about its circumference.[75] Under such circumstances, it was probably not easy to determine the exact shape of the fin. Matham's print appeared naturalistic and resembled earlier woodcuts, and, as Gesner and Belon had posited, the organs of marine animals frequently resembled terrestrial body parts. Why wouldn't the fin look like an ear?

Conclusion

As the example of Matham and Clusius shows, composite animals did not disappear from natural history and the arts when naturalism and empirical knowledge gained an increasingly important foothold in the Netherlands. And it remained a powerful method for constructing plausible animals even as the scientific revolution shifted into a higher gear. Let us look at Maarten de Vos and Adriaen Collaert's *Asia* one last time. As we have seen, this representation of the camel was not unlike Edward Topsell's description of the *nabim* and the *allocamelus*. Nonetheless, Topsell decided not to illustrate his *Historie of four-footed beasts* with an image based on *Asia*. It took several decades before the engraving entered the circulation of natural history through the Polish-Scottish naturalist John Jonstonus's *Historiae naturalis de quadrupetibus libri* (*Historiae de quadrupetibus*) in 1652.

Jonstonus's work is usually considered the last Renaissance encyclopedia in the vein of Gesner and Aldrovandi, produced at a

8

Caspar Merian, *Dromedary and camel*, 1652

engraving, John Jonstonus, *Historiae naturalis de quadrupetibus libri* (Frankfurt am Main, 1652), tab. XLIV, Mannheim University Library, inv. no. Sch 106/343 (photo: http://www.uni- mannheim.de/mateo/camenaref/jonston/vol4/jpg/s140.html).

considerable distance from new scientific developments. Yet Jonstonus traveled through much of Europe in his youth, graduated from Leiden University, and maintained strong contacts with John Amos Comenius (1592-1670) and Samuel Hartlib (c. 1600-1662), among others.[76] The *Historiae de quadripetibus,* and the encyclopedia's other parts, were themselves innovative. It was the first major encyclopedia of natural history to be published with copper engravings, and not woodcuts. For the work, Jonstonus chose Mattha:us Merian (1593-1650), one of the best-known printmakers of the century, and his children.

The *Historiae de quadripetibus* laid a strong emphasis on the variety of nature and on the powers of metonymic composition. In the preface, Jonstonus warned his readers that, in quadrupeds, 'there is a complex and inexpressible variety in all their parts, which could make you stunned upon closer scrutiny'.[77] Some of this variety was actually the result of mating across species. Horses could mate not only with donkeys, for instance, but also with onagers. Female donkeys, in turn, could also bear the offspring of a bull.[78] Nature's complexity could also be observed in camels, which showed extensive geographical variation. Asia was populated by Bactrian, Arabian and Caspian camels, and Africa could boast of another three subspecies – the Hugium, the Becheti and the Ragvahil – not to mention the *allocamelus* in the 'Land of Giants (...) that hath a head, ears, and neck like to a Mule, a body like a *Camel,* a taile like a Horse'.[79] Given the number of subspecies, it is no wonder that Jonstonus decided to provide ten different representations of camels, spread over four folio plates. No single image could do justice to nature's munificence.

Among the large number of illustrations can also be found a copy of the camel in the engraving of *Asia*.[80] In the *Historiae de quadripetibus*, the animal appears without the allegorical figure on top, and the engraver executed some other changes, as well (fig. 8). The elegant waves of the mane are replaced by furry patches of hair, the lips are shortened and the ears can finally be seen on the head. They are small and pointed, just like the ears of a horse. Given the many shapes a camel could take, Jonstonus might have found little reason to exclude an artistic source, even though its credibility was not established. Jonstonus relied on other artistic sources – for example, lions lifted from Rubens – and his encyclopedia even contained an appendix on dubious animals whose existence the author could not ascertain.[81] De Vos and Collaert's print was of sufficiently high quality, appeared naturalistic and used the well-known technique of metonymy. Camels with equine heads might have been a plausible guess for Jonstonus and well worth printing in a universal encyclopedia, especially because they were not all that different from the *allocamelus.* Even if the Flemish artists were wrong, and no subspecies of the camel looked quite like the one in this engraving, there were still nine other images to guide the reader. One of them was surely right.[82]

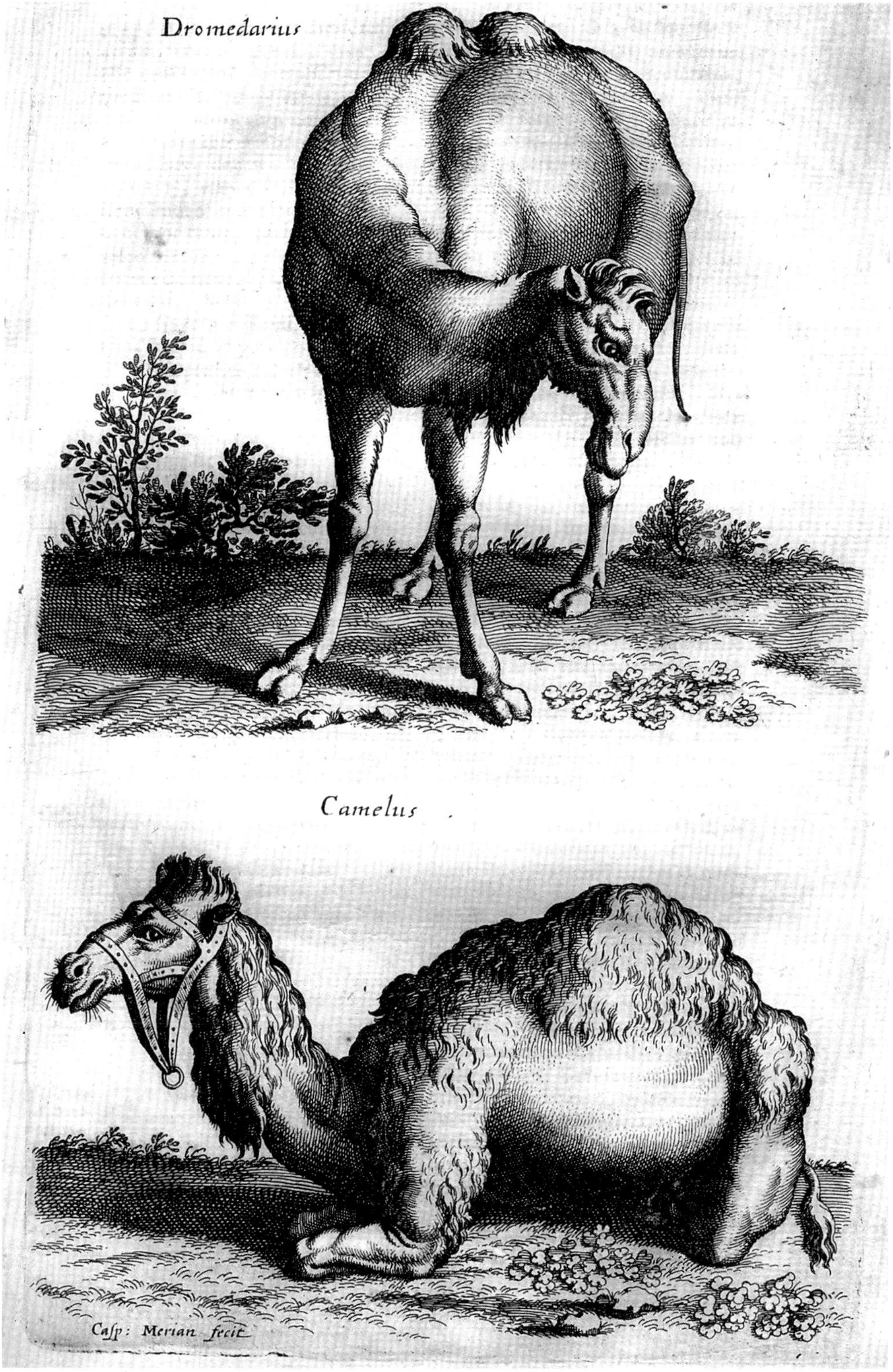
Dromedarius
Camelus
Casp: Merian fecit

Notes

*Versions of this paper have been presented at the Netherlands Yearbook of Art History workshop, the Klopsteg seminar, Northwestern University; the HSS Annual Conference, Phoenix; the Prints and the Production of Knowledge seminar, Harvard University; and the Center for Renaissance Studies, ELTE Budapest. I would like to thank the audiences at these talks, the reviewers, and Ken Alder, Mario Biagioli, Susan Dackerman, Eric Jorink, Iván Horváth, Katharine Park, Bart Ramakers, Pamela Smith, Claudia Swan, Klaas van Berkel, and especially James Clifton for their comments.

1 Du Bartas 1578, 129. On Du Bartas, see Banks 2008.

2 Adriaen Collaert (after Maarten de Vos), *Asia*, c. 1588-1589, engraving, New Hollstein (Collaert dynasty, vol. 6) 1315; part of a series on the *Four continents* (New Hollstein 1314-1317). On the series, see McGrath 2000.

3 On exotic animals, see the series of paintings commissioned by Johann Albrecht I of Mecklenburg, (*Panther, Elephant, Unicorn, Camel*, 1572, all in Schwerin, Staatliches Museum; and *Lion, Deer*, Mainz, Mittelrheinisches Landesmuseum). This early dromedary does not yet sport the features of a horse, suggesting Collaert's active collaboration in the production of De Vos's design for *Asia*. On the planets, see *The seven planets* (engraved by Johannes I Sadeler), 1585, Hollstein (Maarten de Vos) 1380-1387; and *Cathena aurea Platonis* (engraved by Crispijn de Passe), Hollstein 1373-1379. In the painting of the *Seven liberal arts* (Brussels, private collection), the celestial globe closely follows the contemporary astronomical practice in the depiction and positioning of the constellations, suggesting De Vos's familiarity with actual globes. On the cartouches for the engraved map of Abraham Ortelius's *The life and travels of Abraham* (1590, Hollstein 1344), see Bennett 1990.

4 Ortelius was godfather to Collaert's first daughter. Diels 2004-2005, vol. 1, 142.

5 Ważbiński 1999.

6 De Jong 1990.

7 'In tabellis tamen in aes incisis quas Adrianus Collardus de quadrupedibus publicabat, similem fere iconem conspicere memini'. Clusius 1605, 370. On Collaert and De Boodt, see Maselis 1999. For the visual culture of Flanders in this period, see Dupré 2010.

8 In this respect, Collaert follows the tradition set by the giraffe in Von Breydenbach 1486. Similar anatomical errors can also be found in Collaert's *Icones avium* series, for example, the parrots' feet are not zygodactylous, as the Amherst College Library website notes (https://www.amherst.edu/library/archives/holdings/soffer/c).

9 This is not the first camel with a horse's head. See *The camel. Maiolica from the San Paolo Monastery*, 1482, Parma, Galleria Nazionale. An early drawing from 1473 claims to depict a camel from the life, but rather looks like an emaciated dog with oversized hooves (unknown Ulm artist, *The dromedary*, 1473, Stuttgart, Staatsgalerie). Erhard Schoen's woodcut of a *Turk riding a dromedary* (1529, Hollstein 146; see also *Lansquenets with Turkish booty*, c. 1530, Hollstein 147) claims to depict the camels in the Turkish army besieging Vienna and was probably inspired by a camel that was on show in Nuremberg around that time (Timann 1993, 126). Nonetheless, this woodcut's lack of detail and limited circulation probably made it an unlikely model for the engraving of *Asia*.

10 'Eerst aen de behulpsaem moedighe peerden:/ Edel (Seggh' ick), want aen Peerden bevonden/ Zijn veel eyghenschappen'. Van Mander 1604, fol. 38v.

11 Adriaen Collaert (after Jan van der Straet), *The triumph of Caesar*, c. 1612, engraving, New Hollstein (Collaert dynasty) 1191.

12 Mason 2009, 87-123.

13 On the complexities of firsthand observation, see, among many other works, Clifton 1995, Swan 1995, Swan 2005, Parshall 1993, Givens *et al.* 2006, Dickenson 1998.

14 For other examples, see Giulio Romano's winged lions, 'whose feathers are so soft and downy that it seems impossible to believe that the hand of an artist could imitate Nature so closely'. Vasari 1998, 365.

15 Vasari 1998, 288. The anecdote was also known in the Low Countries, at least after the publication of Karel van Mander's *Schilderboeck*, Van Mander 1604, fol. 112v. I thank Claudia Swan for this reference. See also the contribution by Karin Leonhard elsewhere in this volume.

16 Unknown Flemish artist, *The head of Medusa*, c. 1600, Florence, Uffizi. See Turner 1999.

17 Peter Paul Rubens and Frans Snyders, *The head of Medusa*, 1616/1617, Vienna, Kunsthistorisches Museum. The copulating viper couple in the top right is inspired by Pliny, while the double-headed amphisbaena is probably based on an illustration from the circle of Cassiano dal Pozzo. Koslow 1995, 147-149; Freedberg 2002, 292; Mason 2009, 168-170; Cat. Los Angeles & The Hague 2006-2007,180-185. Rubens used a print by Collaert after Stradanus for one of his drawings (Logan & Plomp 2005, 9), and Collaert published a print after Rubens's *Judith beheading Holofernes* (1610-1620, Hollstein 31/II).

18 For an introduction, see Grafton 1992.

19 'Porro non dubium est, istius negotii longe majores progressus futuros, si vestris civibus, qui navigationes in Indiam, Aethiopiam et Americam Deo favente adeo feliciter nunc instituunt, mandaretis (...) Nam tametsi adeo progressa jam sim aetate, ut isto beneficio me fruiturum vix mihi polliceri queam: successuros tamen alios non diffido, qui satis amplam materiam hac ratione nacti, omnium illarum rerum historiam commodius et magis perfecte describere poterunt'. Clusius 1605, ep. ded.

20 Glardon 2009, 129.

21 Ogilvie 2006, 222.

22 On zoology, see Enenkel & Smith 2007, Piñon 1995.

23 Throughout the period, exotica and monsters were frequently treated together. See Daston & Park 1998, Céard 1996. On qualifying Aristotle's own concept of the species, see Lennox 1985.

24 Egmond 2005, 124.

25 Paré 1971, 25.

26 Cardano 1930, 194; Grafton 1999, 50.

27 Céard 2008; Pennuto 2008.

28 'Het is werkelijk alsof je voor een nieuw schouwtoneel van de natuur staat, op een andere aarde bent'. Huygens 1987, 132. See Swan 2005, 5-6; Jorink 2010, 180-182.

29 Ruestow 1996. See also the contribution by Eric Jorink elsewhere in this volume.

30 Campbell 1999.

31 Kepler 1967, 27, and for the dating of the work, xix.

32 Huygens 1698, 23. See also the contribution by Joke Spaans elsewhere in this volume.

33 See, for instance, Gerbi 1985. For the limits of such an analogical thinking, see Elkins 1992.

34 Leitch 2009.

35 Cogley 1986; Huddleston 1967.
36 Rondelet 1558, 361 on monkfish, and 363 on the other marine animals.
37 'Il resemble proprement a un gros Limacs terrestre, n'estoit qu'il est entouré de picqureons. Ceste Pourpre tire ses cornes, comme un Limacs, et chemine en la mesme maniere (...)'. Belon 1555, 420.
38 'Elle est tournee en uiz, mais moindre que celle de la Pourpre (...). Ses aguillons ne sont long comme ceux de la Pourpre (...). Il ha la chair plus durette, que la Pourpre'. Belon 1555, 425. Note that, according to Belon, the Romans thought that the purpura and the murex were the same species. Belon 1555, 420.
39 Unknown artist, *Murex*, 16th century, Vienna, Kunsthistorisches Museum, Kunstkammer inv. no. 5940. See Ferino-Pagden 2007, 197, who describes it as an imitation of a life cast. While contemporaries probably did not mistake the murex for a life cast, the allusion to this genre would not have been lost on them. I thank Pamela Smith for discussing this point.
40 For a description of the process, see Smith & Beentjes 2010.
41 On the issue of variation in Renaissance thought, see Siraisi 1994, and especially Maclean 2000.
42 For a general introduction to the importance of climates, see Wey Gómez 2008, Glacken 1967.
43 For an exhaustive review, see Céard 1996, and Bates 2005, 113-138.
44 Van Nouhuys 1998, 460; Gemma 1575, 82.
45 Cardano 1930, 5, 6.
46 Aristotle argues that hybrids can come into being only when the parental species are similar to each other and have equal gestational periods. Aristotle 2007, IV/3.
47 Topsell 1607, 18, 101, and 261. Not all authors supported this argument, and some of them suggested that only related species could interbreed. On this, and the Biblical complications of hybrids, see Van Ruler 1999.
48 Topsell 1607, 92, 101.
49 On the *allocamelus*, see Mason 2009, 173-196.
50 Jonstonus 1657, 346.
51 Van Helmont 1671, 287.
52 The circulation of zoological illustrations has been studied by, among others, Ashworth 1985, Ashworth 1984, and Krämer 2009.
53 When reproducing Dürer's rhinoceros in his *Historiae animalium*, Gesner specified that the woodcut was designed by Dürer, 'clarissimus ille pictor'. Gesner 1551-1558, vol. 1, 952.
54 Thus Ulisse Aldrovandi took both his amanuensis and a draftsman for his fieldwork on insects. 'Si quid portabatur nomen, naturam, locumque ubi cepissent, inquirebam, ac saepe etiam ipse una cum amanuensibus et pictore, cum ob continue studia fessi essemus, per vineta, agros, paludes, montesqe expatiabar: pictor secum penicillum, amanuenses pugillares et stylum ferebant, ille, si qui caperemus pictu dignum, pingebat, illi quod notatu erat dignum me dictante notabant; atque hoc modo tam variam insectorum supellectilem nancisci contigit'. Aldrovandi 1602, *ad lectorem.*
55 Ogilvie 2006.
56 Belon 1555. While the French version claims that the images are as lifelike as possible, the Latin text claims that the images are taken from the life, insofar as that was possible. I thank James Clifton for calling my attention to this point.
57 'Parquoy i'en laisse asseuré iugement a ceulx qui en pourront plus certainement prononcer, me contentant en cest endroict de pouuoir monstrer le fabuleux pourtraict de l'ancien Cheual de Neptune, tel que les uieilz marbres et medales antiques nous ont enseigné'. Belon 1555, 22, 23.
58 On Gesner, see Ogilvie 2006, Piñon 2005, Kusukawa 2010, Fischel 2010.
59 This was especially true for copies colored in hand, in which, as Gesner admitted, 'those [animals] that are not known to us can only be painted by some sort of conjecture. (Et aliqua nobis incognita non nisi ex coniectura qualicunque pingi possunt.)'. Gesner 1577, 22.
60 'Figura haec talis est, qualis a pictoribus fere hodie pingitur, de qua certi nihil habeo'. Gesner 1551-1558, vol. 1, 781.
61 'Pedes in hoc pisce expressi non placent; quanquam pictura etiam illa (...) pedes ostendit, sed in ea caput tantum ad sceleton ueri capitis factum audio, reliquum corpus ex coniectura aut narratione adiectum'. Gesner 1551-1558, vol. 4, 249.
62 Baldung Grien was a member of the town council of Strasbourg, and Gesner claimed that his walrus was taken from an image in the Strasbourg town hall, suggesting a direct connection between the two. Albrecht Dürer, *Head of a walrus*, 1521, British Museum; Hans Baldung Grien, *Walrus*, in: *The Gebetbuch of Emperor Maximilian I*, 1514-1515, Munich, Bayerische Staatsbibliothek. See also Von der Osten 1983, 314.
63 'Possunt in sceletis, praesertim maiorum piscium, pinnae, ad aliquam pedum unguiumque speciem arte formari'. Gesner 1551-1558, vol. 4, 249.
64 '(...) aiant deux dents semblables à un elephant'. Paré 1971, 108, 109.
65 For a review of these broadsheets, see Faust 2002.
66 'Les prunelles des ses yeux trente livres chacune, Et les oreilles cent cinquante, Le tout pesé es presences de Monsieurs Blanc (...)'. H.H.E., *Description d'un poisson marin*, 1619, Ulm, Stadtbibliothek, Einblattdrucke 922, Faust 2002, 562.1.
67 Jacob Matham (after Hendrick Goltzius), *The beached sperm whale near Berkhey*, 1598, New Hollstein (Matham) 202.
68 Clusius 1605, 131.
69 On Clusius and Dutch collecting in general, see Bergvelt & Kistemaker 1992, Swan 2008; Jorink 2010, 257-346, and esp. 278-289.
70 Kusukawa 2007, and Mason 2007.
71 Clusius 1601, book 2, 256, 257.
72 Roze 1899, 81.
73 Clusius 1605, 2.
74 Jan Saenredam, *Stranded sperm whale at Beverwijck*, 1601, Hollstein 121.II.
75 '(...) crassitudinis ambitum licet nonnulli dicant triginta pedum et unius, fuisse, alij vero longe majorem faciant: exacta tamen mensure sumi non potuit, quia volutatione et agitatione, ante quam interiret, corporis pars quaedam sabulo erat immersa'. Clusius 1605, 131.
76 Hitchens *et al.* 2000; Konior 2003.
77 'In reliquis omnibus multiplex & incomprehensa occurrit varietas, quae tibi, si bene perpenderis, stuporem iniicere poterit'. Jonstonus 1652, 4.
78 Jonstonus 1652, 28.
79 Jonstonus 1652, 101. The *allocamelus* is not mentioned in the *Historiae naturalis,* but in Jonstonus 1657, 210.
80 Jonstonus 1652, tab. XLIV.
81 Jonstonus 1652, tab. LI. The lions on the bottom are taken from Peter Paul Rubens, *Daniel in the lions' den*, c. 1615, Washington, National Gallery of Art. On dubious animals see Jonstonus 1652, 210.
82 On the indeterminacy of scientific illustrations, see Margócsy 2011.

Bibliography

Aldrovandi 1602
U. Aldrovandi, *De animalibus insectis libri septem,* Bologna 1602.

Aristotle 2007
Aristotle, *On the generation of animals* (A. Platt, trans.), Adelaide 2007.

Ashworth 1984
W. Ashworth, 'Marcus Gheeraerts and the Aesopic connection in seven-teenth-century scientific illustration', *Art Journal* 44 (1984), 132-138.

Ashworth 1985
W. Ashworth, 'The persistent beast. Recurring images in early zoological illustrations', in: J. Ackerman (ed.), *The natural sciences and the arts. Aspects of interaction from the Renaissance to the 20th century,* Uppsala 1985, 46-66.

Banks 2008
K. Banks, *Cosmos and image in the Renaissance*, London 2008.

Bates 2005
A. Bates, *Emblematic monsters. Unnatural conceptions and deformed births in early modern Europe*, Amsterdam 2005.

Belon 1555
P. Belon, *La nature et diversité des poissons,* Paris 1555.

Bennett 1990
S.K. Bennett, 'Drawings by Maerten de Vos. Designs to ornament an Ortelius map', *The Hoogstedder Mercury* 11 (1990), 4-13.

Bergvelt & Kistemaker 1992
E. Bergvelt & R. Kistemaker (eds.), *De wereld binnen handbereik. Nederlandse kunst- en rariteitenverzamelingen 1585-1735,* Zwolle 1992.

Campbell 1999
M.B. Campbell, *Wonder and science. Imagining worlds in early modern Europe,* Ithaca 1999.

Cardano 1930
G. Cardano, *The book of my life* (J. Stoner, trans.), New York 1930.

Céard 1996
J. Céard, *La nature et les prodiges*, Geneva 1996.

Céard 2008
J. Céard, 'La notion de prodige selon Cornelius Gemma', in: Hirai 2008, 67-76.

Clifton 1995
J. Clifton, '"Ad vivum mire depinxit". Toward a reconstruction of Ribera's art theory', *Storia dell'arte* 83 (1995), 111-132.

Clusius 1601
C. Clusius, *Rariorum plantarum historia*, Antwerp 1601.

Clusius 1605
C. Clusius, *Exoticorum libri decem*, Leiden 1605.

Cogley 1986
R.W. Cogley, 'John Eliot and the origins of the American Indians', *Early American Literature* 21 (1986), 210-225.

Daston & Park 1998
L. Daston & K. Park, *Wonders and the order of nature 1150-1750*, Cambridge 1998.

De Jong 1990
M. de Jong, '"Viel und mancherley Gefögel". Populaire friezen met vogels uit het eind van de 16de en de eerste helft van de 17de eeuw', *Bulletin van het Rijksmuseum* 38 (1990), 305-313.

Dickenson 1998
V. Dickenson, *Science and art in the portrayal of the New World*, Toronto 1998.

Diels 2004-2005
A. Diels, *De familie Collaert (ca. 1555-1630) en de prentkunst in Antwerpen*, Brussels 2004-2005 (unpub. diss. Vrije Universiteit, Brussels).

Du Bartas 1578
G. du Bartas, *La sepmaine, ou creation du monde*, Paris 1578.

Dupré 2010
S. Dupré, 'Trading luxury glass, picturing collections and consuming objects of knowledge in early seventeenth-century Antwerp', *Intellectual History Review* 20 (2010), 53-78.

Egmond 2005
F. Egmond, *Het Visboek: De wereld volgens Adriaen Coenen 1514-1587*, Zutphen 2005.

Egmond *et al.* 2007
F. Egmond, P. Hoftijzer & R. Visser (eds.), *Carolus Clusius. Towards a cultural history of a Renaissance naturalist*, Amsterdam 2007.

Elkins 1992
J. Elkins, 'On visual desperation and the bodies of protozoa', *Representations* 40 (1992), 33-56.

Enenkel & Smith 2007
K.A. Enenkel & P.J. Smith (eds.), *Early modern zoology. The construction of animals in science, literature and the visual arts*, Leiden 2007.

Faust 2002
I. Faust, *Zoologische Einblattdrucke und Flugschriften vor 1800*, vol. 4, Stuttgart 2002.

Ferino-Pagden 2007
S. Ferino-Pagden, *Arcimboldo 1526-1593*, Milan 2007.

Fischel 2010
A. Fischel, 'Collections, images and form in sixteenth-century natural history. The case of Conrad Gessner', *Intellectual History Review* 20 (2010), 147-164.

Freedberg 2002
D. Freedberg, *They eye of the lynx. Galileo, his friends, and the beginnings of natural history*, Chicago 2002.

Gemma 1575
C. Gemma, *De naturae divinis characterismis*, Antwerp 1575.

Gerbi 1985
A. Gerbi, *Nature in the New World. From Christopher Columbus to Gonzalo Fernandez de Oviedo*, Pittsburgh 1985.

Gesner 1551-1558
C. Gesner, *Historiae animalium*, Zürich 1551-1558.

Gesner 1577
C. Gesner, *Epistolarum medicinalium libri IV*, Zürich 1577.

Givens *et al.* 2006
J. Givens, K. Reeds & A. Touwaide (eds.), *Visualizing medieval medicine and natural history, 1200-1550*, Aldershot 2006.

Glacken 1967
C.J. Glacken, *Traces on the Rhodian shore. Nature and culture in Western thought from ancient times to the end of the eighteenth century*, Berkeley 1967.

Glardon 2009
P. Glardon, 'The relationship between text and illustration in mid-sixteenth-century natural history treatises', in: B. Boehrer (ed.), *A cultural history of animals in the Renaissance*, Oxford 2009, 119-146.

Grafton 1992
A. Grafton, *New worlds, ancient texts. The power of tradition and the shock of discovery*, Cambridge 1992.

Grafton 1999
A. Grafton, *Cardano's cosmos. The worlds and works of a Renaissance astrologer*, Cambridge 1999.

Hirai 2008
H. Hirai (ed.), *Cornelius Gemma. Cosmology, medicine and natural philosophy in Renaissance Louvain*, Pisa 2008.

Hitchens *et al.* 2000
W.J. Hitchens, A. Matuszewski & J. Young, *The letters of Jan Jonston to Samuel Hartlib*, Warsaw 2000.

Huddleston 1967
L.E. Huddleston, *Origins of the American Indians. European concepts, 1492-1729*, Austin 1967.

Huygens 1698
C. Huygens, *The celestial worlds discover'd, or, conjectures concerning the inhabitants, plants and productions of the worlds in the planets*, London 1698.

Huygens 1987
C. Huygens, *Mijn jeugd* (C.L. Heesakkers, ed. & trans.), Amsterdam 1987.

Jonstonus 1652
J. Jonstonus, *Historiae naturalis de quadrupetibus libri*, Frankfurt am Main 1652.

Jonstonus 1657
J. Jonstonus, *An history of the wonderful things of nature* (J. Rowland, trans.), London 1657.

Jorink 2010
E. Jorink, *Reading the book of nature in the Dutch Golden Age*, Leiden & Boston 2010.

Kepler 1967
J. Kepler, *Kepler's somnium. The dream, or posthumous work on lunar astronomy* (E. Rosen, ed. & trans.), Madison 1967.

Konior 2003
A. Konior (ed.), *Jan Jonston w 400 lecie urodzin (1603-2003). 'Europejskość myśli i twórczości naukowej Jana Jonstona po czterech wiekach'*, Leszno 2003.

Koslow 1995.
S. Koslow, '"How looked the Gorgon then ...". The science and poetics of "The head of Medusa"', in: C.P. Schneider, A.I. Davies & W.W. Robinson (eds.), *Shop talk. Studies in honor of Seymour Slive*, Cambridge 1995, 147-149.

Krämer 2009
F. Krämer, *The persistent image of an unusual centaur. A biography of Aldrovandi's two-legged centaur woodcut, Nuncius* 24 (2009) 313-340.

Kusukawa 2007
S. Kusukawa, 'Uses of pictures in printed books. The case of Clusius' *Exoticorum libri decem*', in: Egmond *et al.* 2007, 221-246.

Kusukawa 2010
S. Kusukawa, 'The sources of Gessner's pictures for the *Historia animalium*', *Annals of Science* 67 (2010), 303-328.

Leitch 2009
S. Leitch, 'Burgkmair's *Peoples of Africa and India* (1508) and the origins of ethnography in print', *Art Bulletin* 91 (2009), 134-159.

Lennox 1985
J. Lennox, 'Are Aristotelian species eternal?', in: A. Gotthelf (ed.), *Aristotle on nature and living things*, Pittsburgh 1985, 67-94.

Logan & Plomp 2005
A.-M. Logan & M.C. Plomp, 'Peter Paul Rubens as a draftsman', in: *Peter Paul Rubens (1577-1640). The drawings,* New York, Metropolitan Museum of Art, 2005. (exhibition website, http://www.metmuseum.org/special/rubens/rubens_essay.asp).

Cat. Los Angeles & The Hague 2006-2007
Cat. Los Angeles, The J. Paul Getty Museum & The Hague, Royal Picture Gallery Mauritshuis, *Rubens & Brueghel. A Working Friendship* (A.T. Woollett & A. van Suchtelen), Los Angeles & The Hague 2006-2007.

Maclean 2000
I. Maclean, 'Evidence, logic, the rule and the exception in Renaissance law and medicine', *Early Science and Medicine* 5 (2000), 227-256.

Margócsy 2011
D. Margócsy, 'A museum of wonders or a cemetery of corpses? The commercial exchange of anatomical collections in the early modern Netherlands', in: S. Dupré & C. Lüthy (eds.), *Silent messengers. The circulation of material objects of knowledge in the early modern Low Countries*, Berlin 2011, 185-216.

Maselis 1999
M.-C. Maselis, *The albums of Anselmus de Boodt (1550-1632). Natural history painting at the court of Rudolph II in Prague*, Ramsen 1999.

Mason 2007
P. Mason, '*Americana* in the *Exoticorum libri decem* of Charles de l'Ecluse', in: Egmond *et al.* 2007, 195-220.

Mason 2009
P. Mason, *Before disenchantment. Images of exotic plants and animals in the early modern world*, London 2009.

McGrath 2000
E. McGrath, 'Humanism, allegorical invention, and the personification of the continents', in: H. Vlieghe, A. Balis & C. van de Velde (eds.), *Concept, design and execution in Flemish painting (1550-1700)*, Turnhout 2000, 43-71.

Ogilvie 2006
B.W. Ogilvie, *The science of describing. Natural history in Renaissance Europe*, Chicago 2006.

Paré 1971
A. Paré, *Des monstres et prodiges* (J. Céard, ed.), Geneva 1971.

Parshall 1993
P. Parshall, '"Imago Contrafacta". Images and facts in the Northern Renaissance', *Art History* 16 (1993), 554-579.

Pennuto 2008
C. Pennuto, 'Cornelius Gemma et l'épidémie de 1574', in: Hirai 2008, 77-90.

Piñon 1995
L. Piñon, *Livres de zoologie de la Renaissance*, Paris 1995.

Piñon 2005
L. Piñon, 'Conrad Gesner and the historical depth of Renaissance natural history', in: G. Pomata & N. Siraisi (eds.), *Historia. Empiricism and erudition in early modern Europe*, Cambridge 2005, 241-268.

Rondelet 1558
G. Rondelet, *L'histoire entière des poissons*, Lyon 1558.

Roze 1899
E. Roze, *Charles de l'Escluse d'Arras. Sa biographie et sa correspondence*, Paris 1899.

Ruestow 1996
E.G. Ruestow, *The microscope in the Dutch Republic. The shaping of discovery*, Cambridge 1996.

Siraisi 1994
N. Siraisi, 'Vesalius and human diversity in *De humani corporis fabrica*', *Journal of the Warburg and Courtauld Institutes* 57 (1994), 60-88.

Smith & Beentjes 2010
P.H. Smith & T. Beentjes, 'Nature and art, making and knowing. Reconstructing sixteenth-century life-casting techniques', *Renaissance Quarterly* 63 (2010), 128-179.

Swan 1995
C. Swan, 'Ad vivum, naer het leven, from the life. Defining a mode of representation', *Word & Image* 2 (1995), 353-372.

Swan 2005
C. Swan, *Art, science, and witchcraft in early modern Holland. Jacques de Gheyn II (1565-1629)*, Cambridge 2005.

Swan 2008
C. Swan, 'Making sense of medical collections in early modern Holland. The uses of wonder', in: P.H. Smith & B. Schmidt (eds.), *Making knowledge in early modern Europe. Practices, objects and texts, 1400-1800*, Chicago 2008, 199-213.

Timann 1993
U. Timann, *Untersuchungen zu Nürnberger Holzschnitt und Briefmalerei in der ersten Hälfte des 16. Jahrhunderts*. Münster 1993.

Topsell 1607
E. Topsell, *The historie of four-footed beasts*, London 1607.

Turner 1999
R. Turner, 'Words and pictures. The birth and death of Leonardo's *Medusa*', in: C. Farago (ed.), *Leonardo da Vinci. Selected scholarship*, vol. 2, New York 1999, 287-296.

Van Helmont 1671
J.-B. van Helmont, *Les oeuvres de Jean Baptiste van Helmont*, Lyon 1671.

Van Mander 1604
K. van Mander, *Het Schilderboeck*, Haarlem 1604.

Van Nouhuys 1998
T. van Nouhuys, *The age of two-faced Janus. The comets of 1577 and 1618 and the decline of the Aristotelian world view in the Netherlands*, Leiden & Boston 1998.

Van Ruler 1999
H. van Ruler, 'Waren er muilezels op de zesde dag? Descartes, Voetius en de zeventiende-eeuwse methodenstrijd', in: F. Egmond, E. Jorink & R.H. Vermij (eds.), *Kometen, monsters en muilezels. Het veranderde natuurbeeld en de natuurwetenschap in de zeventiende eeuw*, Haarlem 1999, 133-153.

Vasari 1998
Giorgio Vasari, *The lives of the artists* (J. Conaway Bondanella & P. Bondanella, trans.), Oxford 1998.

Von Breydenbach 1486
B. von Breydenbach, *Peregrinatio in terram sanctam,* Mainz 1486.

Von der Osten 1983
G. von der Osten, *Hans Baldung Grien. Gemälde und Dokumente,* Berlin 1983.

Waźbiński 1999
Z. Waźbiński, 'Adrian Collaert i jego Florilegium. Niderlandzkie zródla wloskiej szesnastowiecznej martwej natury', in: M. Poprz cka (ed.), *Ars longa. Prace dedykowane pamieci profesora Jana Bialostockiego,* Kraków 1999.

Wey Gómez 2008
N. Wey Gómez, *The tropics of empire. Why Columbus sailed south to the Indies,* Cambridge 2008.

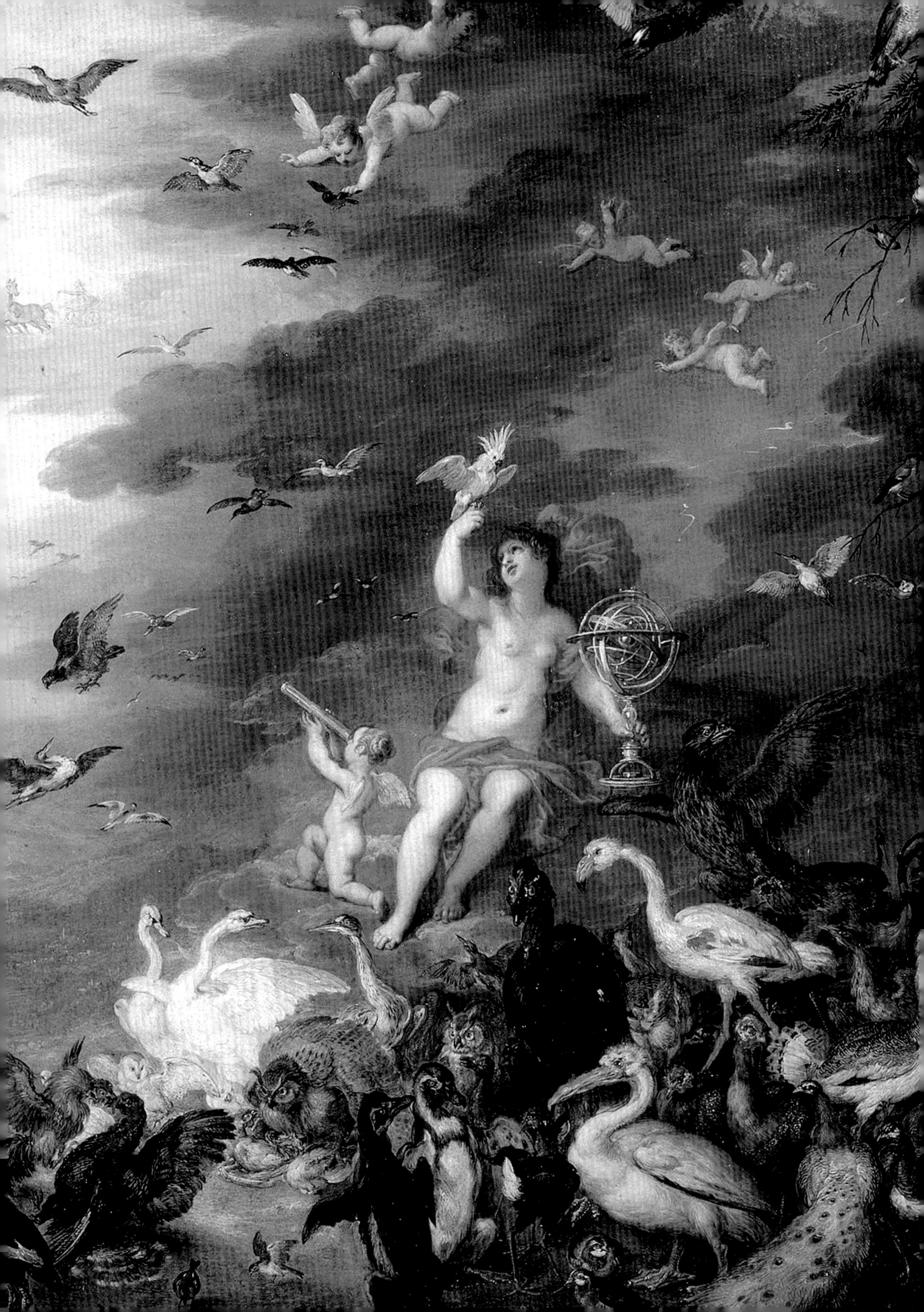

Jan Brueghel's *Allegory of air* (1621) from a natural historical perspective*

Marrigje Rikken and Paul J. Smith

Some paintings are worth the wait. Cardinal Federico Borromeo (1564-1631) must have felt the same when in 1622 he eventually received the *Allegory of air* from Jan Brueghel the Elder (1568-1625) to complete the series of the *Four elements* (fig. 1).[1] The *Musaeum Bibliothecae Ambrosianae* – a sort of museum guide written by the cardinal himself, in which he described the highlights and favourite works of his public art collection that he had accommodated in the Ambrosiana since 1618 – was published in 1625.[2] In that work he enthused about Brueghel's painting:

> He suffused the *[Element of] Air* with every delight as though it were a field of light. If something further needs to be said to compare this painting with the ones discussed earlier, I would note that the artist seems to have expended his last drop of ingenuity on this final painting, thus finishing the whole series.[3]

It is clear from this evidence that Borromeo preferred the *Allegory of air* to the other paintings in the series. This article focuses on the following question: What makes this painting by Brueghel so special? But before going any further, it is first important to consider the genesis of the painting.

The fourth element: the *Allegory of air*

The Counter-Reformationist Federico Borromeo and Jan Brueghel met in Rome around 1593, and Brueghel formed part of the entourage of the cardinal in Milan in 1595-1596.[4] Right from the start Borromeo was charmed by the painter and immediately commissioned paintings, as he would continue to do until Brueghel's death.[5] Although Brueghel returned to Antwerp later in 1596, he kept up an intensive correspondence with Borromeo from which a close friendship emerged.[6] That correspondence is a prime source of information about the genesis of the paintings that Brueghel did for Borromeo.[7]

In a letter of February 1608 Brueghel writes that he had wanted to start on the *Allegory of fire* but was prevented by the cold.[8] Nevertheless, he finished this painting relatively quickly: he writes in a letter of September of that same year that he expected to send the painting within

Detail fig. 1
Jan Brueghel the Elder, *Allegory of air*, c. 1621

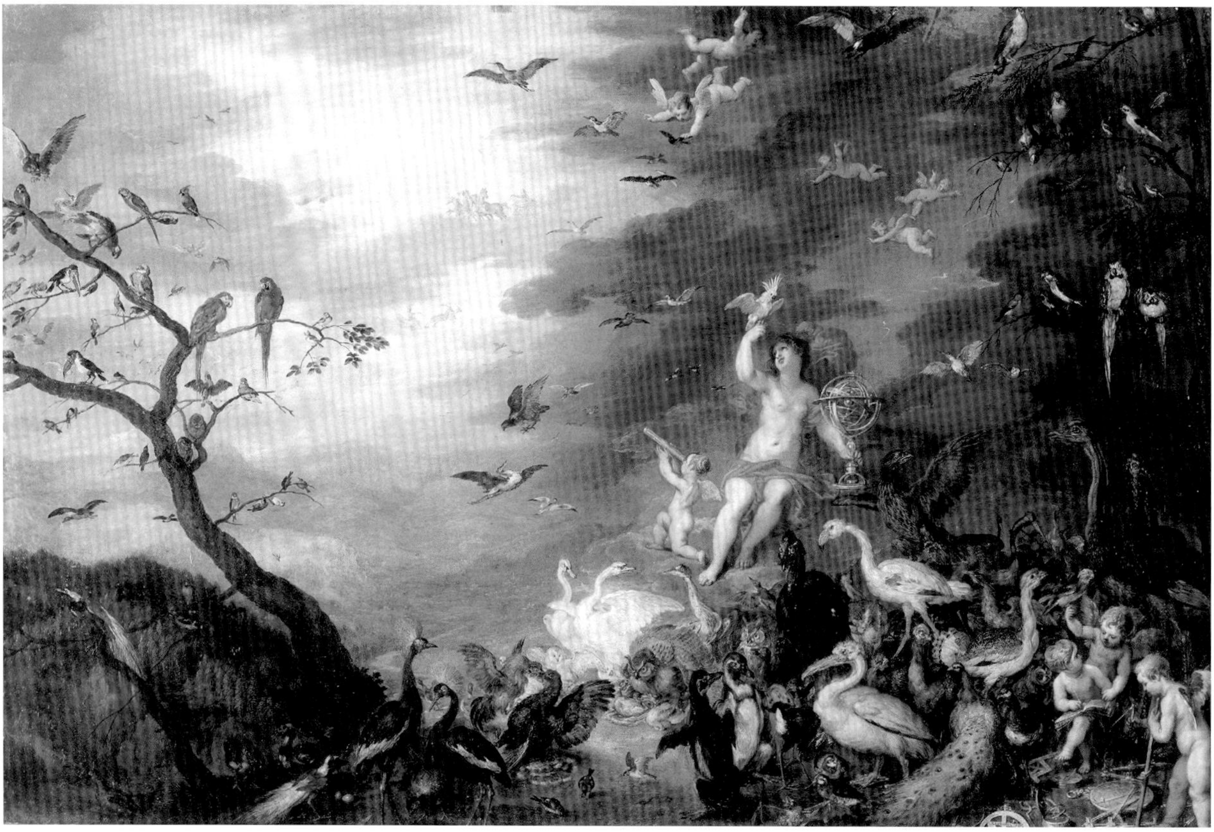

1

Jan Brueghel the Elder, *Allegory of air*, 1621

oil on copper, 46 x 66 cm, Paris, Musée du Louvre.

a few days.[9] The painter dispatched the *Allegory of water* to Milan around March 1614.[10] Exactly when the *Allegory of earth* was produced is more difficult to determine on the basis of the correspondence.[11] At any rate, it was in the cardinal's possession by 1618, because the three elements – fire, water and earth – were a part of the official act of donation of the Ambrosian collection in which the cardinal recorded the works of art that he donated to the library. Borromeo was kept waiting a long time for the fourth element of the series. In 1616 he wrote to Brueghel: 'I await the last Element left, in which I desire and also anticipate greater diligence than is to be found in the others'.[12] It was not until October 1621, more than five years later, that Brueghel announced that the painting was finished.[13] In February 1622 he wrote that he had dispatched the *Allegory of air*.[14] So the complete series was produced between 1608 and 1621.

In his treatises, museum guide and letters, Federico Borromeo repeatedly praised Brueghel for his colourful and detailed style. Moreover, he was also full of admiration for Brueghel's skill in realistically depicting animals and the natural world on a miniature scale. The painter managed to combine natural features and animals that were separated by large distances in reality to create a credible visual representation. This is how Borromeo formulated these qualities:

> In his own genre [of painting in miniature] he was something to be marvelled at: he was able to endow those little bodies with such nobility and liveliness that viewers could reasonably wonder whether those diminutive figures were actually tall or short. Brueghel seems to have wanted to wander, paintbrush in hand, through all of natural creation. As we will see, he painted seas, mountains, caves, and underground caverns. He managed to squeeze all of these natural features into the narrow scope of miniaturist painting, even though in reality they would have been separated by enormous distances.[15]

The cardinal also had great admiration for the overwhelming variety of animal species with which Brueghel populated his works. For instance, he wrote on the *Allegory of water*: 'he [Brueghel] depicted fish catches of such variety that he would seem to have been as skilled at fishing as at painting'.[16]

All these characteristics that so delighted Borromeo in Brueghel's work can also be found in the *Allegory of air*. Slightly right of centre is an allegorical figure on a cloud: Aurora, the goddess of the dawn. She is wrapped in a red drape and holds an armillary sphere in her left hand.[17] A cockatoo is perched on her right hand. Beside her a putto peers through a telescope.[18] More putti float through the air, where we also see the sun god Apollo in his horse-drawn chariot and the moon goddess Diana in hers drawn by deer. The various optical instruments scattered on the ground at the bottom right are rendered so accurately and in such detail that they can all be identified. They include an astrolabe, a regular compass and a proportional compass. Three putti are making use of the instruments. The most striking feature of the painting, however, is no doubt the enormous variety of native and exotic birds that flock around Aurora. Brueghel has depicted the birds in many different natural poses: some are in flight, others are sitting on or hanging from a branch, yet others are walking or standing on the ground. Many of the birds react to one another, sometimes in a hostile way.

The degree of detail with which Brueghel has rendered the birds so realistically seems to invite the viewer to determine the various kinds. The same is true of the many types of fish that he represented in the *Allegory of water*. Originally a list was attached to the rear of the latter painting giving the names of all the fish represented there.[19] The model of this list has been followed in this article on the different kinds of birds; the appendix contains a numbered list of the different species of birds represented by Brueghel in the painting.

The birds in the different versions

By the time Brueghel painted his *Allegory of air* for Cardinal Borromeo in 1621, he already had considerable experience in painting series of the elements and in painting many different species of birds in a single scene.[20] There is an undated series of the *Four elements* – in which the *Allegory of air* is populated by different types of birds – in the Galleria Doria Pamphilj in Rome (fig. 2).[21] Brueghel produced another series of

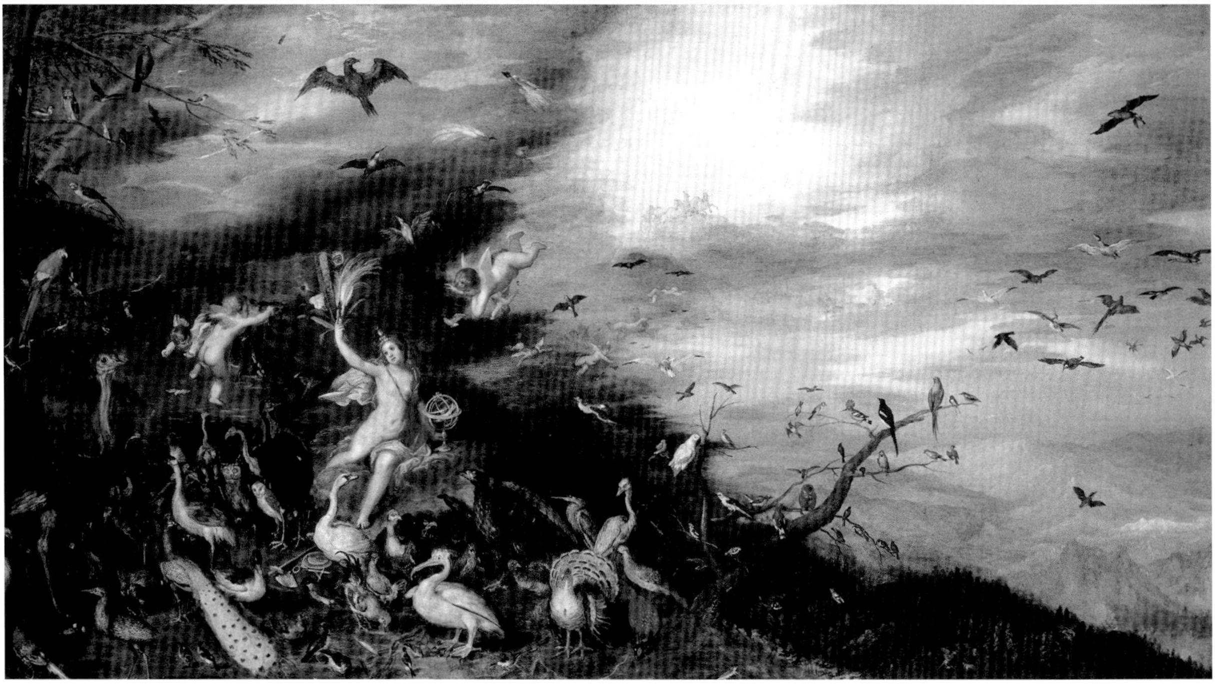

2

Jan Brueghel the Elder, *Allegory of air*, c. 1608
oil on panel, 50,3 x 80,1 cm, Galleria Doria Pamphilj.

the *Four elements* between 1606 and 1611; this series is now in the Musée des Beaux-Arts in Lyon.[22] The *Allegory of air* in the latter series is dated 1611 (fig. 3). There are also a number of extant isolated paintings of the *Allegory of air*.[23] It is a well-known fact that in Brueghel's day it was not uncommon for an artist to produce several versions of the same series. The repetition of motifs or even of complete compositions and series demonstrated the skill and success of the artist with a particular theme.[24] Brueghel, who increasingly specialised in the painting of flora and fauna from 1600 on, was certainly no exception.

Comparison of the versions of the *Allegory of air* in Rome and Lyon with the work that Brueghel executed for Borromeo in 1621 – now in Paris, at the Musée du Louvre – reveals similarities as well as striking differences. Aurora occupies the central position in the Rome and Lyon paintings too, but in these works she holds not only the armillary sphere but also a bunch of feathers, and her head is adorned with a tiara on which a sun and a moon are represented.[25] In these works she is also surrounded by a large variety of birds. Many of these birds can be found in all three paintings, such as the pelican, peacock, eagle and bird of paradise, to name but a few. However, the grouping of the birds is different in each of the three works; in this connection it should be noted that the compositions of the two earlier allegories display significantly more affinity with one another than with that of the *Allegory of air* from 1621. In the latter work Brueghel has not only moved the bare tree from the right to the left, but he has also given many of the birds a different position. For instance, the ostrich is on the far left in the works in Rome and Lyon, while it is represented on the far right in the allegory in Paris.

The same applies to the poses of the birds: while they are similar in most cases in the two early works, they often differ from those in the 1621 painting, in which the pose is inverted on more than one occasion.

These differences in the arrangement of the birds, their poses and the types present in the three versions raise the question of how Brueghel set about making his paintings. In the light of the ornithologically correct representation, the painter must have studied a large number of birds himself. But how did Brueghel come into contact with the birds? We do not know whether he observed birds in their natural surroundings, though it is by no means inconceivable.[26] Most of the exotic sorts, however, he would have observed in the menageries of his day, such as that of Archduke Albrecht of Austria (1559-1621) and the Infanta Isabella of Spain (1566-1633), regents of the Southern Netherlands from 1599, for whom Brueghel worked as court painter from 1606. This function gave Brueghel access to the famous menagerie in Brussels between 1606 and 1621, precisely the period in which he produced the different versions of the *Four elements*.[27] In 1604 Brueghel also paid a visit to the court of Rudolf II (1583-1612) in Prague. Rudolf possessed an impressive menagerie that was particularly known for its large collection of different ornithological species, so Brueghel would have studied birds here as well. In addition Rudolf had a famous collection of curiosities that included various stuffed birds.[28] The rare specimens it contained (that could not yet be observed alive in Europe) would also have aroused Brueghel's interest. For instance, it is not improbable that the red-beaked toucan that Brueghel depicts time and again in his work is based on a stuffed specimen in Rudolf's cabinet of curiosities. Brueghel also sometimes borrowed motifs from illustrations of birds by others. The pelican, for

3

Jan Brueghel the Elder, *Allegory of air*, 1611

oil on panel, 46 x 83 cm, Lyon, Musée des Beaux-Arts.

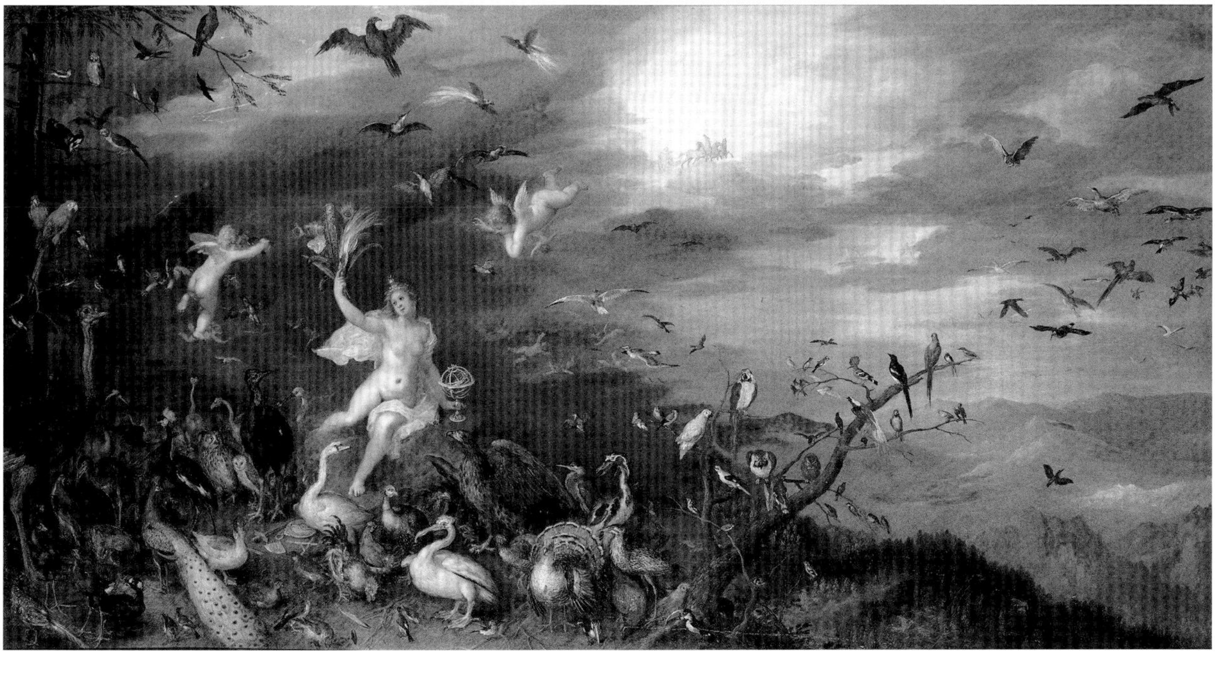

example, one of the few birds that Brueghel depicts clumsily and awkwardly in more or less the same way in all three versions of the *Allegory of air*, displays a striking resemblance to the painting of the bird that the physician Anselmus de Boodt (1550-1632) did in his album for Rudolf II.[29] Copying motifs from others was quite a common artistic practice. An interesting drawing shows a number of individual birds, including their names.[30] These birds are all taken from the work of the Swiss scholar Conrad Gesner (1516-1565). From the sequence of the ornithological sketches it can be deduced that the artist drew not on Gesner's voluminous and erudite standard work *Historiae animalium liber III, qui est de avium natura* (1555), but on the collection of illustrations taken from that work, the more manageable *Icones*, that was specially printed from 1556 for practical use, such as a manual for painters.[31]

The fact that a number of birds are represented in exactly the same pose in the various versions of the *Allegory of air* suggests that Brueghel worked with studies and had a stock of motifs to select from for his paintings. A pen drawing shows that Brueghel sometimes based his birds on live examples: this study of parrots on a branch in different poses is clearly taken from life.[32] The pose of some of these birds corresponds exactly to the parrots that Brueghel depicted in his paintings. The most interesting drawing for present purposes is in the Herzog Anton Ulrich-Museum in Braunschweig (fig. 4). This drawing is a compositional sketch for a

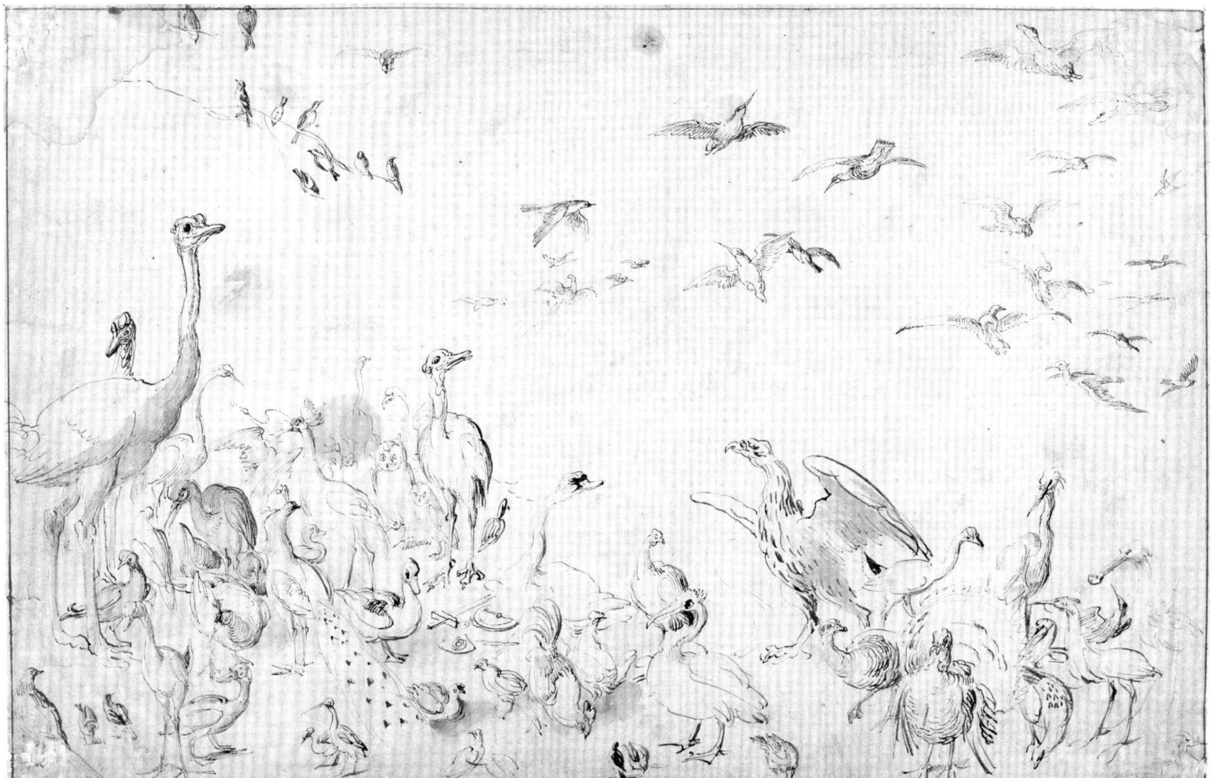

4
Jan Brueghel the Elder, *Ostriches, peacocks and other birds*, c. 1605
pen in brown ink with grey/brown wash, 250 x 380 mm, Braunschweig, Herzog Anton Ulrich-Museum.

painting of the *Allegory of air*. It is noteworthy that the figure of Aurora is absent, though a number of optical instruments lie on the ground beside the spot where Aurora would be introduced in the paintings. There is probably a simple explanation for this absence: it is assumed that the figure of Aurora in the various versions was painted not by Brueghel himself, but by his colleague Hendrick van Balen (1573-1632).[33] Brueghel seems to have done the compositional sketch primarily to determine the positions of the birds. The setting in which the birds are placed displays strong resemblances to the allegories in Rome and Lyon but is not identical with either of them. This might imply that Brueghel painted yet another version with a composition identical to that of the sketch, but it is also possible that he never intended to transfer the sketch exactly to the panel and deliberately introduced variations while he was painting.

Comparison of the sketch with the paintings in Rome and Lyon shows that some motifs can be found in both paintings, while others occur in only one of the two versions. For instance, there are some small birds on a branch in the top left of the drawing that have been moved in both paintings to a much lower tree slightly left of centre. The birds flying in the air in the sketch have been partly incorporated in both paintings but are grouped differently on the panels. There are no birds of paradise in the drawing, but they are to be found in both paintings. The same is true of the parrots. The birds on the ground have been partly taken from the drawing but have sometimes been depicted on another spot. On the other hand, the ostrich and the motif of the eagle owl and barn owl between the black crowned crane and southern cassowary are in the same position in the drawing as in the two paintings. The purple coot standing to the right of the eagle is absent in that position in both paintings, where it has been replaced by a heron. Further to the right (next to the impressive turkey), the drawing contains a pheasant and a bittern. They recur in the painting in Lyon but not in the painting in Rome, where a different sort of pheasant takes the place of the bittern; the bittern has been moved to the lower left, a position occupied by a purple coot and a duck in the drawing and in the Lyon painting. The drawing thus seems closer to the Lyon painting than to the one in Rome.[34] Nevertheless, comparison of the two paintings with the drawing does not lead to conclusions about the chronology of the two paintings.

If the two paintings are compared, there is a striking difference that may well throw light on the chronology and thus on the dating of the *Allegory of air* in Rome. This difference lies in the degree of accuracy of the rendering of the birds. The vast majority of birds in the painting in Rome have a fairly static pose. A number of them are not depicted entirely correctly, or are not depicted correctly at all. For example, the tail feathers of the crowned crane are unnatural. The colours of the same bird in the painting in Lyon from 1611 are better, and the shape of the tail is closer to reality. Likewise, the pose of the eagle owl is more natural in the dated work, and the markings on the head of the barn owl (especially the black eyes and the characteristic, heart-shaped 'face', the veil) are more realistic. Moreover, Brueghel added a number of new bird species to the Lyon painting, such as the two toucans seated on the ground in front of the tree slightly right of centre. One is rendered in profile, the other frontally, which makes it possible to see how narrow the beak is. Brueghel has also added two blue-and-yellow macaws in very natural poses in the tree. Like the toucan, it is relatively difficult to provide a good frontal rendering of this bird.

By comparison with the *Allegory of air* in Rome, the version dated 1611 in Lyon shows clear progress in Brueghel's development. It may therefore be supposed that the painting in Lyon is later, and that a date before 1611 is plausible for the painting in Rome.[35] However, Brueghel's development did not stop in 1611. The *Allegory of air* that he produced in 1621 can be regarded as the definitive step in this line of progress. The natural poses are captured even better, and the interaction between the birds is also much greater. This painting includes an even larger variety of birds than before. Finally, this painting shows a perfection in the correct ornithological depiction of the birds, which suggests that Brueghel was himself reacting to leading authorities in the world of natural history.

Reaction to Clusius

Brueghel's various versions of the *Allegory of air* increasingly display the progress made in the natural sciences. This can be seen, moreover, not only in the various stages of the depiction of the birds. Brueghel underlined his point by representing a number of optical instruments that were naturally necessary for the making of new discoveries.[36] Among those instruments, it was the telescope that appealed most to the imagination. It was a novelty in Brueghel's day: a patent request was filed in 1608.[37] There are five telescopes in Brueghel's oeuvre. The first is probably in the *Landscape with Mariemont Castle* from c. 1608-1612,[38] making Brueghel the first artist to have represented a telescope in a painting.[39] The one represented in this painting has a convenient format with a length of about 40 centimetres and seems to be gilded at both ends.[40] Brueghel painted what is apparently a silver telescope in the *Allegory of sight* in 1617.[41] According to the Italian astronomers Molaro and Selvelli, this telescope, which when fully extended would have an estimated length of 170 to 180 centimetres, corresponds to the model that Johannes Kepler (1571-1630) is supposed to have produced fifteen years later.[42] The telescope in the 1621 *Allegory of air* appears to have gilded ends and is closer to the instrument in the *Landscape with Mariemont Castle* than to the other model (fig. 5). Brueghel may well have opted for the earlier model because it was easier for a putto to handle. By representing a telescope in the *Allegory of air*, Brueghel emphasises the scientific topicality of the painting. The instrument shows how Brueghel keeps on trying to update his paintings: the telescope is absent from the preliminary study and from the earlier versions of the allegory, though they do contain representations of other optical instruments.

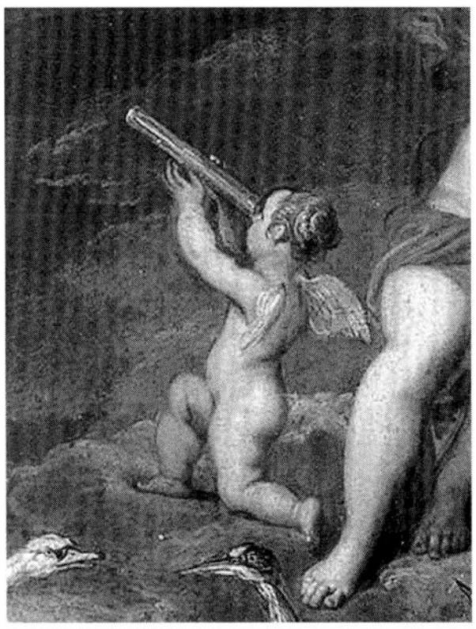

5

Jan Brueghel the Elder, *Allegory of air*, 1621 (detail fig. 1).

The scientific topicality of the painting is expressed above all in the diversity of species of birds that are represented. The growing intensity of trade with distant regions led to a rapid growth in the number of known sorts. This development is well illustrated in the various versions of the

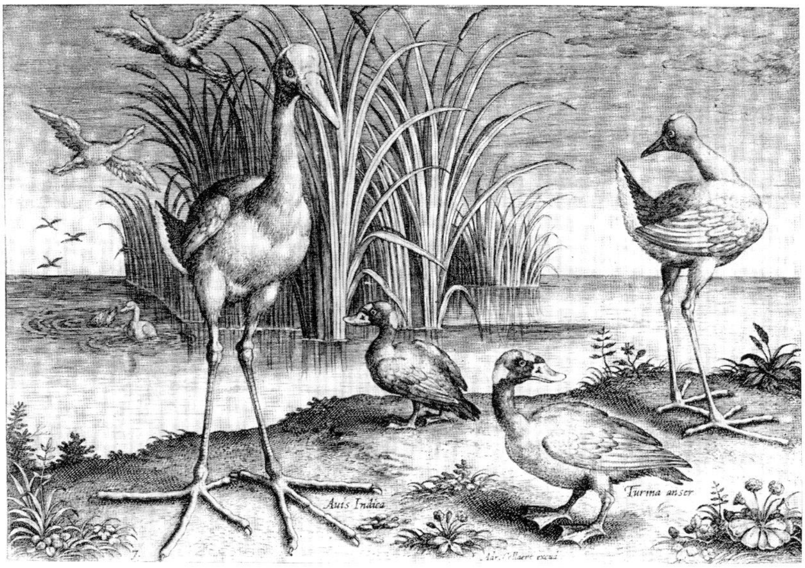

6a
Adriaen Collaert, *Leguatia gigantea*, c. 1580-1590
engraving, 132 x 189 mm, *Avium vivae icones.*

Allegory of air. The compositional sketch includes both a number of European species and a few spectacular, exotic ones that were already known at the time: an ostrich, two cassowaries, a crane, a pelican and two turkeys.[43] The most unusual bird in the drawing is the purple coot, which Brueghel has based on the *Leguatia gigantea*, a sort of giant rail. The only known representation of this extinct bird from Mauritius was done by Adriaen Collaert (1560-1618) from a specimen that could be seen around 1600 in the garden of the Count of Solms in The Hague (fig. 6a).[44] Brueghel borrows the pose from Collaert but reduces the proportions and turns the bird into the familiar purple coot (fig. 6b).[45] It is characteristic of Brueghel's skill as a painter of birds that he manages to play with variations on Collaert's image by representing a second giant rail / purple coot walking naturally.

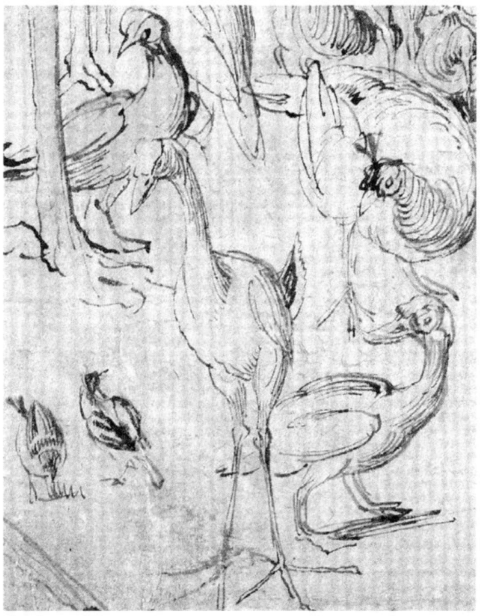

6b
Jan Brueghel the Elder, *Ostriches, peacocks and other birds*, c. 1605
(detail fig. 4).

The drawing lacks some of the types of birds that in the paintings seem to be symbols of the very latest ornithological knowledge. As mentioned above, in terms of being ornithologically up to date, the version in Rome looks earlier than that in Lyon, while the 1621 painting demonstrates an enormous leap forwards. The bird of paradise and the macaw are illustrative of the increase in ornithological knowledge in the different versions. The development of the bird of paradise is the most spectacular. The type represented by Brueghel, the lesser bird of paradise of New Guinea (no. 2 in the Appendix), was known in the sixteenth century from the skins that sailors brought back with them from their voyages in the East. These skins were prepared by the native Papuans without feet or wings, which led to the emergence of the myth that this bird always floated in heaven, lived on dew and could not build a nest because it had no feet. The male was believed to have a cavity on its back in which the female could brood the eggs. The Italian physician Hieronymus Cardanus (1501-1576) was the first to record this myth in writing, and from then on it was endlessly repeated, usually with an illustration of a preserved bird of paradise *sans* feet or wings.[46] Brueghel's work very precisely reflects the exposure of this myth. In the Rome version of the *Allegory of air* the bird of paradise is represented in the air without feet or wings (fig. 7a). In the Lyon painting it is sitting on a branch, and thus has feet, while two others fly in the air, both with clearly visible feet and wings (fig. 7b). The 1621 version shows two birds of paradise, one sitting on a branch, the other on the ground, as though Brueghel wanted to emphasise the fact that they have feet (fig. 7c). By placing this bird of paradise on the ground in the immediate proximity of two eggs, the painter also seems to want to explode the myth of the breeding cavity on its back.

The macaw is also illustrative of the development in Brueghel's ornithological knowledge. He represents two types: the commoner scarlet macaw with yellow wings, and the rarer green-winged macaw. The main difference between them is in the colour of their wings. Until the 1611 *Allegory of air* in Lyon, the macaws in Brueghel's paintings were always scarlet, as in the Rome version. From the Lyon version on, however, the scarlet macaw is regularly replaced by the green-winged macaw (no. 17).[47] Brueghel evidently favoured rarity above contrasting colour.

In representing new sorts of birds, Brueghel showed not only that he was well informed about the state of natural history at the time, but also that he left the professional naturalists far behind him. This becomes clear if the *Allegory of air* from 1621 is compared with the most authoritative work of the time in the field of newly discovered fauna and flora: the *Exoticorum libri decem* by Carolus Clusius (1526-1609) from 1605.[48] Brueghel seems to be reacting to Clusius's accounts and illustrations of the cassowary, crane, bird of paradise, penguin and parrots, which all have a prominent place in the *Allegory of air*.[49] The cassowary (no. 55) seems to have been chosen by Brueghel to demonstrate that he was thoroughly conversant with the latest developments. Clusius was the first to describe the specimen that had been brought from New Guinea in 1597 – which he does at length and in detail – and he included an illustration of it as well.[50] Although Brueghel probably depicted the bird from his own observations, there is a clue that he also had Clusius's book in mind: in the immediate proximity of the cassowary he represents precisely the two birds that Clusius also mentions in this description of the cassowary, namely the ostrich (no. 54) and the great bustard (no. 48).[51] Brueghel's footed birds of paradise can be seen as a piece of one-upmanship vis-à-vis Clusius. The illustration in the latter shows a bird of paradise without feet, but Clusius adds in his description that he has heard from a reliable witness that the bird does have feet in the wild (fig. 7d).[52] Clusius was one of the first to call the myth of the footless bird of paradise into question. Brueghel is the first to illustrate this demystification by representing the bird with feet.

In the case of the black crowned crane (no. 4), Brueghel seems to adopt a more critical attitude towards Clusius. The latter, as he writes himself in his *Exoticorum libri decem*, had only an illustration of the head of the bird at his disposal.[53] Clusius gives a good description of that head with an indication of the colours: a black head and neck, a yellow-

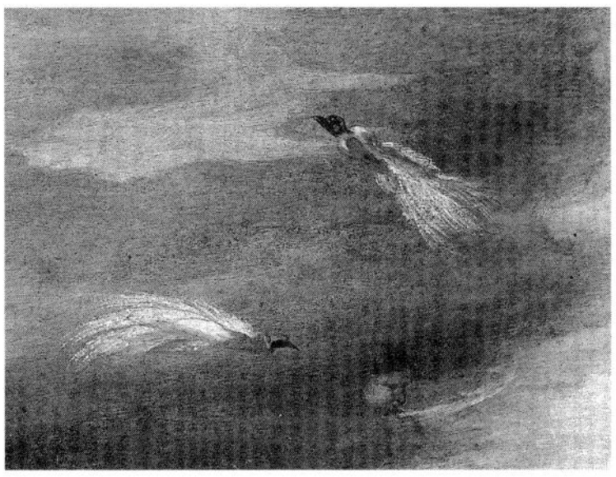

7a

Jan Brueghel the Elder, *Allegory of air*, c. 1608 (detail fig. 2).

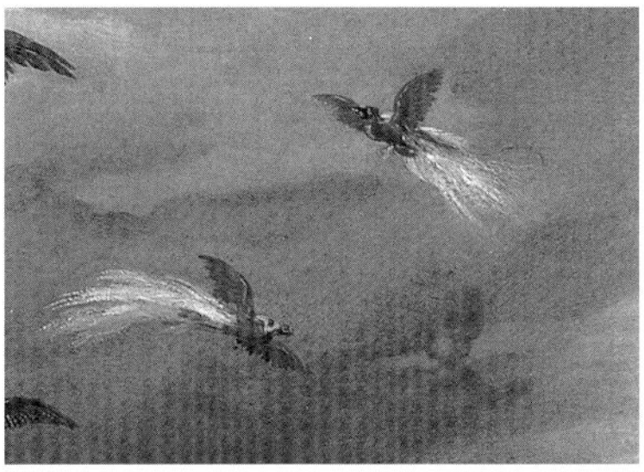

7b

Jan Brueghel the Elder, *Allegory of air*, 1611 (detail fig. 3).

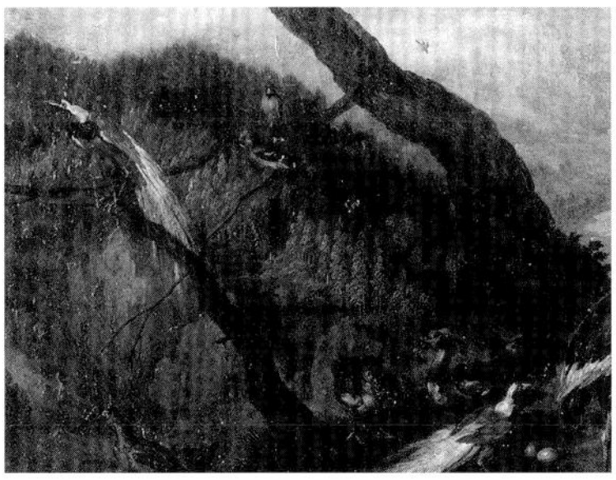

7c

Jan Brueghel the Elder, *Allegory of air*, 1621 (detail fig. 1).

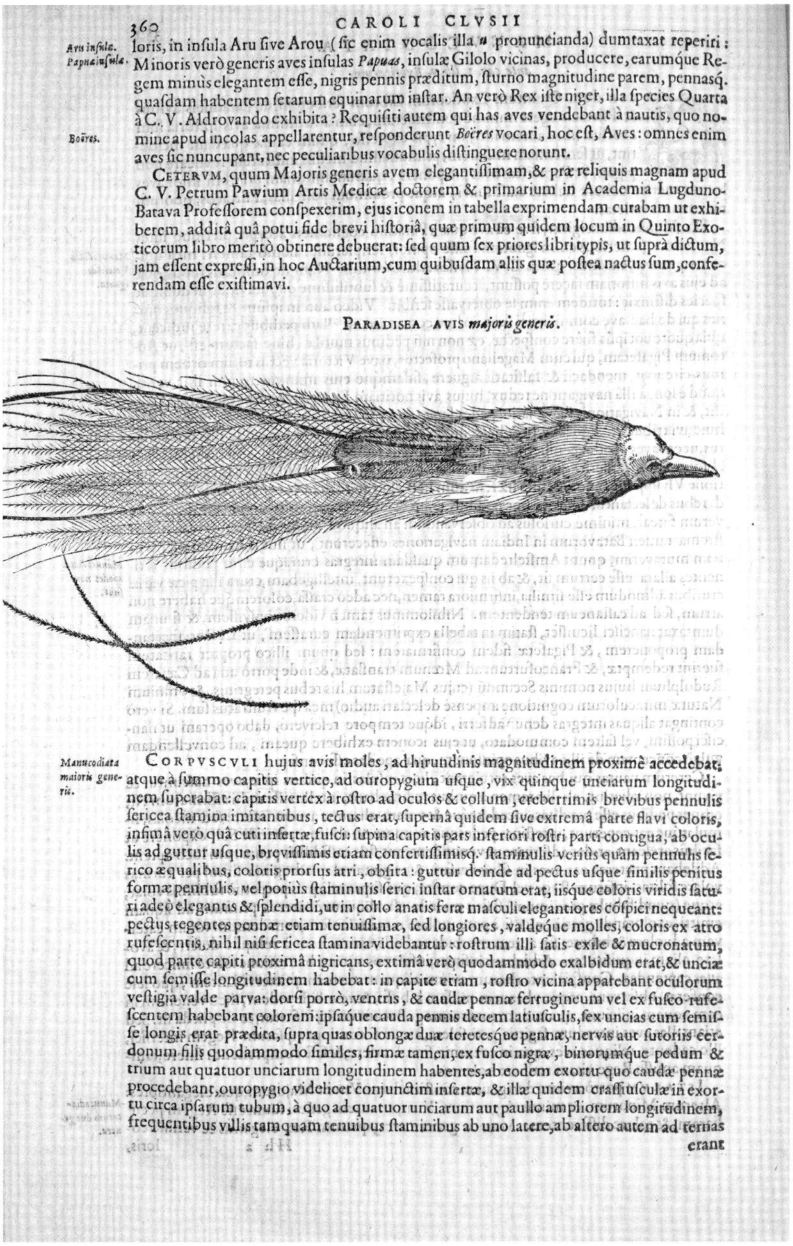

360 CAROLI CLVSII

Aru inſula. loris, in inſula Aru ſive Arou (ſic enim vocalis illa *u* pronuncianda) dumtaxat reperiri: *Papua inſula.* Minoris verò generis aves inſulas *Papuas*, inſulæ Gilolo vicinas, producere, earumque Regem minùs elegantem eſſe, nigris pennis præditum, ſturno magnitudine parem, pennasq. quaſdam habentem ſetarum equinarum inſtar. An verò Rex iſte niger, illa ſpecies Quarta à C. V. Aldrovando exhibita? Requiſiti autem qui has aves vendebant à nautis, quo nomine apud incolas appellarentur, reſponderunt *Boëres* vocari, hoc eſt, Aves: omnes enim aves ſic nuncupant, nec peculiaribus vocabulis diſtinguere norunt. *Boëres.*

CETERVM, quum Majoris generis avem elegantiſſimam, & præ reliquis magnam apud C. V. Petrum Pawium Artis Medicæ doctorem & primarium in Academia Lugduno-Batava Profeſſorem conſpexerim, ejus iconem in tabella exprimendam curabam ut exhiberem, additâ quâ potui fide brevi hiſtoriâ, quæ primum quidem locum in Quinto Exoticorum libro meritò obtinere debuerat: ſed quum ſex priores libri typis, ut ſuprà dictum, jam eſſent expreſſi, in hoc Auctarium, cum quibuſdam aliis quæ poſtea nactus ſum, conferendam eſſe exiſtimavi.

PARADISEA AVIS *majoris generis*.

Manucodiata maioris generis. CORPVSCVLI hujus avis moles, ad hirundinis magnitudinem proximè accedebat; atque à ſummo capitis vertice, ad ouropygium uſque, vix quinque unciarum longitudinem ſuperabat: capitis vertex à roſtro ad oculos & collum, crebertimis brevibus pennulis ſericea ſtamina imitantibus, tectus erat, supernâ quidem ſive extremâ parte flavi coloris, infimâ verò quâ cuti inſertæ, fuſci: ſupina capitis pars inferiori roſtri parti contigua, ab oculis ad guttur uſque, breviſſimis etiam confertiſſimisq. ſtaminulis veriùs quàm pennulis ſerico æqualibus, coloris prorſus atri, obſita: guttur deinde ad pectus uſque ſimilis penitus formæ pennulis, vel potiùs ſtaminulis ſerici inſtar ornatum erat, iisque coloris viridis saturi adeò elegantis & ſplendidi, ut in collo anatis feræ maſculi elegantiores cõſpici nequeant: pectus tegentes pennæ etiam tenuiſſimæ, ſed longiores, valdeque molles, coloris ex atro ruſeſcentis, nihil niſi ſericea ſtamina videbantur: roſtrum illi ſatis exile & mucronatum, quod parte capiti proximâ nigricans, extimâ verò quodammodo exalbidum erat, & unciæ cum ſemiſſe longitudinem habebat: in capite etiam, roſtro vicina apparebant oculorum veſtigia valde parva: dorſi porrò, ventris, & caudæ pennæ ferrugineum vel ex fuſco-ruſeſcentem habebant colorem: ipſaque cauda pennis decem latiuſculis, ſex uncias cum ſemiſſe longis erat prædita, ſupra quas oblongæ duæ teretesque pennæ, nervis aut ſutoriis cerdonum filis quodammodo ſimiles, firmæ tamen, ex fuſco nigræ, binorumque pedum & trium aut quatuor unciarum longitudinem habentes, ab eodem exortu quo caudæ pennæ procedebant, ouropygio videlicet conjunctim inſertæ, & illæ quidem craſſiuſculæ in exortu circa ipſarum tubum, à quo ad quatuor unciarum aut paullo ampliorem longitudinem, frequentibus villis tamquam tenuibus ſtaminibus ab uno latere, ab altero autem ad ternas erant

7d

Artist unknown, *Bird of paradise*, before 1605

woodcut, C. Clusius, *Exoticorum libri decem. Quibus animalium, plantarum, aromatum, aliorumque peregrinorum fructuum historiae describuntur*, Leiden 1605.

coloured crown and a red, naked spot near the eye. He goes on to advance the strange hypothesis that the bird is a 'sea peacock' (*pavo marinus*). This hypothesis is remarkable, because the bird had been described and illustrated earlier. One of those who had discussed it was the Italian naturalist Ulysse Aldrovandi (1522-1605), whom Clusius had apparently not read at that time.[54] Brueghel appears to be wanting to correct Clusius: he depicts one of the two specimens in such a way that the pose of the head corresponds exactly to that of Clusius's 'sea peacock' (figs. 8a, 8b). Brueghel depicts the colours of the crane accurately and supplements Clusius's description by adding the characteristic red shine to the white of the head. The painter also represents the whole bird, true to life, probably on the basis of his own observation.

The Magellanic penguin (no. 40) presents a similar correction of Clusius. This bird was first mentioned and illustrated in an account of the voyage of the Dutch expedition to the Strait of Magellan of 1600.[55] Clusius refers to this source for his illustration of the bird and adds a detailed description that is more precise than the matching illustration (fig. 9a). He was the first to provide a detailed account of this unknown bird. He describes it as '(...) a sort of goose, although the beak is

8a

Artist unknown, *Sea peacock*, before 1605,

woodcut, C. Clusius, *Exoticorum libri decem. Quibus animalium, plantarum, aromatum, aliorumque peregrinorum fructuum historiae describuntur,* Leiden 1605.

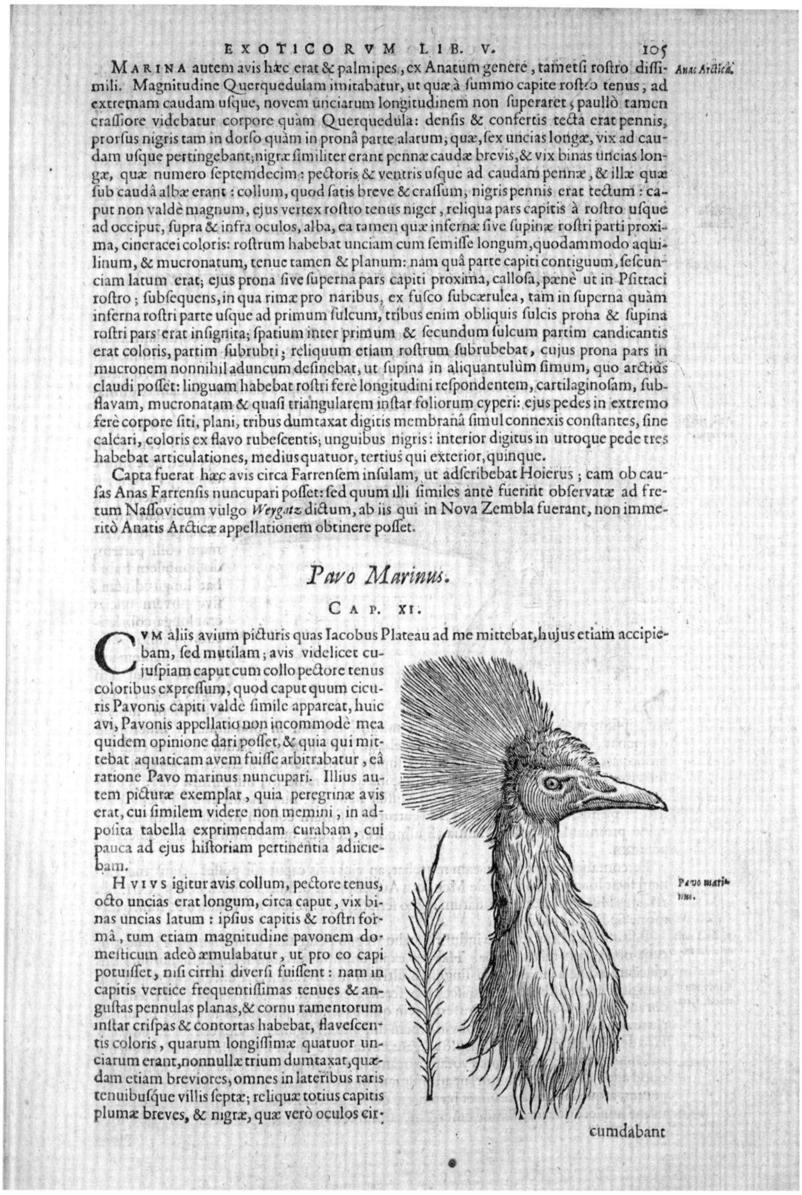

EXOTICORVM LIB. V. 105

MARINA autem avis hæc erat & palmipes, ex Anatum genere, tametſi roſtro diſſimili. Magnitudine Querquedulam imitabatur, ut quæ à ſummo capite roſtro tenus, ad extremam caudam uſque, novem unciarum longitudinem non ſuperaret; paullò tamen craſſiore videbatur corpore quàm Querquedula: denſis & confertis tecta erat pennis, prorſus nigris tam in dorſo quàm in pronâ parte alarum; quæ, ſex uncias longæ, vix ad caudam uſque pertingebant; nigræ ſimiliter erant pennæ caudæ brevis, & vix binas uncias longæ, quæ numero ſeptemdecim: pectoris & ventris uſque ad caudam pennæ, & illæ quæ ſub caudâ albæ erant: collum, quod ſatis breve & craſſum, nigris pennis erat tectum: caput non valdè magnum, ejus vertex roſtro tenus niger, reliqua pars capitis à roſtro uſque ad occiput, ſupra & infra oculos, alba, ea tamen quæ infernæ ſive ſupinæ roſtri parti proxima, cineracei coloris: roſtrum habebat unciam cum ſemiſſe longum, quodammodo aquilinum, & mucronatum, tenue tamen & planum: nam quâ parte capiti contiguum, ſeſcunciam latum erat; ejus prona ſive ſuperna pars capiti proxima, calloſa, pænè ut in Pſittaci roſtro; ſubſequens, in qua rimæ pro naribus, ex fuſco ſubcærulea, tam in ſuperna quàm inferna roſtri parte uſque ad primum ſulcum, tribus enim obliquis ſulcis prona & ſupina roſtri pars erat inſignita; ſpatium inter primum & ſecundum ſulcum partim candicantis erat coloris, partim ſubrubri; reliquum etiam roſtrum ſubrubebat, cujus prona pars in mucronem nonnihil aduncum deſinebat, ut ſupina in aliquantulùm ſimum, quo arctiùs claudi poſſet: linguam habebat roſtri ferè longitudini reſpondentem, cartilaginoſam, ſubflavam, mucronatam & quaſi triangularem inſtar foliorum cyperi: ejus pedes in extremo ferè corpore ſiti, plani, tribus dumtaxat digitis membranâ ſimul connexis conſtantes, ſine calcari, coloris ex flavo rubeſcentis; unguibus nigris: interior digitus in utroque pede tres habebat articulationes, medius quatuor, tertius qui exterior, quinque. *Anas Arctica.*

Capta fuerat hæc avis circa Farrenſem inſulam, ut adſcribebat Hoierus; eam ob cauſas Anas Farrenſis nuncupari poſſet: ſed quum illi ſimiles antè fuerint obſervatæ ad fretum Naſſovicum vulgo *Weygatz* dictum, ab iis qui in Nova Zembla fuerant, non immeritò Anatis Arcticæ appellationem obtinere poſſet.

Pavo Marinus.

CAP. XI.

CVM aliis avium picturis quas Iacobus Plateau ad me mittebat, hujus etiam accipiebam, ſed mutilam; avis videlicet cujuſpiam caput cum collo pectore tenus coloribus expreſſum, quod caput quum cicuris Pavonis capiti valdè ſimile appareat, huic avi, Pavonis appellatio non incommodè mea quidem opinione dari poſſet, & quia qui mittebat aquaticam avem fuiſſe arbitrabatur, eâ ratione Pavo marinus nuncupari. Illius autem picturæ exemplar, quia peregrinæ avis erat, cui ſimilem videre non memini, in adpoſita tabella exprimendam curabam, cui pauca ad ejus hiſtoriam pertinentia adiiciebam.

HVIVS igitur avis collum, pectore tenus, octo uncias erat longum, circa caput, vix binas uncias latum: ipſius capitis & roſtri formâ, tum etiam magnitudine pavonem domeſticum adeò æmulabatur, ut pro eo capi potuiſſet, niſi cirrhi diverſi fuiſſent: nam in capitis vertice frequentiſſimas tenues & anguſtas pennulas planas, & cornu ramentorum inſtar criſpas & contortas habebat, flaveſcentis coloris, quarum longiſſimæ quatuor unciarum erant, nonnullæ trium dumtaxat, quædam etiam breviores, omnes in lateribus raris tenuibuſque villis ſeptæ; reliquæ totius capitis plumæ breves, & nigræ, quæ verò oculos cir- *Pavo marinus.*

cumdabant

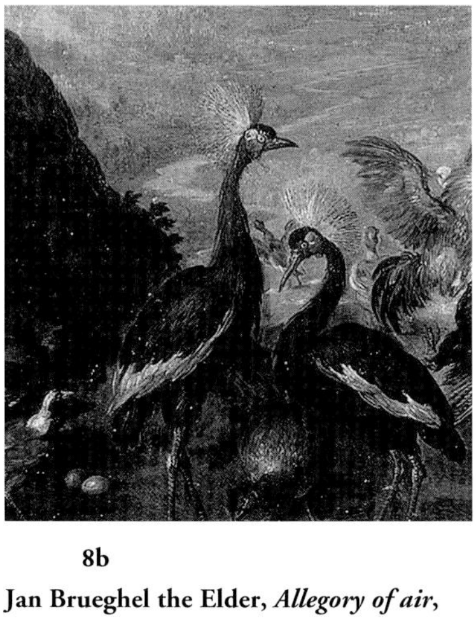

8b

Jan Brueghel the Elder, *Allegory of air*, 1621 (detail fig. 1).

different, living in the sea, has a great deal of fat, and is the size of a very large goose'.[56] He goes on to precisely describe the black and white pattern of its feathers: 'the back was covered with black feathers, the front and belly with white ones; and several had the neck, which is thick and short, ringed with white feathers like a collar'.[57] The woodcut fails to do justice to this accurate description of the colours. In 1621 Brueghel shows that he can do better: he depicts two Magellanic penguins in a lively attitude and with the correct markings (fig. 9b). He probably had to rely on a badly preserved example, because the anatomy of the feet is wrong.[58]

It is possible to reconstruct even more precisely how the penguin entered the oeuvre of Brueghel. To do so, we have to return to the

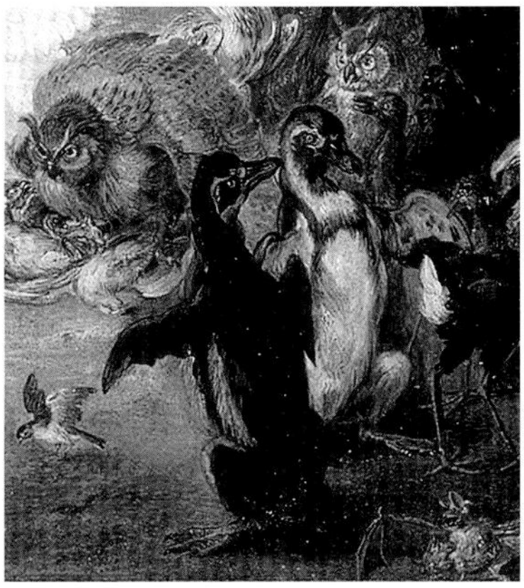

9b

Jan Brueghel the Elder, *Allegory of air*, 1621 (detail fig. 1).

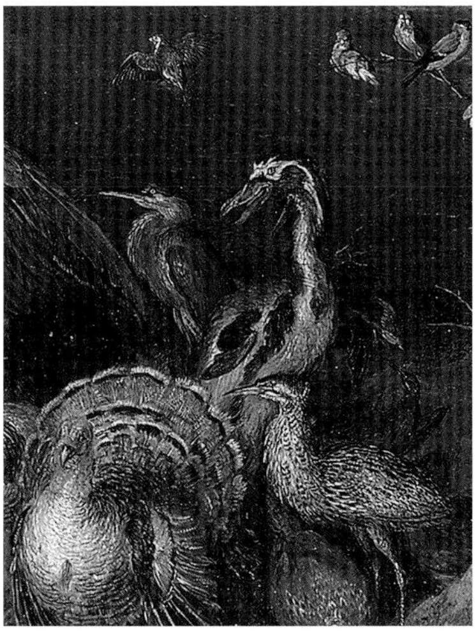

9c

Jan Brueghel the Elder, *Allegory of air*, 1611 (detail fig. 3).

9a

Artist unknown, *Magellanic penguin*, before 1605

woodcut, C. Clusius, *Exoticorum libri decem. Quibus animalium, plantarum, aromatum, aliorumque peregrinorum fructuum historiae describuntur,* Leiden 1605.

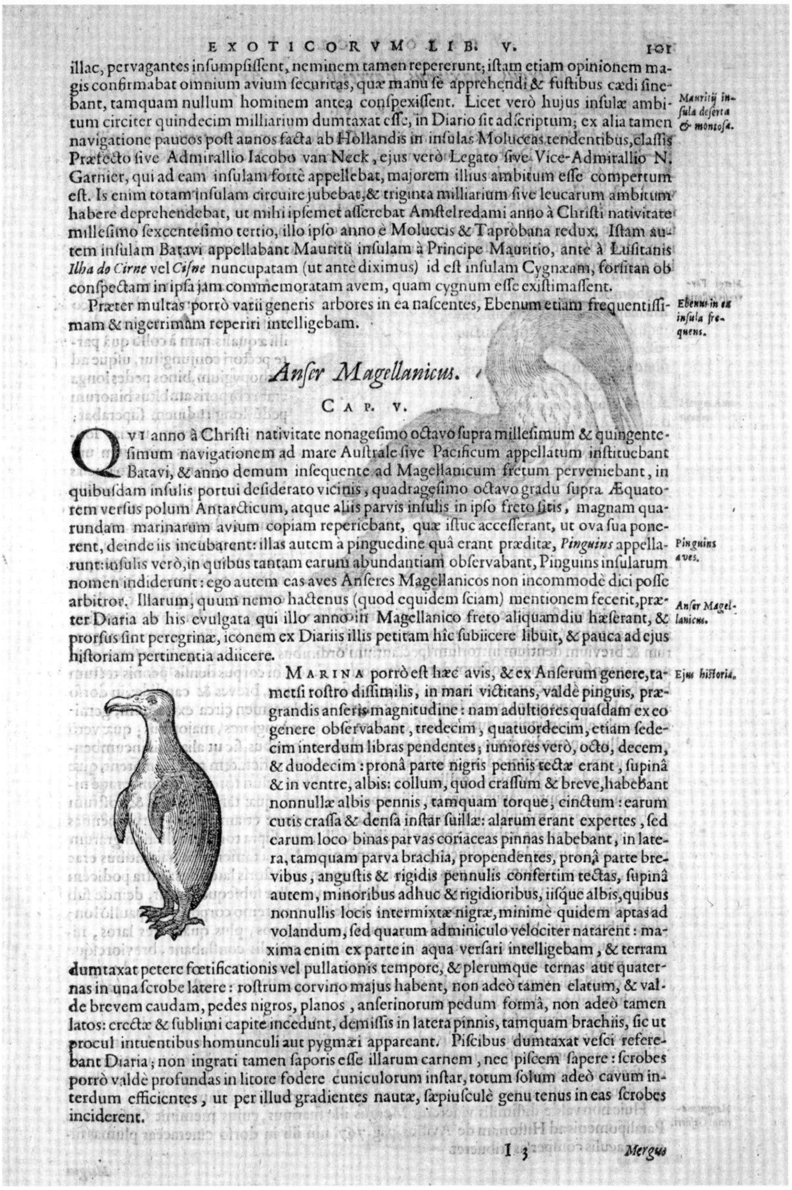

EXOTICORVM LIB. V. 101

illac, pervagantes inſumpſiſſent, neminem tamen repererunt; iſtam etiam opinionem magis confirmabat omnium avium ſecuritas, quæ manu ſe apprehendi & fuſtibus cædi ſinebant, tamquam nullum hominem antea conſpexiſſent. Licet verò hujus inſulæ ambitum circiter quindecim milliarium dumtaxat eſſe, in Diario ſit adſcriptum; ex alia tamen navigatione paucos poſt annos facta ab Hollandis in inſulas Moluccas tendentibus, claſſis Præfecto ſive Admirallio Iacobo van Neck, ejus verò Legato ſive Vice-Admirallio N. Garnier, qui ad eam inſulam fortè appellebat, majorem illius ambitum eſſe compertum eſt. Is enim totam inſulam circuire jubebat, & triginta milliarium ſive leucarum ambitum habere deprehendebat, ut mihi ipſemet aſſerebat Amſtelredami anno à Chriſti nativitate milleſimo ſexcenteſimo tertio, illo ipſo anno è Moluccis & Taprobana redux. Iſtam autem inſulam Batavi appellabant Mauritii inſulam à Principe Mauritio, ante à Luſitanis *Ilha do Cirne* vel *Ciſne* nuncupatam (ut antè diximus) id eſt inſulam Cygnæam, forſitan ob conſpectam in ipſa jam commemoratam avem, quam cygnum eſſe exiſtimaſſent. *Mauritij insula deserta & montosa.*

Præter multas porrò varii generis arbores in ea naſcentes, Ebenum etiam frequentiſſimam & nigerrimam reperiri intelligebam. *Ebenus in ea insula frequens.*

Anſer Magellanicus.

CAP. V.

QVI anno à Chriſti nativitate nonageſimo octavo ſupra milleſimum & quingenteſimum navigationem ad mare Auſtrale ſive Pacificum appellatum inſtituebant Batavi, & anno demum inſequente ad Magellanicum fretum perveniebant, in quibuſdam inſulis portui deſiderato vicinis, quadrageſimo octavo gradu ſupra Æquatorem verſus polum Antarcticum, atque aliis parvis inſulis in ipſo freto ſitis, magnam quarundam marinarum avium copiam reperiebant, quæ iſtuc acceſſerant, ut ova ſua ponerent, deinde iis incubarent: illas autem à pinguedine quâ erant præditæ, *Pinguins* appellarunt: inſulis verò, in quibus tantam earum abundantiam obſervabant, Pinguins inſularum nomen indiderunt: ego autem eas aves Anſeres Magellanicos non incommodè dici poſſe arbitror. Illarum, quum nemo hactenus (quod equidem ſciam) mentionem fecerit, præter Diaria ab his evulgata qui illo anno in Magellanico freto aliquamdiu hæſerant, & prorſus ſint peregrinæ, iconem ex Diariis illis petitam hîc ſubiicere libuit, & pauca ad ejus hiſtoriam pertinentia adiicere. *Pinguins aves.* *Anſer Magellanicus.*

MARINA porrò eſt hæc avis, & ex Anſerum genere, tametſi roſtro diſſimilis, in mari victitans, valdè pinguis, prægrandis anſeris magnitudine: nam adultiores quaſdam ex eo genere obſervabant, tredecim, quatuordecim, etiam ſedecim interdum libras pendentes; iuniores verò, octo, decem, & duodecim: pronâ parte nigris pennis tectæ erant, ſupinâ & in ventre, albis: collum, quod craſſum & breve, habebant nonnullæ albis pennis, tamquam torque, cinctum: earum cutis craſſa & denſa inſtar ſuillæ: alarum erant expertes, ſed earum loco binas parvas coriaceas pinnas habebant, in latera, tamquam parva brachia, propendentes, pronâ parte brevibus, anguſtis & rigidis pennulis confertim tectas, ſupinâ autem, minoribus adhuc & rigidioribus, iiſque albis, quibus nonnullis locis intermixtæ nigræ, minimè quidem aptas ad volandum, ſed quarum adminiculo velociter natarent: maxima enim ex parte in aqua verſari intelligebam, & terram dumtaxat petere fœtificationis vel pullationis tempore, & plerumque ternas aut quaternas in una ſcrobe latere: roſtrum corvino majus habent, non adeò tamen elatum, & valde brevem caudam, pedes nigros, planos, anſerinorum pedum formâ, non adeò tamen latos: erectæ & ſublimi capite incedunt, demiſſis in latera pinnis, tamquam brachiis, ſic ut procul intuentibus homunculi aut pygmæi appareant. Piſcibus dumtaxat veſci referebant Diaria; non ingrati tamen ſaporis eſſe illarum carnem, nec piſcem ſapere: ſcrobes porrò valdè profundas in litore fodere cuniculorum inſtar, totum ſolum adeò cavum interdum efficientes, ut per illud gradientes nautæ, ſæpiuſculè genu tenus in eas ſcrobes inciderent. *Ejus hiſtoria.*

I 3 *Mergus*

compositional sketch. It shows a great white egret on the right between the purple coot and the spoonbill. This bird was copied in the version of the *Allegory of air* in Rome and, initially, in the 1611 version in Lyon. However, the bird in the latter version was overpainted with a bizarre exemplar that at first sight resembles a juvenile swan with short wing stumps (fig. 9c).[59] In fact, this is Brueghel's first attempt to represent a penguin: the strange bird, of which only the head is more or less ornithologically correct, matches Clusius's written account of 'a sort of goose, although the beak is different (...) the size of a very large goose', with black and white markings. There is a significant difference, however, for Clusius describes the neck as 'thick and short'. We might speculate that Brueghel had initially set out from an illustration or specimen of the head of the bird combined with Clusius's account – but overlooked Clusius's description of the neck of the bird – and that he later had a preserved specimen at his disposal for the version in the *Allegory of air* of 1621. Alternatively, the long neck of the penguin may be a remnant of the long neck of the great white egret over which the penguin was painted, and Brueghel may have considered a shorter neck to be awkward in compositional terms.

Clusius's chapter on parrots shows the level of awareness of the rapid developments in ornithological knowledge at the time. Clusius wrote this chapter later and appended it to his *Exoticorum libri decem*. In the meantime he seems to have become acquainted with the work of Aldrovandi, which was not the case with regard to the 'sea peacock' discussed above. Clusius pays full tribute to his Italian colleague and praises him for his detailed description of no less than fourteen kinds of parrot, thirteen of which are illustrated. This certainly is a large number. By contrast, Conrad Gesner and the French ornithologist Pierre Belon (1517-1564) had included only two illustrations in their authoritative reference works of 1555 and simply stated that there were many species. Clusius also makes a further contribution in describing a few new species, including the rare red-fan parrot, which he also illustrates.[60]

It can be seen from the descriptions by Aldrovandi and Clusius that the number of parrot species adduced can be taken as a gauge of the degree of ornithological topicality of their work. It is therefore not surprising that Brueghel rises to the challenge: he adds one or several new sorts to each version of the *Allegory of air*. While not a single parrot is to be seen in the compositional sketch, three sorts are represented in the version in Rome: the yellow-winged macaw, the grey parrot and the rose-ringed parakeet. The 1611 version in Lyon adds the blue-and-yellow macaw and two red-faced lovebirds, and, as mentioned above, the yellow-winged macaw is replaced by the rarer green-winged macaw.[61] The 1621 *Allegory of air* shows a spectacular number of parrots by comparison with the earlier versions.[62] Besides the green-winged macaw (no. 17), the blue-and-yellow macaw (no. 61), the red-faced lovebird (no. 12) and the rose-ringed parakeet (no. 22), it includes a blue-fronted amazon (no. 63) – a type that was already known in the sixteenth century – a black-capped lory (no. 62), which also appeared in the *Allegory of sight*,[63] a yellow-headed amazon (no. 53), a senegal parrot (no. 60) and a canary-winged parakeet (no. 24). As far as we know, this is the first time that the last three birds are represented. Furthermore, Brueghel gives a prominent place to the yellow-crested cockatoo (nos. 33a, 33b); this is the first unmistakable representation of this bird.[64]

The parrots thus stand as symbols of the scientific topicality of the painting. This is accentuated by the putto with a yellow-headed amazon on his hand, and by Aurora with one of two yellow-crested cockatoos on her hand. But Brueghel does not just stress how scientifically up to date the painting is; the work also demonstrates the superiority of painting over printed natural historical observations. The painted image provides much more information than a descriptive text or a printed black-and-white illustration. Moreover, Brueghel also shows with this painting that he is a learned painter – a *pictor doctus*.[65]

There are two complementary sides to the way in which Brueghel depicts the world of the birds: on the one hand he emphasises what is new and unfamiliar, on the other he renders what is known and familiar with greater accuracy, veracity and liveliness.[66] The *Allegory of air* from 1621 represents an unprecedented number of new ornithological species in extremely natural attitudes. Many of them are presented from an unusual angle: not just in profile, but also from the front or the back, or partially (only the head of the surf scoter [no. 47] is visible). Very important for the liveliness of the birds is the increasing interaction between them displayed in the successive versions. The following section will go into the interpretation of these forms of interaction.

Sympathy and antipathy

Although the viewer's attention is drawn first and foremost to the birds in the *Allegory of air*, the painting is one in a series of the *Four elements*. Brueghel does not neglect this layer of meaning and also demonstrates the place of the element of air in nature. To do justice to the originality of Brueghel's allegory of the element, it is important to recall a few aspects of the worldview that generally prevailed at the time. The four elements were connected with the four seasons, the four cardinal points, the four ages of man, the four divisions of the day.[67] This can be represented as follows:[68]

Season	*Cardinal point*	*Part of the day*	*Age*	*Element*
spring	east	morning	childhood	air
summer	south	afternoon	youth	fire
autumn	west	evening	adulthood	earth
winter	north	night	old age	water

Brueghel's *Allegory of air* from 1621 conforms with this system. The rising sun and sinking moon indicate that the direction of the gaze is eastward. This morning coincides with the morning of life and with spring. The latter two aspects are emphasised in the foreground by the bird's eggs and the suckling bat (no. 42). Birdwatchers will recognise a sign of spring in the plumage of the flying great cormorant (no. 39), as this bird has the characteristic white fleck on its thighs only in that season.

Such coherence is less prominent, if present at all, in the allegories of the other three elements that Brueghel painted for Federico Borromeo. The *Allegory of water* only seems to imply the connection between water and old age, as Neptune is represented at the centre of the scene as an old man, while in other versions of this element a young Amphitrite sits at the water's edge. The bare fallen tree in the background seems to be a harbinger of death, and the waning, subdued light could be interpreted in the same way. In the *Allegory of earth* the element is connected with autumn and harvest. There are no such connections at all in the *Allegory of fire*. This may indicate a progression in Brueghel's practice: from 1608 (fire) to 1621 (air), he pays increasing attention to the ramifications of the elements and their connections with other categories. Might this be what Borromeo was referring to when he wrote that Brueghel seemed 'to have expended his last drop of ingenuity on this final painting'?[69]

In the worldview of the time, the four elements were grouped in accordance with the principle of antipathy: fire is opposed to water, and air to earth.[70] Brueghel applies the same principle in his *Allegory of air*. The only element visible in the painting apart from air is earth, the very element that is antipathetic to it. Earth is accentuated as an element by the large number of earthbound birds that are represented: ostrich, cassowary, great bustard and many gallinaceous birds. It is also noteworthy that the many waterbirds are represented here without their element, water, to which they nevertheless refer.[71] Fire too, like water, if present at all, is so only indirectly via the light of the rising sun.

Antipathy is compensated as a regulatory force by sympathy. Brueghel seems to have been increasingly concerned with the representation of universal sympathy and antipathy. He interprets sympathy – in the scientific sense of a force of attraction between objects, organs or bodies – as the force of sexual attraction within the species.[72] In the various versions of the *Allegory of air* Brueghel plays with the theme of sympathy by representing the birds of one type not only in pairs, but also distant from one another so that the viewer is invited to find the matching partners, sometimes at opposite removes of the painting.[73] Examples of this in the 1621 *Allegory of air* are the grey parrot (nos. 21a, 21b), the yellow-crested cockatoo (nos. 33a, 33b) and the great bustard (nos. 48a, 48b). In some cases there is only a single exemplar of a bird present. The viewer discovers this only after having intensively scrutinised the painting. This is a rhetorical device to bind the viewer to the painting.

The antipathy between species is even more striking than the sympathy.[74] Antipathy is not expressed very strongly in the Rome and Lyon versions of the *Allegory of air*. The aggression between animal species that can be found in these early versions is relegated to the margins: in the top right a heron and a bird of prey are fighting in the air, and a shadowy creature – which upon closer inspection proves to be an enormous bat – has caught a bird in its teeth.[75] Visible in the bottom right of these early versions is the struggle between the mythical bird roc and an elephant: the gigantic bird attacks an elephant and flies off, clutching the pachyderm in its claws.[76] The only form of antipathy that is not hidden in the margins is that between two birds at Aurora's feet: the mute swan seems to be reacting to the eagle. The antipathy between these ornithological species was a commonplace of natural history. Pierre Belon wrote in his authoritative *L'histoire de la nature des oyseaux* (1555): 'If the eagle attacks them [sc., the swans], they defend themselves as superior, but they never fight unless attacked first'.[77]

Antipathy is much more explicit in the *Allegory of air* from 1621 than in the two earlier versions, and this is underlined by the distribution of light and shade in the work. The gaze of the viewer is drawn to the light area in the lower centre of the painting, above all to the dazzling white of the two mute swans (no. 35). They seem to be quarrelling with a grey heron (no. 25c).[78] Three other fights between birds are represented around them. In the middle of the painting a large bird of prey attacks a heron in the air (nos. 26b, 25b). (Another struggle between a bird of prey and a heron is represented at top left [nos. 26a, 25a].) A serious fight is also taking place on the ground between a cock and a turkey (nos. 31a, 30). The third combat is between an enormous eagle owl and two gallinaceous birds; the hen seems to be dead, and the cock is trying to defend itself (nos. 34a, 31b, 31c). This scene is carefully observed by a barn owl (no. 32a) and a yellow-crested cockatoo (no. 33a), both of which are difficult to see because their light plumage is outdone by the dazzling white of the swans. These remarkable fights call for further interpretation.

The antipathy between the bird of prey and the heron is well known and is mentioned from the time of Aristotle on. Pierre Belon writes:

> Aristotle said in the first chapter of the ninth book that the eagle attacks the heron and dies in self-defence. When assaulted, the heron tries to win by flying against it instead of fleeing into the distance like some other river birds.[79]

Aldrovandi provides an extensive account of this fight in the sky, which can end badly for the bird of prey because it runs the risk of being pierced in the breast by the heron's sharp beak. Brueghel presents a lively version of the scene at the moment of attack.

The fight between the cock and the turkey was much less familiar for the simple reason that the turkey was not imported into Europe until the sixteenth century.[80] Their mutual antipathy does not figure in the contemporary literature on natural history. The first to illustrate a struggle between the two birds was Marcus Gheeraerts the Elder (c. 1520-c. 1590) in an illustration to a fable in the emblematic collection *De warachtige fabulen der dieren* from 1567, a co-production between Gheeraerts and the poet Eduard de Dene (c. 1505-c. 1578) (fig. 10).[81] The illustrations of this collection were reissued in 1617, this time accompanied by texts by Joost van den Vondel (1587-1679).[82] De Dene and Vondel offer the same interpretation of this fight: it is the Dutch cock that attacks, and the peace-loving foreigner – the turkey – must give way to its violence.[83] It is not possible to conclude from Brueghel's

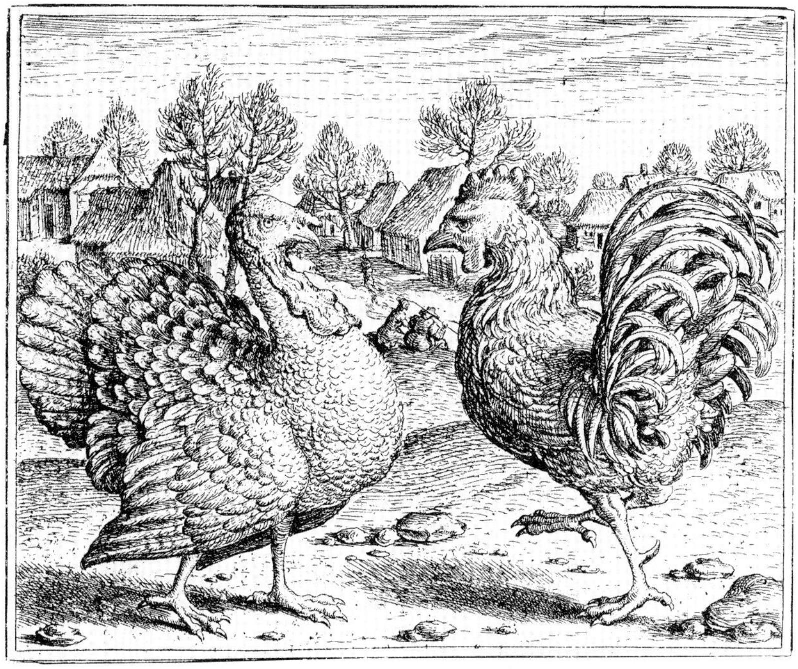

10
Marcus Gheeraerts, *Vlaemsche ende Turksche Hane*, 1567
etching, E. de Dene, *De warachtige fabulen der dieren*, Bruges 1567, 190.

representation of the fight whether he shared their moralising disapproval of violence against foreigners. Perhaps Brueghel wanted to show in the painting how the domestic and the exotic meet and can clash. Moreover, a bird fight was an opportunity for Brueghel to show off his talent as a painter by rendering the birds in an extremely lively way and displaying his skill in depicting their plumage.[84]

The third fight, that between the eagle owl and the cock, is not found in the natural history of Brueghel's day either. It should therefore probably be interpreted symbolically instead of in terms of natural history. The cock was a traditional symbol of sun and light, while the owl is the bird of night and darkness. Brueghel reverses this motif in a sort of paradox: the bird of night is fully illuminated here. Light and dark are also symbolised by the two birds that watch the scene: the yellow-crested cockatoo (no. 33a) symbolises the day, the barn owl (no. 32a) the night.[85] The rivalry between light and dark is also portrayed on a wider scale in the painting: the moon (Diana) has to make way for the rising sun (Apollo). Aurora, who is situated just above the owl and the cock, functions as a *coincidentia oppositorum* in which light and dark, sun and moon coincide.[86]

The 1621 *Allegory of air* thus contains not only interesting cases of sympathy and antipathy among birds, but also more general oppositions between day and night, air and earth, domestic and exotic, as well as paradoxical reversals that underline these oppositions. The bird of paradise and the bat, for instance, which are traditionally associated with the air, are represented on the ground. Waterfowl are represented without a trace of water, either on the land or in the air. Nocturnal creatures such as the bat and the owls are placed in broad daylight. And domestic birds are represented between exotic species all over the painting. This contrasting effect that dominates the scene is further accentuated by the distribution of colour: the contrast between light and dark is continued in the colours of the birds: the light plumage of the pelican (no. 44), flamingo (no. 49) and mute swans (no. 35) contrasts sharply with the dark plumage of the Magellanic penguins (no. 40), cassowaries (no. 55) and cranes (no. 52). It is also noteworthy that the birds with the brightest and most variegated colours – such as the blue-and-yellow macaws (no. 61), the golden oriole (no. 64) and the birds of paradise (no. 2) – are situated in the darkest areas of the painting where their yellow is seen at its best.

In the eyes of Borromeo

After this discussion of the various layers of interpretation of the painting, several important questions still remain. To what extent did Borromeo recognise and appreciate these layers of meaning? Did he also understand the elements and the natural world in terms of sympathy and antipathy? Did the veracity and accuracy of the representations of the many types of birds and the display of being au fait with the latest scientific discoveries mean a lot to him? For it is these very aspects that distinguish the 1621 *Allegory of air* not only from the earlier versions of this theme in Rome and Lyon, but also from the other three elements – fire, earth and water – in the series that Brueghel produced for the cardinal.

Borromeo regarded the whole of nature as a divine creation. In a broad sense the paintings were a manifestation of the divine plan.[87] The elements played an essential role in this plan, because they ensured an orderly universe in which the opposite extremes were linked together in a harmonious whole.[88] Borromeo mentions this opposition or antipathy in the elements in his devotional treatise *Le laudi divine*:

> Now this nature, is she not perhaps the very humble servant and obedient handmaiden of God, who joins the extreme parts of created things by the knot of many suitable and proportioned means? Hence, the pleasantness of the air does much to diminish and moderate the pride of fire. And the harshness and hardness of the earth are sweetened and softened by the softness of the waters, and by their pliant quality and substance.[89]

The enormous variety of natural creatures called for an orderly system to express the formation of God's creation.[90] Animals were deserving of study because they were examples of the divine creation. In a later passage in the same work, Borromeo writes:

> Animals, therefore, must be our teachers, and they will teach us various things. (...) Looking then with attentive study at animals' construction and formation, and at their parts, and members, and characters, can it not be said how excellently divine wisdom has demonstrated the value of its very great works?[91]

The realistic, true-to-nature rendering of the birds that reaches a climax in this painting must have appealed strongly to the cardinal. An unpublished treatise that Borromeo wrote on the keeping of birds indicates that he studied them himself.[92]

The question arises as to whether the topicality of the optical instruments in the painting also contributed to Borromeo's particular appreciation of it. He was well informed about the latest discoveries in the natural sciences and was very interested in findings that were made by means of new optical instruments such as the microscope.[93] The telescope had his interest as well. He purchased one in 1614 and recorded his findings in an unpublished work, *L'occhio celeste*.[94] It may therefore not be fortuitous that Brueghel represented a telescope in the painting. However, the telescope was in fact problematical for Borromeo, a strict anti-Copernican, in so far as it conflicted with the Bible. He also expressed his doubts about the reliability and validity of telescopic observations. Nevertheless, findings that displayed the wonders of the divine creation were deserving of his approval. That is how the presence of the telescope should be understood. In his view, the many recently discovered types of parrots depicted in the painting would have shown the breadth of God's universe. It is therefore not surprising if Borromeo considered that Brueghel had surpassed himself with this painting.

The exceptional nature of Brueghel's *Allegory of air* from 1621 still holds true. This one painting offers insight into various developments in the observation of the natural world of his time. Faber Kolb has already pointed out that Brueghel's paintings were done in the period of the rise of the encyclopaedia of natural history and that his work reflects the growing impulse to catalogue or even to classify and order the (new) species.[95] The shift in Brueghel's work towards a more systematic ordering of the animal species is very well expressed in the paintings of the *Four elements*. As we have seen, however, Brueghel was interested not only in classification, but also in constantly applying the very latest knowledge. The process of Brueghel's assimilation of the most recent ornithological knowledge can be followed step by step in the different versions, reaching a climax in 1621.[96] Brueghel showed himself not only to be a remarkable artist, but also to be in no way inferior to the most authoritative naturalists of his day. This places the relation between artists and scientists in a new light and calls for further investigation. Art and science were not regarded as entirely separate disciplines in Brueghel's day. We hope to have taken a first step towards new research that not only relates the achievement of the natural sciences of that time to artistic production, but also takes into consideration the importance of art for science.

Notes

*This article was written within the framework of the NWO programme *Cultural Representations of Living Nature: Dynamics of Intermedial Recording in Text and Image (c. 1550-1670)*. The English translation is by Peter Mason. We would like to thank the guest editors and an anonymous peer reviewer for their useful comments. We are also grateful to Huib Zuidervaart (KNAW) for his comments on Brueghel's representation of the telescope, and to Sven Dupré (Max Planck Institute for the History of Science) for discussion of Borromeo's complicated attitude towards the telescope.

1 This series is now divided between the Biblioteca Ambrosiana in Milan and the Musée du Louvre in Paris. The *Allegory of fire* and *Allegory of water* are in the Ambrosiana, the *Allegory of earth* and the *Allegory of air* are in the Louvre. All four works are painted on copper and measure 46 x 66 cm.

2 Borromeo 1625 and 2010. For more information on this work, see Quint Platt 1986 and Jones 1993. Borromeo's collection totalled 127 paintings. Almost 30 per cent of them were landscapes, still lifes and Madonnas in a garland of flowers, most of them by northern painters.

3 Borromeo 2010, 168-169: 'Aera ceu campum lucis omni incunditate circumfudit; ac si comparationis causa dicendum est etiam aliquid, extrema haec Tabula omnem Artificis consumpsisse solertiam videtur, sicuti laboris etiam ordine id consumebat opus'. The citations from Borromeo's *Musaeum* are from the edition and English translation in Borromeo 2010.

4 Brueghel went to Italy in 1589 and reached Rome in 1592. He moved back to Antwerp in or shortly before October 1596. Before his return he had received a letter of recommendation from Borromeo addressed to the Bishop of Antwerp. See Crivelli 1868, 3-4.

5 Brueghel produced a total of 29 paintings for Borromeo.

6 For the relationship between Borromeo and Brueghel, see Cutler 2003.

7 Brueghel painted some of his most innovative and most labour-intensive works for Borromeo; Woollett & Van Suchtelen 2006, 12. The letters of Jan Brueghel to Borromeo are published in Crivelli 1868. The correspondence provides a good deal of information about the artistic and diplomatic network in which Brueghel moved. A number of the letters that Brueghel wrote to Borromeo (more than 25) were written by Peter Paul Rubens, who acted as Brueghel's amanuensis, probably because Rubens's Italian was better than Brueghel's; Woollett & Van Suchtelen 2006, 29-30. This is an indication of the close friendship between the two painters.

8 Letter of 1 February 1608. Brueghel does not seem to have grasped the irony of the situation: the cold prevented him from painting fire. Crivelli 1868, 91-92: 'A quello della notto e fuoco non manchero anchora de metter mano quanto prima: et si non fosso stato questo grand.mo freddo, de giga saria cominciato con qualche inventcione del ellemento del fuoco, che spera che piatchera a VS Ill.mo: alla quale bagiando humilmente le mani prego nostro Signor per la lunga conservatione'.

9 Letter of 26 September 1608. Crivelli 1868, 112-113: 'In pochi giorni mandera il quadro Ellementa del fuoco, in quael e de vedere oigni sorte d armeria, metali oro argento e fuoco, abcho l'alchimio et distilattioni, tutti fatti del naturel con grandismo diligenc'. Perhaps Borromeo later returned this work to Brueghel so that he could introduce some changes, as it is not mentioned in the codicil of 1611. See Jones 1993, 237.

10 This can be seen from a letter of 24 December 1614. Crivelli 1868, 220-221: 'I perçiò mi rallegrai molto quando VS Illus.ma si degnò di commandarmi quel quadro del Elemento dell'acqua, i condussi l'opra a quella perfettione che l poco ingegno i debol mano permiserò. Pur al contrario non poco mi afflige doppo l'hauerlo consignato al sig. Ercole Bianchi noue mesi sono, non auerne mai inteso noua alcuna nè del buon ricapito, nè come sia riuscito à gusto di VS Ill.ma alla quale li maggior suoi negocij non permettono tal volta d'attender à bagattille: ma ben sperai ch'il sig. Ercole Bianchi douesse supplir à questo colla solita sua diligenza'.

11 There is mention of this element and the element of water in letters of 1609 and 1611, but it is not entirely certain whether they refer to this series or to another series of the four elements. Ertz claims that these letters from 1609 and 1611 may refer to a second series of the four elements that Brueghel did between 1608 and 1611; Ertz 1979, 364. According to Crivelli, a 'landscape with animals' may have been transformed into this element between 1608 and 1618; see Crivelli 1868, 94-97. Moreover, on more than one occasion Brueghel sent paintings to Borromeo for his approval, who then returned them together with requests for modifications. In that sense it is not unlikely that the *Allegory of earth* may have undergone changes between 1609 and 1618.

12 Quint Platt 1986, 268.

13 Letter of 29 October 1621 to Ercole Bianchi. Crivelli 1868, 274-275: 'Il quader ellemento del ario i fatto con ordini (...)'.

14 Letter of 11 February 1622. Crivelli 1868, 282-283: 'Io ha mandato un quadro fatto per sua ordine, Ellement del ario: un altro mandai senso ordina, un girlande de fiori per fare piatcir: le Madonna fatto diuinamento de mane de Rubbens'.

15 Borromeo 2010, 166-167: 'Fuit in suo genere mirificus, potuitque corpusculis illis inserere tam generosos, vividosque spiritus, ut incertum relinquere videatur in spectantium animis, altane an humili dimensione incluserit tenuium earum figurarum modum. Videtur penicillo suo per cuncta naturae voluisse pervagari. Pinxit enim (...) Maria, Montes, Antra, specusque subterraneos, et omnia ista spaciis disiecta immanibus in angustum coegit, naturam ipsam imitatus non coloribus tantum sed etiam facilitate, quae sicuti naturae, ita etiam artis summum decus est'.

16 Borromeo 2010, 168-169: 'In descriptione Maris et aquarum ita varii generis piscatus expressit, ut non minor piscandi, quam pingendi Artifex fuisse videatur (...)'.

17 The armillary sphere is often represented in allegories of knowledge, as in Marcus Gheeraert's title page of the *Warachtige fabulen der dieren*; De Dene & Gheeraerts 1567. Scholars in the Renaissance also had themselves portrayed with an armillary sphere to show their wisdom and knowledge.

18 The telescope became the icon of science and is as such an important motif in the painting.

19 This is stated in the official act of donation of the Ambrosian collection of 1618; see Quint Platt 1986, 298: 'L'Elemento dell'Acqua di Gio.

Brueghel, alto un braccio, e largo cinque quarti. Et a tergo vi sono i nomi dei pesci e degli altri animali ch'egli con somma diligenza ha ricavato dal naturale, con le cornici d'ebano coi proffili d'oro'.

20 Brueghel also depicted many different species of birds in paintings such as *The entry of the animals into Noah's Ark* (1613), now in the J. Paul Getty Museum (inv. no. 92.PB.82) in Los Angeles, and representations of the Earthly Paradise as in *Adam and Eve in the Garden of Eden* (1615), in the Royal Collection (inv. no. 405512) in London, and *Adam and Eve in Paradise* (c. 1615), painted in cooperation with Rubens and now in the Mauritshuis (inv. no. 253) in The Hague.

21 The series in the Galleria Doria Pamphilj in Rome is painted on panel and measures c. 55 x 95 cm. The series is dated around 1611 by Ertz 1979, 599. For further discussion of this dating, see below.

22 This series in the Musée des Beaux-Arts in Lyon is painted on panels measuring c. 46 x 83 cm. The *Allegory of water* is not dated. The *Allegory of fire* is dated 1606, the *Allegory of earth* 1610. Ertz attributes this series to Jan Brueghel the Younger (1601-1678), the son of our Brueghel. In the catalogue of the museum, Hans Buijs attributes the series to Jan Brueghel the Elder; see Cat. Lyon 1993, 56-58. The RKD also attributes the series to Jan Brueghel the Elder. We share that opinion. There is no reason to doubt the dating, which excludes Jan Brueghel the Younger. Moreover, from 1624 the son signed with 'Breughel', which is not the case in this series. We consider that the style of this series matches that of the other two series. Moreover, the birds are rendered more accurately than in the series in Rome (see below), which makes a copy of the series in Rome by Jan Brueghel the Younger extremely unlikely.

23 The many isolated paintings of one of the four elements in museums, at auctions and on the art market are a sign that Brueghel probably made more series, which have become scattered. Besides, Brueghel's workshop would have produced series of the elements, and Jan Brueghel the Younger copied the work of his father.

24 For more information on repetition and versions in the work of Brueghel, see Paulussen 1997, 11. It was in fact necessary for Brueghel to repeat motifs and compositions, otherwise he would have been unable to satisfy the large demand for his paintings; Woollett & Van Suchtelen 2006, 163.

25 The absence of the bunch of feathers and the tiara in the 1621 version in Paris may be a sign that Brueghel wanted to shift the emphasis from her to the birds.

26 In that case, however, there were limits to his observation of nature. For instance, in the *Allegory of air* in Lyon he represents a breeding great tit on a nest in the fork of a branch, though this bird actually breeds in cavities.

27 In a letter to Bianchi of 5 September 1621, Brueghel wrote that he had painted the animals in a garland of flowers from life on the basis of exemplars that he had seen in the menagerie in Brussels. Crivelli 1868, 272-273: 'Et con le robbe de Enoni, et bone comodita mando un altro quadro, il più bello et rara cosa che habbia fatta in vita mia. Ancho sig. Rubens ha fatta ben mostrande sua virtu in el quadro de megio, essend una Madonna bell.ma. Li oitcelli, et animale son fatto ad vivo de alcuni delli seren.ma Enfanto'.

28 Illustrations of these exemplars were made by hand and collected in albums. See Haupt *et al.* 1990.

29 Maselis *et al.* 1989, 101.

30 The drawing is in the collection of the Metropolitan Museum of Art in New York (inv. no. 2008.178.14a, b) and shows the study of birds on the recto side and a village scene with skaters on the verso side. It is in pen and brown ink, brush and brown wash (recto) and watercolour (verso), and it measures 261 x 182 mm. The RKD attributes the drawing to Jan Brueghel the Elder. Stijn Alsteens, the curator of the drawings and prints in the Metropolitan Museum, has recently attributed the drawing to Albert Flamen (c. 1620-1669), and we would like to thank him for this information. He will shortly publish the new attribution of this drawing.

31 Gesner 1557. The sequence of the birds in the sketch follows the one adopted in Gesner's *Vogelbuch*. In her extensive discussion of Brueghel's relation to natural history, Faber Kolb claims that Brueghel does not seem to refer directly in his work to the encyclopaedia of natural history such as those of Gesner and Aldrovandi, but adopts their methodology; Faber Kolb 2005, 26. We shall argue below that Brueghel does deliberately refer to their works.

32 This pen drawing is in the De Grez Collection in the Koninklijke Musea in Brussels (inv. no. 4060/474). It is in brown ink and coloured wash, 142 x 196 mm. We would like to thank Tine Meganck for the information that the drawing is no longer considered to be an authentic drawing by Brueghel by the Koninklijke Musea. However, it is possible that the drawing is a copy after a lost original by Brueghel. There is a different, smaller study of birds on a branch in the F. Lugt Collection. This is a pen drawing measuring 94 x 139 mm.

33 Brueghel often worked with Van Balen from 1604 on. Both painters moved to the Lange Nieuwstraat in Antwerp in that year, which would certainly have facilitated their cooperation because the paintings could then be transported without much difficulty between the two workshops; Woollett & Van Suchtelen 2006, 157-160. It is noteworthy that there is no mention of the cooperation between Brueghel and Van Balen in the correspondence between Brueghel and Borromeo. The cardinal does not mention Van Balen in his *Musaeum* either, though he does mention a collaboration between Brueghel and Rubens on a painting with a Madonna inside a garland of flowers. Generally speaking, patrons very much appreciated a cooperation between two painters. For more information on cooperation between Brueghel and other painters (especially Rubens), see Woollett & Van Suchtelen 2006.

34 Other examples could be cited to substantiate this thesis. For instance, the spoonbill is depicted on the far right on the sketch and to the left beside Aurora in the painting in Lyon. The spoonbill is entirely absent from the Rome painting.

35 Ertz dated the Rome series around 1611; see Ertz 1979, 599. We consider a slightly earlier date, around 1608, to be more plausible.

36 According to Faber Kolb, the optical instruments demonstrate Brueghel's empirical approach to nature; Faber Kolb 2000, 159.

37 Hans Lipperhey from Middelburg applied for a patent. His request was

turned down because, soon after Lipperhey's demonstration, the art of making a telescope proved to be in several hands.

38 This landscape on canvas measures 84,7 x 130,8 cm and is in the Virginia Museum of Fine Arts, Richmond (inv. no. 53.10). For more information on the telescope in the work of Brueghel, see Molaro & Selvelli 2008.

39 The oldest known sketch of a telescope is in a letter from Giambaptista della Porta to Federico Cesi of 29 August 1609. See Galilei 1808-1811, vol. 10, 252.

40 Brueghel's model was probably the telescope that Archduke Albrecht of Austria purchased 'from the inventor' in 1609.

41 The *Allegory of sight* measures 65 x 109 cm and is in the Museo Nacional del Prado in Madrid (inv. no. P01394).

42 See Molaro and Selvelli 2008. As far as is known, their hypothesis is not confirmed by contemporary sources. If it is correct, the date of Kepler's invention will have to be adjusted.

43 The ostrich and the pelican were represented since the Middle Ages. Among those who represented the crane are Collaert s.a. and Aldrovandi 1599. Images of the turkey began to proliferate after c. 1550; Mason 2000.

44 For more information, see Balis 1968, 31.

45 The purple coot is described but not illustrated by Pierre Belon; Belon 1555, 226.

46 For a survey of the sources, see the commentary by J. Céard in Paré 1971, 194, n. 296. The French surgeon Ambroise Paré writes in his *Des monstres et prodiges* (1573): 'The male has a cavity in the back where the female broods her young' (Paré 1971, 130).

47 Brueghel seems to have had a momentary doubt in *The entry of the animals into Noah's Ark* (1613), in the J. Paul Getty Museum, where the green has a yellow lustre.

48 See Clusius 1605.

49 Unfortunately we have not been able to discover how Brueghel came into contact with the work of Clusius. Rubens owned works on natural history by Aldrovandi and Gesner, but as far as is known no work by Clusius; see Arents 2001, 61. As Brueghel and Rubens were close friends, it is likely that Brueghel would have consulted the book in the house of his learned friend if it had been there. Nothing is known about Brueghel's own library. Clusius worked for many years on his book and asked many people for information. The work therefore enjoyed a great reputation before its publication.

50 Clusius 1605, 98.

51 Clusius compares the head of the cassowary to that of the ostrich. He likens the toes of the cassowary to those of the great bustard ('Otidis sive Tardae pedium modo'); Clusius 1605, 99.

52 Clusius's informant is Johannes de Weely, a merchant from Amsterdam. He even gives the date of the merchant's statement: late June 1605. See Clusius 1605, 361.

53 His correspondent Jacques Plateau, who had sent him this illustration, could not give him any further information; Clusius 1605, 105.

54 Clusius was to read Aldrovandi later, as can be seen from later additions in which Aldrovandi is extensively cited, as though Clusius wanted to remedy his earlier mistake.

55 Potgieter 1600. For more information on this travel account, see Mason 2007, 195-219.

56 Clusius 1605, 101: '(…) ex Anserum genere, tametsi rostro dissimilis, in mari victitans, valde pinguis, praegrandis anseris magnitude'.

57 Clusius 1605, 101: 'pronâ parte nigris pennis tectae erant, supinâ & in ventre albis: collum, quod crassum & breve, habebant nonnullae albis pennis, tamquam torque, cinctum'.

58 The penguin recurs in a painting attributed to Jan Brueghel the Elder, *The Garden of Eden; in the background the Temptation* (panel, 53 x 84 cm, Victoria and Albert Museum, London). The penguin is shown from the back, and once again the feet are not correct. The work is dated around 1600. The presence of the penguin, the birds of paradise with feet and wings and the senegal parrot (see below) lead us to favour a later dating around 1620.

59 The fact that the great white egret is still vaguely visible may be because the layers of paint of the blue sky have become more transparent over the years.

60 Clusius 1605, 365.

61 All of these varieties were already fairly well known. They occur afterwards in Brueghel's *The entry of the animals into Noah's Ark* of 1613 and in *Adam and Eve in Paradise* from c. 1615 painted by Brueghel and Rubens; see note 20.

62 See also Smith 2008, 241.

63 Clusius provides a detailed description of a parrot that may be a black-capped lory. His description is based on a watercolour sent him by his correspondent Jacques Plateau. Plateau saw the bird at the Duke of Aarschot's, his employer at the time.

64 The yellow wash on the bird's ear coverts indicates that it is the yellow-crested cockatoo (*Cacatua sulphurea*) from Indonesia.

65 It should be evident that he had sufficiently mastered Latin to be able to understand natural historical texts (perhaps with some assistance). What Grafton writes about Latin in the early modern period will also have been true of Brueghel: 'Latin (…) was not a parochial dead language but a universal living one, which many could read and understand even if they could not speak and write it fluently and grammatically. The medium of every church service, every university lecture, every science, and even some tourist guides, it prevailed as English does in the modern world. Many who did not fully belong to the learned milieus could nevertheless draw on scholarly books (…)'; Grafton 1992, 75.

66 A good example of ornithological precision is the pair of great tits (no. 10) depicted next to a couple of blue tits (no. 18). Brueghel seems to want to indicate that these closely related sorts should be distinguished from one another, as Pierre Belon also does in his book on birds; Belon 1555, 366-368.

67 For a survey of the doctrine of the elements, see, for example, Dijksterhuis 1975, 23-26.

68 This system is based on a passage in a compendium of ancient wisdom by a certain Pierre Breslay; Breslay 1571. In his final chapter, 'De l'analogie qui est entre les saisons de l'an, & du iour & entre les humeurs du corps humain', he explains: 'Le printemps simbolize avec le Levant, l'esté avec le Sud, & l'autonne répond au Ponant, & l'hiver au Nort. Item le matin est printanier, le midy estival, le vespre autonnier & la nuit hyvernale (...). Item le printemps a correspondance à l'enfance, l'esté à la ieunesse, le declin de l'age à l'autonne, & l'hyver à la froide & grise vieillesse'. A scheme can also be found in Seznec 1972, 47, but it lacks the divisions of

the day, and the cardinal points are linked with other elements.

69 See note 3.

70 The French poet and publisher Simon Goulart explains this in his commentary on *Le grand miroir du monde* by the poet Joseph Duchesne, sieur de la Violette from 1593, who is cited as follows by Michel Foucault in his *Les mots et les choses*: 'L'élément du feu est chaud et sec; il a donc antipathie avec ceux de l'eau qui est froide et humide. L'air chaud est humide, la terre froide est sèche, c'est antipathie. Pour les accorder, l'air a été mis entre le feu et l'eau, l'eau entre la terre et l'air. En tant que l'air est chaud, il voisine bien avec le feu et son humidité s'accommode avec celle de l'eau'; Foucault 1966, 40.

71 In other paintings Brueghel always places water birds in their element. See, for example, *The entry of the animals into Noah's Ark* and *Adam and Eve in Paradise* (see note 20).

72 Sympathy is predominant in Brueghel's *Adam and Eve in Paradise*, which focuses on the love between the animals, and in his *The entry of the animals into Noah's Ark*, in which the animals are led into the Ark two by two (see note 20).

73 This is also true of the *Adam and Eve in Paradise* painted by Brueghel and Rubens (see note 20). For more information on this work, see Smith 2008.

74 This antipathy already played an important role in *Adam and Eve in Paradise* by Brueghel and Rubens, but there it was only latent because a violent outburst of antipathy between the different species would be out of place in the peace of Eden and could only be situated after the Fall; Smith 2008.

75 The size of this bat indicates that it is an exotic species. Clusius provides a survey of the authors who describe large bats in his *Exoticorum libri decem*. Clusius himself is the first to describe and illustrate the largest sort, the Kalong, or flying fox, of Indonesia, though it lives on fruit, not meat.

76 Brueghel may have intended the roc to suggest that the air is superior to the earth, symbolised by the elephant, the largest terrestrial animal. The antipathy between the roc and the elephant was known in the Western world from the time of Marco Polo. This originally oriental theme was often illustrated in scientific works, as in those of Theodore de Bry and Aldrovandi; Wittkower 1938. The absence of the roc in the *Allegory of air* from 1621 is probably because in Brueghel's day naturalists were beginning to call the scientific character of this myth into question.

77 Belon 1555, 151.

78 We have been unable to find any reference to antipathy between these two species in the authoritative works of Belon, Gesner and Aldrovandi.

79 Belon 1555, 190-191.

80 Mason 2000, 63-90.

81 De Dene & Gheeraerts 1567.

82 Vondel 1617, fols. 99r-100v.

83 Here follows the English prose version of the same fable by Arthur Golding (late 16th century) from Golding 1987, 282: 'Of a turkey cock and a dunghill cock; A dunghill Cock being a proud courage, stood so immoderately in his own conceit that he disdained all fowls of his own kind. Wherefore perceiving a Turkey Cock (whom by another name some call an Indian Peacock) to haunt the same places that he did, and to be welcome to the Hens and Young Pullen there for his gentle and quiet behavior: he took such spite at it, that he made cruel and fierce war against him. When the Turkey Cock saw the contention grow so oppressive that it struck a terror even into the Cockerels, and made them keep aloof for fear: he loving quietness, and utterly abhorring such continual battle and discord, forsook his accustomed dwelling place, and chose him another habitation; deeming that body to have lived well, which hath lived unknown'.

84 Brueghel's invention of painted bird fights was clearly a success. This motif was rapidly picked up by other painters such as Frans Snijders and – later in the century – Melchior d'Hondecoeter.

85 The barn owl is even more of a nocturnal bird than the eagle owl. The same paradox applies here: both the barn owl and the eagle owl are in broad daylight.

86 In the Rome and Lyon versions of the *Allegory of air*, this is also symbolised by Aurora's tiara with the sun and moon represented on it.

87 For more information on Borromeo's vision of nature in relation to the paintings of Brueghel, see Faber Kolb 2000, 112-120.

88 Faber Kolb remarks that the four elements keep on cropping up in Borromeo's writings; Faber Kolb 2000, 159. Prosperetti remarks on a painting by Jan Brueghel the Elder in which the four elements are combined (dated 1604, now in Vienna), noting that the scene shows the moment at which the Creation is complete and all the elements are in perfect balance. She thereby establishes a relation between the elements and the Creation that she believes Brueghel to have intended; Prosperetti 2009, 66-67.

89 Borromeo 1632, 15: 'Hoc questa Natura non è ella force la humilissima serua, e l'vbbidente ancella di Deo, il qual congiugne le estremi parti delle cose create col nodo di molti conueneuoli e proportionati mezzi? Quindi è, che la piaceuolezza dell'aere va molto scemando e temperando la fierzza del fuoco; e che le asprezze e durezze della terra sono addolcite ed ammollite dalla morbidezza delle acque, e dalla loro arrendeuole qualità, e costanza'. The syntactic parallelism in this citation that opposes *'che la piaceuolezza dell'aere'* to *'che le asprezze e durezze della terra'* corresponds exactly to Brueghel's practice in the *Allegory of air*.

90 Faber Kolb 2005, 56.

91 Borromeo 1632, 131-132: 'Gli animali adunque douranno essere i nostri maestri, ed hauranno da insegnarci diuerse cose. (...) Riguardando poi con attento studio alla costruttura e formatione loro, ed alle lore parti, e membra, e figure, no si può dire quanto eccellentemente la divina Sapienza dimonstrato habbia il valore delle sue grandissime opere (...)'. For more information about Borromeo's views on nature and painted landscapes, see Jones 1988 and Jones 1993, 79-83.

92 Faber Kolb 2000, 123-124. This treatise, *Memorie per allevare li ucelli et il modo di tenerli in un giardino. In italiano*, is in the Biblioteca Ambrosiana (G 9 inf. [3]).

93 Jones 1993, 83-86. Borromeo expressed his wonder at and use of all kinds of optical instruments in his *Pro suis studiis*; Borromeo 1628, fols. 256r-258v. He also counted Galileo Galilei as one of his intimate circle.

94 This work is part of *Miscellanea varia. Occhiale celeste sive caeles oculus* and is in the Biblioteca Ambrosiana (I 52

suss).

95 Faber Kolb 2000, especially 58-80 (on the relation between Brueghel's paintings and the encyclopaedias of natural history) and 173-191 (on Brueghel's progress from the pictorial description of different animal species to the classification of animals). Also see Faber Kolb 2005, 2, 4. She concludes that the naturalistic description of animals formed the basis of Brueghel's approach, but that his search for a framework in which different species could be combined led to the formulation of a new kind of encyclopaedic representation in which the scientific, religious, courtly and artistic language of his time converged; Faber Kolb 2000, 212, 223; Faber Kolb 2005, 81.

96 We therefore disagree with Faber Kolb's remark that after *The entry of the animals into Noah's Ark* from 1613 Brueghel no longer found it necessary to introduce new animals into his paintings; Faber Kolb 2000, 214.

Bibliography

Aldrovandi 1599
U. Aldrovandi, *Ornithologicae hoc est de avibus historiae libri XII*, Bologna 1599.

Arents 2001
P. Arents, *De bibliotheek van Pieter Pauwel Rubens. Een reconstructie*, Antwerp 2001.

Balis 1968
J. Balis, *Van diverse pluimage. Tien eeuwen vogelboeken*, [Antwerp] 1968.

Belon 1555
P. Belon, *L'histoire de la nature des oyseaux, avec leurs descriptions, & leurs naifs portraicts retirez du naturel* (...), Paris 1555.

Borromeo 1625
F. Borromeo, *Musaeum Bibliothecae Ambrosianae*, Milan 1625.

Borromeo 1628
F. Borromeo, *Pro suis studiis*, Milan 1628.

Borromeo 1632
F. Borromeo, *I tre libri delle laudi divine*, Milan 1632.

Borromeo 2010
F. Borromeo, *Sacred painting. Museum* (K. Rothwell Jr. & P. Jones, eds.), Cambridge 2010.

Breslay 1571
P. Breslay, *L'Anthologie ou recueil de plusieurs discours notables, tirez de divers bons autheurs Grecs & Latins*, Paris 1571.

Clusius 1605
C. Clusius, *Exoticorum libri decem. Quibus animalium, plantarum, aromatum, aliorumque peregrinorum fructuum historiae describuntur*, Leiden 1605.

Collaert s.a.
A. Collaert, *Avium vivae icones*, Antwerp s.a.

Crivelli 1868
G. Crivelli, *Giovanni Brueghel, pittor fiammingo o sue lettere e quadretti esistenti presso l'Ambrosiana*, Milan 1868.

Cutler 2003
L.C. Cutler, 'Virtue and diligence. Jan Brueghel I and Federico Borromeo', *Nederlands Kunsthistorisch Jaarboek* 54 (2003), 203-227.

De Dene & Gheeraerts 1567
E. de Dene & M. Gheeraerts, *De warachtighe fabulen der dieren*, Bruges 1567.

Dijksterhuis 1975
E.J. Dijksterhuis, *De mechanisering van het wereldbeeld*, Amsterdam 1975.

Ertz 1979
K. Ertz, *Jan Brueghel der Ältere (1568-1625). Die Gemälde mit kritischem Oeuvrekatalog*, Cologne 1979.

Faber Kolb 2000
A. Faber Kolb, *Cataloguing nature in art. Jan Brueghel the Elder's Paradise landscapes*, Los Angeles 2000 (unpub. diss. Los Angeles).

Faber Kolb 2005
A. Faber Kolb, *Jan Brueghel the Elder. The entry of the animals into Noah's Ark*, Los Angeles 2005.

Foucault 1966
M. Foucault, *Les mots et les choses. Une archéologie des sciences humaines*, Paris 1966.

Galilei 1808-1811
G. Galilei, *Le opere di Galileo Galilei* (E. Albèri, ed.), Milan 1808-1811.

Gesner 1555
C. Gesner, *Historiae animalium liber III, qui est de avium natura*, Zürich 1555.

Gesner 1557
C. Gesner, *Vogelbuch, darinnen die art, natur und eigenschafft aller vöglen, sampt irer waren Contrafactur, angezeigt wirt*, Zürich 1557.

Golding 1987
A. Golding, *A moral fable-talk* (R.G. Barnes, ed.), San Francisco 1987.

Grafton 1992
A. Grafton, *New worlds, ancient texts. The power of tradition and the shock of discovery*, Cambridge, MA 1992.

Haupt *et al.* 1990
H. Haupt *et al.*, *Le bestiaire de Rodolphe II. Cod. min. 129 et 130 de la Bibliothèque nationale d'Autriche*, Paris 1990.

Jones 1988
P.M. Jones, 'Federico Borromeo as a patron of landscapes and still lifes. Christian optimism in Italy ca. 1600', *Art Bulletin* 70 (1988), no. 2, 261-272.

Jones 1993
P.M. Jones, *Federico Borromeo and the Ambrosiana: Art patronage and Reform in seventeenth-century Milan*, Cambridge 1993.

Cat. Lyon 1993
Cat. Lyon, Musée des Beaux-Arts, *Catalogue sommaire illustré des peintures du Musée des Beaux-Arts de Lyon. Écoles étrangères XIIIe-XIXe siècles. Allemagne, Italie et divers* (H. Buijs), Paris 1993.

Maselis *et al.* 1989
M. Maselis, A. Balis & R. Marijnissen, *De albums van Anselmus de Boodt (1550-1632). Geschilderde natuurobservatie aan het Hof van Rudolf II te Praag*, Tielt 1989.

Mason 2000
P. Mason, 'Of turkeys and men. Towards a historical iconography of New World ethnographic and natural historical representation', in: L. Tongiorgi Tomasi, G. Olmi & A. Zanca (eds.), *Natura-cultura. L'interpretazione del mondo fisico nei testi e nelle immagini,* Florence 2000, 63-90.

Mason 2007
P. Mason, '*Americana* in the *Exoticorum libri decem* of Charles de l'Ecluse', in: F. Egmond, P. Hoftijzer & R. Visser (eds.), *Carolus Clusius. Towards a cultural history of a Renaissance naturalist*, Amsterdam 2007, 195-219.

Molaro & Selvelli 2008
P. Molaro & P. Selvelli, 'The mystery of the telescopes in Jan Brueghel the Elder's paintings', *Memorie della Società Astronomica Italiana* 75 (2008), 282-285.

Paré 1971
A. Paré, *Des monstres et prodiges* (J. Céard, ed.), Geneva 1971.

Paulussen 1997
M. Paulussen, *Jan Brueghel d. Ä. 'Weltlandschaft' und enzyklopädisches Stilleben*, Frankfurt am Main 1997 (unpub. diss. Frankfurt am Main).

Potgieter 1600
B.J. Potgieter, *Wijdtloopigh verhael van tgene de vijf schepen (die int jaer 1598 tot Rotterdam toegherust werden, om door de Straet Magellana haren handel te drijven) wedervaren is* (...), Amsterdam 1600.

Prosperetti 2009
L. Prosperetti, *Landscape and philosophy in the art of Jan Brueghel the Elder (1568-1625)*, Aldershot 2009.

Quint Platt 1986
A. Quint Platt, *Cardinal Federico Borromeo as a patron and a critic of the arts and his 'Musaeum' of 1625*, New York & London 1986.

Seznec 1972
J. Seznec, *The survival of the pagan gods. The mythological tradition and its place in Renaissance humanism and art*, Princeton 1972.

Smith 2008
P.J. Smith, 'Sympathy in Eden. On *Paradise with the Fall of Man* by Rubens and Brueghel', in: C. Göttler & W. Neuber (eds.), *Spirits unseen. The representation of subtle bodies in early modern European culture*, Leiden & Boston 2008, 211-244.

Vondel 1617
J. van den Vondel, *Vorsteliicke warande der dieren. Waer in de zeden-rijcke philosophie, poëtisch, morael, en historiael, vermakelijck en treffelijck wort voorghestelt*, Amsterdam 1617.

Wittkower 1938
R. Wittkower, '"Roc". An Eastern prodigy in a Dutch engraving', *Journal of the Warburg Institute* 1 (1938), 255-257.

Woollett & Van Suchtelen 2006
A.T. Woollett & A. van Suchtelen, *Rubens & Brueghel. Een artistieke vriendschap*, Zwolle 2006.

Appendix

Number	English name	Scientific name
1	undetermined wader	
2a, b	lesser bird of paradise	*Paradisea minor*
3a, b	muscovy duck	*Cairina moschata*
4a, b	black crowned crane	*Balearica pavonina*
5	pheasant	*Phasianus colchicus*
6	hoopoe	*Upupa epops*
7	hawfinch	*Coccothraustes coccothraustes*
8	undetermined goose	
9	woodchat shrike ?	*Lanius senator*
10a, b	great tit	*Parus major*
11a, b	little owl	*Athene noctua*
12a, b	red-faced lovebird	*Agapornis pullarius*
13	barn swallow	*Hirunda rustica*
14a, b	red-billed toucan	*Rhamphastos monilis*
15	european goldfinch	*Carduelis carduelis*
16a, b	undetermined glossy starling	*Lamprotornis spec.*
17a, b	green-winged macaw	*Ara chloroptera*
18a, b	blue tit	*Parus caeruleus*
19a, b	rock thrush ?	*Monticola saxatilis*
20	undetermined bird of prey	
21a, b	grey parrot	*Psittacus erithacus*
22	rose-ringed parakeet	*Psittacula krameri*
23	bohemian waxwing	*Bombycilla garrulus*
24	canary-winged parakeet	*Brotogeris versicolurus*
25a, b, c	grey heron	*Ardea cinera*
26a, b	undetermined bird of prey	
27	brambling (male)	*Fringilla montifringilla*
28	chaffinch (male)	*Fringilla coelebs*
29	undetermined finch	
30	turkey	*Meleagris gallopavo*
31a, b, c	domestic fowl	*Gallus gallus*
32a, b	barn owl	*Tyto alba*
33a, b	yellow-crested cockatoo	*Cacatua sulphurea*
34a, b	eagle owl	*Bubo bubo*
35a, b	mute swan	*Cygnus olor*
36a, b	great spotted woodpecker	*Dendrocops major*
37	eurasian curlew	*Numenius arquata*
38	bullfinch ?	*Pyrrhulla pyrrhulla*
39	great cormorant	*Phalacrocorax carbo*
40a, b	magellanic penguin	*Speniscus magellanicus*
41a, b	purple coot	*Porphyrio porphyrio*
42	undetermined bat	*Chiroptera spec.*
43a, b	red-legged partridge	*Alectoris rufa*
44	pelican	*Pelecanus spec.*
45a, b	helmeted guineafowl	*Numida meleagris*
46a, b	peacock	*Pavo cristatus*
47	surf scoter	*Melanitta perspicillata*
48a, b	great bustard	*Otis tarda*
49	greater flamingo	*Phoenicopterus ruber*
50	great bittern	*Botaurus stellaris*
51	undetermined owl	
52a, b	common crane	*Grus grus*
53	yellow-headed amazon	*Amazona oratrix*
54	ostrich	*Struthio camelus*
55a, b	southern cassowary	*Casuarius casuarius*
56	undetermined eagle	
57	undetermined woodpecker	
58	bullfinch	*Pyrrhulla pyrrhulla*
59	common redstart	*Phoenicurus phoenicurus*
60	senegal parrot	*Poicephalus senegalus*
61a, b	blue-and-yellow macaw	*Ara ararauna*
62	black-capped lory	*Lorius lory*
63	blue-fronted amazon	*Amazona aestiva*
64	golden oriole	*Oriolus oriolus*
65	common kingfisher	*Alcedo atthis*
66	northern goshawk	*Accipiter gentilis*
67	eurasian jay	*Garrulus glandarius*

Appendix
Jan Brueghel the Elder, *Allegory of air*, 1621
Numbered image of the most important birds depicted.

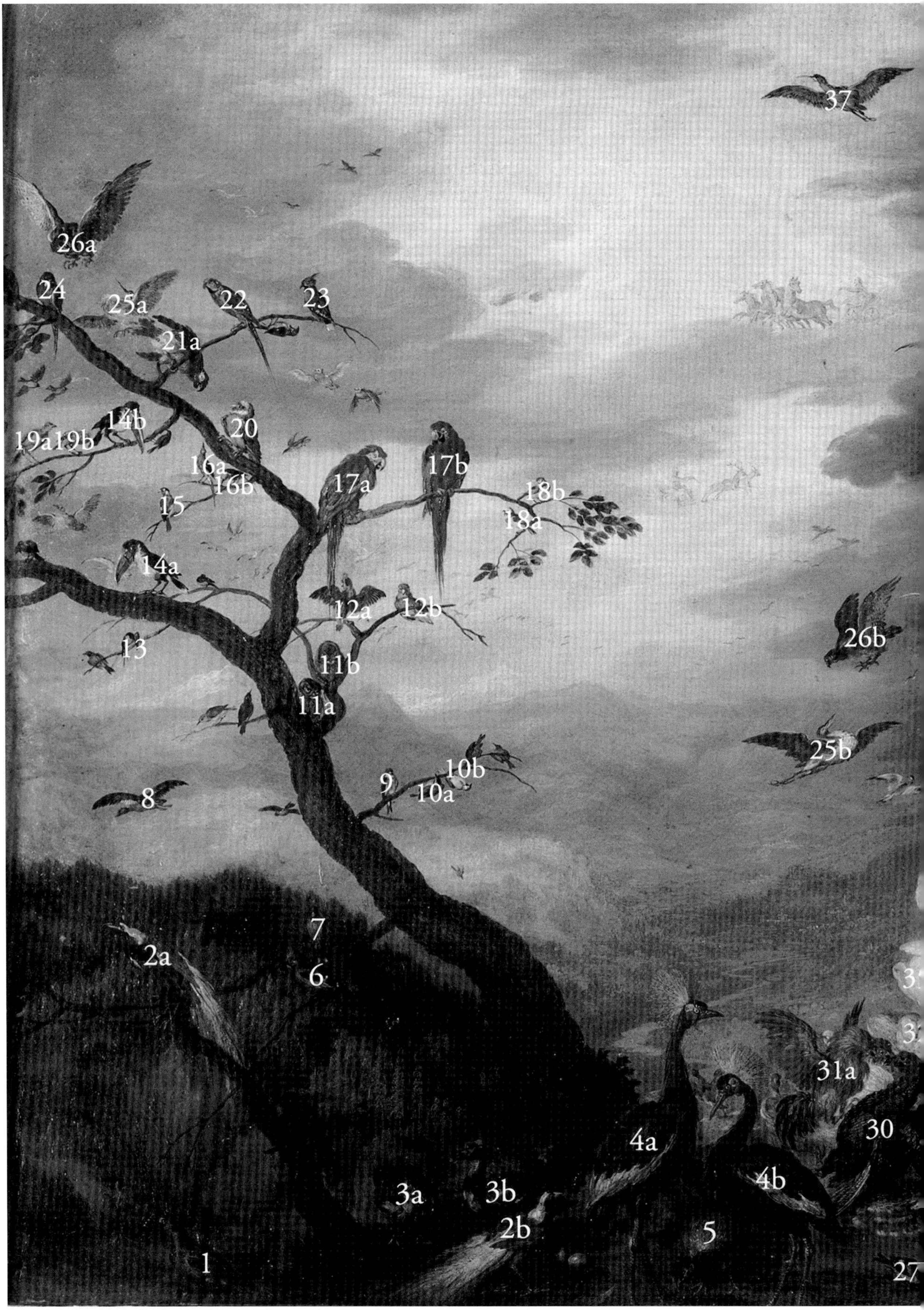

37
26a
24
25a
22
23
21a
14b
20
19a19b
16a
16b
17a
17b
18b
18a
15
14a
12a
12b
26b
13
11b
11a
25b
10b
9
10a
8
7
2a
6
31a
30
4a
3a
3b
4b
2b
5
1
27

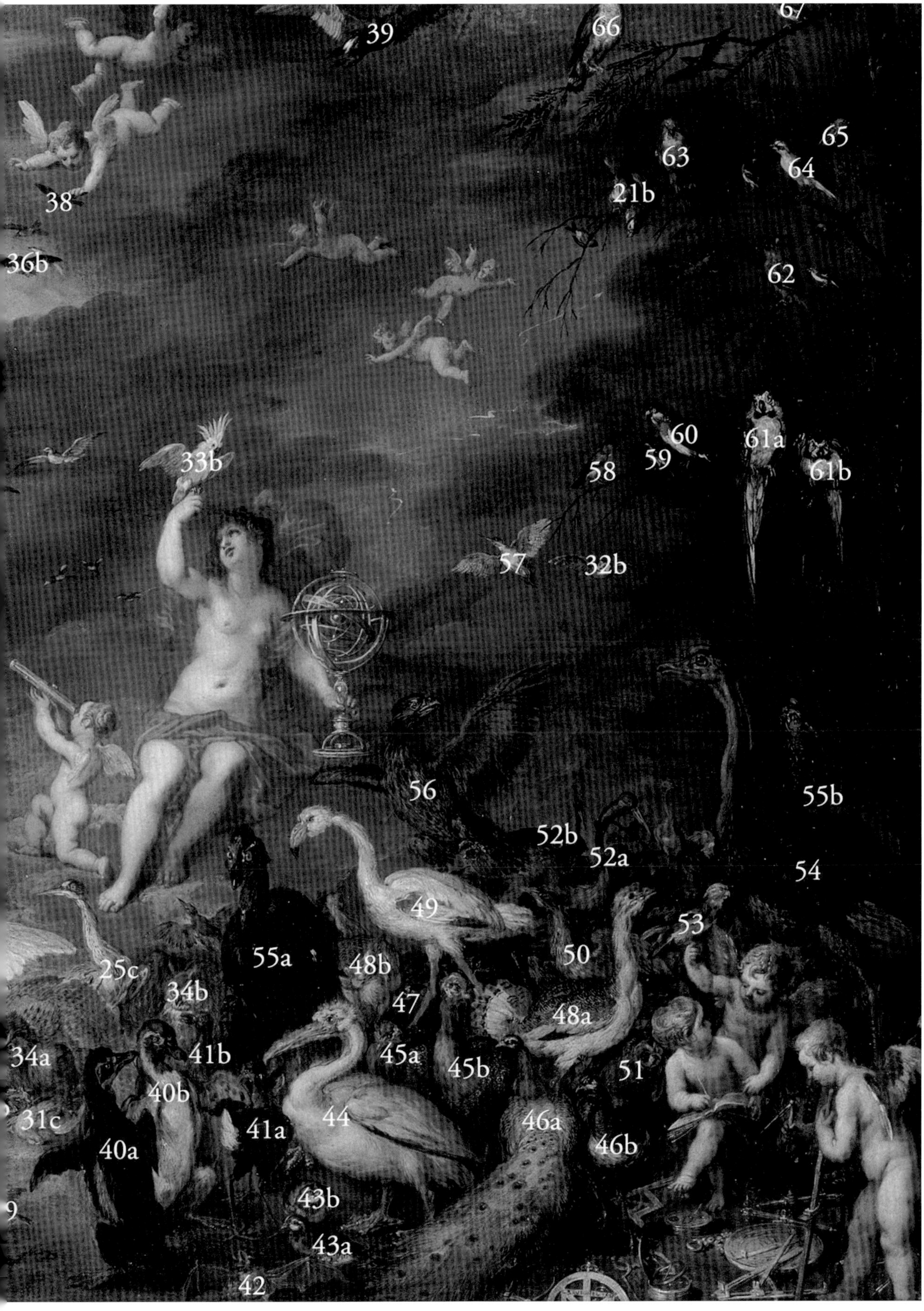
39
66
67
38
65
63
64
21b
36b
62
60
61a
33b
59
58
61b
57
32b
56
55b
52b
52a
54
49
53
50
25c
55a
48b
34b
47
48a
34a
41b
45a
45b
51
40b
31c
44
41a
46a
46b
40a
43b
9
43a
42

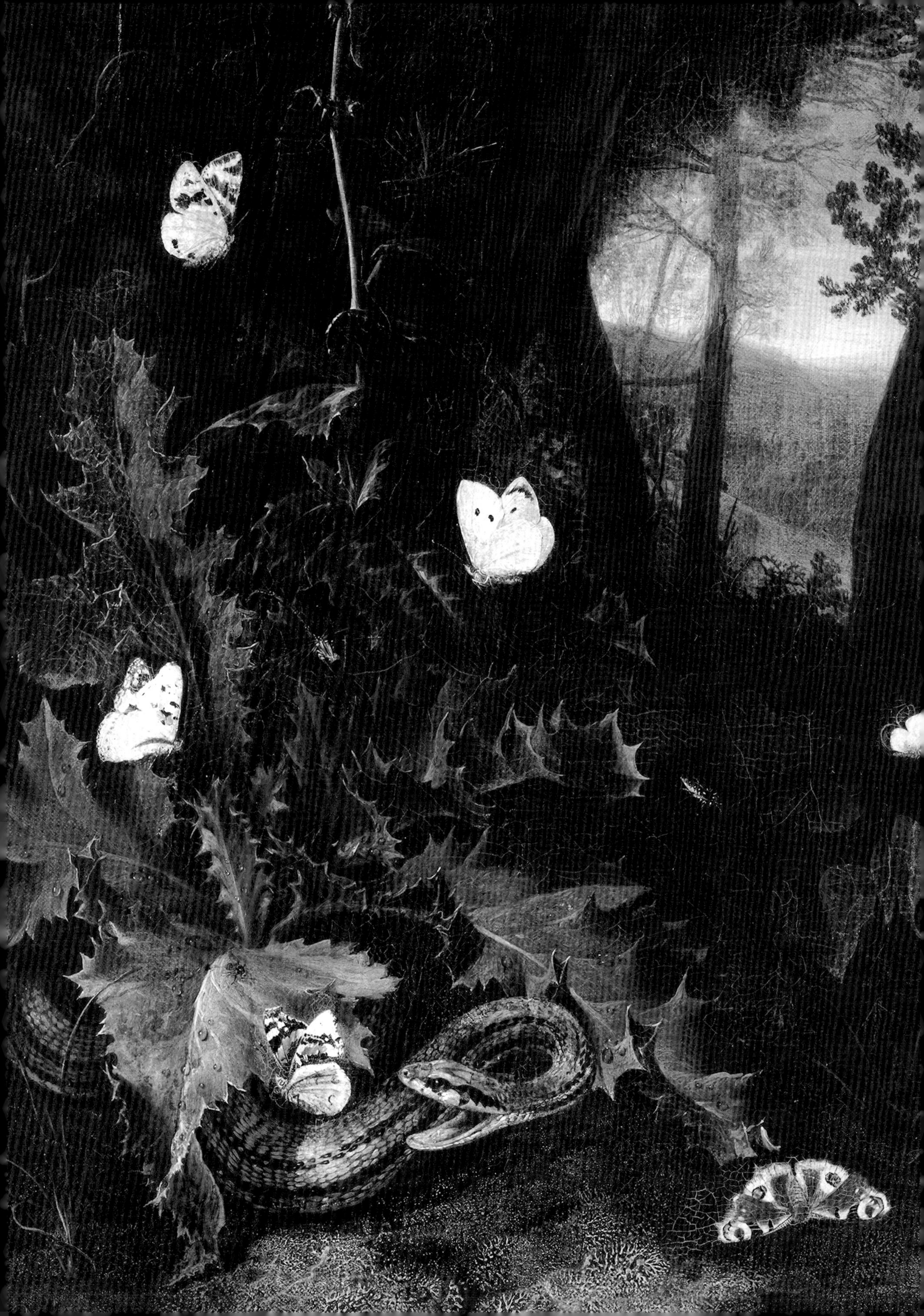

Painted poison
Venomous beasts, herbs, gems, and Baroque colour theory

Karin Leonhard

Her face still betrays its former classical features, though they appear somewhat rough and masculine. The Gorgon's severed head lies on the ground motionless or petrified, with a frowning forehead and half-opened eyes and mouth. Some carefully trimmed locks of golden hair curl at the temple, but otherwise the scalp is covered with thick strands of snakes writhing together in an agitated mass wherever they touch the ground. The bleak setting depicts the entrance to a grotto or the dark cleft of a rock. A small pool of blood has seeped onto the ground where some chthonic creatures – a lizard, a rat, a bat, some frogs – have emerged and now encircle the scene, mimicking the silent observers at a deathbed (fig. 1).

For a long time, the painting was thought to have been the work of Leonardo da Vinci (1452-1519). In his *Vita di Leonardo* (1568), Giorgio Vasari (1511-1574) gives an account of two famous versions painted by the artist at an early stage of his career.[1] Both examples are well known and inspired several early seventeenth-century painters. The account describes how a peasant had presented a wooden shield to Leonardo's father, having sought him out to have it painted. The father, in turn, asked Leonardo to complete the work, and the son decided on depicting a horrible monster:

> To do what he wanted Leonardo carried into a room of his own, which no one ever entered except himself, a number of green and other kinds of lizards, crickets, serpents, butterflies, locusts, bats, and various strange creatures of this nature; from all these he took and assembled different parts to create a fearsome and horrible monster which emitted a poisonous breath and turned the air to fire. He depicted the creature emerging from the dark cleft of a rock, belching forth venom from its open throat, fire from its eyes and smoke from its nostrils in so macabre a fashion that the effect was altogether monstrous and horrible. Leonardo took so long over the work that the stench of the dead animals in his room became unbearable, although he himself failed to notice because of his great love to painting.[2]

Vasari's description of Leonardo's second painting with a monstrous subject, then, is much shorter though not without significance for adaptations:

Detail fig. 9
Otto Marseus van Schrieck, *Sottobosco with grass snake, lizard and butterflies*, ca. 1665.

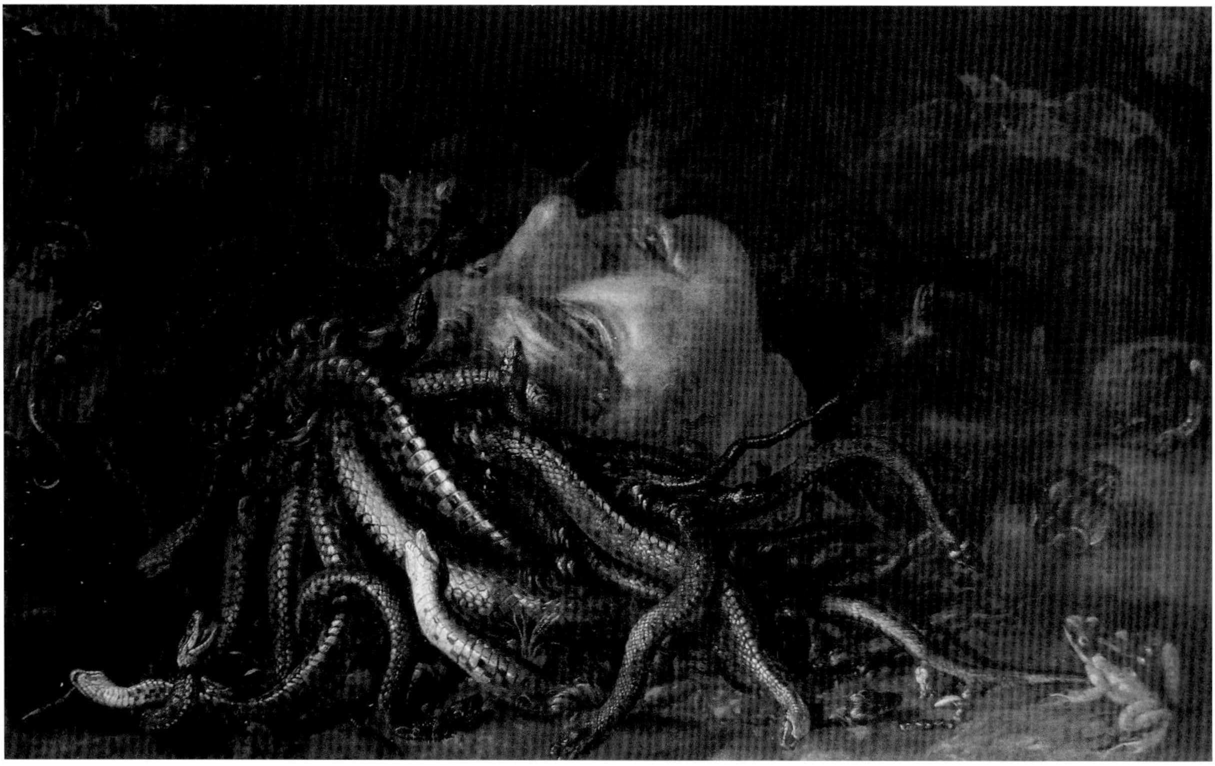

I

Artist unknown (Flemish), *Head of Medusa*, c. 1600
oil on wood, 49 x 74 cm, Florence, Galleria degli Uffizi.

> The fancy came to paint a picture in oils of the head of Medusa, with the head attired with a coil of snakes, the most strange and extravagant invention that could ever be imagined, but since it was a work that took time, it remained unfinished, as happened with almost all his things. It is among the rare works of art in the Palace of Duke Cosimo...

We now know that our painting is in fact of Flemish origin. Documents show that it was bequeathed to Ferdinando II de' Medici (1610-1670) by his former footman and *segretario di camera*, Ippolito de Vicq of Bruges, and entered his collection only in 1668.[3] What once was reckoned to be young Leonardo's masterpiece is now lost. Peter Paul Rubens (1577-1640) (presumably in cooperation with Frans Snyders [1579-1657]) and Caravaggio (1571-1610) painted their own versions of the subject, possibly to compete with the famous shield, though what they owed to Leonardo's painting is uncertain (figs. 2, 3).[4] Interestingly enough, as it provides us with an early record of the sinister attraction of the motif, in his autobiography of 1629-1630, Constantijn Huygens (1596-1687) gives an account of how he saw what must have been a copy of Rubens' famous *Head of Medusa* in the collection of Nicolaas Sohier (1590-1642) in Amsterdam.[5] In the early years of the seventeenth century, then, the subject matter of the 'head of Medusa' was clearly in high fashion. Besides its obvious reference to Leonardo's famous version of the motif, it was not least motivated by a growing interest in herpetology and toxicology in Europe's main cultural and scientific centres.

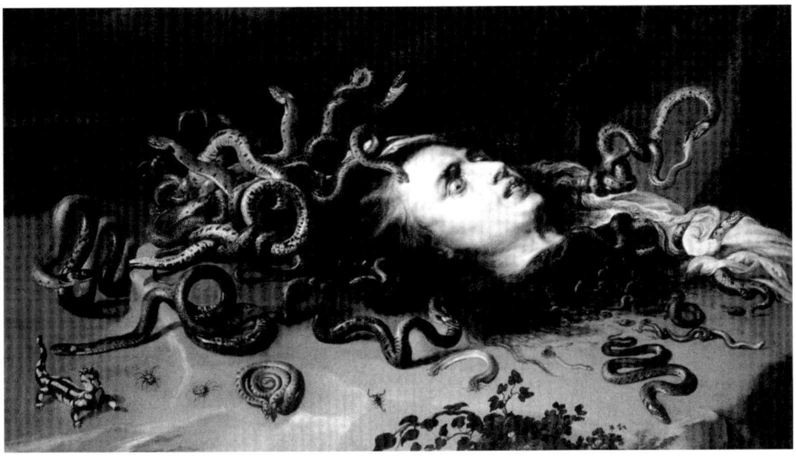

2
Peter Paul Rubens / Frans Snyders, *Head of Medusa*, ca. 1615-1616
oil on canvas, 69 x 118 cm, Vienna, Kunsthistorisches Museum.

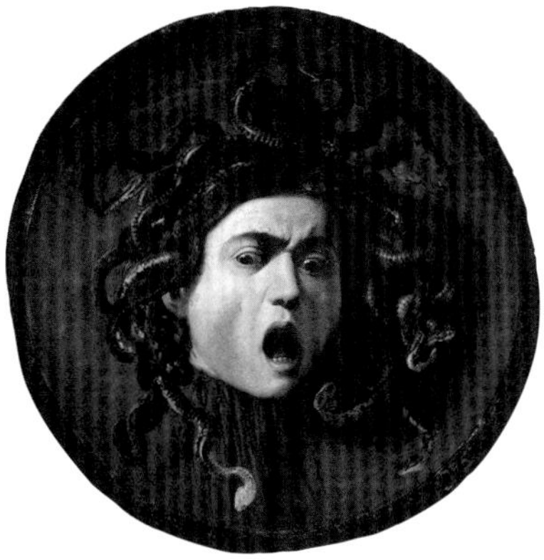

3
Caravaggio, *Medusa*, 1597
oil on wood covered with canvas, diameter 55 cm, Florence, Galleria degli Uffizi.

In early modern science, these two disciplines ranged, somewhat vaguely, somewhere between zoology and medicine. By around 1600, they had gained an immense actuality following for reasons still to be discovered. In Italy – or rather in Rome, Naples, Ferrara, Bologna, but especially at the Medici court in Florence and within the Accademia del Cimento – venomous beasts became a prime subject for investigation. Only later in the century would the toxicological discussion be taken on by the royal societies of Paris and London, as a science whose scope was only beginning to be defined, slowly turning into a discipline of its own. It is important to note that this new interest in toxicology could be observed both in the sciences and in the arts, in part as a result of a new empirical approach towards nature. But there is more at stake; things are not that easy.

Painting as 'pharmakon'

'The fear of death is what gives all witchcraft, all occult medicine, a hold. The *pharmakeus* (sorcerer, magician) is banking on that fear'.[6] – I quote from an essay on 'Plato's Pharmacy' by Derrida. In Plato's *Phaedrus*, Socrates compares the written texts Phaedrus has brought along to a drug (*pharmakon*) and explains that 'one thing held against the invention of the pharmakon is that it substitutes the breathless sign for the living voice and can no more answer for itself than a sculpture or inanimate painting'.[7] The pharmaceutical remedy is not only beneficial, but also 'harmful because it is artificial. The pharmakon goes against natural life'.[8]

In this essay, I will investigate the idea of painting as *pharmakon*, concentrating on some aspects of the complex relationship between painting, toxicology and pharmacology. It immediately brings into play other contexts of the sixteenth and seventeenth centuries in which poisons and their remedies were used – such as herpetology, herbology or gemology – thus mapping onto ways of 'acting' or 'affecting' the healing or damaging properties of natural substances both in medicine and in art. Such notions will also touch upon the ambivalent qualifications of what Derrida calls a 'god-doctor-pharmacist-magician'[9] (and I will add: –artist) who finds his place in the in-between, or dialectics, of life and death.[10]

We know that the word *pharmakon* in Greek can mean both 'remedy' and 'poison', but it also 'means to paint, not a natural colour but an artificial tint, a chemical dye that imitates the chromatic scale given in nature'.[11] Socrates uses the word to designate pictorial colour: pigments are *pharmakeia* 'which resemble the things imitated, and out of which the picture is composed' (Phaedr., 434a-b). As its polysemy permits the rendering of the same word by 'remedy', 'recipe', 'poison', 'drug', 'potion', 'philtre' or 'paint', one has already come to the conclusion that painting itself can act as a *pharmakon*, or, vice versa, that *pharmakon* is 'the drug of representation' or the 'doctoring of art'. I would like to stick to this very general notion to see how far we can develop it in respect of seventeenth-century image production and the use of 'paint' and colour. I have chosen some examples of an altogether neglected genre, the so-called *sottobosco* genre, once developed by Otto Marseus van Schrieck (c. 1620-1678) and others in exactly the context of discussion that concentrated on toxicological and pharmacological subjects covering the whole spectrum of science and magic. Sottoboschi depict venomous beasts, herbs and fruits – snakes, toads and fungi, for example – in eerie setting, such as dark underwood or clefts of rocks, and were, as a sub-genre, invented in Italy in the 1650s.[12] I will argue that the genre not only formally developed out of depictions of the female chthonic monster's head of Medusa – as it was, for example, famously depicted by Leonardo or Rubens – but that its iconography derives from pharmacological origins as well. The medical and mythical spheres converge in the heated contemporary debate on poisons and their antidotes, leading to questions of how images function in (and with) the natural world.

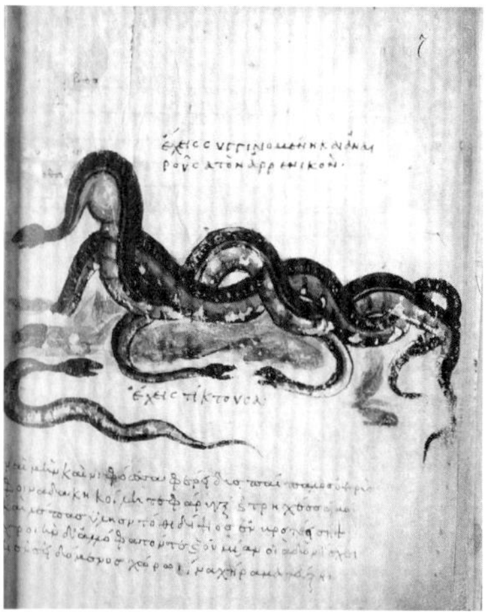

4
Poisonous creatures sprung from the slain Typhon
Nicander of Colophōn, *Theriaka and Alexipharmaka*, 10th-century manuscript (fol. 7r), Paris, Bibliothèque nationale de France.

Chthonic darkness and Apollonian light[13]

In Greek mythology, Python was the earth-dragon of Delphi and was represented as a serpent. He became the chthonic enemy of the later Olympian deity Apollo, who slew him, abandoned the place to purify himself and returned to Delphi crowned. From then on, the priestess of the oracle became known as the Pythia, after the place-name Pytho, which the Greeks explain as being named after the rotting (πύθειν) of the slain serpent's corpse in the powerful heat of Hyperion.[14] It is noteworthy, however, that older accounts even mention two dragons, confusing the dragon of Delphi, Python, with Typhon, the adversary of Zeus.[15] It is Nicander of Colophōn, a Hellenistic Greek poet of the second century BC, who states, in his poem on poisonous animals, that poisonous creatures and other plagues sprung from the slain Typhon (fig. 4).[16] But by contrast and in a much better-known version, in Apollonios Rhodius' *Argonautica*,[17] Ovid's *Metamorphoses*[18] and Lucan's *Pharsalia*[19], all sorts of venomous beasts, such as the vipers of the Sahara, were said to be born from drops of Medusa's blood. Some also say that Athena, having received the Gorgon's head from Perseus, gave Medusa's magical blood to the physician Asclepius, some of which was a deadly poison, and some of which had the power to raise the dead. With these conflated stories in mind we turn back to Medusa's severed head by the Flemish master in the Uffizi (once thought to be Leonardo) or by Rubens and wonder how far the ancient myths have embedded their iconographic imagery into new genres of Baroque painting.[20]

Man's fight against the serpent – symbolized in Python, Typhon and Medusa – strikes against the dominance of chthonic elements and the subterranean region; that is, against the union between Gaia and Tartarus, which brings mortality, death and decay to man. But 'no viper, nor harmful spiders, nor deep-wounding scorpion inhabit the glades of Clarus, for Apollo veiled its deep grotto with ash-trees and purged its grassy floor of noxious creatures'.[21] The passage is taken from a fragment by the aforementioned Nicander, of whose numerous works only two didactic poems survive: the *Theriaca* (on poisonous animals), and *Alexipharmaca* (on poisons and their antidotes). Nicander's poems are the oldest extant works on the subject to survive from antiquity, though they are thought to derive from Apollodorus, who wrote on poisons early in the third century BC. *Theriaca* deals with the bites of venomous snakes, spiders and scorpions, and how to remedy them. In his work, more than a dozen snakes and other creepy-crawlies (including the two-headed amphisbaena as depicted in Rubens' *Medusa*) as well as the effects of their poisons are described. Written in hexameters, the poem may have been dedicated to Attalus III, King of Pergamum, whom Galen ranked with Mithridates as an experimenter of poisons, which he administered to criminals in order to test the efficacy of their antidotes.[22] Nicander would become one of the firsthand references on the subject of poison, its prophylactics and remedies, from antiquity far into the seventeenth century, along with Dioscorides and Galen, who were both giving special attention to the pharmaceutical use of all materials, including herbs and minerals, against snake bites. Together with Pliny they also form the main reference for the toxicological iconography of sottobosco painting (figs. 5a, 5b).

Venomous beasts

Drops of Medusa's blood produced a generation of serpents. But how are we to understand the production of images of this kind? Does the archaic bloodletting also lead to any offspring in art? If I am not mistaken, the Flemish *Head of Medusa* in the Uffizi and the Rubens-Snyders painting of the same subject rank among the first paintings to initiate a new genre – the sottobosco painting of the 1650s – also understood as the depiction of a *Niederwelt*.

Let us see how the relationship could be retraced (fig. 2). The Rubens-Snyders head of Medusa was surrounded by a mass of coiled grass snakes (*Natrix natrix*) as well as a smooth snake (*Coronella austriaca*) – both of them non-venomous species that, nevertheless, are depicted in a frenzied scene, weaving about, rising up and down, their struggle to escape by attacking themselves making for a gruesome sight.

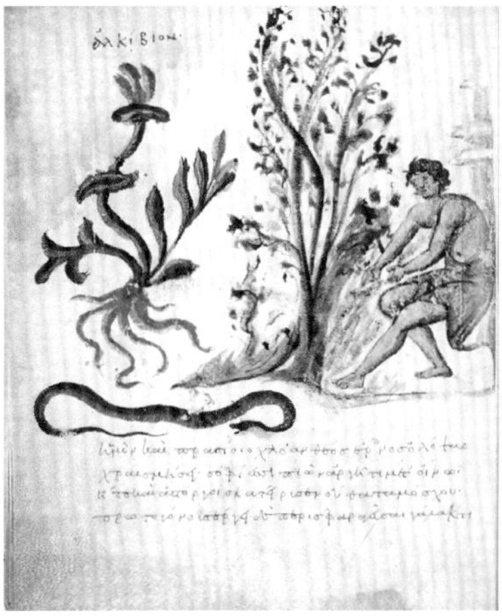

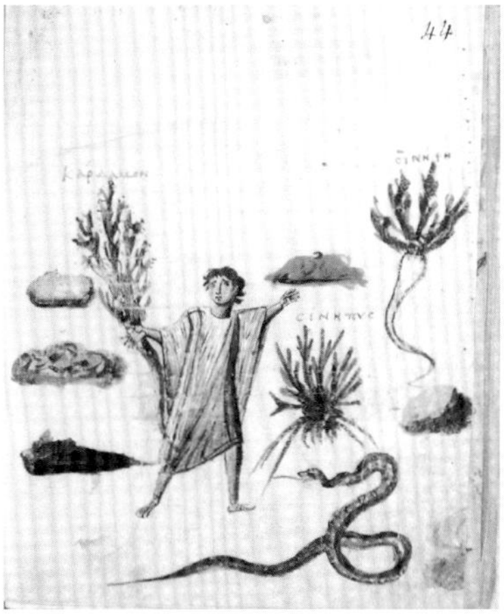

5a and 5b

Herbs and minerals against snake bites

Nicander of Colophōn, *Theriaka and Alexipharmaka*, 10th-century manuscript (fol. 16v and fol. 44r), Paris, Bibliothèque nationale de France.

Still there are more venomous creatures to observe. The yellow striped fire salamander (*Salamandra salamandra*) possesses toxic glands that usually coincide with the coloured sections of its skin. In addition, two spiders remind us of the poisonous bite of the tarantula.[23] They are joined by a black rock scorpion (*Urodacus manicatus*), and even by a two-headed amphisbaena, a fabulous creature depicted in Gesner's natural history, amongst others. So far, the painting features the classical repertoire of venomous beasts as reported in the toxicological texts by Nicander, Dioscorides, Galen, Pliny and their respective followers. We can clearly see how the creatures are generated from Medusa's blood – blood that slowly coagulates and transforms itself into snaking bodies. Poison as the acme of human suffering and disease derives from the slain head of the Gorgon (or Python or Typhon). We recall that it signifies a chthonic revolution against the sun-like authority of Perseus/Apollo/Zeus. The latter, on the other hand, have to act in the capacity of physicians, liberating and healing the diseased or, figuratively, chasing away the

6
Otto Marseus van Schrieck, *Sottobosco with fungi and saturnal animals*, 1655
oil on canvas, 38 x 47.8 cm, Florence, Palazzo Pitti, Inv. 5268 (provenance: Collection Cosimo III de' Medici, Galleria degli Uffizi 1687. Villa La Topaia 1699).

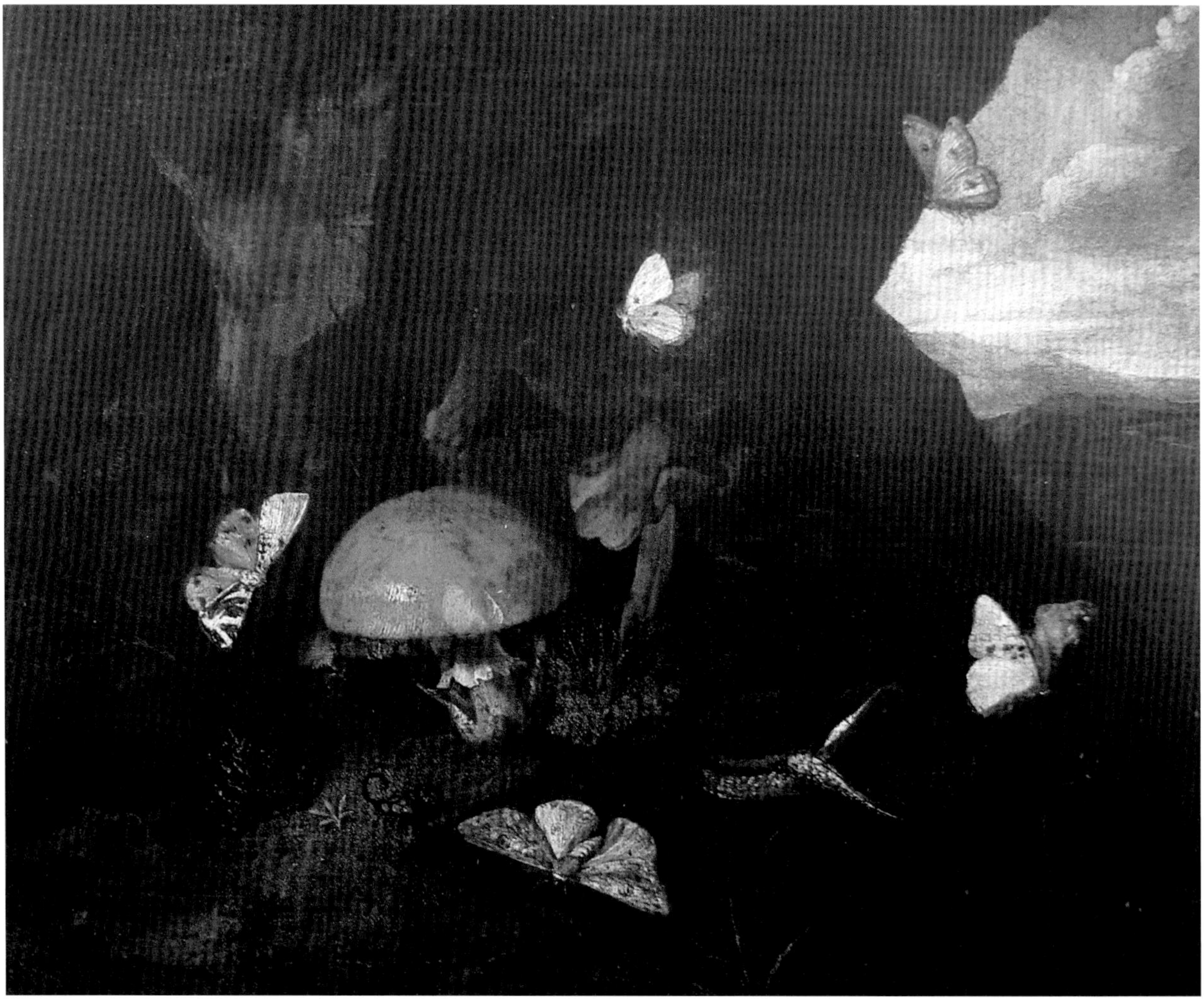

darkness from a clouded sky. Sharing this Apollonian task, art is bound to pharmacy.

Not just in medicine but also in art has Medusa left behind her kin. Otto Marseus van Schrieck's *Sottobosco with fungi and saturnal animals* – which is signed 'fecyt in Roma 1655 Ly to aug.' and once belonged in the collection of the keen naturalist Cosimo III de' Medici (1642-1723) at the Villa La Topaia – features a whole range of noxious creatures (fig. 6). The setting, again, is a small grotto or a cleft in a rock high up in the mountains, where a four-lined ratsnake (*Elaphe quatuorlineata*) slithers down the stones, its mouth already open, ready to catch the hovering moth before it. Four more moths circle around a cluster of mushrooms (*Amanita caesarea* and perhaps *Inocybe*) that have sprung up from the grotto's barren ground, some of them looming in the dark like animated beings, hardly recognizable. In addition, a black rock scorpion uses the damp grotto as its dwelling place. And while a glimpse through the cave's mouth signals rather fine weather, the conditions inside are cold, dank and unhealthy. The contrast is mirrored in the colouring as well. The rock forms a ragged demarcation line that separates light from darkness.

Even a short summary analysis of sottobosco painting suffices to show how much of its constitutive structure is still owed to ancient myths and traditional imagery. But herpetologists now also wonder whether it was the artist's skills in observation and the depiction of naturalism in art that advanced scientific illustration, such as in early modern natural histories. But how can we prove a relationship between archaic belief and keen naturalism? Do they converge at any point?

In 1648 Otto Marseus van Schrieck from Nijmegen had left the Low Countries for Rome. In his very early years as a painter, Otto Marseus practised on flower pieces in the style of the Utrecht school. However, in Rome he turned away from the production of colourful floral bouquets towards the depiction of a *Niederwelt*, 'denigrating' his flower bouquets by placing them on forest grounds or rocky soil and swamping them with insects and low-life creatures such as venomous snakes, spiders, lizards and toads. In 1652, Samuel van Hoogstraten (1627-1678) met Otto Marseus in Rome. In his *Inleyding tot het hooge schoole der schilderkonst* (Rotterdam 1678) he later recalled his stay and reports that Marseus was nicknamed the '*snuffelaer*' (ferreter) by the Roman *bentvuegels*, as he 'was always sniffing about everywhere for strange coloured and speckled snakes, lizards, caterpillars, spiders, butterflies and strange plants and herbs'.[24] Van Hoogstraten also wondered about his keeping a vivarium with all sorts of reptiles and amphibians.[25] His obsession was obviously already well developed at that point of his career. After his return to Amsterdam in the early 1660s, Marseus bought himself a house and piece of lowland which he called Waterryck and where he once more gathered venomous animals and insects for observation. His description of natural phenomena had now become as precise and accurate as scientific illustration.[26]

The genre of sottobosco soon enough became popular in Naples, where Paolo Porpora (1617-1673) and his teachers Giuseppe and Giovanni Recco (1634-1695; c. 1615-c. 1660) adopted the new style of low-life painting. In the Netherlands, it was taken on by artists such as Rachel

Ruysch (1664-1715), Elias van den Broeck (1649-1708) and Abraham Begeyn (1637-1697). But it was the Medici court in Florence around Cardinal Leopoldo de Medici (1617-1675) and Grand Duke Ferdinando II, and later around his successor Cosimo III, that was most deeply interested in the new subject.[27] Death by poisoning had been a threat for the Medici family for generations, and an interest in lethal substances and poison was cultivated at an early stage. Already around 1600, the keeping of venomous beasts had been fashionable at court, and Ferdinando I (1549-1609) was brought a pair of highly poisonous Libyan vipers (*Cerastes cerastes*), which were depicted by Jacopo Ligozzi (1547-1627) for Ulisse Aldrovandi's natural history. Within a short time the Tuscan dukes had also gathered the most prominent images of Medusa for their collection – paintings that were as much admired for their artwork as for their magical properties as personal protectors. Ferdinando II had then taken up the family interest by dealing in antidotes. In a pharmaceutical treatise – *Elixir Vitae* (1624) – he was called 'the new Mitridate' because of his keen involvement in toxicological matters.[28] Later, he initiated investigations at the Medicean Experimental Academy of Science that aimed to locate and assess the toxicity of viper's bile. The experiments were closely connected to the ceremony of theriac production at the Tuscan court – a medicament administered in pills or small doses as a prophylactic remedy against a poisoning attack, the plague, epilepsy or some other disease that could be supposedly prevented by the frequent use of the illustrious antidote. Within the Florentine context, a well-known reference to *theriaca* can already be found in Marsilio Ficino's *Three Books on Life* (*De vita libri tres*, 1489) where the drug was recommended to cleanse and prepare the body for the reception of astral virtues, especially for those inclined to melancholy or a sedentary and reflective life. As Beecher put it recently:

> But *theriaca* in any considered or analytical medical context was anything but a tried and proven remedy for simple maladies. It was and would remain (...) not only a pharmaceutical icon, but a drug under empirical siege, and a site for philosophical debate over the nature of the occult or unexplained pharmaceutical actions, as opposed to those deemed mechanical in accordance with the received doctrines concerning the humors. (...) *Theriaca*, itself, was composed of ingredients, each of which carried its own potential for astral fortification, and so might very well perform in the same talismanic way as the words, music, stones, jewelry, amulets, medals, and the multitude of other magically iconic objects which the *gens de bien* right up to the Medici kept around them in fifteenth-century Florence.[29]

Things had not much changed when Francesco Redi (1626-1697) – member of the Accademia del Cimento, court physician and superintendent of the pharmacy of the Medici court – started his famous investigations into the toxicity of viper bile in the early 1660s. Each year in June, hundreds of vipers were brought from Naples to Florence to be slaughtered and roasted to ashes for the preparation of *theriaca* or *panacea*. Redi had been assigned to watch over the preparation of the antidote when a suggestion came to him from the grand duke to make use of the situation and investigate the nature of the envenomation of poisonous serpents. Redi's experiments became a milestone in toxicology and were intended to be used as an instrument against the fabulous stories reported by ancient authors such as Dioscorides, Pliny, and so forth. Amongst them was the idea that leaves of certain trees and plants are shunned by vipers; Redi could not prove it. He also dispelled the myth that vipers drank wine and shattered wine flasks, that human saliva poisoned vipers and that venom's strength was determined by the viper's diet, amongst other age-old beliefs. In 1664, his *Observations on vipers* (*Osservazioni intorno alle vipere*), a document 'unmasking of the truth', as Redi triumphantly called it, appeared in print.

At this point, we might have come to the conclusion that ancient myths and traditional beliefs had been left behind and that investigations into these matters directly led towards a scientific discipline, namely toxicology, or herpetology. True enough, Redi effectively applied the Accademia del Cimento's experimentalism to the natural sciences, and, of course, in so doing he acted as a great reformer. The modern reliance on experiments and the power of observation, however, did not mean that older beliefs had been abolished. This is demonstrated by subsequent events, since Redi's ideas did not gain full acceptance until one century later.[30] But it is not so much the opposition to progressive thought by some members of the scientific community, but the heterogeneity of cultural tradition itself in which great changes are always in some way connected to the legacy of the past that helps us to correct the predominant image of a slow but steady march of empiricism in the seventeenth century. Redi himself worked as much *against* as *with* a foil of historic traditions that were transported in oral or written language, or by the imagery, such as in herbals or natural histories, etc.[31] In a fulminant article, Jay Tribby offers a close reading of Redi's *Osservazioni* and analyses the conflicting relationship between an aurally experienced past and a visually experienced present, then quite openly displayed at the Tuscan court (fig. 7).[32] In fact, Redi played with and against the word-pictures offered to him by the statements of the ancients. When he actively tested the relationship between ancients and moderns by restaging the circumstances under which certain assertions had been made and by comparing their results to his, what happened was not so much the glimpse of a future science, but the close exchange between past and present. Redi is at least as much debating with Nicander, Dioscorides, Pliny or Galen as with his fellows at the Royal Society in London or with Moyse Charas (1619-1698), his opponent at the Académie royale des sciences in Paris.[33]

If we have paused on a figure of exchange, it is because it is significant for an understanding of the importance of visual experience in early modern sciences in general, and at the Tuscan court in particular. Visual experience, then, does not simply mean that 'what you see is what you get', because there is no scientific or indeed cognitive image resulting from pure optical perception. Rather, it involves a comparative

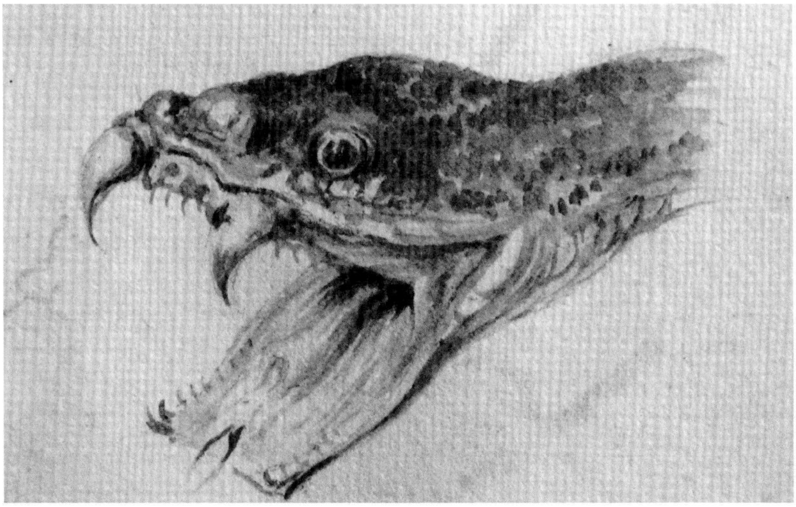

7
Stefano della Bella, *A viper's head*, c. 1660s
watercolour on paper, Florence, Gabinetto Disegni e Stampe degli Uffizi.

perception that perpetually shifts between historical images and natural objects, or between past and present experiences. But does it also teach us about the relationship between art and science? We may ask ourselves, for example, if the genre of sottobosco painting forms an intricate fusion of the old and the new. When it comes to the disquieting topic of human danger and mortality – expressed in imageries of poison and their antidotes or man's fight against the serpent – it hovers between myth and science as its two main semantic points of origin. If we seek to situate the genre, then it occupies a place somewhere between a tradition of images and a new cultural configuration in which it partakes and which it actively accounts for.[34]

Painted poison

The way in which Baroque painting makes use of venomous creatures soon enough develops into a formula. Since a double structure of tradition and perception – myth and science – is part of the concept or even initiated it, the creatures themselves quite aptly linger in some sort of limbo between darkness and light (fig. 8). Marseus' Brunswick painting, for example, is well known for its eerie atmosphere and sinister setting. On the other hand, one cannot help but recognize the sharp observational skill with which the painter describes the plants, reptiles and amphibians, hence showing an educated attitude and a scientific interest in biological (low) life forms. In this new light, they become rather awesome creatures and turn into aesthetic objects, and it is not least because of the patterns of the snakeskin, the vermilion red of a mushroom's cap or the colour of a poisonous flower or berry that venomous beasts and plants naturally become 'painted' and redoubled in art. In the Apollonian realm, danger is transformed into attraction, and a toxic colour into an aesthetic value.

Art and medicine, painting and pharmacy make fine partners. We know that discussions of the specific qualities of herbs and gems mostly

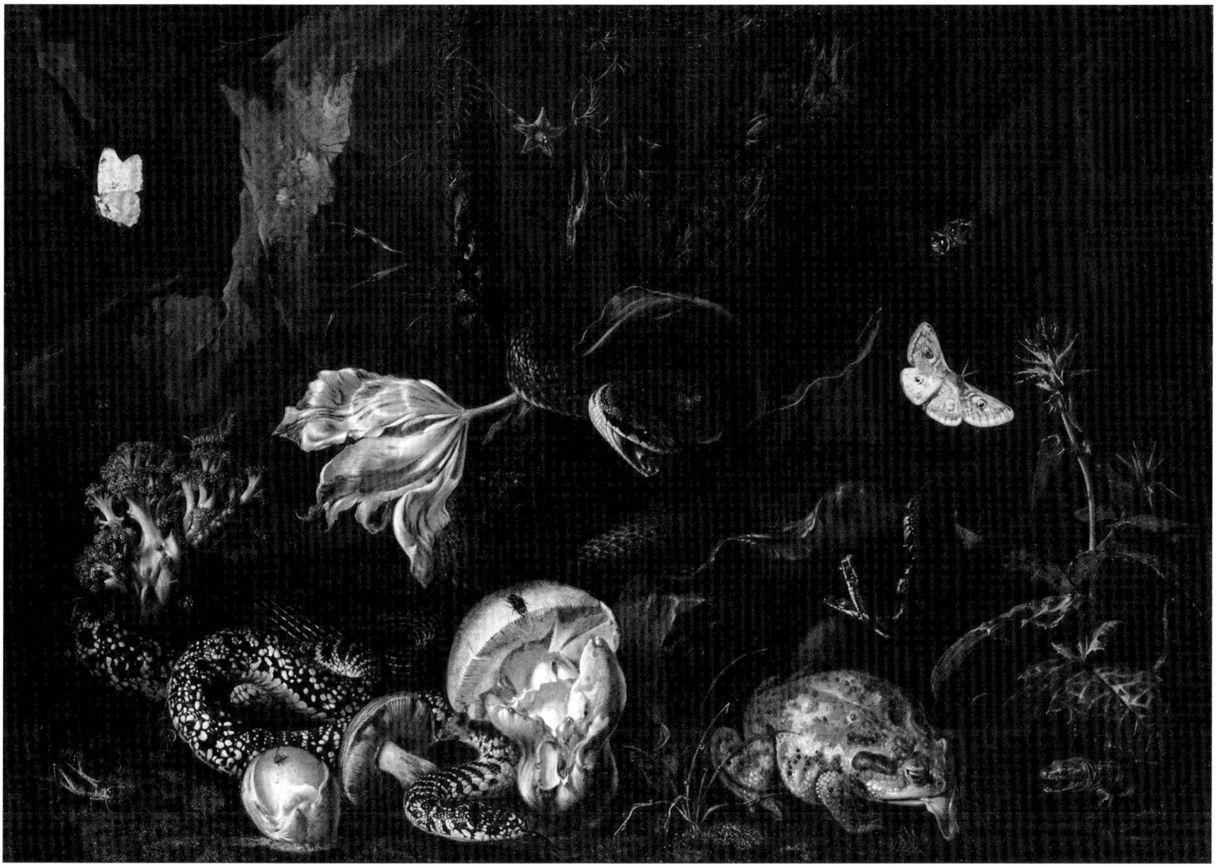

8

Otto Marseus van Schrieck, *Sottobosco with fungi, toad and snakes*, 1662
oil on canvas, 50.7 x 68.5 cm, Brunswick, Herzog Anton Ulrich-Museum, Inv. 431.

took place in a medical context, as the physician had to understand the meaning of health as a means of balance between extremities of the four humours (or basic substances). This commonly held view, again, was closely related to the theory of the four elements, so that earth and fire, water and air played important roles in prophylactic and curative treatments. Possible remedies also included the substantial transformation of former venoms into medication, as a seventeenth-century natural scientist noted:

> As for example, a toad for itself is mere poison, yet if first dryed in the sun, (...) and to burn to ashes, if these ashes be laid upon a place stung or venomously bit, or the like, you shall see one venom attract the other, and the place freed from venom. (...) So antinomy, though full of venomous arsenic qualities, after his poison is taken away, may be transmuted into medicine, which never more for the future retaineth poison, but be so able and efficatious as to attract the poison of diseases to itself and expel it out of the body, with it self[35]

– according to Dioscorides' dictum, 'The serpent, drunk in wine, heals its own bite'.[36]

Such conclusions confirm the hierarchy of the natural order, but simultaneously turn it upside down, poison and low-life creatures in turn becoming themselves the antidotes. This is entirely in accordance with homeopathic principles, that poisonous substances are rendered preventions and cures when administered in small or diluted doses. Ancient writers, therefore, have allowed even the most primitive or noxious creatures a place within medical treatment.[37]

One point at which art and medicine, painting and pharmacy meet is through colour – colour understood as the property of a substance, or a *pharmakon*. The decisive point at which all these issues coincide is expressed in a passage by Pliny:

> We cannot sufficiently admire, and, indeed, adore, the wondrous providence displayed by Nature. She had given us, as already shewn, plants soft to the touch, and agreeable to the palate; in the flowers *she has painted the remedies for our diseases with her varied tints,* and while commingling the useful with the delicious, had attracted our attention by means of the pleasures of the eye.[38]

It is in this light that we must now turn to colour and its natural pigments, which are obtained from plants, herbs and gems by grinding or distilling. Matteo Zaccolini (1574-1630), for example – whose *De Colori* (1618) was one of the earliest treatises devoted entirely to the principles of colouring in painting and who was also writing on the use of colour as an antidote in the tarantella dance – started his treatise by stating exactly this: that the generation of colours in nature and in art is to be discussed together, as colours are as much generated naturally by the mixture of the elements as they can be recovered from nature for the painters' use. The earth and the minerals ('de quali serve alli Pittori per dipingere l'operi loro, che toccarli tutto il giorno con Mani') make fine examples, but also ('vicino della superficie della Terra') in the vegetation on the surface of the earth, in herbs, plants, trees and fruits, colours can be seen and found. Because painting colour is extracted and ground from organic and inorganic substances, it is at the same time natural colour. He then describes the generation of colours in the four elements, beginning with the earth and the minerals that he equates with painting pigments, and continuing with water, air and fire. When applying colour to the canvas, a painter also has to deal with substances and their qualities (and with respect to Socrates we could state, with *pharmakeia*). Art and medicine have indeed contracted an inextricable alliance. They both draw upon the various properties of minerals and plants that, amongst others, are displayed in their colours and hues. The relationship is confirmed by the fact that earth, mineral and biological pigments were as much used for medical drugs, potions or plasters as for the preparation of tempera or oil colour on the painter's palette. We know that in Florence, for example, until the end of the sixteenth century, painters belonged to the Guild of Doctors and Apothecaries (*Arte dei Medici e Speciali*), as they bought their pigments from the apothecaries: the guild's banner depicted the Lady and child – Saint Luke was believed to be have been a doctor and also the painter of the first portrait of the Virgin Mary.[40] In establishing an analogy between pharmaceutical and aesthetic values, Pliny, Zaccolini, Marseus and others bring the healing (and damaging) properties of painting into focus, which, at its extreme, turns into apotropaism or antidotism.[41]

Herbs

Let us discuss this in an example that we have met before. In Marseus' *Sottobosco with fungi and saturnal animals* (fig. 6) we notice a contrast of colouring that is also marked by the ragged outline of the cave's mouth. As if a frame, the rock gives way to a view towards fading hillsides and a blue summer sky that is already obscured by clouds. Inside the grotto, the colours are toned down to tenebrous tinges of brown and black, with one exception: the *Amanita*'s orange cap provides a vivid colour accent. Commonly known as 'Caesar's mushroom', it ranks amongst the most highly prized edible fungi (its name derives from being the favourite of the Roman emperors), though it can easily be confused with other deadly members of the *Amanita* family – for example, the poisonous fly agaric (*Amanita muscaria*) – so that the painting's iconography falls back into ambivalence.[42] On the whole, and to the naked eye, there is no clear-cut distinction between edible and poisonous fungi. Dioscorides, whose book on poison (*Materia Medica*, VI) was keenly read throughout the sixteenth and seventeenth centuries, thought that it could also be the location or surrounding substances that made mushrooms unpalatable or poisonous.[43] Pliny has repeated his dictum at length:

> In general, these plants are of a pernicious nature, and the use of them should be altogether rejected; for if by chance they should happen to grow near a hob-nail, a piece of rusty iron, or a bit of rotten cloth, they will immediately imbibe all these foreign emanations and flavours, and transform them into poison. (…) There are other circumstances, too, which render them noxious; if they grow near the hole of a serpent, for instance, or if they should happen to be breathed upon by one when just beginning to open; being all the more disposed to imbibe the venom from their natural affinity to poisonous substances.[44]

We can also cite Nicander's influential text on the avoidance of poisonous and mouldy places:

> Let not the evil ferment of the soil injure a man; it will often swell up in his chest, at other times it will choke him, when it is fostered over the Viper's coil deep in its lair, sucking up the monster's venom and the noxious breath from its mouth. This is the evil ferment which they call Fungi in general, for to different kinds different names have been assigned.[45]

Overall, the body of fungi was characterized as cold, spongy and corrupt, and Baroque herbals and natural histories often paraphrased Dioscorides' dictum of their besetting nature when taken: 'Most of them do suffocate and strangle the eater'.[46] Galen, too, takes a rather negative view of fungi in the diet, with the possible exception of Caesar's mushroom (*Amanita caesarea*), which is edible but wich, because it belongs to a genus with so many toxic species, is best avoided by all but the most expert mycologists.[47] We have already seen that the *Amanita* features prominently in Marseus' sottobosco. Part of its iconography centres on the *Amanita*'s orange cap that is a sign both for pleasure and well-being, and for menace, death and decay, depending on its location and immediate environment. Within this context, it is interesting to see that the mushroom's cap is not only vermilion in colour but literally made of vermilion – a bright orange-red pigment permanent in oils, which was said to become redder the more it was ground, and suffered from its hazardous poisonous nature.[48] Together with orpiment (otherwise known as king's yellow or royal yellow) – an arsenic sulphide, which could be quite lethal for the careless artist – the pigment was used to build up the lustre of the Caesar's mushroom.

Such notions offer us a first glimpse into the complex relationship between poisonous substances and Baroque colour theory – colour being not only an outward appearance but also the indicator of an inner quality:

> All sorts of mushrooms are nothing but the swellings of the earth, occasioned by a viscous humidity of the earth and trees, for Luna is Lady of all Moisture. (...) Some think that the pale coloured are the most hurtful because they have the more Lunar poison in them.[49]

For a long time, mushrooms were considered a plant and therefore found their way into ancient and early modern herbals. But they were also regularly listed in manuals for dye work, since the organic compounds of mushrooms produce strong and vivid colours. Dyeing with mushrooms and lichen has a long history and was still very much popular in the sixteenth and seventeenth centuries; see, for example, Giovanni Ventura Roseto's *Libro di tentoria intitolato Plicto* (Venice 1540 and 1565). When used for dyeing wool and other natural fibres, the whole range of the colour spectrum could be achieved. Only the invention of synthetic colours put an end to mushroom dyes.[50]

During the seventeenth century, a huge output of pharmaceutical literature stimulated the market, from expensive herbals to cheap 'daily companions' showing 'the magical and physical vertues of trees and herbs, and their vertues for curing of most diseases'.[51] We further register a boom in the publication of 'books of secrets' that explained the arts of drawing, colouring, painting and engraving, of ink-making, gilding, or lacquering. Often enough, they were bound together with books on dyeing and distilling: 'Notwithstanding, the so great variety of Colour in the Herbs, Roots, Leaves and Flowers from whence they were distilled'.[52]

> Colours are either merely tinctures of vegetables, or substances of minerals, or both: (...) Vegetables are rootes, juces, berries, and such like things as grow out of the earth. Minerals are such as are dig'd out of the earth, as earth, and stones, &c.[53]

Herbals informed their readers about the specific virtues of plants and herbs, and their 'cracht en werckinghe' (power and effects), as we read, for example, in Rembert Dodoens' *Cruyde-boeck* (1554), of which we know Otto Marseus owned a copy.[54] A 'table wherein is conteyned the nature, vertue, and dangers, of all the Herbes, Trees, and Plantes' completes this book. Besides six entries on how 'to drive away serpents' (by burning, chewing or rubbing in aromatic herbs), almost seventy herbal remedies are listed 'against the biting of serpents', 'against poison of vipers and serpents' or 'against all poison'.[55] Amongst them, plants such as the willow-herb, the elderberry (*Sambucus*), watercress

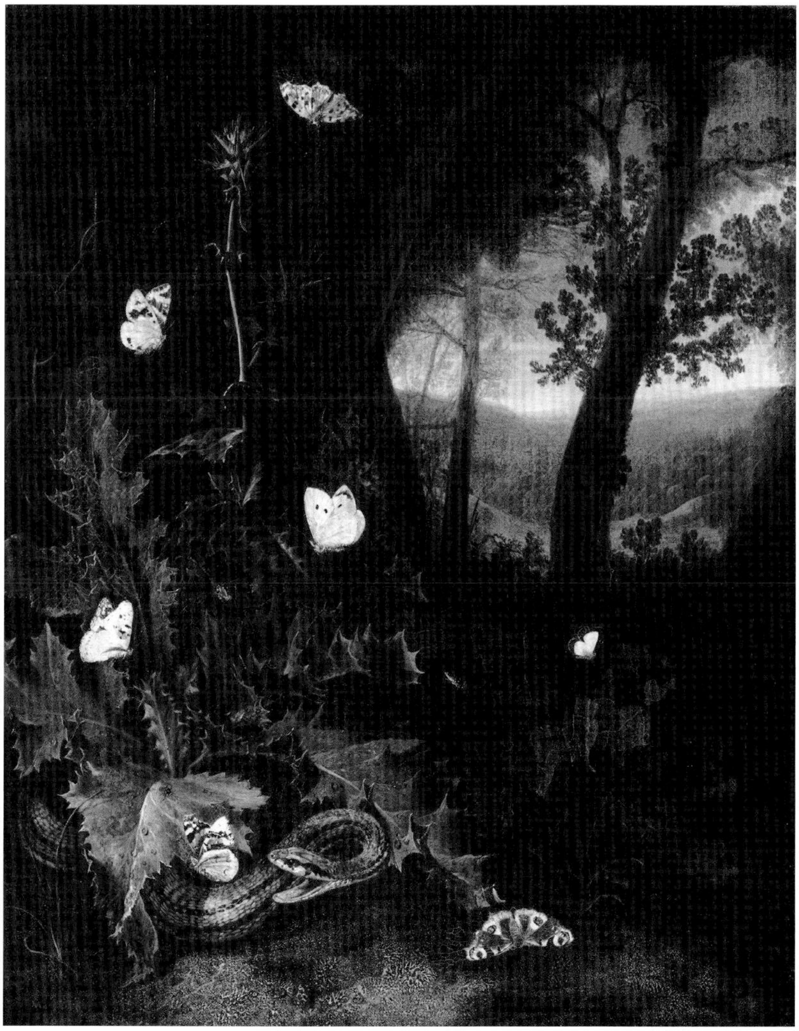

9
Otto Marseus van Schrieck, *Sottobosco with grass snake, lizard and butterflies*, ca. 1665
oil on canvas, 68.4 x 53 cm, Schwerin, Staatliches Museum, Gemäldegalerie.

(*Nasturtium*), or the monk's pepper tree (*Vitex agnus-castus*) 'driveth away serpents, and all other venomous beastes creeping on the ground,'[56] so that 'no venomous beastes have the power to hurt'.[57]

Otto Marseus van Schrieck must have been a keen reader of this and other herbals. Dodoens' *Cruyde-boeck* tells us that cyclamen plants are said to help against snake bites when they are dried and their powder drunk in wine, and indeed they feature in Marseus' *Sottobosco with grass snake, lizard and butterflies* in Schwerin (fig. 9).[58] We also read that the root sap of the arum and the bistort, themselves ominous shade plants, would be avoided by snakes when it is applied to the hands: Marseus paints them several times, the poisonous arum most prominently featured in the Schwerin painting of 1676 in dark and earthen tones (fig. 10).[59] The mayflower we find in a sottobosco, signed and dated 1668, that once belonged to Cosimo III (fig. 11);[60] the toxic digitalis, or foxglove, popularly known as dead man's bells or witches' gloves, in a painting in Göttingen (fig. 12);[61] and the red poppy, together with the arum (that now reveals its lily's cluster of poisonous berries) in an early painting in Florence (fig. 13).[62] In the *Sottobosco with orache, digitalis, poppy and morning glory*, once in the Dresden collection, a now recognizable assembly of chthonic creatures looms around a flower bouquet of notably dangerous plants: once ingested, they will induce oblivion, sleep, death and decay. Only the bright morning glory promises a glimpse of heavenly blue, despite it struggling hard to free itself from the meshed entanglement.[63] There are many more examples to be found. Marseus has

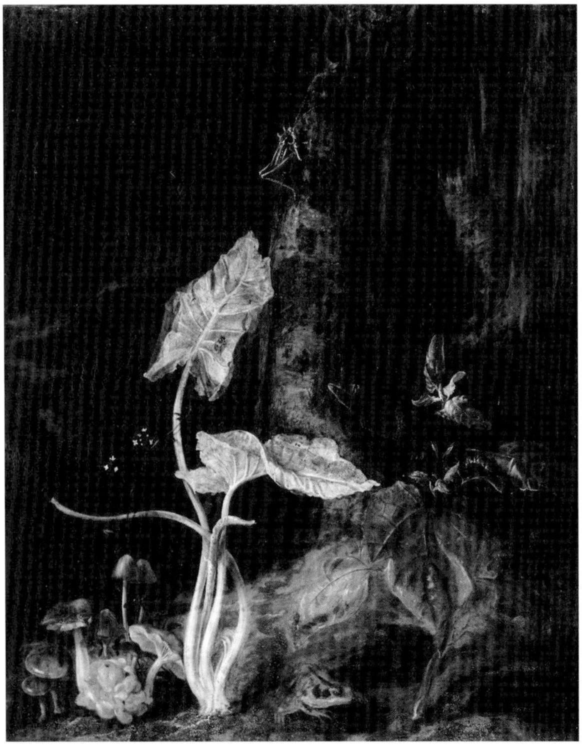

10
Otto Marseus van Schrieck, *Sottobosco with arum*, 1676
oil on canvas, 54 x 47 cm, Schwerin, Staatliches Museum, Gemäldegalerie.

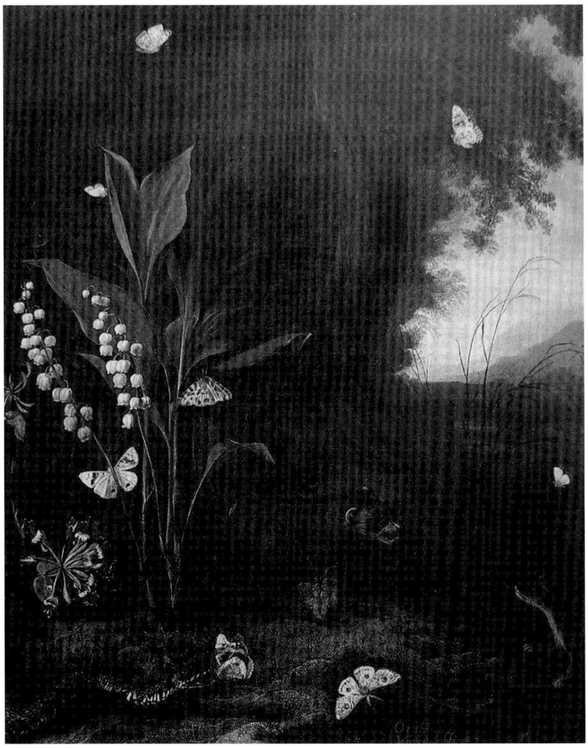

11
Otto Marseus van Schrieck, *Sottobosco with mayflower*, 1668
oil on canvas, 81.5 x 67.5 cm, Florence, Galleria degli Uffizi (provenance: Cosimo III de' Medici).

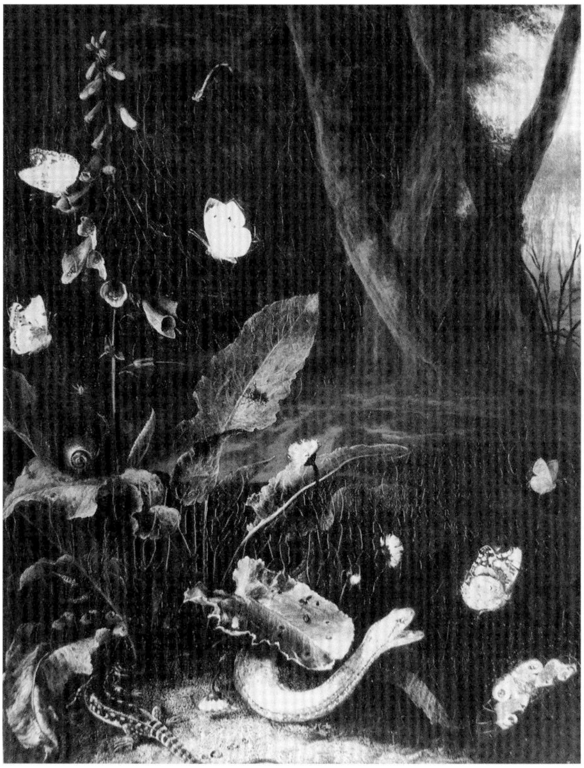

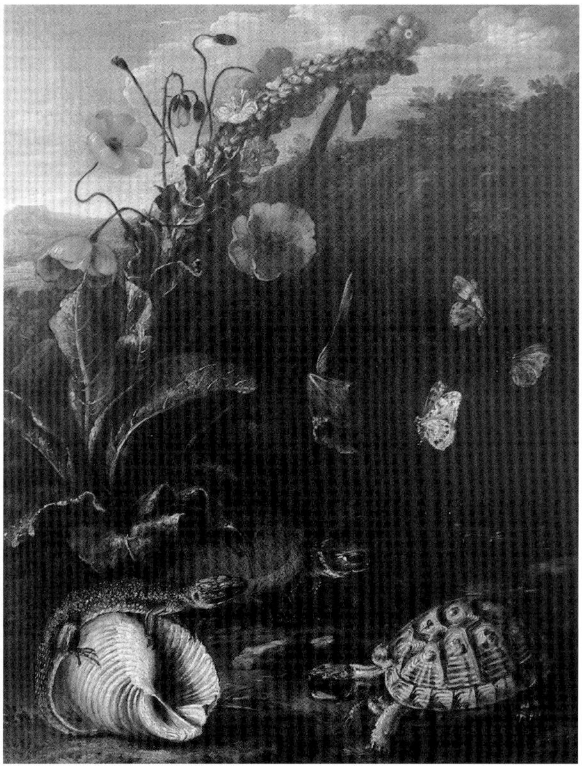

12

Otto Marseus van Schrieck, *Sottobosco with toxic digitalis*, 1667
oil on canvas, 58 x 45.5 cm, signed OTTO Marseus 12.7.1667, Göttingen, Kunstsammlung der Universität Göttingen, Gemäldegalerie.

13

Otto Marseus van Schrieck, *Sottobosco with red poppy and the arum*, 1655-1660
oil on canvas, 64.5 x 49 cm, Florence, Palazzo Pitti.

arranged poisonous plants and their herbal cures in endless combinations, just like a pharmacist who knows about the different properties of herbal plants and is determined to make use of them.

Amongst the hot and astringent herbal remedies for snake bites, however, the lady's thistle plays the central part. According to an inventory drawn after his death, we find that more than ten paintings showing 'een Maria distel' accompanied by a 'slang and hagedissen' (snake and lizards), both finished and unfinished, remained in Marseus' house. The same source provides information about the aforementioned 'herbarium van Doddenees', and also about 'een hooft van Medusa' (a head of Medusa) that formed part of the painter's estate (fig. 14). In addition, the inventory repeatedly lists paintings that prominently feature the amaranth, a popular herb whose bitter decoction is, as we read in the *Cruyde-boeck*, helpful against the bite of venomous beasts.[64] We find a similar entry concerning the lady's thistle: 'The seede is given with great profite to suche as are bitten with serpentes, or other venomous beastes'.[65] The orache plant, in turn, makes another regular feature in Marseus' sottoboschi. Dodoens describes its moist and cold nature and cautions the reader against regular use, as it was 'causing divers spottes, freckles and pimples'.[66] As with mushrooms, the colouring properties of plants and herbs were much discussed at that time. Since the Middle Ages, the orache was regularly used as a green hue, the amaranth and the thistle as a yellow colourant. Strong dyestuff was borne from lichen and moss. It was extracted from shells, snails and insects as well. All of them feature in sottobosco paintings.

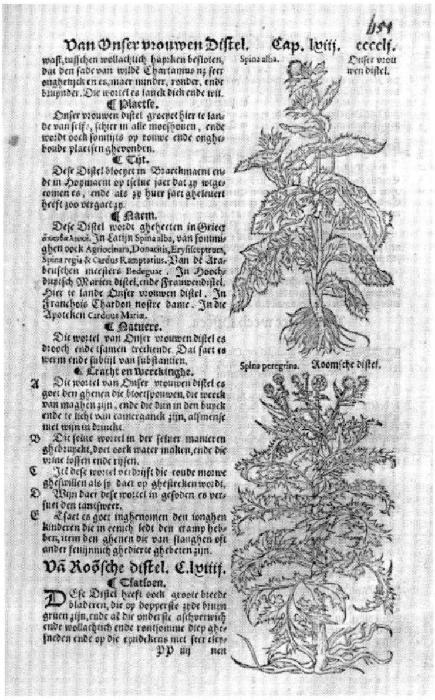
654

Van Onser vrouwen Distel. Cap. lviij. cccclj.

wast, tusschen wollachtich hayrken besloten, dat den sade van wilde Chartamus nz seer ongheliick en es, maer minder, ronder, ende bruynder. Die wortel es lanck dick ende wit.

Plaetse.

Onser vrouwen distel groeyet hier te lande van selfs, schier in alle moeshoven, ende wordt oock somtijts op rouwe ende ongheboude plaetsen ghevonden.

Tijt.

Dese Distel bloeyet in Braeckmaent ende in Hoymaent op tselue jaer dat zy wtgecomen es, ende als zy huer saet ghelevert heeft zoo vergaet zy.

Naem.

Dese Distel wordt gheheeten in Grieck ἄκανθα λευκή. In Latijn Spina alba, van sommighen oock Agriocinara, Donacitis, Erysisceptrum, Spina regia & Cardus Ramptarius. Van de Arabeuschen meesters Bedeguar. In Hoochduytsch Marien distel, ende Frawwendistel. Hier te lande Onser vrouwen distel. In Franchois Chardon nostre dame. In die Apoteken Carduus Mariæ.

Natuere.

Die wortel van Onser vrouwen distel es drooch ende tsamen treckende. Dat saet es werm ende subtijl van substantien.

Cracht en Werckinghe.

A Die wortel van Onser vrouwen distel es goet den ghenen die bloetspouwen, die weeck van maghen zijn, ende die dun in den buyck ende te licht van camerganck zijn, alsmense met wijn in drincht.

B Die selue wortel in der seluer manieren ghebruyckt, doet oock water maken, ende die vrine lossen ende rijsen.

C Item dese wortel verdrijft die coude morwe gheswillen als sy daer op ghestreken wordt.

D Wijn daer dese wortel in gesoden es versuet den tantsweer.

E Tsaet es goet inghenomen den ionghen kinderen die in eenich ledt den cramp hebben, item den ghenen die van slanghen oft ander fenijnnich ghedierte ghebeten zijn.

Vā Roōsche distel. C.lviiij.

Tfatsoen.

DEse Distel heeft oock groote breede bladeren, die op dopperste zyde bruyn gruen zijn, ende al die onderste aschverwich ende wollachtich ende rontsomme diep ghesneden ende op die eyndekens met seer cley-

PP iiij nen

14

Cruyde-boeck

Antwerp, Rembert Dodoens 1563, p. 493.

Gems

The stunning depiction of a *Toad and a Morning Glory*, signed and dated 'O. MARSEUS. D S 1660', is our next example.[67] Once again it zooms into the dark realm of the underworld, where only a glimpse of light in the upper background gives us the sense of space and air, familiar in traditional landscape painting (fig. 15). To imitate the moss structure in the foreground, the artist has used a special printing technique (the so-called fat press[68]), applying oil colour on a tool and pressing it onto the canvas, pulling it off quite rapidly so that a ridged surface remains that is afterwards painted over with a different colour and carefully wiped out again. He was aiming, undoubtedly, at a double-effect, and we are meant to see both the mimetically presented features of moss and lichen *and* the sticky and moist oil colour that helps to generate the naturalistic forms. The floral world literally grows out of the thickly applied oil colour on the canvas, an idea that stresses the material basis of painting. The toad seems to creep out of the soil or an earthen burrow or, perhaps more accurately, out of the brown pigment and sticky paint of the oil colour. Marseus chose to depict the animal leaving its subterranean hiding place, presenting itself on the ground, and in doing so demonstrating a momentary state of incompleteness of form. We observe, as a result, how animation takes place, with the toad's slow coagulation into form, and the morning glory's burst of vital energy into different stages of bloom in increasingly radiant and brilliant colours.

It is an unusual painting with an odd iconography, though there is an earlier version by Otto Marseus at the Royal Collection in London, showing the tendril of an *Ipomoea*, a splendid blue bell in its centre, in combination with a snake.[69] Butterflies flutter round the winding flower stalks, while the twines, in turn, are formally repeated in the thin and artificially coiled body of the snake. The leaves are heavily damaged and pitted by insects and caterpillars. Somehow the painting raises the profound question of biological and moral growth and decay, and maybe it stages a type of *psychomachia*, but it does not convey any more of a specific meaning. Actually, we should reflect upon this fact, and now turn our attention to one of the painting's most striking features: that of colour and 'paint'.

Before moving on, however, I would like to pause on some seventeenth-century references to the fabulously blue trumpet-shaped flowers of the morning glory.[70] As any gardener knows, many species of *Convolvulaceae* are problematic weeds, because they take over more valuable plants by strangling them. It is exactly this parasitic nature that is discussed in sixteenth- and seventeenth-century herbals.[71] But the tone changes completely when it comes to the blue species, which, for example, was described as 'rarissima & pulcherrima' in Lobelius' *Stirpium adversaria nova* (1571).[72] More interestingly, in Rembert Dodoens' *Cruyde-boeck* we find the following entry:[73] 'The blue Withiwinde hath slender branches and small, by whiche it clymbeth up, and wrappeth or windeth it selfe about trees and poles. (...) The flowers are fashioned like belles, blewe and hollowe',[74] to which is added: 'The

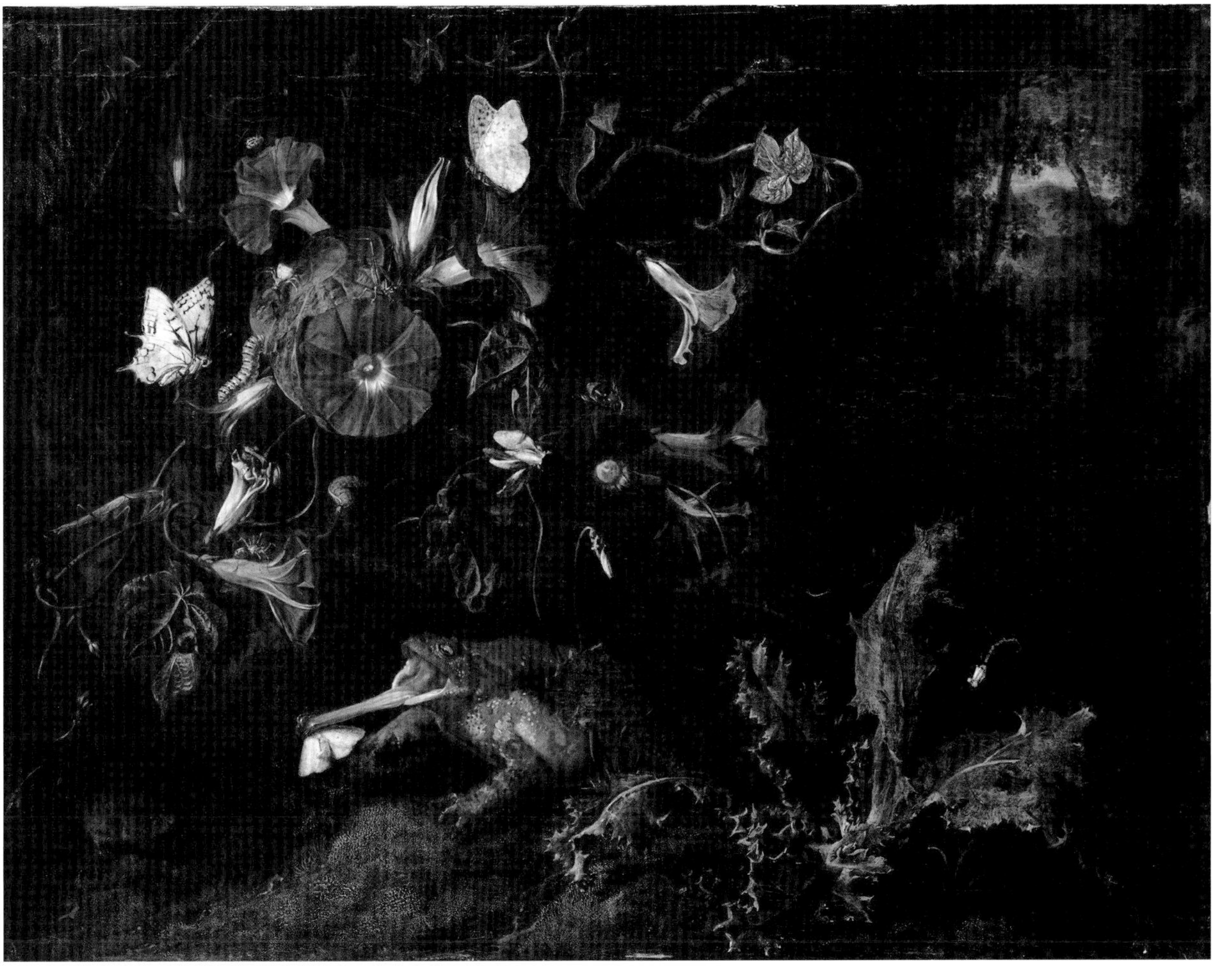

15
Otto Marseus van Schrieck, *Toad and a morning glory*, 1660
oil on canvas, 53.7 x 68 cm, Schwerin, Staatliches Museum, Gemäldegalerie, Inv. 154.

Nature: Bindeweede or Withiwinde is of a hoate and drie qualitie or nature'.[75] The most interesting implication of this assertion is not so much the idea of the flower's entangling or strangling nature, but of its hot and dry qualities. Obviously, the plant is introduced into the picture as a strong opponent to the phlegmatic nature of the toad and the cold humidity of its earthen burrow. To avert the symbolic danger of the noxious beast that has already caught a butterfly (*psyche*), the artist has drawn upon the principles of sympathy and antipathy, contrasting cold with warm, wet with dry and the identical with the different, like a physician curing a poisonous attack or threatening disease.

Lobelius once provided a special name for the blue species of Convolvulaceae. He called it 'Asureus, sive caeruleus Convulvulus Hederaceus':[76] *asureus* meaning, according to contemporary colour theory, made of lapis lazuli or azurite. Throughout history, people have attributed beneficial healing powers to this stone. Legend has it that it brought the blue of the sky – and with it the divine light – down to earth. It was meant to work against melancholy and depression, and when ground up it was used for medicinal purposes and as a cosmetic, not to mention as a source for the pigment ultramarine in tempera and oil-based paints. It is therefore significant that Otto Marseus painted the flower cups hovering above the toad's bulky body in brilliant lapis lazuli, applying it in pure form, almost as a decorative glaze. The blue flower makes a striking feature, especially as Marseus hardly ever used the pigment in other paintings. Let us dwell on some passages in Agricola's *De natura fossilium* (1546) and other lapidaries: 'Lapis-lazuli especially resembles the heavens because of the golden points which represent the stars'.[77] In his *Lapidary, or, the history of pretious stones* (1652), Thomas Nicols, paraphrasing Boetius, states that its stones are of a dry nature, 'and the very best of these are very comfortable to the eyes, if they be often looked on'.[78] It is a 'fair, ceruleous, sky-coloured stone', hence called in Latin 'Cæruleus lapis, in Italian Azuro ultra-marino'. And, 'Of this stone is made the excellent azure that is more pretious then gold'. It also withstood the stiffest test: 'If this stone being put into the fire doth not change its colour'.[79] Dioscorides recommended it as an antidote to snake bites and 'saith that this stone hath a repercussive faculty and that it is good in all melancholy diseases'.[80] Nicols also informs us that 'other Physicians have found experience that it hath in it a purgative facultie; and that it is good in all melancholy diseases'.[81] He then adds a recipe by Antonius Musa Brassavolus (1500-1555), apothecary of Ferrara,[82] describing pills of ground lapis lazuli, camphor, anise, cinnamon and other purgative ingredients. 'As such, it is unlikely that a material such as ultramarine would have been seen solely as a *colore*, but simultaneously as capital, as a colouring agent, as a medicine, and as a gem'.[83]

Within the tradition of images, the vibrant blue colour was often used for the garments of the Madonna because of its high value. It might also have carried associations with the main healing property of lapis lazuli, which was thought to be protective care. It would not be more than a suspicion to relate the morning glory to Marian symbolism if Marseus had not also chosen the lady's thistle to act as the main feature in so many of his other paintings.[84] Butterflies spiral around its thorny leaves that, again, were used as medicine in ancient and early modern pharmacy because of the thistle's hot and burning nature, while venomous snakes are lurking on the forest ground or slipping down some clammy rock. We do not know enough about Marseus' religious beliefs, but there is evidence of an early interest in his paintings amongst Rome's highest circles of clergy and aristocracy.[85] We may also relate the idea of a celestial Mary crushing a serpent ('that in her the seed of the serpent may have no part'[86]), figuratively repeated in the morning glory's struggle against chthonic forces, with a similar reading of theriac preparation in books on medicine, where it is said that the treacle also was a moral curative, that is, a medicine 'to make a man reject the poisonous sin which has entered into his soul'.[87] It would mean that Otto Marseus blended main narrative patterns of salvific history into the more secular dramas of natural history, and it would also explain the slight irony with which he put every figure into place, not least picking up narratives of old fairy tales. Salvation happens when the princess kisses the toad on its mouth or provides another potent antidote. And suddenly, the picture's featured encounter between a toad and morning glory becomes similarly dichotic: it holds for both a literal and a figurative reading; it constitutes *and* demonstrates, presents *and* represents the antagonistic forces in nature's great realm of matter and form, or substances articulated in colours. In so doing, it brings together material, aesthetic and symbolic values on the picture's plane, hence carefully fusing the domains of art and nature, art and science.

A similar statement can be made about the famous butterfly wing imprints that had been a stumbling block for contemporary criticism. Marseus pasted real butterflies onto the canvas rather than painting them, blurring the border between art and nature's formative abilities to 'paint' and 'design'. During the seventeenth century, contemporary thought became resolutely conscious of the fact that Man's creativity still had to be tested, that it was always challenged by the great abundance and variety of God's creation.[88] Marseus surely paid credit to the traditions of image-making and their highly developed iconographical traditions. But he also decided to insert 'matter itself' into his paintings, inviting nature to participate in substance in his work of art.

There is at least one more aspect to consider. It shows the lively occupation during the seventeenth century with visual perception and physical explanations of colour and light, which will lead into the optical investigations of the next century. In 1674, the naturalist, court botanist, alchemist and physician to Elector Friedrich Wilhelm of Brandenburg (1620-1688), Johann Sigismund Elsholtz (1623-1688), announced that

> the change of light doth also cause a new colour to appear. So we have observed, the flowers of *Convolutus* or *Bind-weed* (that sort with the Blew flowers and ivy-shaped leaves, with sharp points) which Lobelius calls Asureum, in the day time is of a delicate *Blew*, but at night by candle-light it appears of a *Violet colour*.[89]

In so saying, Elsholtz draws our attention to the frequencies of visible light and the dependence of colour perception to day- or night-time, so that the phenomenon of colour is no longer tied to being simply a material 'substance' but in the variable circumstances of its perception. With this, the old belief of colour as a fixed inherent quality reaches its limit. In contrast, colour perception is not stable at all, but subject to time and space. It signals the beginning of the end for earlier concepts such as Otto Marseus' indebtedness to colour symbolism, which, at the same time, was realism, or substantialism.

Touched by light

For the time being, Elsholtz's thoughts on how perceptions of colour are dependent on the immediate illumination can be nothing but a side-note. But then his views on this are extremely significant in further developments of colour theory. Contemporary thought is much involved in questions of visual perception and the real or apparent properties of colour. In the seventeenth century, references to gemstones and plants appear in many writings on colour and colour production, and it is no accident that it is within a lapidary that a three-colour primary system as the basis for colour mixing appears for the first time.

The landmarks in lapidaries during the second half of the sixteenth and early seventeenth century were Georgius Agricola's comprehensive texts on mineralogy and gemology *De re metallica* (1530 and 1541) and *De natura fossilium*, and Anselm de Boodt's *Gemmarum et lapidum historia* from 1609. Both authors were physicians and keenly aware of the traditional belief in the natural (healing or damaging) properties of stones and gems, and though they recognized the remedial effect of mineral potions, they also tried to adjust them to empirical observation and sometimes chemical and physical experimentation. While Agricola also extensively deals with chthonic creatures in *De animantibus subterraneis liber* (1549), which is therefore of some significance for the sottobosco genre, Anselm de Boodt's book on gems is particularly relevant because he already speculates about the existence of a distinct atomic structure in minerals: for him, crystal has no real colour itself, but is reducible to fine powder through grinding. Being an amateur watercolourist himself, De Boodt propagates a three-colour primary system that derived from the use of gems as pigments. In fact, the lapidaries of the sixteenth and seventeenth centuries pioneer modern colour theory:

> The *primarii colores*, or especiall colours which arise not from commixtion of other colours are these: White, black, blue or skie colour, yellow, red. (...) Almost all kinds of colours are caused by the various mixtures of these colours. But in their own originals they are not mixt with others.[90]

The seventeenth century's quest to provide a modern foundation for painting culminated in the discovery that colour mixing was governed by a simple law: that all colours may be made from just three, the painters' primaries, together with white and black. Robert Boyle (1627-1691) wrote his *Essay about the origine and virtues of gems* in 1672, and it was considered the most scientific work possible of its time, dealing with the formation of minerals and their crystals from solutions as reported by direct observation. Minor remarks were made at the time on the impossibility of using gems for medicinal purposes due to their insolubility. This line of argument was taken up by the London physician Robert Pitt (1653-1713) in 1703. In *The craft and frauds of physick expos'd* Pitt reiterates the insolubility of gems and the fact that this would make it impossible for them to affect the body. As Lorenzo de' Medici died after having taken pulverized gemstones and diamonds – as much an expensive as a fatal cure – he would have approved of it.[91]

Burst into colour

The epistemic shift from substantialism to subjective perception is one of the most thrilling developments in seventeenth-century colour theory. The colour ultramarine, for example, would no longer be explained as the inherent property of an entity, but as a visual perceptual property: as a sensation, though it still remains the responsibility of the entity. An object such as a rock of lapis lazuli is then said to have the colour of light leaving its surface, though the surface must have distinct properties. It is no accident that this knowledge was to some extent developed and displayed in early modern lapidaries, while gems and stones, in turn, provided pigments for colours. It triggered thoughts about the interactivity of the physical world with the beholder's eye and led to a totally revised version of chromatic theory: that there is no real colour in nature, and that all colour is apparent colour. In the longer term, the discussion changed the perception of painting as well, because it affected its material basis and rendered it a precondition of subjective vision. Yet we cannot presume that the new theories were widely known before the turn of the century.

In the case of Otto Marseus van Schrieck and sottobosco painting, we have to step back a little. We can say for certain that he is deeply interested in colour values and colour changes in nature and art. He chooses to depict the chameleon and the chameleon herb (a species of thistle with speckled foliage), the feathers of parrots, the chromatic change of autumn leaves or leaves slightly dried and curling. The blue bell of the morning glory provides a brilliant colour effect amid the shady realm of the earth. The arum's lily presents glaringly bright red berries. But above all he was intrigued by the idea of painting rock with rock, earth with earth and herbs with herb dye: the vermilion mushroom cap with vermilion, the heavenly blue blossom with lapis lazuli, the emerald lizard's scales in colours accordingly, and butterflies by pasting real butterfly wings onto the canvas. The structure of moss and lichen is reproduced with sticky paint. It indicates that was thinking of dealing with substances when applying his colours, thus loading the picture plane with the gravity of essentialism.

The iconography of a flower, a plant or a creature is therefore never an end in itself. It always transcends the mere identification of an object because of its substantial realization in colour, which presents *and* represents the antagonistic forces in nature's great realm of matter and form. In sottobosco painting, the dualism is repeated in the central topic of poisons and their antidotes, or the powerful agencies of sympathy and antipathy. For us it is important to note that it is displayed simultaneously on the level of the signifier and the signified. Conflict is shown iconographically – in the animal world for example – but it also brings out a main feature pictorially: the painter is forced to see how compatible or incompatible the colours are likely to be with each other. As they express substantial properties or qualities, they, too, make fine protagonists, or opponents, or partners; they may act as ground natural substance, or artificial paint, or drug, or poison, or counterpoison. 'Painting as *pharmakon*' effectively tackles the problematics of presentation and representation, with the painter replacing living nature by its artificial double and, hence, entering the realm of signs, or representation, or *pharmakeia*.

Joachim von Sandrart (1606-1688) once reflected upon the painter's Apollonian task. Each day his brush, he says, had to chase off the darkness that loomed over the world's creation – like Apollo or Aurora who had to drive away the shadows of the night and uncover Nature's splendours and riches:

> In the beginning / when the wise creator had produced all things / that we see with our eyes / he had firstly / by the creation of light / uncovered the chaos or the mingled lump / from which all has been created. / In this chaos all colours existed at the same time and were mixed together / until they were separated. But even after the things attained their form / they are repeatedly obscured by the darkness of the night / and have to be made visible by the day light.[92]

When trying to retrieve hidden forms out of the dark grounding of the picture plane, the artist depicts the virtue of daylight that warms the soil and encourages plants and creatures to rise up and spread out. The blossoms and leaves of flowers and herbs, the colourful caps of mushrooms, the patterned skin of salamanders, lizards and snakes, the glistening bodies and wings of beetles and dragonflies all sparkle against the dark background like gemstones recovered from the depths of the earth. It is because they are equally touched by light and by the tip of the painter's brush that they come to life and animate the picture plane. They are sketched in gradual stages of vitality and visibility, thus unfolding their unique properties in a range – a burst – of colours between chthonian darkness and Apollonian light.

Notes

1 Vasari 1987, 259-260. See also Vasari 1963, vol. 2, 158-159.

2 The story ends in a moment of perfect illusion: Having finished the work, Leonardo 'adjusted the buckler in a good light on the easel, and put to the window, in order to make a soft light, and then he bade [his father] come in to see it. Ser Piero, at the first glance, taken by surprise, gave a sudden start, not thinking that that was the buckler, nor merely painted the form that he saw upon it, and, falling back a step, Leonardo checked him, saying, "This work serves the end for which it was made; take it, then, and carry it away, since this is the effect that it was meant to produce"' (Vasari 1987, 260). On emotional responses to Medusa paintings and their theoretical reflection in the 17th century see Heinen 2001, Heinen 2007, Heinen 2010, and Koslow 1995. On Leonardo and the concept of 'art versus ugliness' see e.g., Rubin 1990, and Pfisterer 2003, 271; on similar discussions in 17th-century Dutch art theory see Weststeijn 2009, 344-345. See also the contribution by Dániel Margócsy earlier in this volume.

3 Cat. Florence 2008, 66.

4 Caravaggio's patron Cardinal Francesco Maria Del Monte commissioned him to paint Medusa as the figure on a ceremonial shield presented in 1601 to Ferdinando I de' Medici, Grand Duke of Tuscany – Del Monte was the Medici agent in Rome, and the grand duke was then reorganizing his personal armoury. The poet Marino claimed that it symbolized the Duke's courage in defeating his enemies. In 1568 Vasari had written of the Medusa shield by Leonardo in the grand duke's collection. If it still existed in 1597, Del Monte would have known it, and in giving this commission to Caravaggio he was setting his protégé against the man recognized even then as the greatest of all painters.

5 'Of the many paintings [that I have seen] I always seem to call to mind one which our friend Nicholas Sohierus of Amsterdam once showed to viewers among his magnificent furnishings: Medusa's head, which is entangled with snakes born from her hair, has been cut off. In it he mixes the still pleasing sight of a woman, who recently was very beautiful, and a sight horrid with recent death and the most disgusting reptiles with such ineffable diligence that the viewer is struck with sudden terror (indeed the painting is usually covered) and delighted by the vivid and charming cruelty of the matter. Yet while I prefer to eulogise this [painting] rather in the house of friends than my own, the witty saying about the envoy of the Teutons came to my mind, who, when a precious painting of an old Shepherd with his staff was exhibited on the Forum Romanum, was asked what he thought was the value of it; he responded (forsooth with the frankness of his people) that he would rather not have the original of someone like him even, at a gift [Pliny, *Historia naturalis*, 35, 8]. Away with you, you experts, who appreciate beautiful things just because of their horror; if someone loves to sing to me about murder and blood with the

same grace of voice by which he is capable to convey mirth, jokes and laughter, should I not ask him to delight me with his saying as well? Someone may represent beautiful things less elegantly, but horrible things never pleasantly'. Huygens's Latin manuscript is in the Bibliotheek der Koninklijke Akademie, Royal Library Den Haag, *Huygens-handschriften, Catalogus van Handschriften* no. XL.VIII; see also Worp 1897, 73, and Worp 1891, 119–120. The English translation by Heinen 2008. For a discussion on 'painting the ugly' see Aristotle, *Poetics* 1448b 10-12, Plutarch, *Moralia* I 18, and Thomas von Aquin, *Summa Theologiae* I qu 39 ar 8. In 17th-century Dutch art theory, see Van Hoogstraten 1678, 77, and Junius 1641, 65. Rubens painted the picture around 1617 – possibly in collaboration with the animal specialist Frans Snyders (1579–1657). The version that Huygens saw must have been a studio copy after the original that was listed in the estate of the Duke of Buckingham (1592–1628) in 1635 and is now in Vienna. On the different versions of the painting see W. Prohaska in Cat.Vienna 1977, no. 23, 81–83, and P.C. Sutton in Cat. Boston 1994, 245–247.

6 Derrida 1981, 120.

7 Ibid., 92.

8 Ibid., 99-100.

9 Ibid., 90.

10 Socrates has been taken out of the city into the countryside by Phaedrus. They have stretched out on the ground, and Phaedrus wants to reconstitute the argument and design of a text by one of the most talented writers of their day, Lysias. However, he has not learned the speech by heart; he has hidden the text under his cloak. It is the moment when Socrates tells him about Theuth, the god of writing, who discovered how to represent or replace the living voice by signs, and whose gift was criticized by the king because it could supplement memory but also destroy it (so writing is good because it helps us memorize, and it is bad because it destroys our memory: we now rely on signs, we have externalized our memory). Socrates therefore calls Theuth a pharmacist, as medicine supports and irritates the natural course of living beings, as it also has contrary values.

11 Derrida 1981, 129.

12 Sottoboschi depict botanical and zoological life in dark underwood or at the humid margins of pools, and were as a sub-genre invented in Italy in the 1650s by the Dutch artist Otto Marseus van Schrieck and taken on by artists such as Rachel Ruysch, Elias van den Broeck, Abraham Begeyn and Franz de Hamilton. The genre has been investigated by Steensma 1999, and Hildebrecht 2005. I will use the term sottobosco throughout the article. Hildebrecht decided to call them 'nature pieces' to give them a broader meaning (Hildebrecht 2005, 1, 7-14, 84-141). I, however, will concentrate on certain aspects only, referring to the contemporary discussions on Nature's subterranean forces and potencies. The German name *Niederwelt* was coined by Habicht 1923; the term *bosstilleven* or *Waldbodenstilleben* by Bol 1969. For the terminology and an overview of the history of research see Steensma 1999. Further reading: De Mirimonde 1968; De Mirimonde 1970; Bergström 1974; Marcus 1974; Franchini Guelfi 1977; Bol 1982; Cat. Bergamo 1986; Cat. Florence 1989; Weber 1993; Gemar-Koeltzsch 1995.

13 For metaphors on light and darkness, see *Homeric Hymn to Apollo* ll. 348-374.

14 Every four years, the Pythian Games and Delphic Festivals were organized, with sacrifices and a performance of the Sacred Drama, whose main theme was the duel between Apollo and Python. On the Pythian Games and Delphic Festival see Deane 1833, 329, Fontenrose 1959, Kerenyi 1980, 135-136, and Rohde 1890-1894. There are various versions of Python's birth, and death at the hands of Apollo. A version of the *Homeric Hymn to Apollo* related by Gaius Julius Hyginus (c. 64 BC-AD 17) (*Fabulae* 140) holds that when Zeus lay with the goddess Leto, and she was to deliver Artemis and Apollo, Hera sent Python to pursue her throughout the lands, so that she could not deliver her babies anywhere that the sun shone. Thus when Apollo grew up, he drove the python from Mount Parnassus to the oracle of Gaia at Delphi where he dared penetrate the sacred precinct to kill her with his arrows beside the rock cleft where the priestess sat on her tripod. The myth has it that Zeus instituted the Pythian Games for Apollo to preside over as penance for his act and to purify himself for the sacrilege.

15 Kerenyi 1980, 136: The enemy dragoness 'actually became an Apollonian serpent, and Pythia, the priestess who gave oracles at Delphi, was named after him. Many pictures show the serpent Python living in amity with Apollon and guarding the Omphalos, the sacred navel-stone and mid-point of the earth, which stood in Apollon's temple'. See Watkins 1995, 448-459. Typhon, in turn, was Gaia's last son, fathered by Tartarus. As Hesiod narrates, he attempted to destroy Zeus at the behest of Gaia. Hesiod has it that he was finally conquered by Zeus and cast into Tartarus, while others tell us that he was confined beneath Mount Aetna where he is still the cause of its eruptions, Typhon being the chthonic figuration of volcanic forces, just as Hephaestus is their civilized Olympian manifestation; see Hesiod, *Theogony* 820-822. In an alternative account of his origin, the *Homeric Hymn to Apollo* makes the monster Typhon at Delphi a son of archaic Hera, produced out of herself. For a detailed comparison see Miller 1986, Pindar, *Pythian Ode* 1.19-20, and Aeschylus, *Prometheus Bound* 370.

16 Nicander, *Theriaca* 8-20. Gow & Scholfield 1997, 170, comment that 'no such lore is to be found in Hesiod, cite Acusilaus for the statement that "panta ta daknonta", i.e. "anything that bites" sprang from the blood of Typhon, who, if not a Titan, belongs to that order of beings'. On Nicander's toxicology see Scarborough 1979; on his use of minerals and stones in potions and drugs see Mottana 2006.

17 Apollonios Rhodius, *Argonautica* 4.1515: 'For when over Libya flew godlike Perseus Eurymedon (...) bearing to the king the Gorgon's head newly severed, all the drops of dark blood that fell to the earth, produced a brood of those serpents. Now Mopsus stepped on the end of its spine, setting thereon the sole of his left foot; and it writhed round in pain and bit and tore the flesh between the shin and the muscles. And Medea and her handmaids fled in terror; but Canthus bravely felt the bleeding wound, for no excessive pain harassed him. Poor wretch! Already a numbness that loosed his limbs was stealing beneath his skin, and a thick mist was spreading over his eyes. Straightway his heavy limbs sank helplessly to the ground and he grew cold; and his

comrades and the hero, Aeson's son, gathered round, marvelling at the close-coming doom. Nor yet though dead might he lie beneath the sun even for a little space. For at once the poison began to rot his flesh within, and the hair decayed and fell from the skin'.

18 Ovid, *Metamorphoses* 4.770: 'And as he o're the Libyan Deserts flew; the blood, that dropt from Gorgon's head, streight grew to various serpents, quicken'd by the ground: With these, those much infested climes abound'.

19 Lucan, *Pharsalia* 9.1070-1088: 'Where western waves on farthest Libya beat, warm'd with the setting sun's descending heat, dreadful Medusa fix'd her horrid seat. (...) It was from this monster to afflict mankind, that nature first produc'd the snaky kind: on her, at first, the forky tongues appear'd; from her their dreadful hissings first were heard. Some wreath'd in folds upon her temples hung, some backwards to her waist depended long; some with their rising crests her forehead deck; some wanton play, and lash her swelling neck: And while her hands the ending vipers comb, poisons distil around, and drops of livid foam'.

20 The three Gorgon sisters – Medusa, Stheno and Euryale – were children of the ancient marine deities Phorcys and his sister Ceto or, sometimes, Typhon and Echidna, in each case chthonic monsters from an archaic world.

21 Nicander, *Ophiaca* (= frag. 31).

22 Justin relates that the king 'employed himself in digging and sowing in his garden, mixing noxious herbs with harmless ones, and sending them all indiscriminately, moistened with poisonous juices, as special presents to his friends' (*Epitome* XXXVI.4).

23 The tarantula became an important topic in the seventeenth century. There is a connection between the tarantella dance and Baroque colour theory, e.g., in the treatises of Athanasius Kircher and Matteo Zaccolini. I am preparing a paper on this subject.

24 Van Hoogstraten 1678, 358: 'omdat hy allerwegen naar vremd gekleurde of gespikkelde slangen, hagedissen, rupsen, spinnen, flintertjes, en vremde gewassen en kruiden omsnuffelde'.

25 Ibid., 69: 'Maer zeker onzen Otho Marseus (alias Snuffelaer) heeft zijn bequaemheit in de konst, ent tot wat deel hy neygde, genoeg laten blijken: want als ik in 't jaar 1652 by hem te Romen was, verwonderde ik my over zoo veel gedrochten, als hy onderhielt en voede: welkers natuere hy ook zoo wonderlijk deurgronde, als hy hare gedaentens levendich heeft uitgebeelt'.

26 In Amsterdam, Marseus was acquainted with Jan Swammerdam and Johannes Hudde. For a description of the scientific climate, esp. of the Dutch interest in entomology and the microscope, see Jorink 2010, 181-256, and the contribution by Eric Jorink elsewhere in this volume.

27 The number of sottoboschi done by Marseus in the possession of the Medici comes as no surprise: Baroque herpetological investigation was moving fast towards its peak when Otto Marseus started depicting the *Niederwelt*. From his early career in the 1650s until his death in 1678 the Medici bought twelve paintings from him, and Cosimo III even visited the painter's studio during his journey through the Netherlands in 1668. Hildebrecht 2005, 58: 'The presence of such a large number of Marseus' pictures in Florence made the Medici and visitors to the Granducal court the most important audience of the artist's career'.

28 D'Eremita, an 'aromatico e medico di Rocca d'Evandro, dell'Ordine dei Predicatori', and author of the treatise *Elixir Vitae* (1624) writes: '(...) al pari dei suoi predecessori e imitatori di Mitridate, re del Ponto, impiega tesori nelle vere preparazioni di preziosissimi antidoti, e per tal opra appresso di sé mantiene gran numero di artefici, pratici, e intendenti di misteri dell'occulta filosofia, da' quali mi glorio anch'io d'aver, ne' primi anni appreso il modo d'avanzarmi in questa professione'. Cited in Torresi 2004, 22-23.

29 Beecher 2002, 243.

30 It was not until a 1781 publication by Felice Fontana that they were generally accepted. (True poison was the spirit formed in the mind of the enraged reptile; see Knoefel 1988, xiii).

31 Knoefel 1988, 23; Tribby 1991. For Otto Marseus van Schrieck and Francesco Redi see e.g., Hildebrecht 2005.

32 Tribby 1991.

33 Moyse Charas took up the profession of apothecary and was declared Master by Frederick Henry, Prince of Orange. Later, in Paris, he had contacts with the Dutch diplomat Constantijn Huygens and his son, the scientist Christiaan Huygens, and with the English physician and philosopher John Locke. After shorter visits to London and the Royal Society he lived there for a certain time, being called to contribute to cure King Charles II. In 1692 he was appointed member of the Académie royale des sciences (established in 1666). Charas was a fierce opponent of Francesco Redi.

34 For a similar approach towards a pharmacological picture theory, with a special focus on the role of imagination in Baroque response theory, see Barker 2004, and Weststeijn 2008; in general: Schaub & Suthor 2005.

35 Thrasher 1669, 129-130.

36 Dioscorides, *Materia medica* VII, lx.

37 Pliny, *Historia naturalis* XXII, 7: 'Nature who, as we will prove beyond a doubt, has never failed in coming to the assistance of man, and has implanted (inseruisse) remedies for our use even in the most despised' plants or animals.

38 Pliny, *Historia naturalis* XXII, 7.

39 Dragonsblood is a resinous product of climbing plants in the Indies and Canaries and could be used as medicine, as pigment and as dye. In ancient times it was thought to be the congealed blood of dragons and elephants mixed together as they fought to the death.

40 The Compagnia di San Luca, once a separate entity from the guild system, developed into the Florentine Accademia e Compagnia delle Arti del Disegno in 1563, which was then formally incorporated into the city's guild system in 1572; see Hughes 1986, 3-10. For the use of pigments see DeLancey 2003, esp. 142-143: 'The Florentine *speziale* occupied a position at a nexus between the – to the modern mind – seemingly disparate professions of doctor and artist, and the disparate fields of medicine and art. And yet, our knowledge of the Renaissance mind tells us that these fields were not unrelated at the time. (...) The professional associations among the members of the *Arte dei Medici e Speziali* suggest close ties, not only in ways of moving, making and working with materials, but also in the way the professionals were bound into a network, with the apothecary, as supplier of goods and services to both, at the center. That these same goods frequently served both doctor and artist

simply reinforced that bond'. See also ibid., 148: 'By examining apothecaries and their trade, we can better understand the position that pigments most likely held, not only in the mind and business of *speziali*, but also, through them, in Florentine culture more widely. In addition to their use by painters as the *sine qua non* of their craft, they served doctors, alchemists, sculptors, and dyers. As such, it is unlikely that a material such as ultramarine would have been seen solely as a *colore*, but simultaneously as capital, as a coloring agent, as a medicine, and as a gem, and therefore of particular value to its main merchant, the *speziale*'. See Jack 1976.

41 For discussions on the enmity and friendship of colours in early modern colour theory, see e.g., Fehrenbach 2003.

42 *Amanita caesarea* was first described by Italian mycologist Giovanni Antonio Scopoli in 1772 as *Agaricus caesareus* (Scopoli 1772, 419), before later being placed in the genus *Amanita* by Persoon in 1801 (Persoon 1801, 252). The Romans called it Bōlētus, derived from the Ancient Greek βωλιτης, for this fungus as named by Galen (Ramsbottom 1953, 6). Interestingly enough, the word has apparent analogies in Dutch, 'padde(n)stoel' (toad-stool/chair), and German, 'Krötenschwamm' (toad-fungus). With this, I refer to the storytelling of the day and to the play on words that was certainly familiar to the Dutch artist.

43 Dioscorides, *Materia medica* IV, Lxxix: 'Either they are edible, or they are poisonous, and come to be so on many occasions, for either they grow amongst rusty nails or rotten rags, or ye holes of serpents'.

44 Pliny, *Historia naturalis* XXII, 47.

45 Nicander, *Alexipharmaca* 522-527.

46 Gerard 1597.

47 Dugan 2008.

48 The identification of the pigments was provided by several restorers specializing on 17th-century painting. According to Cennini, if you ground it every day for 20 years it would still show improvement; Cennini 1899, chap. XL.

49 Hiebner 1698, 103: 'All sorts of mushrooms are nothing but the swellings of the earth, occasioned by a viscous humidity of the earth and trees, for Luna is Lady of all Moisture. (...) They give, according to the nature of all Lunars a pale colour. (...) Some think that the pale coloured are the most hurtful because they have the more Lunar poison in them'. For a connection to Baroque theories on spontaneous generation, see Leonhard 2010.

50 However, the bright-coloured *Amanita* species like *Amanita muscaria* have no dyeing ability at all. See also Perkins 1986, and Kok 1966.

51 Durant 1697, frontispiece.

52 Elsholtz 1677, 17.

53 Bate 1634, 120.

54 It was translated first into French in 1557 by Charles de l'Écluse (*Histoire des plantes*), into English (via l'Écluse) in 1578 by Henry Lyte, and later into Latin in 1583. In his time, it was the most translated book after the Bible. It became a work of worldwide renown, used as a reference book for two centuries.

55 Many of these recipes were taken from Nicander's *Theriaca* and *Alexipharmaca* and Dioscorides's *Materia medica*, which contained a whole book on poison that was successfully repeated in Pietro Andrea Mattioli's *Commentarii*; from Pliny's *Natural history*, with its numberless entries upon the protection from poisons and venoms; or Galen's two writings *De antidotis* and *Theriake*, reissued in Paris in 1535. Galen's writings contain a collection of over 130 antidotes, theriacs and drugs, again culled from many physicians. See Watson 1966, 4: 'Theriake (theriac) was the name given to the kind of antidote which was specially meant to counteract the bites of venomous creatures, serpents, scorpions, spiders, bees, mad dogs, and the like. To the composition of both antidotes and theriacs were herbs, animal substances, mineral substances'.

56 Dodoens 1563, 299g.

57 Ibid., 307d.

58 Steensma 1999, BI. 60. For the cyclamen see e.g., Dodoens 1563, III. xi.

59 Steensma 1999, BI. 103.

60 Ibid., BI. 58.

61 Ibid., BI. 57.

62 Ibid., BI. 3.

63 Ibid., BI. 90.

64 'Good against the bytings and stingings of venomouse beastes' (Lyte 1578, 89). Dioscorides lists it under the name Ageratum.

65 Dodoens 1563, IV, Lxii.

66 Ibid., V, i.

67 See Leonhard 2010.

68 Beier 1987; Weber 1989, 35f; Steensma 1999, 71-75.

69 Steensma 1999, BI. 5.

70 E.g., the often-mentioned *Volubilis minor*, or *Helxine cissampelos*.

71 Dioscorides, *Materia medica* IV, xxxvii; at more length in Matthioli, *Commentarrii* IV, xxxvii.

72 Pena & Lobelius 1571, 271-274. Lobelius studied medicine in Montpellier. After long travels through France, Italy, Germany and England, he practised in Antwerp and Delft. In England, he became physician to James I.

73 Rembert Dodoens was born in Malines in 1517, studied medicine at Louvain, became physician to the emperors Maximilian II and Rudolph II and in 1582 professor of medicine at Leyden, where he died three years later. Otto Marseus van Schrieck owned a copy of his *Cruyde-boeck*; see his inventory in Steensma 1999, 213: 'Vier stucx schilderijen getekent a.b.c.d. die S Evert Marseus toekomen neffens een herbarium van Doddenees, dwelcke beleent syn door den overleden met 250 gulden' (the herbarium of Dodoens which was pawned by the deceased for 250 guilders). See Hildebrecht 2005, 81; and also Hildebrecht 2005, 513-533 (526); Marseus also possessed a snake-stone.

74 Lyte 1578, III, 392.

75 Ibid., 393: 'There are two sortes of Bindeweede or Withiwinde, the one bearing a blue flower, the other a white, whereof one is great, the other small. (...) The blue Withiwinde hath slender branches and small, by whiche it clymbeth up, and wrappeth or windeth it selfe about trees and poles. The leaves be large and cornered, like to the old leaves of Ivye, saving that they be not so harde. The flowers are fashioned like belles, blewe and hollowe'. To which is added the trailer: 'The Nature: Bindeweede or Withiwinde is of a hoate and drie qualitie or nature. The Vertues: Withiwinde or Bindeweede is not fit to be put in medicine, as Galen and Pliny witnesseth'. Pliny, in turn, had compared the *convolvulus* with the beautiful form of a lily (*Historia naturalis* XXI, 11).

76 Pena & Lobelius 1571, 274: 'Asureus, sive caeruleus Convulvulus Hederaceus, vel smilaceus, exoticus: Haec omni huius classis rarissima & pulcherrima,

nec nisi hortis Italie, Galliæ, & Belgi nota, ubi amœna purpura porrectis attextísq; secus parietes & perticas funibus viticeis,gracilibus, umbracula cócinannt, atq; fornices, visu periucúda: vicende quippe floribus, è pediculorú calyce exeútibus, Iasmini, aut Vincæ pervincae similibus, sed ora quinis stellatim prominentibus angulis, nol cærulæ vel asureæ effigie: unde Campana Lasura Italis. Folia ternis angulis singola, in teneriusculis pediculis, Ciclamen, Hederámve annosam imitata, nonnihil hirsuta: longa & gracilis radix, nulla sua parte in Medicorum commercia venit'.

77 Agricola: 'Sapphirus (lapis-lazuli) and cyanus (sapphire) are dark blue, hence the name of the latter stone. (...) Lapis-lazuli is enlivened by small golden points. Both gems are blue as the heavens but lapis-lazuli especially resembles the heavens because of the golden points which represent the stars'.

78 Nicols 1652, 84. Albertus Magnus said that polished lapis put into the eye healed diseases of the eye (Perry 1984, 49).

79 Nicols 1652, 152. See Hartley 1982, 44: 'Both lapis lazuli and the pigment made from it could be tested by heat, because, whereas the genuine article was unaffected and remained blue, azurite which might be substituted for the stone would turn black'.

80 Nicols 1652, 152.

81 Ibid., 152.

82 Taken from Brassavolus 1555, 180.

83 DeLancey 2003, 148.

84 A thistle is also added in the Schwerin painting.

85 Hildebrecht 2005.

86 Johnson 2004, 259-270.

87 Henry Grosmont, 1st Duke of Lancaster, *Livre de seyntz medicines*, 1354.

88 On butterfly imprints see Weber 1989, 36, and Weber 1993, 24-29.

89 Elsholtz 1677, 12. Already in the ps.-Aristotelian treatise *De coloribus* there are passages on apparent colour. Kemp 1992, Bell 1992, Leonhard 2011.

90 Nicols 1652, 14, paraphrases Anselm de Boodt's *Gemmarum et lapidum historia*.

91 Aakhus 2008. For a 16th- and 17th-century discussion of the translucency of glass and gems see the article by Sven Dupré, and of the invention of three-colour printing, the article by Fokko Jan Dijksterhuis in this volume.

92 Von Sandrart 1675-1679, I, 3, 86: 'Der Anfang / als der weise Schöpfer alle Dinge / was wir mit den Augen sehen / hervorgebracht / hat er erstlich / durch hervorruffung des Liechtes / das Chaos oder den vermängten Klumpen / aus welchem alles erschaffen worden / entdecket. In diesem Chaos, waren damals alle Farben beysammen und durch einander vermänget / bis sie von einander gesondert worden. Nachdem auch alle Dinge ihre Gestalt bekommen / werden sie doch durch die Nacht-Finsternis wieder verdecket / und müssen täglich durch das Tagliecht wieder sichtbar werden'.

Bibliography

Aakhus 2008
P. Aakhus, 'Astral magic in the Renaissance. Gems, poetry, and patronage of Lorenzo de' Medici', *Magic, Ritual, and Witchcraft* 3 (2008), 185-206.

Agricola 1955
G. Agricola, *De natura fossilium (Textbook of mineralogy)*, (M.C. Bandy and J.A. Bandy, trans.), New York 1955.

Apollonius Rhodius 1990
Apollonius Rhodius, *Argonautica*, Cambridge 1990.

Barker 2004
S. Barker, 'Poussin, plague, and early modern medicine', *Art Bulletin* 86 (2004), 659-689.

Bate 1634
J. Bate, *The mysteryes of nature and art*, London 1634.

Beecher 2002
D. Beecher, 'Ficino, theriaca, and the stars', in: M.J.B. Allen, V. Rees & M. Davis (eds.), *Marsilio Ficino. His theology, his philosophy, his Legacy*, Leiden & Boston 2002, 243-256.

Beier 1987
B. Beier, 'Contre-Épreuves in der barocken Stillebenmalerei', *Restauro* 93 (1987), 35-39.

Bell 1992
J. Bell, 'Aristotle as a source for Leonardo's theory of colour perspective after 1500', *Journal of the Warburg and Courtauld Institutes* 56 (1993), 100-118.

Cat. Bergamo 1986
Cat. Bergamo, Galleria Lorenzelli, *Orbis pictus. Natura morta in Germania, Olanda e Fiandra, XVI-XVII secolo*, Bergamo 1986.

Bergström 1974
I. Bergström, 'Marseus, peintre de fleurs, papillons et serpents', *L'Oeil* 233 (1974), 24-29.

Bol 1969
L.J. Bol, *Holländische Malerei des 17. Jahrhunderts, nahe den großen Meistern. Landschaften und Stilleben*, Brunswick 1969.

Bol 1982
L.J. Bol, 'Goede Onbekenden'. *Hedendaagse herkenning en waardering van verscholen, voorbijgezien en onderschat talent*, Utrecht 1982.

Cat. Boston 1994
Cat. Boston, Museum of Fine Arts, *The Age of Rubens*, Boston 1994.

Brassavolus 1555
A.M. Brassavolus, *De medicamentis tam simplicibus*, Lugduni 1555

Celoria 1992
F. Celoria, *The metamorphoses of Antoninus Liberalis. A translation with a commentary*, London & New York 1992.

Cennini 1899
C. Cennini, *The book of the art (Libro dell' arte)*, Firenze 1899.

Deane 1833
J.B. Deane, *The worship of the serpent*, London 1833.

DeLancey 2003
J.A. DeLancey, 'Dragonsblood and ultramarine. The apothecary and artists' pigments in Renaissance Florence', in: M. Fantoni, L.C. Matthew & S.F. Matthews-Grieco (eds.), *The art market in Italy. Fifteenth-sixteenth centuries*, Modena 2003, 141-150.

De Mirimonde 1968
A.P. de Mirimonde, 'Psyche e le papillon', *L'Oeil* 168 (1968), 2-11.

De Mirimonde 1970
A.P. de Mirimonde, 'Une nature morte enigmatique de Paolo Porpora', *La Revue du Louvre* 20 (1970), 145-154.

D'Eremita 1624
D. d'Eremita, *Elixir Vitae*, Naples 1624.

Derrida 1981
J. Derrida, *Dissemination* (B. Johnson, trans.), Chicago 1981.

Dodoens 1563
R. Dodoens, *Cruyde-boeck*, Antwerp 1563.

Dugan 2008
F.M. Dugan, *Fungi in the ancient world*, St. Paul, 2008.

Durant 1697
J. Durant, *Art and nature joyn hand in hand, or, the poor man's daily companion*, London 1697.

Elsholtz 1677
J.S. Elsholtz, *The curious distillatory, or, the art of distilling coloured liquors, spirits, oyls, etc. from vegitables, animals, minerals and metals*, London 1677 (*Destillatoria curiosa, sive ratio ducendi liquors coloratos...*, Berolini 1674).

Fehrenbach 2003
F. Fehrenbach, 'Calor nativus – color vitale. Prolegomena zu einer Ästhetik es 'Lebendigen Bildes' in der frühen Neuzeit', in: U. Pfisterer & M. Seidel (eds.), *Visuelle Topoi. Erfindungen und tradiertes Wissen in den Künsten der italienischen Renaissance*, Munich 2003, 151-170.

Cat. Florence 1989
Cat. Florence, Gallerie e Musei Statali di Firenze, *I dipinti olandesi dei Seicento e del Setttecento* (M. Chiarini), Rome 1989.

Cat. Florence 2008
Cat. Florence, Gallera degli Uffizi, *Medusa. Il mito, l'antico e i Medici*, Florence 2008.

Fontenrose 1959
J.E. Fontenrose, *Python. A study of Delphic myth and its origins*, Berkeley, Los Angeles & London 1959.

Franchini Guelfi 1977
F. Franchini Guelfi, 'Otto Marseus van Schrieck a Firenze I', *Antichità Viva* 2 (1977), 15-26; part II, *Antichità Viva* 4 (1977), 13-21.

Gemar-Koeltzsch 1995
E. Gemar-Koeltzsch, *Holländische Stillebenmaler im 17. Jahrhundert*, Lingen 1995, 3 vols.

Gerard 1597
J. Gerard, *The herball or generall historie of plantes*, London 1597.

Gow & Scholfield 1997
A.S. Gow & A.F. Scholfield, *Nicander. The poems and poetical fragments*, London 1997 (1953).

Habicht 1923
K. Habicht, 'Ein vergessener Phantast der holländischen Malerei', *Oud Holland* 41 (1923), 31-37.

Hartley 1982
R.H. Hartley, *Artists' pigments c. 1600-1835*, London 1982 (1970).

Heinen 2001
U. Heinen, 'Haut und Knochen – Fleisch und Blut. Rubens' Affekt-malerei', in: U. Heinen & A. Thielemann (eds.), *Rubens Passioni. Die Kultur der Leidenschaften im Barock*, Göttingen 2001, 70-109.

Heinen 2007
U. Heinen, 'Zur bildrhetorischen Wirkungsästhetik im Barock. Ein Systematisierungsversuch nach neurobiologischen Modellen', in: J. Knape (ed.), *Bildrhetorik*, Baden-Baden 2007, 113-158.

Heinen 2010
U. Heinen, 'Huygens, Rubens and Medusa. Reflecting the passions in paintings, with some considerations in the neuroscience in art history', in: S. Dickey & H. Roodenburg (eds.), *The passions in the arts of the early modern Netherlands*), Zwolle 2010 (= *Nederlands Kunsthistorisch Jaarboek* 60 [2010]), 151-178.

Hiebner 1698
I. Hiebner, *Mysterium sigillorum, herbarum & lapidum*, London 1698.

Hildebrecht 2005
D.R. Hildebrecht, *Otto Marseus van Schrieck (1619/20-1678) and the nature piece. Art, science, religion and the seventeenth-century pursuit of natural knowledge*, Ann Arbor 2004 (unpub. diss. University of Michigan).

Homeric Hymns 1976
The Homeric Hymns (A.N. Athanassakis, trans.), Baltimore 1976.

Hughes 1986
A. Hughes, '"An academy for doing" I. The Accademia del Disegno, the guilds and the principate in sixteenth-century Florence', *Oxford Art Journal* 9 (1986), no. 1, 3-10.

Jack 1976
M.A. Jack, 'The Accademia del Disegno in late Renaissance Florence', *Sixteenth Century Journal* 7 (1976), no. 2, 3-20.

Johnson 2004
T. Johnson, '"That in Her the seed of the serpent may have no part". The Agredan visons and the Immaculate Conception of the Virgin in early modern Spain and Germany', in: R.N. Swanson (ed.), *The Church and Mary*, Woodbridge 2004, 259-270.

Jorink 2010
E. Jorink, *Reading the book of nature in the Dutch Golden Age, 1575-1715*, Leiden & Boston 2010.

Junius 1641
F. Junius, *De schilder-konst der Oude*, Middelburg 1641.

Kemp 1992
M. Kemp, 'In the beholder's eye. Leonardo and the "errors of sight" in theory and practice' (Hammer Prize Lecture), *Achademia Leonardi Vinci* V (1992), 153-162.

Kerenyi 1980
K. Kerenyi, *The gods of the Greeks*, London 1980 (1951).

Knoefel 1988
P.K. Knoefel, *Francesco Redi on vipers*, Leiden & New York 1988.

Kok 1966
A. Kok, 'A short history of the orchil dyes', *The Lichenologist* 3 (1966), 248-272.

Koslow 1995
S. Koslow, '"How looked the Gorgon then ...". The science poetics of "The Head of Medusa "by Rubens and Snyders', in: C.P. Schneider (ed.), *Shop talk. Studies in honor of Seymour Slive*, Cambridge MA 1995, 147-149.

Leonhard 2010
K. Leonhard, 'Pictura's fertile field. Otto Marseus van Schrieck and the genre of sottobosco painting', *Simiolus* 34 (2010), 95-118.

Leonhard 2011
K. Leonhard, 'White earth, or, how to cultivate color in the field of painting. Still life and Baroque color theory', *The object as event seminar I. Vision and its instruments in the early modern period* (= *The object as event. Max Planck seminars*, vol. I), New Haven 2011 (in preparation).

Lucan 1720
M.A. Lucanus, *Pharsalia* (N. Rowe, trans.), London 1720.

Lyte 1578
H. Lyte, *A new herball, or historie of plantes* (translation of Rembert Dodoens, *Cruyde-boeck*, Antwerp 1556; and, enlarged, Antwerp 1563), London 1578.

Marcus 1974
C.G. Marcus, 'Otto Marseus van Schrieck et les peintres de reptiles, insectes et sous-bois', *Art et Curiosité*, Paris 1974, 80-83.

Mottana 2006
A. Mottana, 'Nicander on stones and inorganic materials', *Rendiconti Lincei. Scienze fisiche e naturali*, series 9, 17 (2006), 333-353.

Miller 1986
A.M. Miller, *From Delos to Delphi. A literary study of the Homeric Hymn to Apollo*, Leiden 1986.

Nicols 1652
T. Nicols, *A lapidary, or, the history of pretious stones*, Cambridge 1652.

Ovid 1632
P. Ovidius Naso, *Metamorphoses* (George Sandys, trans.), London 1632.

Pena & Lobelius 1571
P. Pena & M. Lobelius, *Stirpium adversaria nova*, London 1571.

Perkins 1986
P. Perkins, 'Ecology, beauty, profits. Trade in lichen-based dyestuffs through Western history', *Journal of the Society of Dyers and Colourists* 102 (1986), 221-227.

Perry 1984
W.N. Perry, 'The alchemy in homeopathy', *Studies in Comparative Religion* 16 (1984), nos. 1-2.

Persoon 1801
C. Persoon, *Synopsis methodica fungorum*, Göttingen 1801.

Pfisterer 2003
U. Pfisterer, 'Erste Werke und Autopoiesis. Der Topos künstlerischer Frühbegabung im 16. Jahrhundert', in: U. Pfisterer & M. Seidel (eds.), *Visuelle Topoi, Erfindung und tradiertes Wissen in den Künsten der italienischen Renaissance*, Munich & Berlin 2003, 263-302.

Pliny 1952
Pliny the Elder, *Natural history*, Cambridge 1952.

Ramsbottom 1953
J. Ramsbottom, *Mushrooms & toadstools*, London 1953.

Rohde 1890-1894
E. Rohde, *Psyche*, Leipzig 1890–1894, 2 vols.

Roseto 1540-1565
G.V. Roseto, *Libro di tentoria intitolato Plicto*, Venice 1540 and 1565.

Rubin 1990
P.L. Rubin, 'What men saw. Vasari's life of Leonardo da Vinci and the image of the Renaissance artist', *Art History* XII (1990), 34-46.

Scarborough 1979
J. Scarborough, 'Nicander's toxicology I. Spiders, scorpions, insects and myriapods', *Pharmacy in History* 21 (1979), no. 1, 3-34; part II in: *Pharmacy in History* 21 (1979) 73-92.

Schaub & Suthor 2005
M. Schaub & N. Suthor (eds)., *Ansteckung. Zur Körperlichkeit eines ästhetischen Prinzips*, Munich 2005.

Scopoli 1772
G.A. Scopoli, *Flora carniolica exhibiens plantas carnioliae indigenas et distributas in classes, genera, species, varietates ordine Linnaeano*, Vienna 1772.

Steensma 1999
S. Steensma, *Otto Marseus van Schrieck. Leben und Werk*, Hildesheim 1999.

Thrasher 1669
W. Thrasher, *The marrow of chymical physick*, London 1669.

Torresi 2004
A.P. Torresi, *Il ricettario Medici. Alchimia, farmacopea, cosmesi e tecnica artistica nella Firenze del seicento*, Ferrara 2004.

Tribby 1991
J. Tribby, 'Cooking (with) Clio and Cleo. Eloquence and experiment in seventeenth-century Florence', *Journal of the History of Ideas* 52, (1991), 417-439.

Van Hoogstraten 1678
S. van. Hoogstraten, *Inleyding tot de hooge schoole der schilderkonst. Anders de zichtbaere werelt*, Rotterdam 1678.

Vasari 1963
G. Vasari, *The lives of the painters, sculptors and architects* (H.B. Hinds, trans.), London 1963, 4 vols.

Vasari 1987
G. Vasari, *The lives of the artists* (1568), Middlesex 1987.

Cat. Vienna 1977
Cat. Vienna, Kunsthistorisches Museum, *Peter Paul Rubens. 1577–1640. Ausstellung zur 400. Wiederkehr seines Geburtstages*, Vienna 1977.

Von Sandrart 1675-1679
J. von Sandrart, *Teutsche Akademie der edlen Bau-, Bild- und Malereikünste*, Nuremberg 1675-1679.

Watkins 1995
C. Watkins, *How to kill a dragon. Aspects of Indo-European poetics*, Oxford 1995.

Watson 1966
G. Watson, *Theriac and mithridatum. A study in therapeutics*, London 1966.

Weber 1989
G.J.M. Weber, *Stilleben alter Meister in der Kasseler Gemäldegalerie*, Melsungen 1989.

Weber 1993
G.J.M. Weber, 'Stilles Leben am Waldboden', *Kunst und Antiquitäten* 1/2 (1993), 24-29.

Weststeijn 2008
T. Weststeijn, '"Painting's enchanting poison". Artistic efficacy and the transfer of spirits', in: C. Göttler & W. Neuber (eds.), *Spirits unseen. The representation of subtle bodies in early modern European culture*, Leiden & Boston 2008, 141-178.

Weststeijn 2009
T. Weststeijn, *The visible world. Samuel van Hoogstraten's art theory and the legitimation of painting in the Dutch Golden Age*, Amsterdam 2009.

Worp 1891
J.A. Worp, 'Constantijn Huygens over de Schilders van zijn Tijd', *Oud Holland* 9 (1891), 106–136.

Worp 1897
J.A. Worp, 'Fragment eener autobiographic van Constantijn Huygens', *Bijdragen en Mededeelingen van het Historisch Genootschap* 18 (1897), 1-121.

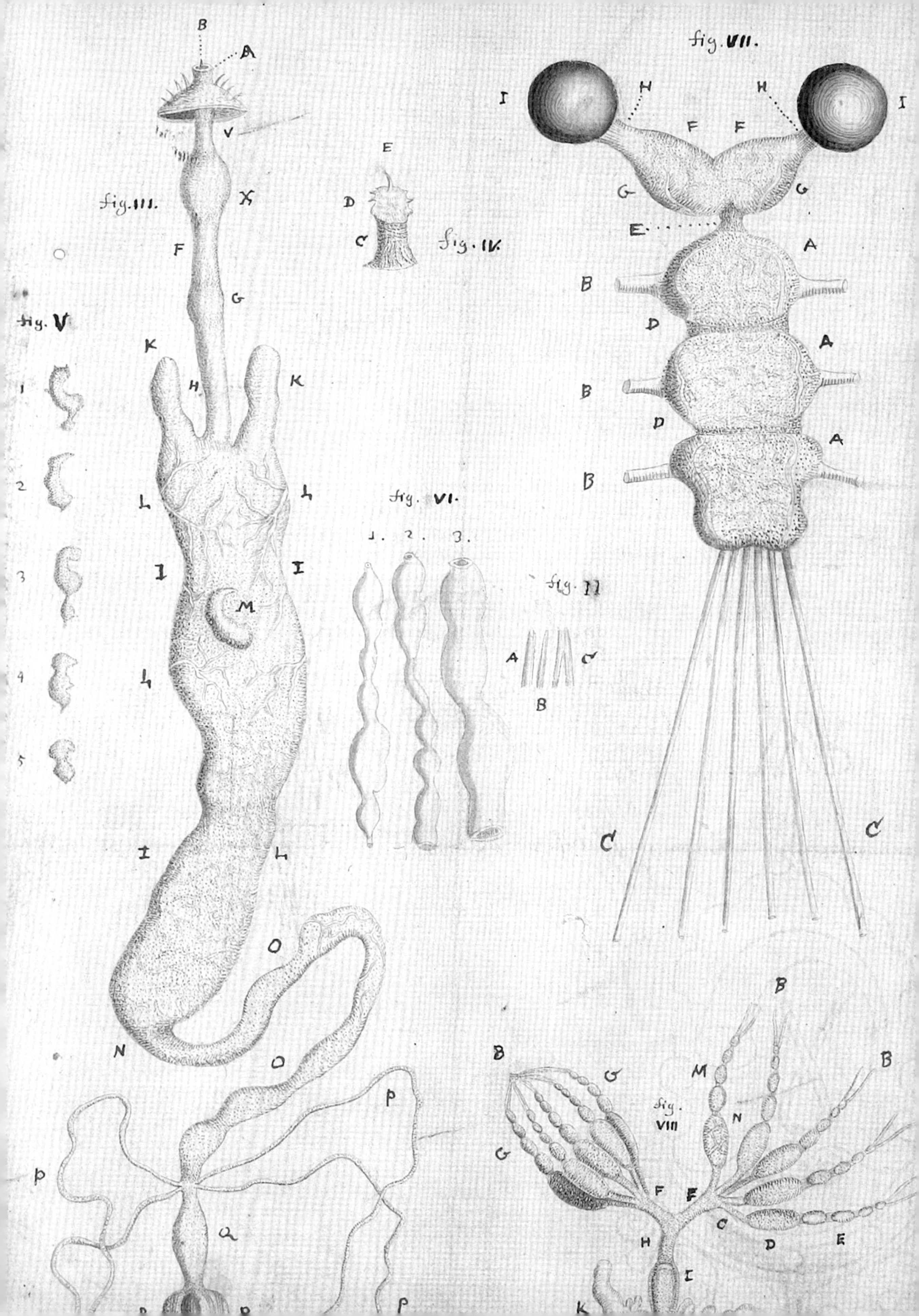
fig. III
fig. IV
fig. V
fig. VI
fig. II
fig. VII
fig. VIII

Beyond the lines of Apelles
Johannes Swammerdam, Dutch scientific culture and the representation of insect anatomy*

Eric Jorink

In April 1678 Amsterdam microscopist Johannes Swammerdam (1637-1680) sent a fascinating research report to his French patron Melchisédech Thévenot (c. 1620-1692). 'I hereby present to you', began the accompanying letter,

> the Almighty Finger of GOD in the Anatomy of a Louse, in which You will find miracles heaped upon miracles (...) The lines of Apelles are admired by all the world, but here you will discover in one part of one line the complete structure of all the most ingenious Animals in the entire universe together, as contained in one concise concept. What people, my lord, are capable of understanding this? Yet what artist can there be other than GOD who could in any way investigate and depict it?[1]

Behind these lyrical opening lines lay considerable self-awareness, since the only person at that moment who was capable of investigating and depicting such divine miracles was Swammerdam himself. One of Europe's pioneers in the field of microscopic research, he had succeeded, through much laborious effort, in dissecting and drawing the intestines of a louse. Proudly, he refers to the legendary Greek painter Apelles, whose paintings were so fine and so lifelike that both humans and animals were confused by them. More specifically, Swammerdam refers to the story, as narrated by Pliny the Elder, of Apelles expressing his admiration for the immensely laborious and infinitely meticulous work of Protogenes. Apelles visited the artist to make his acquaintance, and finding not him there but only an old woman, he walked over to Protogenes's easel and painted an extremely fine line on the panel. He instructed the woman to tell that it was painted by him. Upon returning, Protogenes drew an even finer line in another colour above Apelles's line. When Apelles returned, '(...) ashamed to being beaten, [he] cut the lines with another in a third colour, leaving no room for any further display of minute work'.[2] The implication of Swammerdam's reference to this well-known story was that he not only could at least equal Apelles in the lifelikeness of his representation of the natural world, but even could outdo him in the drawing of microscopically fine lines.

Detail fig. 2
Original drawing by Swammerdam of the intestines of the louse (1678), later published in his posthumous *Bybel der nature* (1737-1738).

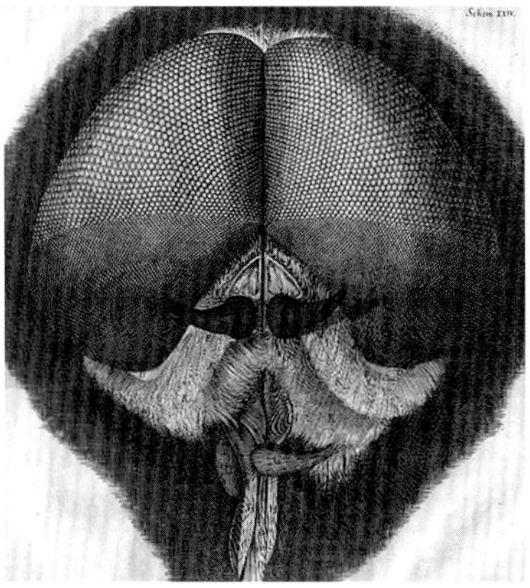

1
The multi-faceted eye of a fly, as depicted in Robert Hooke's *Micrographia* (1664)
Universiteitsbibliotheek Amsterdam, OTM: OG 80-19.

In 1608, in the Dutch Republic, an application was made for a patent on a revolutionary new instrument, the telescope, and some ten years later the inventor and natural philosopher Cornelis Drebbel (1572-1633) developed its sister instrument, the compound microscope, also consisting of two lenses fit into a tube of metal.[3] Through a variety of factors this new piece of equipment was not deployed in scientific research until the 1660s.[4] The publication of Robert Hooke's *Micrographia* (1664) – with its spectacular engravings of the minute structures of moulds and hairs and, in particular, the external anatomy of insects – fired imaginations all across Europe (fig. 1). 'Good figures. Flea and louse as big as a cat,' noted Christiaan Huygens (1629-1695).[5] A short time later, Swammerdam, using techniques of preparation and observation he had invented himself, managed to carry out research into the internal anatomy of insects.

With the autopsy he performed on a tiny louse, Swammerdam had reached the limits of human capability and imaginative power. His letter to Thévenot was accompanied not only by a 29-page research report but also by his own drawings, in pencil and Indian ink, that offered an impression of both the external and the internal anatomy of the louse (fig. 2). In words and images, Swammerdam reported on anatomical details of the insect that no one before him had ever observed: the stomach, the heart, the nervous system, the gastrointestinal tract and even the genitals were minutely described. In these descriptions, text and illustrations belong together. In his text, Swammerdam continually refers to his drawings; conversely, the remarkable pictures are quite incomprehensible without the explanatory text. A third component in the presentation of his work was the specimens he kept in his collection of *naturalia*. Using conservation techniques he had developed for the purpose, he could store his specimens and show them to curious or sceptical colleagues.

Swammerdam is known mainly as a pioneer in the field of microscopy, as well as for the religious crisis he suffered in and around 1675.[6] In this article I will examine two linked facets of his achievements that have as yet been subjected to surprisingly little scrutiny: the part played in his work by drawings and specimens, and his conception of nature as God's work of art. After a short biographical sketch, in which Swammerdam is situated in the flourishing culture of seventeenth-century Amsterdam, I will focus on his scientific work and the role images and anatomical preparations played in it. Ideas borrowed from René Descartes (1596-1650) about the order and structure of nature were his underlying motive for charting a previously unexplored domain.

Swammerdam had an extraordinary graphic talent. Many original drawings by him are preserved at the University Library in Leiden, but they have barely been studied until now. He took care to sign them 'delineavit auctor'.[7] He used engravings based on these drawings to illustrate his works, of which the most important was published posthumously. That Swammerdam produced his own visual research reports was fairly unusual, since most of his colleagues, such as Antoni van Leeuwenhoek (1632-1723), lacked the necessary skills and had to rely on professional draughtsmen.[8] Swammerdam's work is all the more remarkable for the fact that there was no pictorial tradition for him to fall

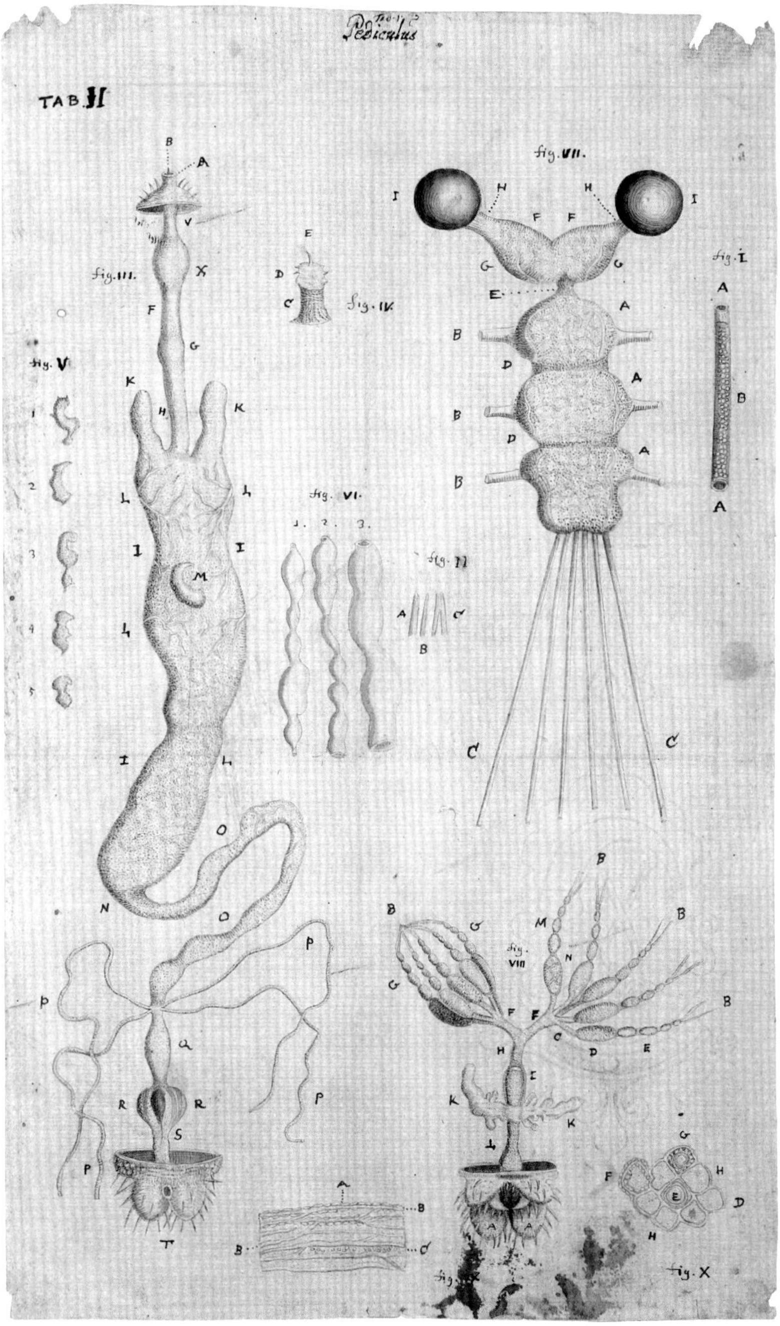

2

Original drawing by Swammerdam of the intestines of the louse (1678), later published in his posthumous *Bybel der nature* (1737-1738)

Fig. iii to the left shows the gastrointestinal tract; fig viii to the lower right represents the female uterus.

Universiteitsbibliotheek Leiden, Ms BPL 126 C f 3r.

back on. The challenges he faced were even greater than those encountered by illustrators attempting to portray the hitherto unknown life-forms of the East and West Indies. In his study of the microscope, Ruestow asserts that 'whereas it might be hard not to draw analogies between humans and the higher animals, insects demanded decidedly more imagination'.[9] The strategy of 'metonymic composition', or the construction of the unknown based on fragments of known life-forms, was impossible here.[10] Unlike Van Leeuwenhoek, his contemporary, Swammerdam attached great importance to the use of images in his work and in so doing displayed not only admirable practical skills but also a capacity for profound theoretical speculation.

Swammerdam also reflected at length on the epistemological and rhetorical functions of drawings and specimens, on the problems surrounding representation and authority, and on the way in which previously unseen body parts and bodily structures should be portrayed 'from the life' ('naer het leven'). This latter point touches upon a question that has for decades been the subject of fierce debate among art historians – namely, the degree to which we should attribute a deeper significance to the countless seventeenth-century images that portray nature (or elements of nature), such as still lifes or landscape paintings.[11] I hope that this contribution demonstrate the plausibility that the lifelike visual representation of nature in the context of the culture of the Dutch Golden Age is far more complex and nuanced than being either a value-free 'art of describing' on the one hand or a symbolic rendering of the *vanitas* motif on the other. However, before examining these aspect in more depth, let us first take a look at the world in which Swammerdam operated: seventeenth-century Amsterdam, where the arts and sciences merged seamlessly into one another.

Between *naturalia* and *artificialia*

Swammerdam was born on 12 February 1637, the son of Jan Jacobszoon Swammerdam (1606-1678) of Amsterdam, owner of an apothecary's shop called De Star, close to the Montelbaanstoren and the harbour on the IJ. On the first floor of his house, beyond the semi-public realm of the shop, the apothecary kept all kinds of objects from all over the world, both art objects and *naturalia*. He was one of many Amsterdam citizens who assembled cabinets of curiosities, acquiring exotic *naturalia* and *artificialia*, antiquities, artworks and scientific instruments and keeping them in collections that were, on the face of it, haphazardly arranged.[12] Other famous Amsterdam collectors include Gerard Reynst (1599-1658), Roetert Ernst (1616-1685), Johannes Wtenbogaert (1608-1680), Frederik Ruysch (1638-1731), Nicolaes Witsen (1641-1717) and Albertus Seba (1665-1736).[13] Rembrandt too had such a collection.[14]

Cabinets like the one belonging to Swammerdam's father had an important part to play in the seventeenth-century information explosion, as the focus of a community of discourse (see fig. 1, page 7).[15] They existed at the junctures of art and science, written knowledge and empiricism, faith in authority and scepticism. Initially, the presence of, for example, a long twisted horn or a huge bone seemed to illustrate the existence of the unicorns and giants referred to in the Bible.[16] This changed during the course of the seventeenth century. Perceptive visitors examining such cabinets learned never to accept any story as true but to look at all the evidence with their own eyes and analyze it with their own intellects. Johannes Swammerdam would later become a fervent propagator of this lesson. The objects he brought together formed the basis for a complex process of generation, transmission and verification of knowledge, which could be disseminated orally, in writing or visually. For his book about the *naturalia* and medicine of the East and West Indies, the influential physician Willem Piso (1611-1678), for many years deacon of the Collegium Medicum in Amsterdam, drew the skull of the babirusa – a member of the pig family that is found only on Sulawesi – in the possession of 'Mr. Swammerdam, apothecary in Amsterdam who is well-versed in exotica'.[17] In later life the artist and author Romeyn de Hooghe (1645-1708) remembered having seen a remarkable oriental depiction of the cosmos, 'from the hand of the late D. Swammerdam, being in the collection of rarities of his father'.[18]

Changing the order of nature: Swammerdam's network and the love for the lowest creatures

In accordance with an influential tradition that went back to antiquity, Johannes Swammerdam regarded creation as an artwork from the hand of God. Both in the Bible and in the work of pagan authors such as Cicero, Pliny and Galen, nature was seen as a product of the almighty power of a divine creator, and the beauty and efficiency of creation as proof of His existence. Swammerdam was a passionate adherent to this belief. He frequently described God in terms such as 'all-eyed Artist' ('algeoogde Kunstenaar') or 'Artist of artists' ('Kunstenaar der kunstenaars') and the works of His hand as an artistic masterpiece or an astonishingly beautiful book.[19] Swammerdam, who was not only a skilled researcher with a gift for drawing but a talented writer and poet as well, was able to write lyrical reports about everything he observed.

Scientific culture of the Dutch Golden Age is generally regarded as businesslike, pragmatic and rationalistic.[20] This impression, disseminated for many years by historians of science, has had a considerable influence on art historians too, leading to the misapprehension that 'science' was a purely descriptive and value-free undertaking.[21] On closer inspection, the descriptive-mathematical approach of scholars like Simon Stevin (1548-1620), Isaac Beeckman (1588-1637) and Christiaan Huygens turns out to be far less typical of the intellectual culture of the Dutch Golden Age than we tend to assume.[22] Here it will suffice to observe that the faith-tinted way in which Swammerdam observed nature was common in his day rather than exceptional. Colleagues, including Frederik Ruysch, Johannes Goedaert (1620-1668), Maria Sybille Merian (1647-1717) and Simon Schijnvoet (1652-1727), explicitly underwrote the a priori notion that nature was God's work of art.[23]

The innovative character of Swammerdam's achievements arises from the fact that he took this belief all the way to its ultimate conclusion. Partly as a result of the influence of Aristotle's hierarchy of forms of existence, those who adhered to the traditional conception of nature paid little attention to the lowest forms of life: insects, reptiles and worms, let alone toadstools and other damp and slimy living things.[24] It was accepted, partly on Aristotle's authority, that these lesser creatures had no internal anatomy and were produced by spontaneous generation – in other words, out of rotting organic waste. Fleas could originate from plants or from dirty hair or excrement. A dead horse could give rise to beetles, maggots and wasps. The nature of such creatures meant they were unworthy of serious study. Until well into the seventeenth century, countless species of insect were completely ignored by both artists and naturalists. There were a few exceptions, such as the bee, the butterfly, the spider and the grasshopper, which became symbols beloved of the artists who created illuminated manuscripts, paintings and emblems, partly because of their biblical connotations.[25] For centuries the butterfly, assumed to arise out of the corpse of a dead caterpillar, symbolized the Resurrection. The beehive and the community of bees within it – which was thought to be led by a king – were regarded as a metaphor for Christian society.

Partly under the influence of Pliny's famous dictum *rerum natura nusquam magis in minimis tota sit* (nature reveals itself nowhere more clearly than in its smallest products), Humanist scholars like Conrad Gesner (1516-1565) and Ulysse Aldrovandi (1522-1605) began to turn their attention to a far broader range of insects. Aldrovandi's *De insectis*, illustrated with fairly unsophisticated woodcuts, appeared in 1601, while Gessner's work would be included in Moffet's *Theatrum insectorum* (1634). Both scholars were outstanding representatives of the Humanist ideal of knowledge. The description and depiction of the external features of each species are but one facet of an endlessly complex enumeration of names, characteristics, references, correspondences, analogies, significances, sympathies and antipathies.[26] With reference to the Bible, the classics and the *Physiologus*, insects were regarded above all as symbols.

The example set by Gesner and Aldrovandi was followed by their kindred spirit Joris Hoefnagel (1542- c. 1600), who depicted countless insect species and other small creatures in his famous *Archetypa* series (1592). His pictures are extremely realistic, and the different species are placed in sophisticated ensembles in a religious-emblematic context.[27] Hoefnagel's example was followed by Jacques de Gheyn II (1565-1629).[28] Through Constantijn Huygens (1596-1687), his neighbour in The Hague, De Gheyn encountered Drebbel's microscope, and Huygens tried to persuade him that he should publish a collection of observations he had made using the instrument. It is precisely in this 'New World' of the smallest of creatures that we encounter the dedication of a divine Architect, a jubilant Huygens said, 'and everywhere we will come upon this same ineffable Majesty'.[29] De Gheyn died before his plan could be realized. In the end it was Swammerdam who gave shape to Huygens's idea.

It has often been claimed that Hoefnagel and De Gheyn helped to create the specifically Dutch genre of the still life with flowers, shells and small living creatures – one famous example being the work by Johannes Goedaert (1620-1668).[30] This pious painter from Middelburg earned his living with still lifes and worked in the same tradition as Ambrosius Bosschaert the Elder and Balthasar van der Ast (fig. 3).[31] He was also active as an observer and describer of insects. Moreover, we have only recently

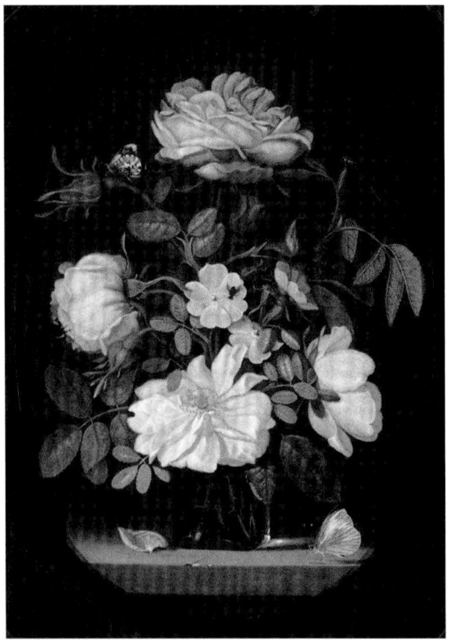

3
Johannes Goedaert, *A Bouquet of Roses in a Glass Vase*, date unknown.
oil on canvas, 50 x 37 cm. Zeeuws Museum Middelburg, inv. no. M96-031.

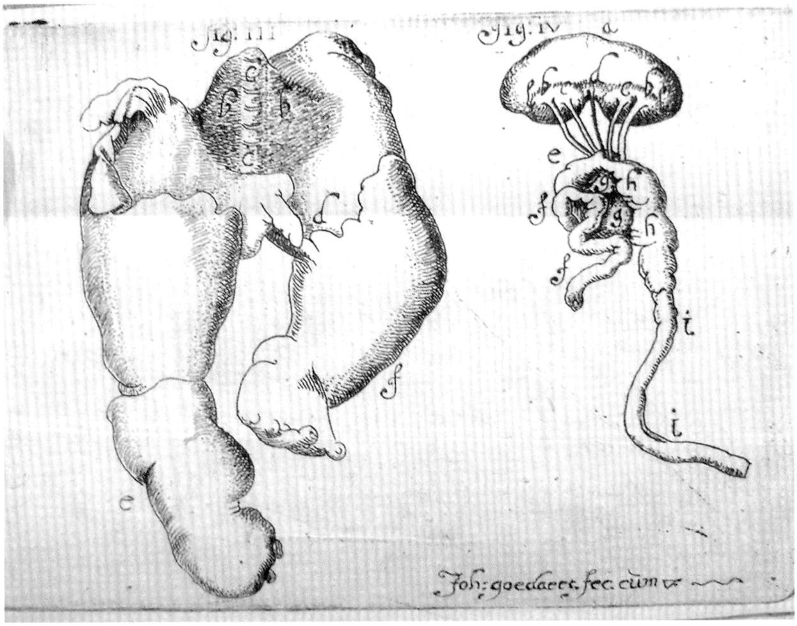

4
Illustration by Johannes Goedaert of the monstrous birth dissected at the Theatrum Anatomicum in Middelburg in 1662.
From A. Everard, *Lux è tenebris affulsa* (1662)
Universiteitsbibliotheek Leiden, 517 G 5:3.

become aware that he was commissioned to produce a picture of the foetal 'monster' dissected in Middelburg in 1662 by the local college of physicians (fig. 4).[32] In about 1635, Goedaert became one of the first scholars in Europe to begin a systematic study of the life and genesis of insects. He described around 150 species in total, ranging from bees to moths and from butterflies to blowflies. Goedaert was an enthusiastic advocate of the theory of spontaneous generation. He described watching with his own eyes as two identical caterpillars died and different insects then emerged from their remains. From one little corpse rose a beautiful butterfly, while out of the other came a swarm of 'flies'. Goedaert noted that he was 'extremely astonished' on seeing this.[33] It must indeed have been a remarkable sight, although the contemporary reader will detect in it not so much the hand of God as the activities of the ichneumon wasp. The entire process of metamorphosis was interpreted by Goedaert as the coming into existence of a life-form (caterpillar, grub), which then died and from whose remains new life (butterfly, fly) emerged by spontaneous generation. After a while these died too, 'until a new resurrection'.[34] Goedaert's observations were published between 1660 and 1669 in the three-part, richly illustrated *Metamorphosis naturalis*, but however important Goedaert's work may have been, by the time of its publication it was already outdated as a result of a quick succession of developments in natural science. Unlike Robert Hooke (1635-1703), for instance, Goedaert had not used a microscope, and his choice of the woodcut technique for the illustrations meant it was almost impossible to make out the details, which were essentially the insect book's *raison d'être* (fig. 5). Above all, Goedaert's observations were coloured by his religious-emblematic thinking. The butterfly was a symbol of the Resurrection, the

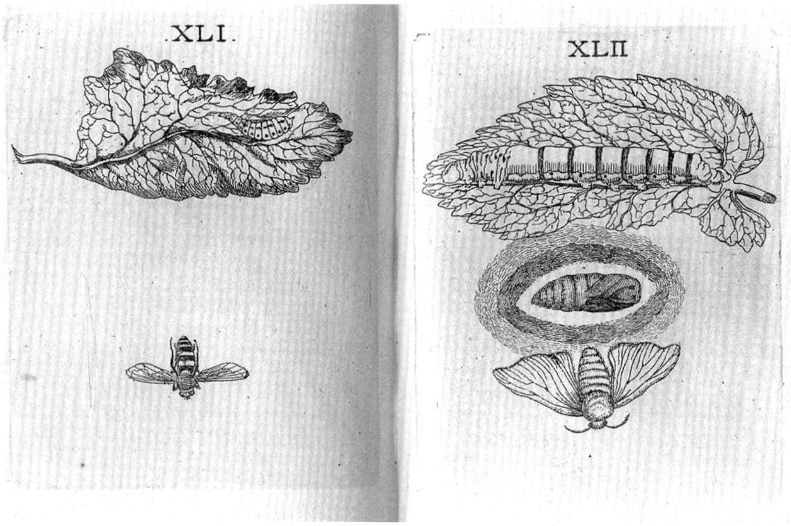

5
Two woodcuts depicting insects, taken from Goedaert, *Metamorphosis naturalis* I (1660)
Koninklijke Bibliotheek, The Hague, 2215 H 32.

ant was an emblem of diligence and the grasshopper was a reference to Old Testament plagues.

The theoretical starting point for Swammerdam's study of the world of the tiniest creatures lay not in the biblical Humanist tradition but somewhere else entirely – namely, in the philosophy of René Descartes. As Rienk Vermij sets out elsewhere in this volume, the Frenchman, who lived in the Dutch Republic from 1628 to 1649, was a man of great influence in his second homeland.[35] Descartes regarded the universe as a vast mechanism in which the only factors that truly mattered were material and motion. Planets, clouds, people and animals consisted purely of moving particles (fig. 6). Everything in the cosmos obeyed a limited number of fundamental natural laws.

Several other implications of Descartes's natural philosophy will be briefly outlined here. Firstly, the mechanistic model gave an enormous boost to anatomical research. The concept of the body as a machine, mechanism or laboratory created a revolutionary new interpretative framework that allowed Harvey's theory of the circulation of the blood (*De motu cordis*, 1628) to be given a philosophical basis and encouraged countless other studies of anatomy and physiology.[36] Secondly and even more importantly, from Descartes's point of view nature was stable, orderly and uniform, governed by laws.[37] Those same laws applied at all times and in all places, and there was no ontological difference between the various forms of life. The traditional distinctions between high and low, large and small, beautiful and ugly, rare and common, or between male and female, were dispensed with. Everything was merely matter-in-motion; the cosmos was essentially a meaningless mechanism.

The Cartesian concept of nature as a mechanism raised a number of questions. If people and animals are machines, where do baby machines come from? If the laws of nature are the same always and everywhere, can spontaneous generation be ruled out? With the aid of a microscope, might it not be possible to see the minute particles of which nature was now thought to consist? In Amsterdam, from about 1660 onwards, there was an active network of men who regularly posed questions like these. Inspired in part by the new Cartesian natural philosophy and the challenges it posed, they committed themselves to nature study in general and the investigation of the problem of reproduction in man and the animals in particular. Among the members of this informal group, who included Swammerdam, were the young student of philosophy and medicine Burchardus de Volder (1643-1709); physician Mattheus Sladus (1628-1689), since 1650 a medical practitioner at the Binnengasthuis; Nicolaes Witsen, a son of the ruling elite; and the painter Otto Marseus van Schrieck (c. 1620-1678), creator of his very own genre of paintings figuring reptiles, amphibians, insects and toadstools in ominous formations.[38] All such creatures were generally assumed to emerge by spontaneous generation, and they were consequently seen as the lowest link in the 'great chain of being'. Marseus had developed his unique style during a stay in Italy, and in about 1660 he settled near Amsterdam.[39] In his country house, called Waterrijk, in the marshy area around Diemen, Marseus had a collection of *naturalia* and artworks, and he bred all kinds of snakes, reptiles and amphibians for use as models for his paintings. At Waterrijk he continued his work and succeeded in creating a niche for his art, even though only *virtuosi* ('liefhebbers') were capable of understanding and properly appreciating it.

The major source of inspiration for the activities of this network was a man who later became the city's mayor and director of the East India Company, regent Johannes Hudde (1628-1704). In the 1650s he had studied philosophy and medicine in Leiden, and he was enthralled by Descartes's philosophy.[40] Hudde was fascinated by the microscope, the new instrument that, along with the telescope and the air pump, would emerge as an icon of the Scientific Revolution.[41] In the early 1660s Hudde succeeded in manufacturing simple microscopes using melted balls of glass, with the help of which Swammerdam and later Van Leeuwenhoek made their revolutionary discoveries. Hudde generously shared his invention with a wide circle of like-minded enthusiasts, and by 1662 at the latest, both Otto Marseus and Johannes Swammerdam were in possession of Hudde microscopes.[42] Hudde was also responsible for the research agenda that the informal group around him worked through in the 1660s, focusing on microscopical details and the problem of procreation. Swammerdam, Marseus and Sladus devoted themselves, both individually and collectively, to unravelling the mystery of reproduction.

Marseus made the provocative decision to position the lowest forms of life at the centre of his compositions. In his dynamic narrative paintings, insects and fungi were often depicted at various stages of their development. Marseus deliberately made the dividing line between nature and its representation the subject of controversy by pressing real butterflies into the still-wet paint. Later his friend Swammerdam would describe in detail how 'one could ornament a painting with the wings of a Butterfly'.[43] Swammerdam and others also recount that Marseus's ambitions went beyond those of the *pictor et inventor*. He was labouring

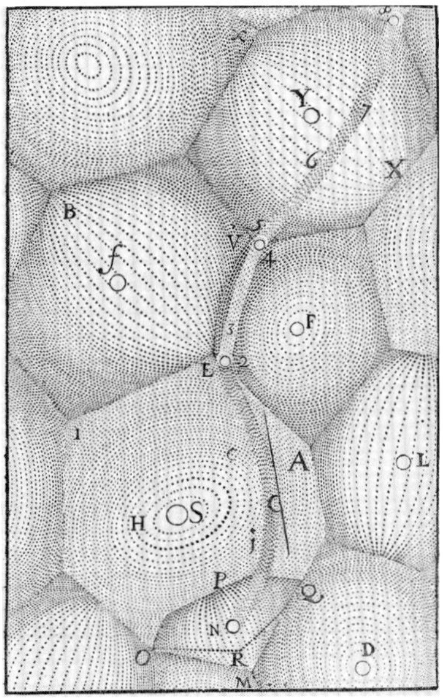

6

Descartes' model of the universe: particles collide and coagulate, forming planets and comets. The cosmos is just matter in motion, without any deeper meaning or hidden significance. From R. Descartes, *Principia philosophia* (1644) Universiteitsbibliotheek Leiden, 616 C 13.

away at a scientific work, never published, that included pictures and descriptions of bloodless animals.[44]

Swammerdam expressed an even more explicit preference for the 'least esteemed creatures' than Marseus did. He too studied amphibians, reptiles, fungi and insects (fig. 7). Time and again he stressed that they had an admirable anatomy whose minuscule size actually made it all the more admirable. The assumption that they were produced by spontaneous generation was not only untrue, it was blasphemous. After all, such a theory implied that part of nature had exempted itself from the order and laws laid down by God. Belief in spontaneous generation was a 'direct route to Atheism'.[45] Nature was orderly; there could be no such thing as chance.[46] With explicit reference to Descartes, Swammerdam translated this belief into natural history.

Once he came to this conviction, Swammerdam would devote his entire scientific career to disseminating the notion that those forms of life that had traditionally been the least valued were in no sense inferior to higher forms. Natural history as practised by his predecessors and contemporaries, he claimed, was deeply rooted in Humanism. Practitioners blindly followed the commentaries, annotations and marginalia of earlier authorities without using their own eyes. In a telling comparison he claimed that the image his contemporaries had of nature was that of a 'decent painting', which because of 'our flights of fancy' and

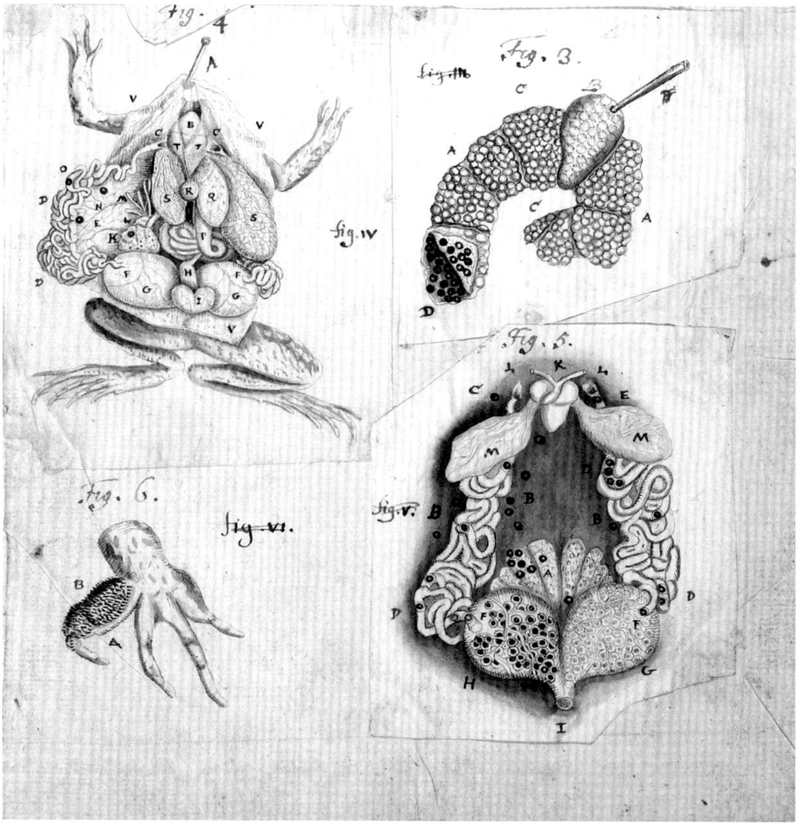

7
Like his friend Otto Marseus, Swammerdam was fascinated by toads and frogs, and studied their anatomy. **Fragment of an original drawing by Swammerdam,** now in Universiteitsbibliotheek Leiden, Ms BPL 126 C f 48^r.

'harmful traditions has become as if befouled and defiled'.[47] Swammerdam's aim was to remove all the baked-on varnish and dirt by means of his own research, so that the painting would be restored 'in its rightful sheen and its own beauty'.[48]

Drawings and specimen: the importance of visual representations and material evidence

Like several of his friends and colleagues – Witsen, Huygens and Spinoza, for example – Swammerdam was exceptionally good at drawing. He probably took lessons as a child at one of Amsterdam's many artists' studios. The illustrations he made of the objects of his research served a number of purposes, some of which overlap. For one thing, drawing was a first step in the cognitive process. By giving shape to what he saw, in his head and with his hand, he could abstract from reality and offer an initial interpretation of the things he observed.[49] Secondly, illustrations helped him to compile his report. For example, they enabled him to look back later to see whether or not he had observed a particular organ of a given species correctly, and to compare analogous organs in different species.[50] The drawings were also the source material for the engravings he used to illustrate his observations and descriptions in his published work. Last but not least, in the shadowy zone that lay between private research and publication in the scientific arena, the drawings served as silent witnesses that could be called upon in case of any dispute over precedence. In the extraordinarily dynamic scientific culture of the seventeenth century, discoveries followed each other in quick succession. As a result, establishing the nature of a specific observation and attributing precedence over the claims of others became epistemological problems of the first order.[51] Were there craters on the moon or not? Was it in fact true that an air pump could create a vacuum? Were those bulges around Saturn, as seen through the telescope, really rings? Questions of a similar kind arose from the virtually unexplored territory on which Swammerdam was active. To generate irrefutable proof, he produced dated drawings of important observations and gave them to colleagues who were widely acknowledged to be honourable, such as Sladus and Witsen, to look after. Should a conflict arise with third parties, the drawings could serve as evidence.

Swammerdam explicitly strove to make drawings 'naer het leeven' of the things he observed. He was well aware of the inevitable problems, both theoretical and practical, of this approach. Behind his apparently clearly defined project lurked an epistemological debate that was as relevant to art theory as it was to the practitioners of research into natural history.[52] Putting aside the largely irrelevant question of whether the supposedly realistic depictions of all kinds of animals by artists such as Albrecht Dürer, Joris Hoefnagel and Otto Marseus belong to the domain of art or science, we need to ask whether 'from the life' refers to specific animals or plants with their own highly individual characteristics (flaws, injuries, degrees of asymmetry) or whether it involves a certain amount of idealization ('the' rhinoceros, 'the' camel, 'the' human being).

Swammerdam reflected deeply on this issue and took a pragmatic stance. In line with the tradition of anatomical illustrations, he generally abstracted and emphasized the shared features of a species, but if there were reasons for depicting exceptional cases, he would do so. The earliest reference to a drawing by Swammerdam concerns a (since lost) picture of a hermaphrodite dissected in the anatomical theatre at the University of Leiden on 12 October 1662.[53]

Swammerdam's career as an anatomist can be divided into a number of phases that gradually overlapped, from his research into humans and mammals to his dissection of the very smallest of insects. He must have got to know Hudde and his circle in about 1660 before going on to study in Leiden from 1661 to 1667. After gaining his doctorate he went back to live with his father in Amsterdam and continued his studies there until his father died in 1678, with a short interruption in 1675-1676 because of his religious crisis. By the time of his death in 1680 he had, in effect, carried out the projects formulated by Huygens and Hudde of studying nature's smallest creatures by means of the microscope.

During his studies in Leiden, Swammerdam excelled at anatomical dissection. Along with three other students – Niels Stensen (better known as Steno, 1637-1689), Reiner de Graaf (1642-1673) and Frederik Ruysch – he was part of a brilliant quartet that made innumerable discoveries under the leadership of tutors Franciscus de le Boë Sylvius (1614-1672) and Johannes Van Horne (1621-1670).[54] Of particular importance was the technique for preparing specimens that had been developed by Swammerdam around 1663.[55] Previous anatomical research had always been hampered by the inevitable decay of the material, which meant among other things that body parts and structures became difficult to see and discoveries could not be preserved. Swammerdam's technique involved soaking specimens or parts of specimens in turpentine for a long period. This caused the various bodily fluids to flow out and the veins and arteries to soften. Warm coloured wax was then injected into them, making all kinds of structures visible in detail (fig. 8). Once embalmed, the specimens could be preserved successfully. Over the years, Swammerdam developed his technique further and shared it with his fellow students. He kept his own specimens in his rapidly expanding cabinet, which was essentially a research collection, complementing and overlapping his collection of drawings. Based on Swammerdam's approach, Frederik Ruysch was able to put together his world-famous collection of anatomical specimens.[56]

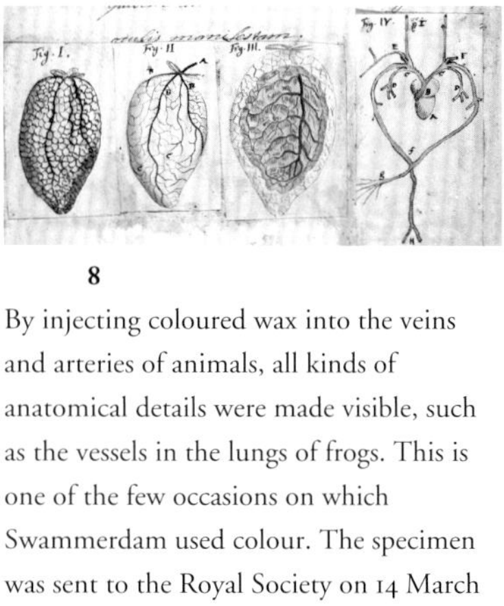

8

By injecting coloured wax into the veins and arteries of animals, all kinds of anatomical details were made visible, such as the vessels in the lungs of frogs. This is one of the few occasions on which Swammerdam used colour. The specimen was sent to the Royal Society on 14 March 1673.
Archive Royal Society Ms . S I, no. 119, reproduced with permission.

Meanwhile Swammerdam was turning the main focus of his attention to the world of insects. Here too, using his refined preservation techniques, he was increasingly able to study not just external features but also internal organs. Until the publication of his *Historia generalis insectorum* (1669) he made little use of the microscope for this kind of work, but between 1670 and 1679, challenged by the richness of his field of research and aided by his own developing skills and continually improving techniques, he concentrated mainly on the internal features of his insect specimens, culminating in his dissection of the louse. In 1678 he was able to send Thévenot the description cited at the beginning of this article and several drawings of the louse, and he promised that eventually he would even be able to provide a specimen of the creature's ovary (!).[57]

It was in this context that Swammerdam's talent for drawing proved so important. His illustrations enabled him to show in detail what he had seen. In *Historia generalis insectorum* he had presented mainly detailed descriptions and drawings of the external appearance of insects. In this field there was hardly anything that could be described as traditional. Swammerdam wrote derisively about Goedaert, whose approach meant that he 'seems to be writing a novel rather than a true History'.[58] He was full of praise for Hoefnagel, however, and he hugely admired the pioneering work of Hooke. At the same time, Swammerdam entered into a debate with the Englishman not just verbally but also pictorially.[59] Hooke had portrayed the freshly discovered microscopic world in all its oddity; the flea, the louse, the water flea and, most famously of all, the compound eye of the fly were shown entirely without context, so that the viewer could only guess at their scale and relative proportions. The engravings that Swammerdam included in his work were explicitly intended to supplement and improve upon Hooke's efforts. The Englishman had drawn the larva of a mosquito in isolation, whereas Swammerdam's engraving was not only more detailed but also showed the creature in its natural habitat, under the surface of the water. He offered two representations of the creature in one illustration, one on a scale true to life and another as seen through a magnifying glass at around fifteen times its natural size (fig. 9). Later, Swammerdam would use Hooke's illustration of the decontextualized eye of a fly as a point of departure for his own, more elaborate depictions of the organ and its parts (fig. 10; cf. fig. 1).

Swammerdam used his illustrations to show the reader previously unseen

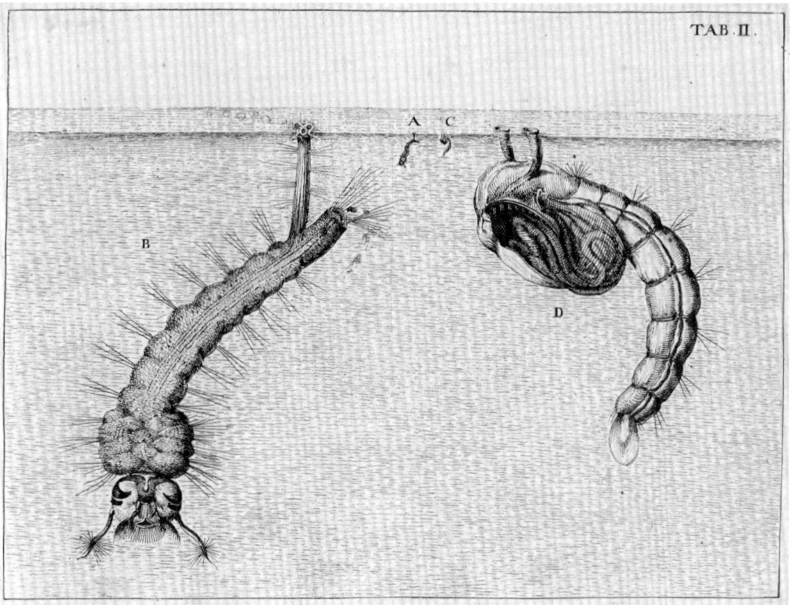

9

Representations of the larva of the mosquito in its natural habitat, under the surface of the water. In figs. A and C the creatures are depicted life-sized; in B and D magnified about fifteen times. **From Swammerdam *Historia insectorum generalis* (1669)** Koninklijke Bibliotheek, The Hague 842 E 92.

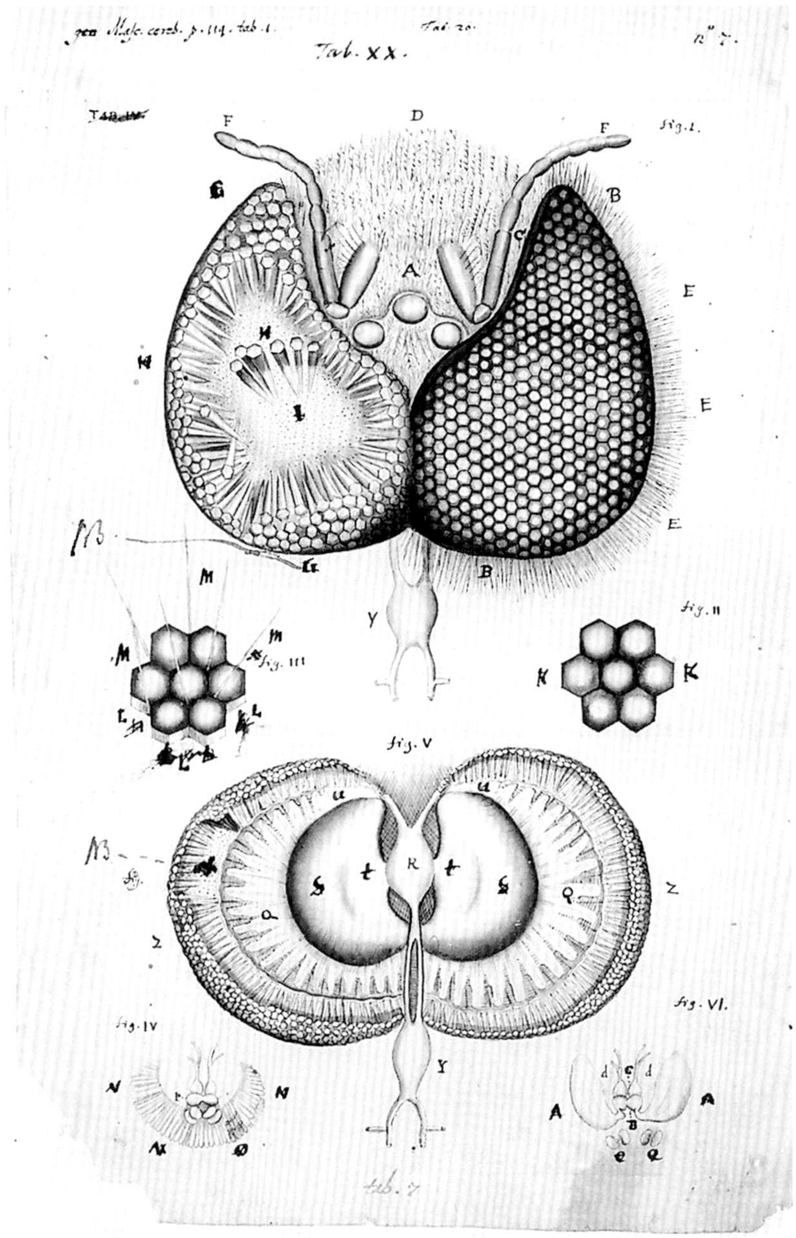

10

Swammerdam engaged in a visual dialogue with Hooke, using the Englishman's engravings as a point of departure for his own illustrations. In this original drawing, Swammerdam shows not only the outward appearance of the fly's eye, but also its internal details (cf. fig. 1). Universiteitsbibliotheek Leiden Ms BPL 126 C f21r.

details; furthermore, they were part of a visual narrative that mirrored his scientific theories. At the heart of *Historia generalis insectorum* is the axiom that insects do not arise from spontaneous generation or by means of a mystical metamorphosis but, like other species of animal, from a female egg, or ovum. Moreover, in words and images, Swammerdam made the comparison between the growth of plants and the generation of animals, thereby demonstrating the uniformity of nature in a very powerful way (fig. 11). The process of metamorphosis in insects was purely a matter of change in form and scale. A louse grew out of a minute egg. A

11

Visual representation and demonstration of the identical processes of generation of animals and plants, in this case the frog and the carnation. **From Swammerdam *Historia insectorum generalis* (1669)** Koninklijke Bibliotheek, The Hague 842 E 92.

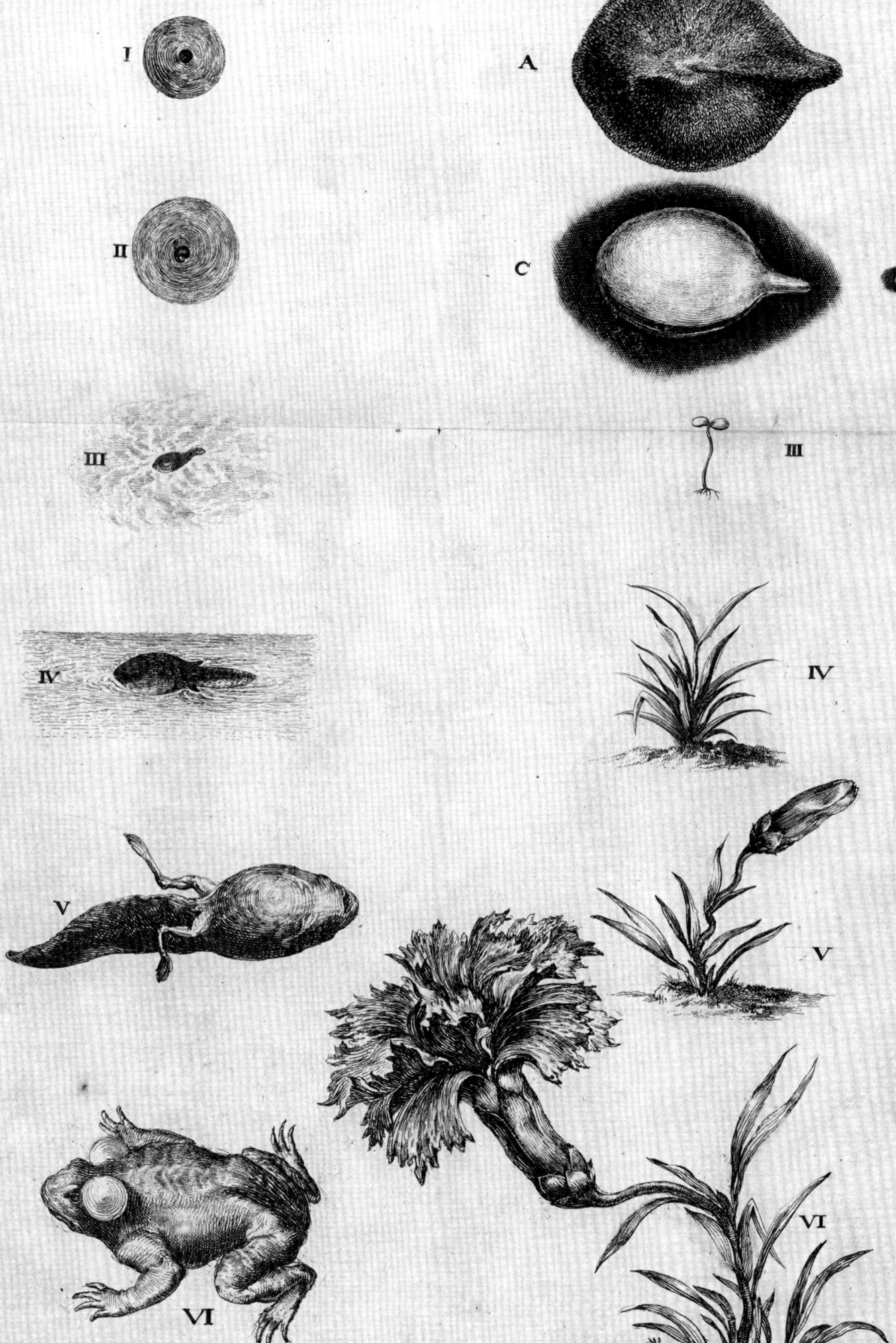
I
II
III
IV
V
VI
A
C
B
I
II
III
IV
V
VI

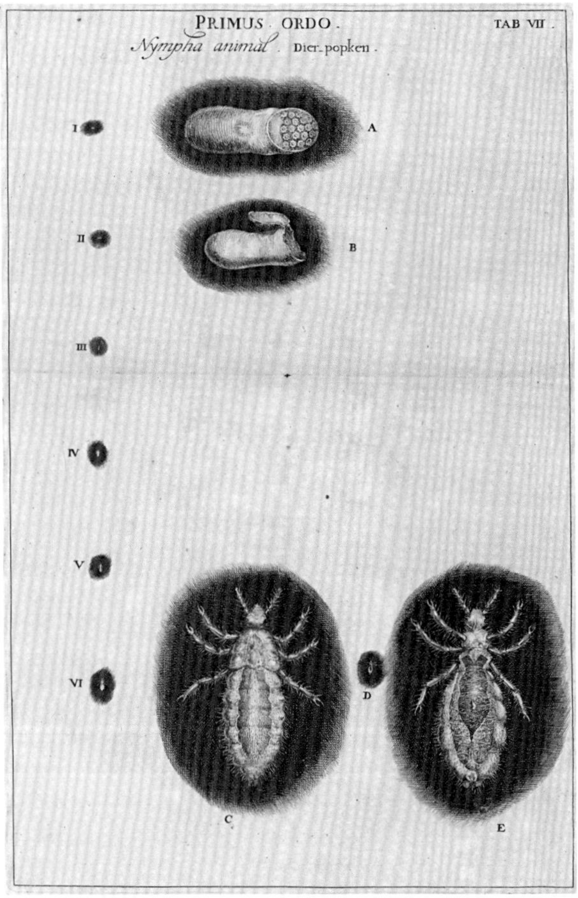

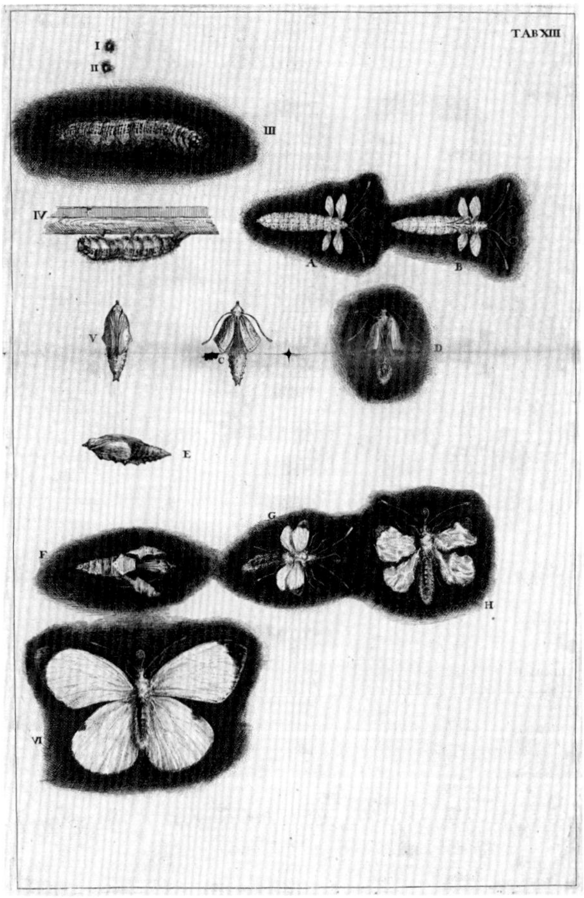

12 and 13

Swammerdam depicted the genesis of various insects and their subsequent stages of development. Here we see the growth of a louse from the egg stage and a caterpillar turning into a butterfly. In the original engravings the size of the eggs and the full-grown insects corresponds to reality. Note the use of a black background to make anatomical details more easily visible. **Taken from Swammerdam, *Historia insectrorum generalis* (1669).** Koninklijke Bibliotheek, The Hague 842 E 92.

caterpillar became a butterfly not as a consequence of a puzzling 'inadvertent transformation, abandonment of form, known as death and resurrection',[60] but rather of a process of growth that was precisely the same as that of 'a Chicken, which does not *change* into a hen but grows in its limbs and so *becomes* a hen'.[61] In his engravings in *Historia generalis insectorum*, Swammerdam shows step by step how the various insect species change their shape and size, starting with a drawing of an egg. The realistic effect is enhanced by the fact that in the original engravings, the entire process is shown at life-size (figs. 12 and 13). The reader sees not just the egg from which the louse or the caterpillar emerges and the caterpillar that results but also the parts of the butterfly that are already present in the caterpillar, along with the pupa and the butterfly itself. Every one of his engravings tells a story and can be read as a narrative, each depiction within it representing a chapter.

Swammerdam was very much aware of the artistic problems involved. He rejected the use of colour since it would distract attention from the depiction of anatomical structures. He realized that certain details – the white hairs on some small animals, for example – were best depicted against a black background, and he promised his readers that all his drawings were 'naer het leeven', by which he actually meant they were slightly idealized, emphasizing the features specific to each species, since the depiction of anatomical details demanded a certain degree of abstraction.[62] He explicitly mentioned that some details were shown on a slightly larger scale in relation to the rest of the body than was the case in reality, since it meant they were easier to see and given greater emphasis.

The crucial role that drawings played for Swammerdam in presenting his research to the public becomes evident in his difficulties in having them translated in equally accurate reproductions. Swammerdam was certainly conscious of his talents as an anatomist, taxidermist, observer and draughtsman, but he had to depend on others to transform his drawings into engravings. From his correspondence it becomes obvious that he had difficulty all his life in finding engravers who could transfer to copper the drawings he made with such meticulous care.[63] We do not know who was responsible for the engravings published in *Historia generalis insectorum*, but it is clear that the author knew Romeyn de Hooghe well and had high regard for his skills. In October 1670 Swammerdam toyed with the idea of having his book translated, writing that in that case

> one must have Romijn cut the plates again, leaving out the brown backgrounds. There are improvements to be made to the plates here and there as well. (...) This might be taken in hand very competently by Romijn, and not incompetently by others who wish to do the work. I could profitably have let it all be done [at my own expense], but since I receive from my father nothing but room and board, that is impossible for me.[64]

In other words, Swammerdam could not actually pay De Hooghe, who was both extraordinarily skilful and part of the circle that included Hudde and Swammerdam. He had created, for example, the frontispieces for Witsen's *Aeloude Scheepsbouw en bestier* (1670). De Hooghe had also been responsible for the title pages to the Amsterdam edition of Francesco Redi's *De insectis* (1672), as well as a volume of letters written by Antonie van Leeuwenhoek and the illustrations to the description of Levinus Vincent's cabinet of *naturalia*, entitled *Wondertooneel der nature* (1706).[65] In 1678 Swammerdam seems again to have wanted to use the services of De Hooghe. Meanwhile Dirk Bosboom (1641-1702), another member of the circle, had made engravings for him,[66] but every now and then Swammerdam grumbled about the slowness and above all inaccuracy of his engravers, accusing them of creating illustrations that were 'against nature, and against painting' and of working purely for monetary gain.[67]

Such outbursts only increased as Swammerdam's research became more complex. Several months before the publication of *Historia generalis insectorum*, Marcello Malpighi's *De bombyce* (1669) was published, in which he described the internal anatomy of the silk worm using no fewer than 48 engravings. Swammerdam, who refused to believe anything until he had seen it with his own eyes, decided to replicate the Italian's observations – and discovered mistakes in the way the anatomy was depicted.[68] He claimed publicly that Malpighi's illustrations 'seem to have originated in his own mind'.[69] His words appear to have been carefully chosen with the aim of accusing Malpighi of not always having drawn from the life. Malpighi's work encouraged Swammerdam to study the internal anatomy of insects systematically, using a powerful single-lens microscope. From 1670 onwards he would examine, as no one else had ever done, the insides of insects, including the mayfly, the cheese mite and that least esteemed of all creatures, the louse. Since Hooke and Malpighi had moved on to other subjects and Antonie van Leeuwenhoek had not yet entered the scientific arena, Swammerdam held a monopoly on this area of study for several years.

Swammerdam quickly had to develop his own dissection technique. There was no tradition for him to follow, since Malpighi had been frustratingly silent about his methods. Swammerdam suffocated or drowned the creatures in spirit of wine or urine and then cut them open with a tiny pair of scissors (the use of a knife would have damaged the structure too severely), after which bodily fluids and fat were removed by soaking the specimens in turpentine. The internal organs were then studied one by one under the microscope. Swammerdam usually removed individual organs and inflated them with a tiny glass tube or injected them with coloured wax, before fixing them to a piece of glass or to a background in a contrasting colour. He observed the water flea in a tiny half-globe of glass filled with water:

> And should it happen that the little animal could not be seen properly on a white ground, we would change the white to yellow, green, blue and so on, applying to those aforementioned pieces of glass blue dyes, black dyes, vermillion and other stains.[70]

Observations were always replicated and verified, since

> the spectacle is always clearer in one than in another, but everything depends upon the eye and the hand with which it is seen and dissected.[71]

Here again we see the importance attributed to reproducibility of observations and experiments in scientific culture from this time onwards, most explicitly advocated by the Royal Society.[72]

Swammerdam created as accurate a picture as possible of every body part or organ he observed. He understood better than anyone what an inordinately difficult task this was, comparing it, for example, to drawing the sun with a piece of charcoal.[73] As he penetrated more and more deeply into the mysteries of nature, his means of representation changed. Formerly, when depicting the external and, to a limited extent, some of the internal parts of insects, he had aimed for contextualization, but as he stepped into unknown territory it became impossible to give any context. He could show the extraordinary forms and structures of insect eyes, nervous systems, hearts and reproductive organs only separately, or in relation to each other (cf. figs. 2, 14 and 17). External reference points or

14
This collection of original drawings of beetles, ca. 1673, nicely illustrates both how Swammerdam mastered the technique of representing creatures 'from the life' and how he sought ways to represent their previously unexamined intestines. Universiteitsbibliotheek Leiden, Ms BPL 126 C f 31r.

tab. 30 ad hist. scarab. Tab. XXX.

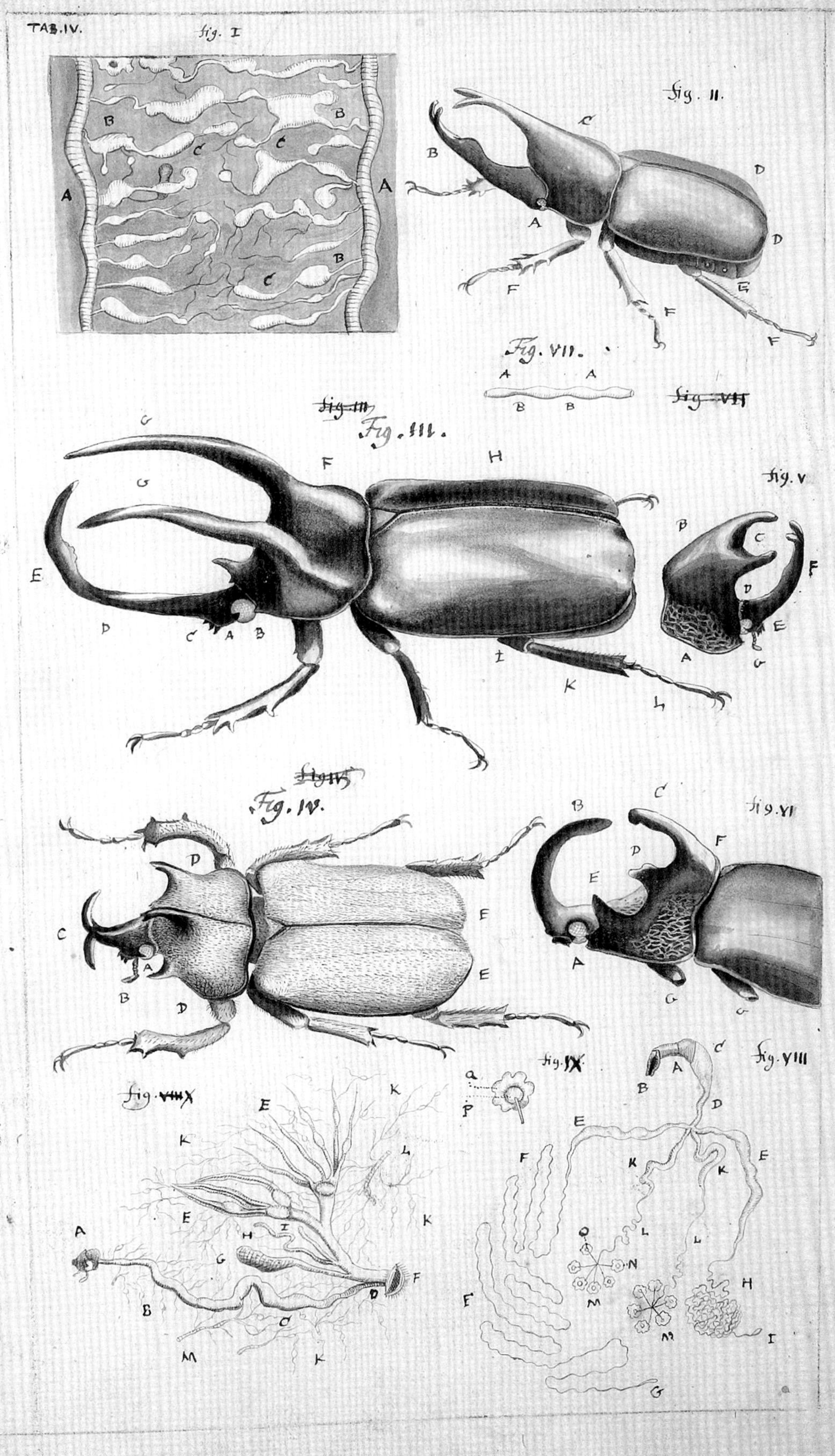

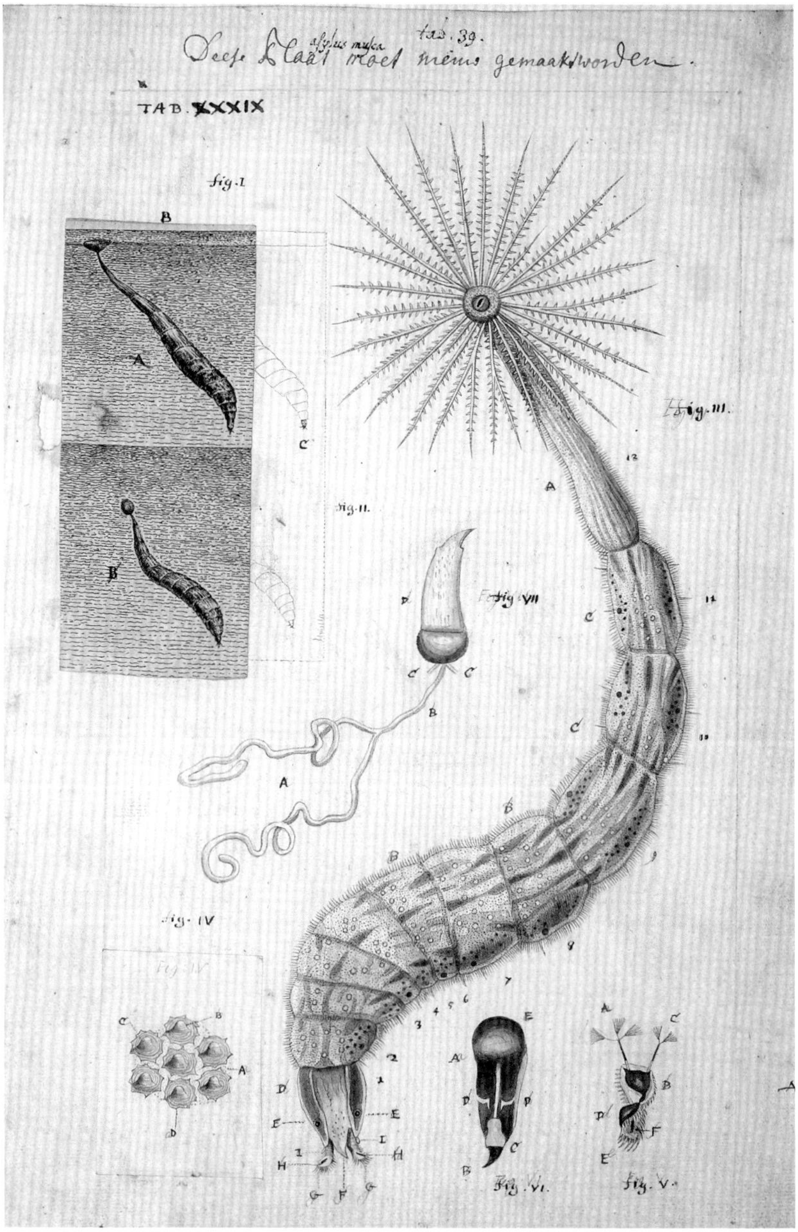

15
On the basis of the engraving published in the *Historia generalis insectorum*, Swammerdam added new observations and drawings, in this case of the larva of a particular kind of fly (the *Stratiomyia furcata*). The drawings were made around 1678, and part of the published engraving from 1669 was clued onto it. It was published in **Swammerdam, *Bybel der Natuure* (1737-1738)**
Universiteitsbibliotheek Leiden, Ms BPL 126 C f20^{r}.

indications of scale were now absent. His earlier illustrations had to a certain extent spoken for themselves; now they could not be understood without elucidation. Swammerdam's drawings were given comprehensive legends, as if they were maps or plans of unknown territories.

The drawings he made from 1670 onwards – most of which were published posthumously more than half a century later in his 'great work', the *Bybel der natuure* (1737-1738) – have another distinguishing feature. Showing almost exclusively anatomical details, with no reference to anything external, Swammerdam now also abandoned the artistic techniques he had once employed, such as the use of a black background or the careful addition of shadows to produce a realistic effect. He adjusted the engravings made for *Historia generalis insectorum* when he came to reuse them in preparation for publishing his great work (fig. 15). Lettering was now included, and he dispensed with the black backgrounds. Nature was presented in its bare essence.

Occasionally Swammerdam would once again show an entire insect in all its complexity. Perhaps the most ambitious drawing he ever made was of the mayfly, first published in 1675 (fig. 16).[74] The insect was dissected organ by organ, then each part was depicted and finally an image of the whole was put together based on the separate parts. (The method used to achieve this is reminiscent of the way Dutch still life painters constructed their compositions.) Letters were placed next to each part to indicate where the detailed explicatory text could be found.

Everything Swammerdam saw through the microscope he interpreted and presented in terms borrowed from art, reminding us of the flourishing culture of collecting and painting of the time. Hidden things such as the anatomy of the mayfly or the louse, or the virtually invisible spores of ferns, turned out to be 'genuine showpieces', 'confronted with which all the lines of Apelles and all the sophistries of human intelligence must be regarded as Folly'.[75] Swammerdam associated the order he observed through the lens with 'invention' and 'art', and again and again he described parts of insects as 'ingenious' or 'artistic'.[76] He compared the inside of a gall with a 'still life' and described the wings of a butterfly as 'brilliant and dazzling mother of pearl (...) that has been polished'.[77] He was astonished not only by the multiplicity of caterpillar species, 'different and different again in their contrivance, and in their markings indescribable', but also, and above all, by the beauty of butterflies, which put even peacock feathers in the shade:

> So are its wings studded, in an orderly manner, as if with Pearls and diamonds, which seem to catch yet more of a gleam from untold Sapphires, Turquoises and Rubies; the mother-of-pearl shells and the silver places on its wings, with their sparkling reflection of rays, outstrip the tints of rainbows.[78]

Repeatedly he spoke of the astonishing 'handiwork' or 'embroidery' of the tiniest creatures, terminology we find again later in the work of his colleague Ruysch.[79]

Nature was not an autonomous artwork; rather, it pointed to the divine artist. Human hands were always outdone by the perfection of God's creation:

> In its competition with nature, art always renders up losers. Therefore, however skilfully the great sculptor Phidias chiselled his own depiction of Minerva onto a shield, should that shield be broken every image by Phidias would be lost. Artworks always bear the stamp of imperfection, so immediately after they emerge they may be reduced to nothing, whereas by contrast the works of nature (oh admirable, oh

Tab. XIV

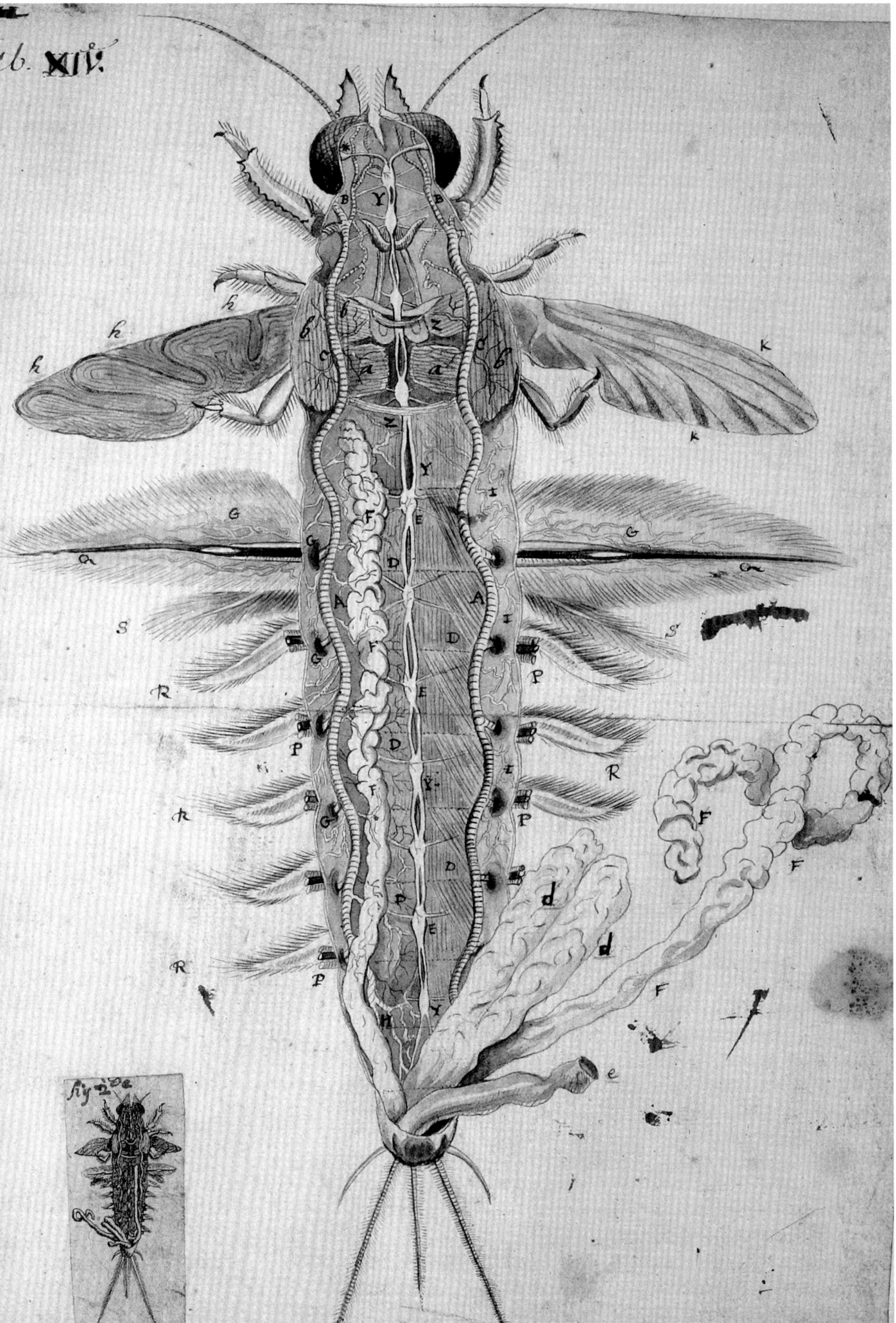

Delineavit auctor

adorable knowledge!) are of such perfection that through the whole of a body, and through the separate parts of that body, they demonstrate the supremacy of the work of art, and as the almighty creator of this universe is almost visible in all His works, so one can almost reach out and touch Him in its very smallest parts.[80]

The changing meaning of the representation of insects

Swammerdam's discoveries were of great importance to knowledge about insects and to debates about reproduction.[81] His observations also had enormous consequences for the traditional understanding of insects. They and other small animals were now de facto detached from the biblical-emblematic context explicitly given them by artists such as Hoefnagel and Goedaert. Swammerdam forthrightly claimed that as a result of his discoveries, centuries-old analogies ceased to be tenable. The famous metaphor of the beehive, for example, did not hold water now that its inhabitants had suddenly turned out to be led by a female (fig. 17).[82] When he discovered the organs of a butterfly inside a caterpillar, he remarked:

(...) so we see clearly here the error of those who have tried to use these natural and comprehensible changes to prove the resurrection of the dead, whose power, evidenced by the order we find in nature, not only far exceeds theirs but has no equal at all in that same.[83]

The devout Goedaert had taken spontaneous generation to be a fact, regarding it as an inexplicable divine miracle and attributing to it all kinds of pious significances. Swammerdam took for granted the axiom that everything in nature obeys the same 'rules and order'.[84] There was no difference between 'high' and 'low' forms of life. All were, as he demonstrated, equally complex in their structures, including – indeed especially – the very smallest of species. For Swammerdam the source of piety was no longer the analogy between a caterpillar's metamorphosis into a butterfly and the Resurrection but the unfathomable anatomy and genesis of such creatures. In their entrails the might of the divine Creator was 'depicted' both visibly and legibly.[85]

In a more general sense it is possible to conclude that for Swammerdam nature was no longer a complex collection of largely random symbols that derived their significance from scriptural tradition. His creatures all pointed *linea recta* in one direction – namely, towards God, the divine Architect. Nature was God's artwork. Swammerdam continually praised 'the incredible Miracles of the embroidered Entrails of the large or smallest animals'.[86] Describing and depicting them so faithfully to nature was not just a scientific requirement, it was above all a religious duty. This was a business as serious as the reading of Scripture. Swammerdam regarded mortals who read the 'Bible of Nature' carelessly as outright sinners and those who denied that the laws of nature created by God applied everywhere as atheists.

The transformation of a symbolic-emblematic concept of nature

16

The anatomy of the mayfly, as depicted by Swammerdam, again life-sized (bottom left), and enlarged at the right side, showing the seminal vesicles of the male (FFF) removed from the body. Universiteitsbibliotheek Leiden, Ms BPL 126 C f 15r.

into one in which order and underlying patterns were central was propagated by Swammerdam not just in speech and in writing but also in his illustrations, which, as we have seen, allocate an increasingly crucial place to the infinitely refined, tiny structures of the 'least esteemed creatures'. In *Historia generalis insectorum* he depicted the astonishing external characteristics of countless insects, and in the ten years that followed he concentrated mainly on microscopic anatomical details. With the exception of the anatomy of the mayfly – published in 1675 during his religious crisis – his observations were known only to a small though influential circle that included Hudde, Witsen, Huygens, Thévenot, Steno, Redi, members of the Royal Society and philosophers such as Nicolas Malebranche (1638-1715) and Gottfried Wilhelm Leibniz (1646-1716).[87] A broader readership was able to learn of Swammerdam's microscopic observations only after Herman Boerhaave published the richly illustrated manuscript in 1737-1738, under the title *Biblia naturae – Bybel der natuure.* Christian-classicist thinking on nature as an artwork, fysico-theology and the 'argument from design' became extraordinarily popular in the late seventeenth century, partly as a result of the kind of research Swammerdam had carried out.[88]

There was another means by which Swammerdam's ideas penetrated the world of scholars and 'enthusiasts': his collection of specimens. As his studies progressed his collection of *naturalia* grew. It consisted of a wide variety of objects, including embalmed human and animal organs, dried plants and shells and, above all, his countless exotic and native insects (or carefully prepared specimens of parts of them). In contrast to collections of rarities such as the one assembled by his father, his own collection was centred on the orderly and ordinary, rather than the exceptional, foreign or exotic. Swammerdam did not collect unicorn horns, Indian headdresses or stuffed armadillos but instead butterflies, beetles and mites. The way in which he organized his collection was a departure from tradition as well. In cabinets of rarities, *naturalia* and *artificialia* were mixed together, generally in an associative arrangement. Swammerdam took a quite different approach. He put similar species next to each other and displayed each of them in various stages of its development. This narrative approach made his collection more than simply a tactile accompaniment to the illustrations in *Historia generalis insectorum*; it became and remained an actual part of nature instead of a representation of it. In the eyes of Swammerdam, living insects, the drawings he made of them 'naer het leeven' and the specimens he kept in his collection had the same ontological and epistemological status. Later, his friend Ruysch would argue that anatomical preparations were more trustworthy than drawings, while his adversary Govaert Bidloo, author of the first great anatomical atlas since Vesalius, maintained the contrary.[89] To Swammerdam all had the same status, since all pointed to God, the divine Architect, who was both the creator of Nature, as well as man's abilities to study and depict nature with the aid of science and art.

Thus, the collection had a deeply metaphysical meaning as well, as it made the developmental processes of insects, dictated by natural laws, visible and tangible. As he wrote in 1674:

17
Swammerdam discovered that the so-called king bee was actually a queen. In this drawing, executed around 1674 and published in 1737
We see the female reproductive system, most prominently the ovaries. Without a legend and an explanatory text, the figure would be incomprehensible.
Universiteitsbibliotheek Leiden, Ms BPL 126 C f 31r.

anat. fæm. ad p. 102 · Tab. 19 · n° 6.

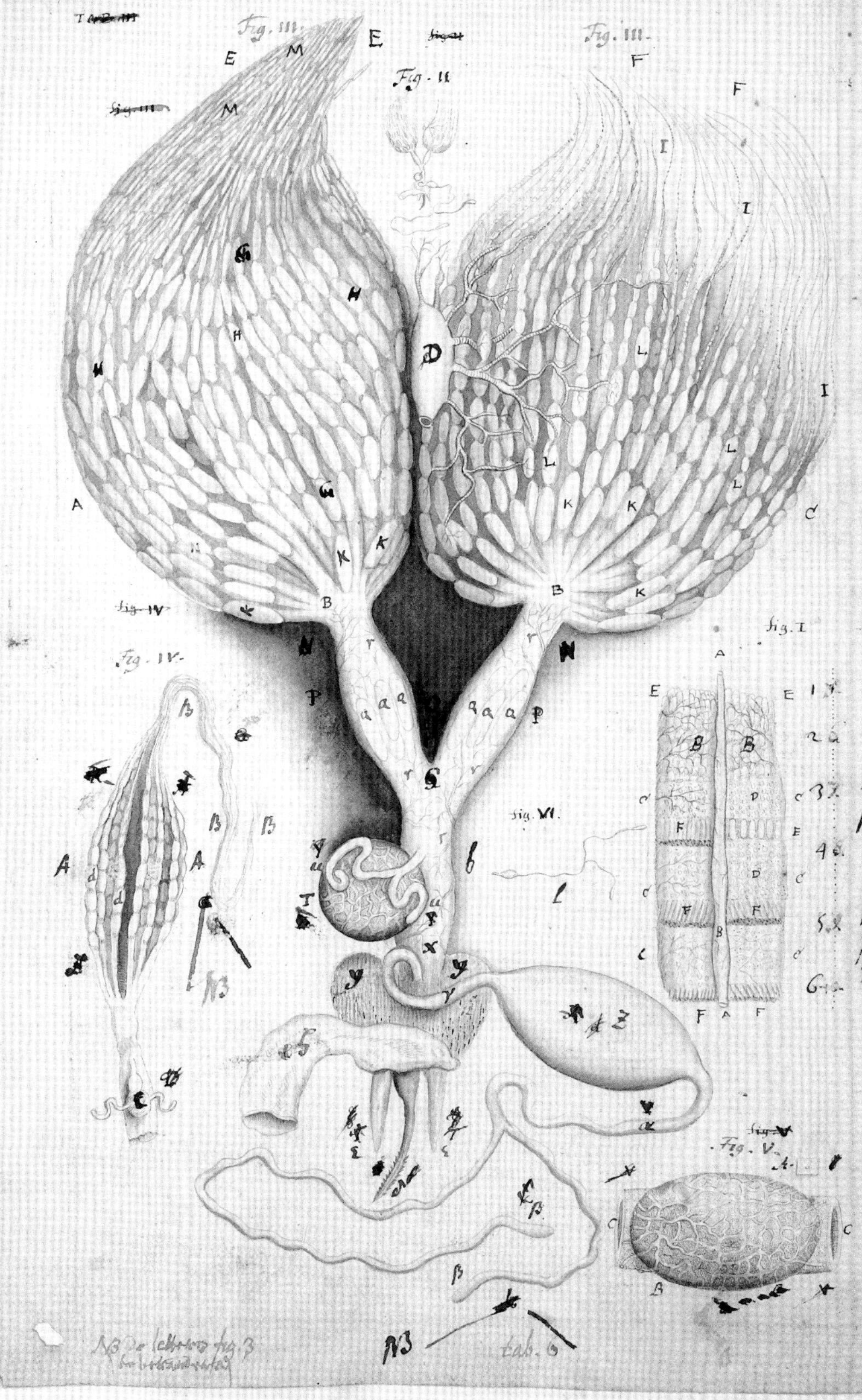

NB · tab. 6

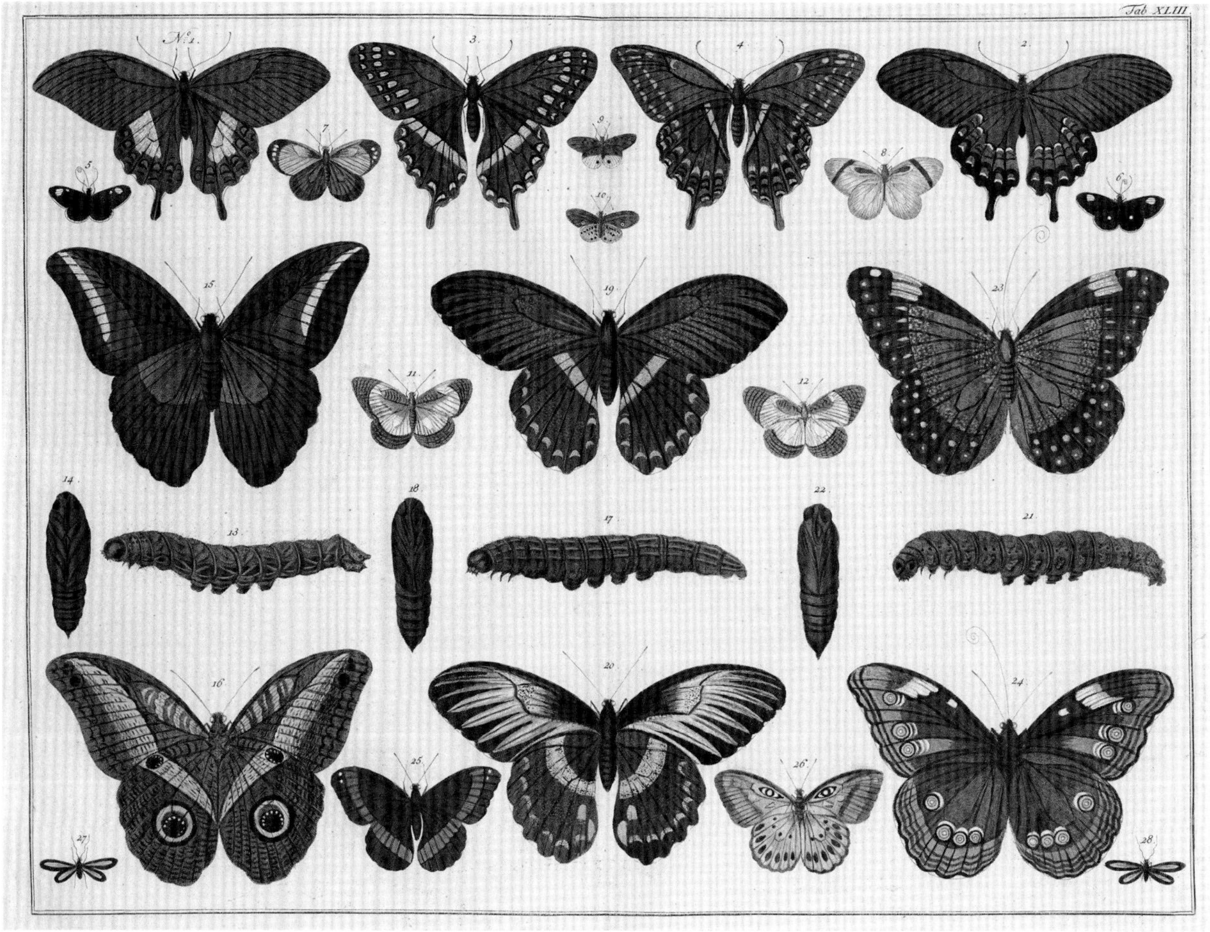

18
In the wake of Swammerdam's work, collecting insects became extremely popular in the Dutch Republic. This ensemble of butterflies was published in volume IV of Albertus Seba's *Thesaurus* (1734-1765), Koninklijke Bibliotheek, The Hague, 394 B 26.

> All the rarities that I have researched up to now I keep in their entirety on pieces of Glass, to some time entertain a few Friends, who love such Studies and desire to set forth this pleasant and excellent investigation, which shows to us in every moment the Divine and adorable architect of the universe.[90]

Swammerdam's collection is one example of the seventeenth-century transition from cabinets of rarities to collections of *naturalia*.[91] The former had been made up of extremely diverse objects with countless connotations and layers of meaning. Swammerdam's collection too was a system of symbols laden with religious significance, but in his case the textual tradition was comparatively unimportant, and all that counted were external manifestations and underlying patterns. The differences between beautiful and ugly, rare and everyday, preternatural and natural were no longer relevant, and neither were biblical connotations, symbolic meanings or emblematic interpretations.

This downgrading of a complex system of significances went hand in hand with a spectacular increase in the number of insects in Dutch collections. In 1660 they had hardly been collected at all, with the exception of items typified by the 'locust of the sort that St. John the Baptist ate in the wilderness' spotted by an English traveller in a Delft cabinet in 1663, yet by about 1680 they were the most spectacular aspect of the collections.[92] Collectors like Witsen, Levinus Vincent, Henricus d'Acquet and Albertus Seba, often with explicit reference to Swammerdam's vast collections, now displayed exotic and to an increasing extent indigenous insects (figs. 18 and 19).

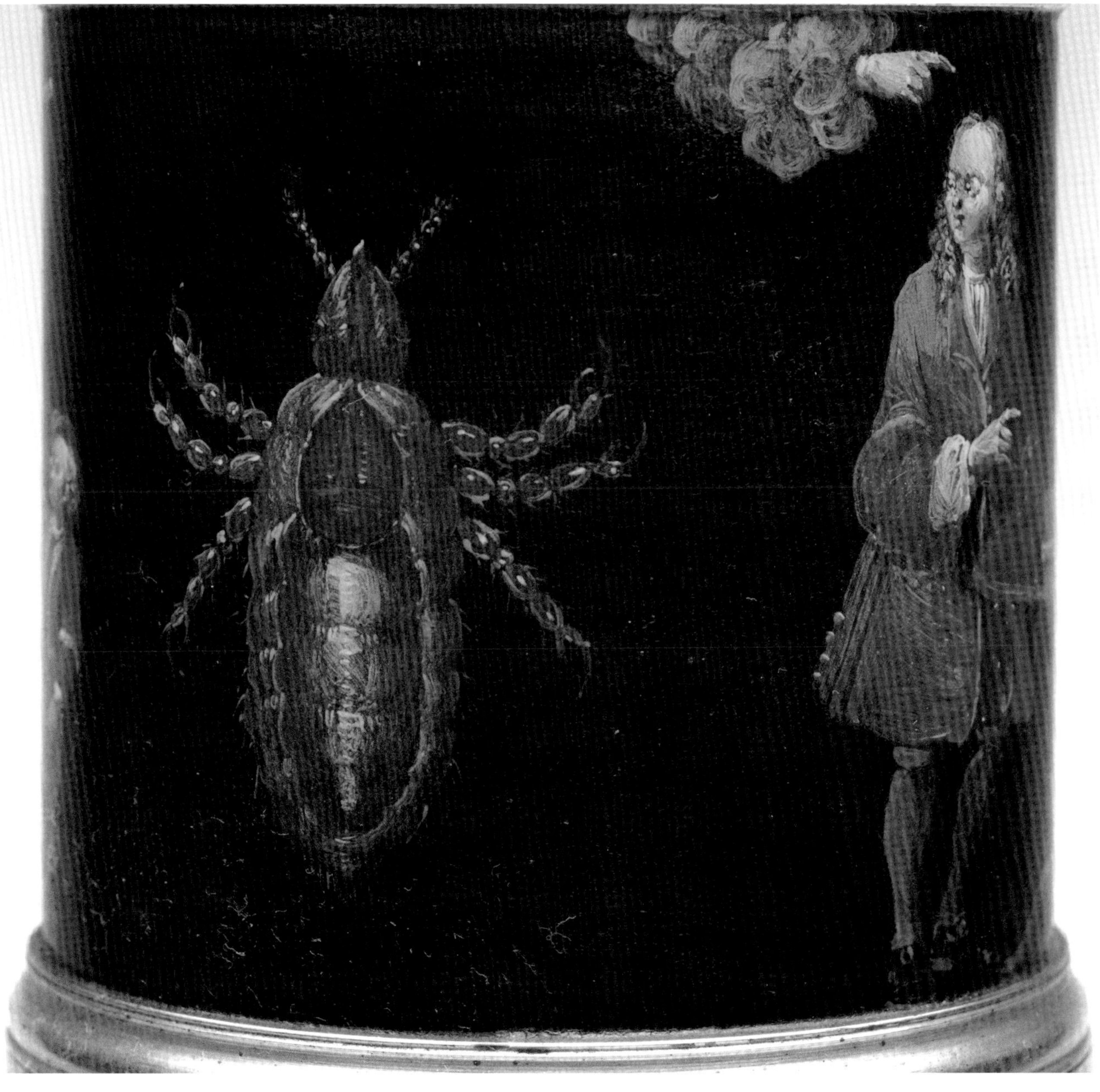

19
Art and the sciencific instrument: detail of a decorated compound microscope manufactured by Lucas Cramer, ca. 1740 (see fig 6, page 15). The louse and the finger are undoubtedly allusions to Swammerdam's treatise on the anatomy of the louse, published in 1737, Museum Boerhaave Leiden, inv. no. V07083.

Conclusion

Although it is difficult to determine exactly what direct influence Swammerdam had on his contemporaries, his work is clearly characteristic of a changing attitude to nature. James Bono has described the changes as 'the turn of science in the seventeenth century from texts to things, from language to laboratory, from nature emblematized to nature laid bare'.[93] Swammerdam removed insects from the margins of discourse and gave them a central place. They were no longer seen as symbols but observed, cut open, studied under the microscope and drawn. The growing emphasis on rules and patterns in the second half of the century, and the growing resistance to religion and reliance on chance, can in essence be traced back to Descartes's influence. It is probably no coincidence that in about 1670 Swammerdam launched an attack on the belief in spontaneous generation, while Spinoza targeted faith in miracles – in both cases the new assumption was that nature could not be contingent, or, in other words, that parts of creation operated independently of the almighty power of God.[94] Nature was uniform and stable. Elsewhere in this volume, Van de Roemer contends that it was precisely in these years that the phenomenon of chance in art theory and practice came to be seen in a bad light. Moreover, can it be pure coincidence that the two Dutch scholars who were most influenced by Descartes – Hudde and Huygens – began studying theories of probability in about 1670, demonstrating that a mathematical law lay at the root of apparent chance?[95]

Under the influence of Cartesianism, nature was regarded as a mechanism and stripped of its symbolic significance. This is not to say, however, that the study of nature was deprived of all meaning. On the contrary, the Cartesian notion of the 'laws of nature' breathed new life into ancient concepts of nature as a work of art and God as an artist, the supreme Creator or Architect. Voyages of discovery and the invention of the telescope and the microscope showed that creation was infinitely richer and more complex than had been assumed based on the authority of the ancients.[96] As Swammerdam put it, 'the more the works of nature are opened up and divided into separate parts, the more their peerless structure is discovered and acknowledged'.[97] Swammerdam was an influential representative of the 'argument from design', the conviction that order and structure in the natural world could be attributed only to God. In this he is probably a far more typical representative of scientific culture in the Dutch Republic than we tend to assume. On closer examination it turns out that Dutch scientific culture in the early modern period was greatly inspired by metaphysics and far less pragmatic and down-to-earth than is usually assumed. No clear dividing line can be drawn between 'faith' and 'knowledge', nor indeed between 'art' and 'science' or representation 'naer het leeven' and symbolic and religious meanings. Observation, study, representation and the attribution of meaning were elements in a broad spectrum of theoretical reflections and practical activities, rather than polar opposites.

Notes

*Translated by Liz Waters. I would like to thank Boudewijn Bakker, Esther van Gelder, Marrigje Rikken, Bert van de Roemer, Claudia Swan, Rienk Vermij and Huib Zuidervaart, as well as my colleagues of the board of the *NKJ*, for their comments on earlier versions of this paper.

1 Swammerdam 1737-1738, 67: 'Ik presenteer U ED alhier den Almaghtigen Vinger GODS, in de Anatomie van een Luys; waar in Gy wonderen op wonderen op een gestapelt sult vinden (...) De linien van Appeles doen al de wereld admirereeren, maar alhier sult gy in een gedeelte van eene linie de gansche structuur van alderkunstigste Dieren, van het geheele univers te samen, als in een kort begrip opgesloten vinden. Wat menschen, my heer, syn capabel dit te begrijpen? Maar wat kunstenaar kan daar buyten GODT ook weesen, die dit eenighsins sou kunnen navorsschen en uytbeelden'. The original, slightly different letter is held by the Universitätsbibliothek Göttingen, partly published in Lindeboom 1975, 104-105. A copy of the manuscript in Swammerdam's own hand, as well as the drawing reproduced here, is in the Universiteitsbibliotheek Leiden, BPL 126 B I (no pagination) and BPL 126 C f 3ʳ.

2 Pliny, *Historia Naturalis* XXV.79-100; esp. 89-82, here quoted after Pliny 1938-1952, vol. 9, 321-323.

3 Van Helden *et al.* 2010; Ruestow 1996, 6-104.

4 Fournier 1996; Ruestow 1996, 104-280; Vermij forthcoming.

5 Christiaan Huygens to Hudde, 4 April 1665, Huygens 1888-1950, vol. 5, 305. On the impact of Hooke, and especially the images of *Micrographia*, see Harwood 1989, and Wilson 1995, 75-86. Hooke's book was written in English, a language few Dutchmen had mastered at the time. See, for example, Johannes Hudde to Christiaan Huygens, 5 April 1665, Huygens Huygens 1888-1950, vol. 5, 305-311: ''T is mij zo leet, dat ik nu geen Engelsch kan, dat, zo mij geen gewichtiger dingen belette, ik zouw expres engelsch gaan leeren, al was t maar alleen om (...) deze Micrographia van Hook te lezen' (309).

6 The traditional, somewhat romanticized image of Swammerdam as a tormented zealot goes back to the short biography Herman Boerhaave included in Swammerdam 1737-1738, A1-I2, and is, for example, repeated in Schierbeek 1946, Lindeboom 1975, 3-33, Ruestow 1996, 105-145, and Cobb 2002. A more balanced view and additional sources can be found in Nordenström 1954, De Baar 2004, 292-293, 329-335, 376-376, 401-405, Jorink 2010, 229-246, 324-331 and *passim*. I am currently working on a new biography of Swammerdam. It was Boerhaave who introduced the forename 'Jan'; Swammerdam signed all his letters, works and other document 'Johannes'.

7 See, for example, Universiteitsbibliotheek Leiden BPL 126 C f 15ʳ.

8 Cf. Daston & Galison 2008, 86.

9 Ruestow 1996, 108.

10 Cf. Mason 2009, 87-123. See also the contribution by Dániel Margócsy earlier in this volume.

11 See, for example, some contributions in Franits 1997; Bakker 2004a and 2004b; and Stumpel 2004. Cf. Van de Roemer 1998.

12 Bergvelt & Kistemaker 1992; Van de Roemer 2005; Cook 2007, 1-225; Jorink 2010, 257-346. For the international context, see Impey & MacGregor 1985; Pomian 1987; Schnapper 1988; Olmi 1992; Grote 1994; Findlen 1994; Bergvelt *et al.* 2005 and MacGregor 2007.

13 Logan 1979; Van Veen 1992; Van de Roemer 2008.

14 Van den Boogert 1999.

15 Findlen 1994, 97-154; Jorink 2010, 257-346; Cf. the remarks on a similar community in Goldgar 2007.

16 Jorink 2011.

17 Bontius 1931, 239.

18 De Hooghe 1735, 122-123: 'uyt de handen van D. Swammerdam Zaliger, in zijns vaders Rariteytkamer zynde'. I owe this reference to Jo Spaans; see also her contribution elsewhere in this volume.

19 See, for example, Swammerdam 1737-1738, 61, 169, 300, 367, 394.

20 See, for example, Dijksterhuis 1970; Van Berkel 1983; Cohen 1990; Van Berkel *et al.* 1999, and Cook 2007.

21 See, for example, Alpers 1985 and Taylor 1995.

22 Jorink 2010.

23 Wettengl 1998; Van de Roemer 2004; Van de Roemer 2005; Van de Roemer 2008; Huisman 2009.

24 Lovejoy 1936, 236-240; Slaughter 1982, 33-35.

25 Freeman 1962; Lehman-Haupt 1973; Ritterbush 1985.

26 See Harms 1985; Harms 1989; Ashworth 1990, and the contributions by Rikken & Smith and Van de Roemer elsewhere in this volume.

27 Vignau-Wilberg 1994; Vignau-Wilberg 2007; Hendrix 1995.

28 Swan 2005.

29 Huygens 1987, 132-133.

30 See, for example, Bergström 1956; Bergström 1985 and Bol 1982.

31 Bol 1982, 30-35; cf. Jorink 2010, 201-209.

32 Everard 1662; see Zuidervaart 2009, 89-90.

33 Goedaert 1660-1669, vol. 1, 4, 45.

34 Goedaert, 1660-1669, vol. 2, 140. See also vol. 1, 7-8.

35 See also Verbeek 1992; Van Ruler 1995 and Van Bunge 2001, 34-92.

36 Lindeboom 1974.

37 Vermij 1999; Jorink 2010, 85-88.

38 On this circle, see Peters 2010, 30-53, and Jorink forthcoming, additional remarks in Vermij forthcoming. On Marseus, see Steensma 1999; Hildebrecht 2004; Leonhard 2010; and the contribution by Karin Leonhard earlier in this volume.

39 Hildebrecht 2004, 47-59, 248-282.

40 Vermij 1995; Vermij & Atzema 1995.

41 Ruestow 1996, 6-60.

42 De Monconys 1666, vol. 2, 152-164; Borch 1983, vol. 1, 128; Swammerdam 1669, 81.

43 Swammerdam 1669, 129: 'Datmen met de vleugelen van Capellen een Schilderie sou kunnen stofferen'.

44 De Monconys 1666, vol. 2, 178; see also Tolmer 1941: 'L'on m'a dit qu'il y avait un Flamand (...) qui nourrissait des lézards, des crapauds et mille vilaine bètes qui faisaient ses délices, dont il examinait la nature et qu'il peignant par excellence, mais peu savant pour en faire la relation'. From the context of this source, it is clear that this is a statement by Swammerdam, and that he is referring to Marseus.

45 Swammerdam 1737-1738, 669: 'regte weg tot het Atheismus'.

46 See also the contribution by Gijsbert van de Roemer elsewhere in this volume.

47 Swammerdam 1737-1738, 56-57: 'een nette schilderey [die] door onse inbeeldingen [en] quade overleveringen, is als bevuylt ende verontreinigt geweerden'.

48 Ibid.: '(...) in haar regte glans ende eygen schoonheid hersteld werd'.

49 Cf. Bredekamp 2000.
50 Much to Swammerdam's chagrin, he lost some of these research reports, including drawings. See, for example, Swammerdam 1675, 85-86: 'Eenige andere aanteykeningen heb ick omtrent deese kuuwens ende hare vaatkens noch gedaan, die ick niet en kan dencken, waar dat sy gebleven sijn. Gelijck als oock de memorie my daar van ontvallen is'.
51 See, for example, Shapin & Schaffer 1985, Shapin 1994, Bredekamp 2000, and Biagioli 2006.
52 See, for example, Swann 1995, and Daston & Galison 2008. On the term 'Ad vivum' see, most recently, Bakker 2011.
53 Borch 1983, vol. 2, 215.
54 Lindenboom 1974]; Beukers 1987; Scherz 1988; Smith 1999; Huisman 2009; Ragland 2008; Roemer 2008.
55 Swammerdam 1672; Cook 2002.
56 Van de Roemer 2008.
57 Swammerdam to Thévenot, 20 June 1678, Lindeboom 1975, 114.
58 Swammerdam 1737-1738, 487: 'Goedaert fabuleert seer aardig (...) soo dat hy eer een roman, dan een waare Historie schijnt te beschrijven'.
59 See also Meli 2010; Meli 2011, 215-234.
60 Swammerdam 1669, 27: 'verkeerdelijk vervorming, gedaanteaflegging, doot ende opstanding genoemt'.
61 Ibid., 9: 'een Kuyken, het welke niet *verandert* in een hoen, maar aangroeiende in leedematen soo *wort* het een hoen', emphasis added.
62 Much has been published in recent years about scientific illustrations vis-à-vis the representation of 'reality'; see, for example, Crombie 1985; Baigrie 1996; Kemp 2007; Enenkel & Smith 2007, and Daston & Galison 2008.
63 See also Leonhard 2007.
64 Swammerdam to Thévenot, 30 October 1670, Lindeboom 1975, 56: '(...) soo most men van Romijn de plaaten opnieuw laate snijden, ende de bruijne gronden agter laaten. Soo most men meede hier en daar wat in de plaaten verbeeteren (...) Het selve sou van Romijn seer bequaam, ende niet onbequam van andere, die dat werk wilde, by de hant neemen, gedaan kunnen werden. Al het selve soude ik tot mijn profijt selve kunnen laaten doen, dan alsoo ik van mijn vader niet als de kost en een kamer heb, soo is het mij onmogelyk'. 'Again' (opnieuw) may refer to the fact that either De Hooghe had made the earlier engravings and would now produce them again, or he would be asked to improve on somebody else's work.
65 Verkruijsse 2008.
66 Swammerdam 1672; Swammerdam 1675. On Bosboom, see Van de Roemer 2005, Peters 2010, 204-205, 320-322, and the contribution by Van de Roemer elsewhere in this volume.
67 Swammerdam to Thévenot, 1680, Lindeboom, 1975, 128: 'Jay reçeu le microscope et les figures, mais c'est une brutalite du graveur, qu'il a fait dans la IV Tab. fig. VII dans la mesme mouche, deux ailes differentes, c'est contre la nature, et pienture: il faut cela sur tout corriger; et aussi le pectus, ne convient en chaque costé. les autres fautes, se corrigent facilement. Quand il fait beau jour j'esprouveray le miscroscope, et j'attendrois les deux autres avec les figures correctes. je ne suis avec vous, tout a fait sa (...) jespere que mes autres figures seront beaucoup plus et mieux faitez, que celle que vous m'aviez envoyé'.
68 See also Cobb 2002, Meli 2010, and Meli 2011, 214-234.
69 Swammerdam 1672, 16: '(...) sed figuram etiam ipsam mentem concepisse videtur (...)'.
70 Swammerdam 1669, 81: 'Ende of het quam te gebeuren, dat het dierken op een witte grond niet wel gezien konde werden: Soo hebben wy het wit in geel, groen, blaauw, en so voorts verandert: settende van gelijken onse genoemde glaaskens, op blaausel, swartsel, vermiljoen, ende ander veruwen'.
71 Ibid., 20: 'the spectacle is always clearer in one than in another, but everything depends upon the eye and the hand with which it is seen and dissected'. This seems to be a paraphrase of Hooke's famous dictum '(...) with a sincere hand and a faithful eye (...)'. Cf. Hooke 1665 A2[v].
72 Cf. Shapin & Schaffer 1985; Cobb 2002, 140; Daston & Galison 2008, 11-16, 55-114; Thomas 2011, 4.
73 Swammerdam 1737-1738, 177: 'Waarom ik ook myn gansche beschryinge voor niets anders uytgeef, als of met een houte kool de Son wilde uytbeelden' .
74 The original drawing, with the significant signature 'delineavit auctor' is in the Universiteitsbibliotheek Leiden, 126 B f 15[r]; two different engravings are to be found in Swammerdam 1675 (by Dirk Bosboom) and Swammerdam 1737-1738 (by Jacobus van der Spyk, 1696-1763).
75 Swammerdam 1737-1738, 908: 'waarachtige pronkstukken (...) waar voor alle linien van Apelles, en alle spitsvondigheden der menselyke verstanden, voor een Sotheyt moeten geacht worden'.
76 Ibid., 232, 583, 708, 713, 886.
77 Ibid., 764, 784: 'stralend ende schitterde parlemoer (...) dat gepolyst is'.
78 Swammerdam 1669, 134: '(...) die anders ende anders van maaksel sijn; ende haare tekeningen onbeschrijfelijk (...) Soo sijn haare vleugelen als met Peerlen, ende diamanten, geschiktelijk besaait;de welke als van onnoemelijke Saphhiren, Turkoisen, ende Robijnen meerder glans ontfangende, de Peerle moer doppen, ende de silveren plaatsen, haarer vleugelen, door een tintelende weerkaatsing van straalen, de veruwen der Reegenbogen doen overtreffen'.
79 Swammerdam 1669, passim; Swammerdam 1675, passim. Cf. Van der Roemer 2008.
80 Swammerdam 1672, 1: 'Ars naturam aemulando suos semper defectus prodit. Quare, ut ut dextrè magnus ille statuarius Phidias propriam suam imaginem Minervae clypeo insculpserat, fracto tame clypeo omnis Phidae imago periisset. Usque daeo artis opera imperfectionis notam semper secum ferunt, ut statim ac nata sunt redigi in nihilum possint: cum econtra Nature opera (O admirandam! O adorandam sapientiam!) tantae perfectionis sint, ut toto corpore & singulis sui corporis partibus majestatem artificis ostentent: & quomodo summus ille plastes hujus universi in omnibus quasi visibilis est; ita in minimutissimis eorundem partibus quasi mani palpari posit'.
81 See, for example, Bowler 1971; Wilson 1995, passim, Fournier 1996, passim, Ruestow 1996, 201-259, and Cobb 2006, 94-154.
82 Swammerdam 1669, 136.
83 Ibid., 28: '(...) soo blijkt hier clarelijk de dwaaling van die geenen, de welke uyt deze natuurelijke ende verstaanbare veranderingen; de opstanding der dooden hebben willen bewijsen, dewelke de kragt, vande order in de natuur bemerkelijk, niet alleen geheel te boven gaat: maar ook, gans geen gelijkenis inde selve vindende'.
84 Ibid., 1-3.
85 Ibid., 56-57.
86 Swammerdam 1675, 322-323: 'de

onbegrijpelijcke Wonderen der geborduurde Ingewanden van de groote ofte de kleenste dierkens'.

87 Jorink 2010, 236-239.

88 Vermij 1991; Van de Roemer 2005; Van de Roemer 2008.

89 Margóscy 2011. On the epistemological significance of the relation between texts and objects see also Thomas 2011.

90 Boccone 1744, 157: 'Alle de zeldzaemheden die ik tot nog toe onderzocht hebbe, bewaere ik in hun geheel op stucken Glas, om zomtyds enige Vrienden te onderhouden, die de Studiën beminnen, en deze aengename en voortreffelyke naspeuringen begeeren voort te zette, welke de Goddelyke en aenbiddelyke bouwmeester van 't geheel al ons ieder ogenblik vertoont'. See also Ibid., 159: 'Ik zal, zo het UE. gelieft, deze zeldzaemheden [microscopische preparaten van koralen] bewaren, om hen te vertonen aen die behaegen scheppen Godt in zyne onnaspeurelyke wonderen te verheerlyken.'

91 Van de Roemer 2008; Jorink 2010, 256-347.

92 Ray 1673, 25. Cf. Matthew 3:4; Mark 1:6.

93 1995], 272.

94 Jorink 2003.

95 See, for example, Hudde to Huygens, 16 August 1671, Huygens 1888-1950, vol. 7, 95-98.

96 See also the contributions by Margócsy, Rikken & Smith and Spaans elsewhere in this volume.

97 Swammerdam 1672, 1: 'Etenim, quo naturae opera magis confringuntur, & in partes dividuntuntur, eo inimitabilis eorundem structura magis magisque detegitur & innotescit'.

Bibliography

Alpers 1983
S. Alpers, *The art of describing. Dutch art in the seventeenth century*, Chicago 1983.Alpers 1983

Ashworth 1990
W. Ashworth, 'Natural history and the emblematic worldview', in: D.C. Lindberg & R.S. Westman (eds.), *Reappraisals of the Scientific Revolution*, Cambridge 1990, 303-323.

Baigrie 1996
B.S. Baigrie (ed.), *Picturing knowledge. Historical and philosophical problems concerning the use of art in science*, Toronto 1996.

Bakker 2011
B. Bakker, 'Au vif – anaar 't leven – ad vivum. The medieval origin of a Humanist concept'?, in: A. Boschloo *et al.* (eds.), *Aemulatio. Imitation, emulation and invention in Netherlandish art from 1500 to 1800. Essays in honour of Eric Jan Sluijter*, Zwolle 2011.

Bakker 2004a
B. Bakker, *Landschap en wereldbeeld van Van Eyck tot Rembrandt*, Bussum 2004.

Bakker 2004b
B. Bakker, 'Some notes on Method. Regarding Jeroen Stumpel's review of my *Landschap en wereldbeeld van Van Eyck tot Rembrandt*', *Simiolus* 31 (2004), 260-262.

Bergström 1956
I. Bergström, *Dutch still-life painting in the seventeenth century*, London 1956.

Bergström 1985
I. Bergström, 'On Georg Hoefnagel's manner of working with notes on the influence of the *Archetypa* series of 1592', in: G. Cavalli-Björkman (ed.), *Netherlandish Mannerism*, Stockholm 1985, 177-188.

Bergvelt & Kistemaker 1992
E. Bergvelt & R. Kistemaker (eds.), *De wereld binnen handbereik. Nederlandse kunst- en rariteitenverzamelingen 1585-1735*, Zwolle 1992.

Bergvelt *et al.* 2005
E. Bergvelt, D. Meijers & M. Rijnders (eds.), *Kabinetten, galerijen en musea. Het verzamelen en presenteren van naturalia en kunst van 1500 tot heden*, Zwolle 2005.

Beukers 1987
H. Beukers, 'Clinical teaching in Leiden from its beginnings until the end of the eighteenth century', *Clio Medica* 21 (1987) 139-152.

Biagioli 2006
M. Biagioli, *Galileo's instruments of credit. Telescopes, images, secrecy*, Chicago 2006.

Boccone 1744
P. Boccone, *Natuurkundige naspeuringen op proef- en waernemingen gegrond*, Amsterdam 1744.

Bol 1982
L.J. Bol, *'Goede onbekenden'. Hedendaagse verkenning en waardering van verscholen, voorbijgezien en onbekend talent*, Utrecht 1982.

Bono 1995
J.J. Bono, *The Word of God and the languages of man. Interpreting nature in early modern science and medicine. Vol. 1. Ficino to Descartes*, Madison 1995.

Bontius 1931
J. Bontius, *On tropical medicine*, Amsterdam 1931.

Borch 1983
O. Borch, *Olai Borrichii Itinerarium 1660-1665* (H. Schepelern, ed.), Copenhagen 1983.

Bowler 1971
P. Bowler, 'Preformation and pre-existence in the seventeenth century. A brief analysis', *Journal of the History of Biology* 4 (1971), 221-244.

Bredekamp 2000
H. Bredekamp, 'Gazing hands and blind spots. Galileo as draftsman', *Science in Context* 13 (2000), 423-463.

Cobb 2002
M. Cobb, 'Malpighi, Swammerdam and the colourful silkworm.

Replication and visual representation in early modern science', *Annals of Science* 59 (2002), 111-147.

Cobb 2006
M. Cobb, *The egg and sperm race. The seventeenth-century scientists who unravelled the secrets of sex, life and growth*, London 2006.

Cohen 1990
H.F. Cohen, 'Open and wide, yet without heights or depth', *Traxtrix. Yearbook for the History of Science, Medicine, Technology and Mathematics* 2 (1990), 159-165.

Cook 2002
H. Cook, 'Time's bodies. Crafting the preparation and preservation of naturalia', in: P. Smith & P. Findlen (eds.), *Merchants and marvels. Commerce, science, and art in early modern Europe*, New York & London 2002, 223-247.

Cook 2007
H. Cook, *Matters of exchange. Commerce, medicine, and science in the Dutch Golden Age*, New Haven 2007.

Crombie 1985
A.C. Crombie, 'Science and the arts in the Renaissance. The search for truth and certainty, old and new', in: J.W. Shirley & F.D. Hoeniger (eds.), *Science and arts in the Renaissance*, Washington 1985, 15-26.

Daston & Galison 2008
L. Daston & P. Galison, *Objectivity*, New York 2008.

Daston & Park 1998
L. Daston & K. Park, *Wonders and the order of nature 1150-1750*, New York 1998.

De Baar 2004
M. de Baar, *'Ik moet spreken'. Het spiritueel leiderschap van Antoinette Bourignon (1616-1680)*, Zutphen 2004.

De Hooghe 1735
R. de Hooghe, *Hieroglyphica of merkbeelden der oude volkeren, namentlyk Egyptenaren, Chaldeeuwen, Feniciers, Joden, Grieken, Romeynen, enz. Nevens een omstandig Bericht van het Verval en voortkruypende Verbastering der Godsdiensten door verscheyde Eeuwen; en eyndelyk de Hervorming, tot op deze Tyden toe vervolgt. Overzien en Beschaaft door Arn. Henr. Westerhovius*, Amsterdam 1735.

De Monconys 1666
B. de Monconys, *Journal des voyages*, Lyon 1666.

Dijksterhuis 1970
E.J. Dijksterhuis, *Simon Stevin. Science in the Netherlands around 1600*, The Hague 1970.

Enenkel & Smith 2007
K.A.E. Enenkel, & P.J. Smith (eds.), *Early modern zoology. The construction of animals in science, literature and the visual arts*, Leiden 2007.

Everard 1662
A. Everard, *Lux è tenebris affulsa*, Middelburg 1662.

Findlen 1994
P. Findlen, *Possessing nature. Museums, collecting, and scientific culture in early modern Italy*, Berkeley 1994.

Fournier 1996
M. Fournier, *The fabric of life. Microscopy in the seventeenth century*, Baltimore 1996.

Franits 1997
W. Franits (ed.), *Looking at seventeenth-century Dutch art. Realism reconsidered*, Cambridge 1997.

Freedberg & De Vries 1991
D. Freedberg and J. de Vries (eds.), *Art in history, history in art. Studies in seventeenth-century Dutch culture*, Chicago 1991, 377-427.

Freeman 1962
R. Freeman, 'Illustrations of insect anatomy from the beginning to the time of Cuvier', *Medical and Biological Illustration* 12 (1962), 174-183.

Fuchs & Breen 1916
R. Fuchs & J.C. Breen, 'Aus dem *Itinerarium* des Christian Knorr von Rosenroth', *Jaarboek Amstelodamum* 14 (1916), 201-256.

Goedaert 1660-1669
J. Goedaert, *Metamorphosis naturalis, ofte historische beschrijvinge van den oirspronk, aerdt, eygenschappen ende vreemde veranderingen der wormen, rupsen, maeden, vliegen, witjens, byen, motten en diergelijke dierkens meer; niet uit eenige boeken, maar aleenlyck door eygen ervaretntheyt uytgevonden, beschreven ende na dekonst afgeteykent*, Middelburg s.d. [1660]-1669, 3 vols.

Goldgar 2007
A. Goldgar, *Tulipmania. Money, honor, and knowledge in the Dutch Golden Age*, Chicago 2007.

Grote 1994
A. Grote (ed.), *Macrocosmos in Microcosmo, die Welt in der Stube. Zur Geschichte des Sammelns, 1450 bis 1800*, Opladen 1994.

Harms 1985
W. Harms, 'On natural history and emblematics in the 16th century', in: J. Ackerman (ed.), *The natural sciences and the arts. Aspects of interaction from the Renaissance to the 20th century*, Uppsala 1985, 67-83.

Harms 1989
W. Harms, 'Beteutung als Teil der Sache in zoologische Standartwerken der frühen Neuzeit (Konrad Gesner, Ulysse Aldrovandi)', in: H. Boockmann *et al.* (eds.), *Lebenslehre und Weltentwürfe im Ubergang vom Mittelalter zur Neuzeit*, Göttingen 1989, 531-553.

Harwood 1989
J.T. Harwood, 'Rhetoric and graphic in *Micrographia*', in: M. Hunter & S. Schaffer (eds.), *Robert Hooke. New studies*, Woodbridge 1989, 119-147.

Hendrix 1995
Hendrix, L., 'Of hirsutes and Insects: Joris Hoefnagel and the Art of the Wondrous', *Word & Image* 11 (1995) 373-390.

Hildebrecht 2004
D.R. Hildebrecht, *Otto Marseus van Schrieck (1619/20-1678) and the nature piece. Art, science, religion and the seventeenth-century pursuit of natural knowledge*, Ann Arbor 2004 (unpub. diss, University of Michigan, Ann Arbor).

Hooke 1665
Hooke, R., *Micrographia: or Some Physiological Descriptions of Minute Bodies made by Magnifying Glasses. With Observations and Inquiries there-upon,* London 1665.

Huisman 2009
T. Huisman, *The finger of God. Anatomical practice in 17th-century Leiden*, Leiden 2009.

Huygens 1888-1950
C. Huygens, *Oeuvres complètes de Christiaan Huygens*, The Hague 1888-1950, 23 vols.

Huygens 1987
C. Huygens, *Mijn jeugd* (C.L. Heesakkers, transl.), Amsterdam 1987.

Impey & MacGregor 1985
O. Impey & A. MacGregor (eds), *The origins of museums. The cabinets of curiosities in sixteenth- and seventeenth century Europe*, Oxford 1985.

Jorink 2003
E. Jorink, '"Outside God, there is nothing". Swammerdam, Spinoza and the Janus-face of the early Dutch Enlightenment', in: W. van Bunge (ed.), *The early Enlightenment in the Dutch Republic, 1650-1750. Selected papers of a conference held at the Herzog August Bibliothek, Wolfenbüttel, 22-23 March 2001*, Leiden 2003, 81-108.

Jorink 2010
E. Jorink, *Reading the Book of Nature in the Dutch Golden Age, 1575-1715*, Leiden & Boston 2010.

Jorink 2011
E. Jorink, 'Noah's Ark restored (and wrecked). Dutch collectors, natural history and the problem of Biblical exegesis', in: S. Dupré & C. Lüthy (eds.), *Silent messengers. The circulation of material objects of knowledge in the early modern Low Countries*, Berlin 2011, 185-184.

Jorink forthcoming
'Swammerdam, hoveling? Moeilijkheden en mogelijkheden in de carrière van een wetenschappelijk onderzoeker', *De zeventiende eeuw* 28 (2012).

Kemp 2007
M. Kemp, *The human animal in Western art and science*, Chicago 2007.

Lehman-Haupt 1973
H. Lehman-Haupt, 'The microscope and the book', in: J. Benzing *et al.* (eds.), *Festschrift für Claus Nissen. Zum siebzigsten Geburtstag 2. September* 1971, Wiesbaden 1973, 471-502.

Leonhard 2007
K. Leonhard, 'Kritik an der Hand. Zum Verhältnis von Wissenschaftler und Zeichner in der frühen Mikroskopie', in: G. Wimböck, K. Leonhard & M. Friedrich (eds.), *Evidentia. Reichweiten visueller Wahrnehmung in der Frühen Neuzeit*, Münster 2007, 235-260.

Leonhard 2010
K. Leonhard, 'Pictura's fertile field. Otto Marseus van Schrieck and the genre of sottobosco painting', *Simiolus* 34 (2010), 95-118.

Lindeboom 1974
G.A. Lindeboom, *Descartes and medicine*, Amsterdam 1974.

Lindeboom 1975
G.A. Lindeboom (ed.), *The letters of Jan Swammerdam to Melchisedec Thévenot*, Amsterdam 1975.

Lindeboom 1980
G.A. Lindeboom, *Het cabinet van Jan Swammerdam (1637-1680)*, Amsterdam 1980.

Logan 1979
A.-M. S. Logan, *The 'cabinet' of the brothers Gerard and Jan Reynst*, Amsterdam, Oxford & New York, 1979.

Lovejoy 1936
A. Lovejoy, *The great chain of being. A study of the history of an idea*, Cambridge, MA 1936.

MacGregor 2007
A. MacGregor, *Curiosity and enlightenment. Collectors and collections from the sixteenth to the nineteenth century*, New Haven 2007.

Margócsy 2011
D. Margócsy, 'A museum of wonders or a cemetery of corpses? The commercial exchange of anatomical collections in the early modern Netherlands', in: S. Dupré & C. Lüthy (eds.), *Silent messengers. The circulation of material objects of knowledge in the early modern Low Countries*, Berlin 2011, 185-216.

Mason 2009
P. Mason, *Before disenchantment. Images of exotic animals and plants in the early modern world*, London 2009.

Meli 2010
D. Bertoloni Meli, 'The representation of insects in the seventeenth century. A comparative approach', *Annals of Science* 67 (2010), 405-429.

Meli 2011
D. Bertoloni Meli, *Mechanism, experiment, disease. Marcello Malpighi and seventeenth-century anatomy*, Baltimore 2011.

Nordenström 1954
J. Nordenström, 'Swammerdamiana. Excerpts from the travel journal of Olaus Borrichius and two letters from Swammerdam to Thévenot', *Lychnos* 15 (1954), 21-65.

Ogilvie 2006
B.W. Ogilvie, *The science of describing. Natural history in Renaissance Europe*, Chicago 2006.

Olmi 1992
G. Olmi, *L'inventario del mondo. Catalogazione della natura e luoghi del sapere nella prima età moderna*, Bologna 1992.

Peters 2010
M. Peters, *De wijze koopman. Het wereldwijde onderzoek van Nicolaes Witsen (1641-1717), burgemeester en VOC-bewindhebber van Amsterdam*, Amsterdam 2010.

Pliny 1938-1952
Pliny, *Naturalis historia* (H. Rackham, ed. & transl.), London & Cambridge, MA 1938-1952.

Pomian 1987
K. Pomian, *Collectioneurs, amateurs et curieux. Paris, Venise: XVIe-XVIIe siècle*, Paris 1987.

Ragland 2008
E. Ragland, 'Experimenting with chymical bodies. Reinier de Graaf's investigations of the pancreas', *Early Science and Medicine* 13 (2008), 615-664.

Ray 1673
J. Ray, *Observations, topographical, moral & physiological made in a journey through parts of the Low-Countries, Germany, Italy and France*, London 1673.

Ritterbush 1985
P.C. Ritterbush, 'The organism as symbol. An innovation in art', in: J.W Shirley & F.D. Hoeniger (eds.), *Science and arts in the Renaissance*, Washington 1985, 149-167.

Ruestow 1996
E.G. Ruestow, *The microscope in the Dutch Republic. The shaping of discovery*, Cambridge 1996.

Schierbeek 1946
A. Schierbeek, *Jan Swammerdam (12 februari 1637-17 februari 1680). Zijn leven en werken*, Lochem 1946.

Schnapper 1988
A. Schnapper, *Le géant, la licorne, la tulipe. Collections françaises au XVIIe siècle. Vol. 1. Histoire et histoire naturelle*, Paris 1988.

Segal 1989
S. Segal, *A prosperous past. The sumptuous still life in the Netherlands 1600-1700*, The Hague 1989.

Shapin 1994
S. Shapin, *A social history of truth. Civility and science in seventeenth-century England*, Chicago 1994.

Shapin & Schaffer 1985
S. Shapin & S. Schaffer, *Leviathan and the air-pump. Hobbes, Boyle and the experimental life*, Princeton 1985.

Slaughter 1982
M. Slaughter, *Universal language and scientific taxonomy in the seventeenth century*, Cambridge 1982.

Smith 1999
P.H. Smith, 'Science and taste. Painting, the passions, and the New Philosophy in seventeenth-century Leiden, *Isis* 90 (1999), 420-461.

Steensma 1999
S. Steensma, *Otto Marseus van Schrieck. Leben und Werk*, Hildesheim 1999.

Stumpel 2004
J. Stumpel, review of Bakker 2004, *Simiolus* 31 (2004), 115-124.

Swammerdam 1667
J. Swammerdam, *Tractatus physico-anatomico-medicus de respiratione usuque pulmonum*, Leiden 1667.

Swammerdam 1669
J. Swammerdam, *Historia generalis insectorum ofte Algemeene verhandeling van de bloedeloose dierkens*, Utrecht 1669.

Swammerdam 1672
J. Swammerdam, *Miraculum naturae sive uteri muliebris fabrica*, Leiden 1672.

Swammerdam 1675
J. Swammerdam, *Ephemeri vita of Afbeeldingh van 's menschen leven vertoont in de wonderbaarlijcke en nooyt gehoorde historie van het vliegent ende een-dagh-levent Haft of Oever-aas*, Amsterdam 1675.

Swammerdam 1737-1738
J. Swammerdam, *Bybel der natuure of historie der insecten. Biblia naturae, sive historia insectorum*, Leiden 1737-1738.

Swan 1995
C. Swan, 'Ad vivum, naer het leven, from the life. Defining a mode of representation', *Word & Image* 2 (1995), 353-372.

Swan 2005
C. Swan, *Art, science, and witchcraft in early modern Holland. Jacques de Gheyn II (1565-1629)*, Cambridge 2005.

Taylor 1995
P. Taylor, *Dutch Flower Painting 1600-1720*, New Haven & London 1995.

Thomas 2011
J. Thomas, 'Compiling 'God's great book [of] univseral nature'. The Royal Society's collecting strategies', *Journal of the history of collections* 23 (2011) 1-13.

Tolmer 1941
L. Tolmer, 'Une page d'histoire des sciences 1661-1669. Vingt-deux lettres inédites d'André de Graindorge à P.D. Huet', *Mémoires de l'Académie des sciences, arts et belles-lettres de Caen* 10 (1941), 245-337.

Van Berkel 1983
K. van Berkel, *Isaac Beeckman (1588-1637) en de mechanisering van het wereldbeeld*, Amsterdam 1983.

Van Berkel *et al.* 1999
K. van Berkel, A. van Helden & L. Palm (eds.), *A history of science in the Netherlands. Survey, theme and reference*, Leiden 1999.

Van Bunge 2001
W. van Bunge, *From Stevin to Spinoza. An essay on philosophy in the seventeenth-century Dutch Republic*, Leiden 2001.

Van den Boogert 1999
B. van den Boogert (ed.), *Rembrandts schatkamer*, Zwolle & Amsterdam 1999.

Van de Roemer 1998
G.M. van de Roemer, 'God en het rariteitenkabinet. Het religieuze motief van Noord-Nederlandse rariteitenverzamelaars einde zeventiende en begin achttiende eeuw', *Theoretische geschiedenis* 25 (1998), 242-255.

Van de Roemer 2004
G.M. van de Roemer, 'Neat nature. The relation between art and nature in a Dutch cabinet of curiosities from the early eighteenth century', *History of Science* 42 (2004), 47-84.

Van de Roemer 2005
G.M. van de Roemer, *De geschikte natuur. Theorieën over natuur en kunst in de verzameling van zeldzaamheden van Simon Schijnvoet (1652-1727)* Amsterdam 2005 (unpub. diss., University of Amsterdam).

Van de Roemer 2008
G.M. van de Roemer, 'Het lichaam als borduursel. Kunst en kennis in het anatomisch kabinet van Frederick Ruysch', in: *Body and embodiment in Netherlandish Art* (A.-S. Lehmann & H. Roodenburg, eds.), Zwolle 2008 (=*Nederlands Kunsthistorisch Jaarboek/Netherlandish Yearbook of the History of Art* 58 [2008]), 216-240.

Van Helden *et al.* 2010
A. van Helden, S. Dupré, R. van Gent & H. Zuidervaart (eds.), *The origins of the telescope. Proceedings of the conference 'The invention of the Dutch telescope. Its origin and impact on science, culture and society, 1550-1650' held in Middelburg in September 2008*, Amsterdam 2010.

Van Hoogstraten 1678
S. van Hoogstraten, *Inleyding tot de hooge schoole der schilderkonst. Anders de zichtbaere werelt*, Rotterdam 1678.

Van Lieburg 1992
M.J. van Lieburg, 'Reinier de Graaf en zijn plaats in het fysiologisch onderzoek van de zeventiende eeuw', *Gewina* 15 (1992), 73-84.

Van Nierop *et al.* 2008
H. van Nierop *et al.* (eds.), *Romeyn de Hooghe. De verbeelding van de Gouden Eeuw*, Zwolle 2008.

Van Ruler 1995
J.A. van Ruler, *The crisis of causality. Voetius and Descartes on God, nature and change*, Leiden 1995.

Van Veen 1992
J. van Veen, 'Dit klain vertrek bevat een Weereld vol gewoel. Negentig Amsterdammers en hun kabinetten', in: E. Bergvelt & R. Kistemaker (eds.), *De wereld binnen handbereik. Nederlandse kunst- en rariteiten-verzamelingen, 1585-1735*, Zwolle & Amsterdam 1992, 232-258; 313-334.

Verbeek 1992
T. Verbeek, *Descartes and the Dutch. Early reactions to Cartesian philosophy, 1637-1650*, Carbondale 1992.

Verkruijsse 2008
P. Verkruijsse, 'Door Romeyn de Hooghe geschreven en geïllustreerde boeken', in: Van Nierop *et al.* 2008, 258-290.

Vermij 1991
R. Vermij, *Secularisering en natuurwetenschap in de zeventiende en achttiende eeuw. Bernard Nieuwentijt*, Amsterdam 1991.

Vermij 1995
R. Vermij, 'Bijdrage tot de bio-bibliografie van Johannes Hudde', *Gewina* 18 (1995), 25-35.

Vermij 1999
R. Vermij, 'Een nieuw concept. De wetten der natuur', in: F. Egmond, E. Jorink & R. Vermij (eds.), *Kometen, monsters en muilezels. Het veranderende natuurbeeld en de natuurwetenschap in de zeventiende eeuw*, Haarlem 1999, 105-119.

Vermij forthcoming
R. Vermij, 'Instruments and the amking of a philosopher: Spinoza's career in optics', *Intellectual history review* 22 (2013)

Vermij & Atzema 1995
R. Vermij & E. Atzema, '*Specilla circularia*. An unknown work by Johannes Hudde', *Studia Leibnitiana* 27 (1995), 104-121.

Vignau-Wilberg 1994
T. Vignau-Wilberg, *Archetypa studiaque patris Georgii Hoefnagelii 1592. Natur, Dichtung und Wissenschaft in der Kunst um 1600*, Munich 1994.

Vignau-Wilberg 2007
T. Vignau-Wilberg, '*In minimis maxime conspicua.* Insektendarstellungen um 1600 und die Anfänge der Entomologie', in: K.A.E. Enenkel & P.J. Smith (eds.), *Early modern zoology. The construction of animals in science, literature and the visual arts*, Leiden 2007, 217-244.

Weststeijn 2009
T. Weststeijn, *The visible world. Samuel van Hoogstraten's art theory and the legitimation of painting in the Dutch Golden Age*, Amsterdam 2009.

Wettengl 1998
K. Wettengl (ed.), *Maria Sybilla Merian. Kunstenares en natuuronderzoekster 1647-1717*, Haarlem 1998.

Wilson 1995
C. Wilson, *The invisible world. Early modern philosophy and the invention of the microscope*, Princeton 1995.

Zuidervaart 2009
H.J. Zuidervaart, 'Het in 1658 opgerichte theatrum anatomicum te Middelburg. Een medisch-wetenschappelijk en cultureel convergentiepunt in een vroege stedelijke context', in: *Archief. Mededelingen van het Koninklijk Zeeuwsch Genootschap der Wetenschappen*, Middelburg 2009, 73-140.

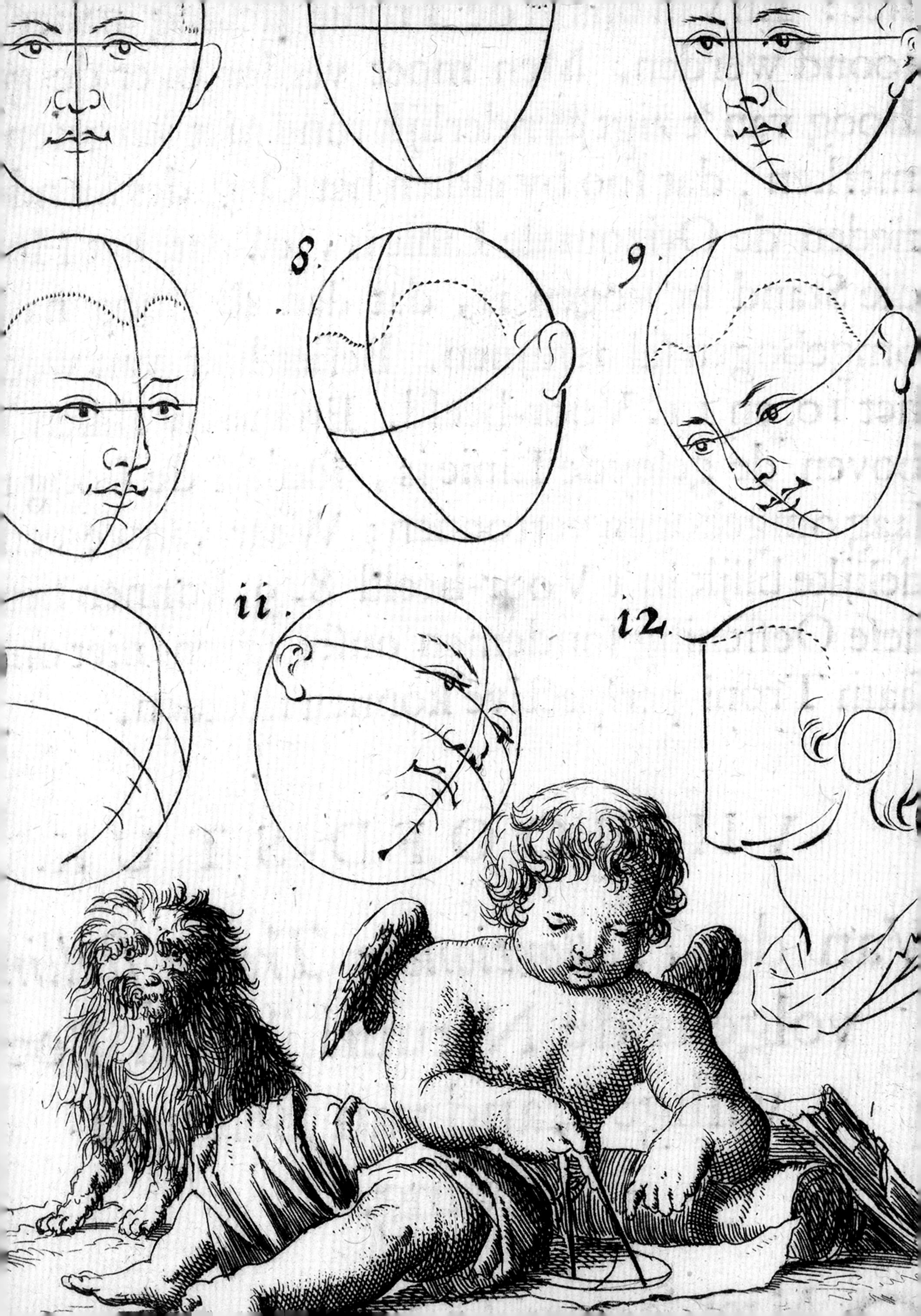
8.
9.
11.
12.

Regulating the arts
Willem Goeree versus Samuel van Hoogstraten*

Gijsbert M. van de Roemer

Introduction

In 1670 Willem Goeree (1635-1711) published a small treatise titled *Inleyding tot de practijck der al-gemeene schilder-konst* (Introduction to the practice of the general art of painting). Eight years later Samuel van Hoogstraten (1627-1678) published his lengthier *Inleyding tot de hooge schoole der schilderkonst* (Introduction to the academy of painting). Given the subject of these works, it is not surprising that both authors are mainly studied by art historians; however, in light of the present volume, the question may be asked if this does justice to the full range of their endeavours. Their interests were certainly not limited to the visual arts. After publishing several titles on art, Goeree directed his attention to other fields of knowledge, such as theology and world history.[1] Van Hoogstraten wrote, next to his treatise on painting, subtitled *De zichtbaere werelt* (The visible world), a counterpart titled *De onzichtbare werelt* (The invisible world). The exact content remains unknown, since the book was never published, but experts suspect it dealt with philosophical matters.[2] Like most men and women of some education, Goeree and Van Hoogstraten had interests that widely surpassed present-day disciplinary boundaries. A more comprehensive approach to their work seems therefore desirable, if not obligatory.

The interest art historians have paid to the two authors has been somewhat unequal. Recently Van Hoogstraten was rewarded with four extensive monographs, while Goeree is barely studied as a subject in his own right.[3] The only extensive work written on the versatile man is the annotated text edition by Michael Kwakkelstein of his earlier publication, *Inleydinge tot de al-ghemeene teycken-konst*. Apart from that, there have been articles about Goeree's writings on architecture and world history, and he is mentioned in studies on draughtsmanship and art education.[4] A reason for this limited attention might be that his work is considered to be not very original. Also, Goeree was not an artist himself, which might make him less appealing to art historians. Furthermore, the fact that his writings on art are somewhat scattered probably hinders systematic research. Goeree did not write a coherent *magnum opus*, like Van Hoogstraten did, but published five separate works between 1668 and 1682.[5] Apart from the *Inleyding (...) schilder-konst*, he published works on drawing, watercolour, architecture and the human body. Nevertheless,

Detail fig. 1
Artist unknown, *The human head in different positions*

Goeree's earlier ambitions regarding the arts were as inclusive as those of Van Hoogstraten. He considered his writings as part of a grander work, which he first envisioned in 1670. This plan was never executed, but the publications were reissued in different collections in 1697, 1704 and 1705. The *Inleyding (...) schilder-konst* can be considered the general introduction to the greater work he hoped to write, and the other publications as substantial parts of the whole.[6]

As they are of a similar time frame, and since Dutch writings on art from the seventeenth century are comparatively rare, the two writers – when discussed together by art historians – are often treated as if they were of one accord. Hans-Jörg Czech assumes that Van Hoogstraten's elaborate chapter division inspired Goeree to do likewise.[7] Lyckle de Vries discusses the writings of Goeree as an 'important link between Van Hoogstraten and De Lairesse'.[8] Thijs Weststeijn calls Van Hoogstraten's work 'more ambitious in terms of both theoretical scope and volume' and uses many utterances by Goeree to support his interpretation of Van Hoogstraten's text.[9] Long before that, Jan Emmens included the works of Van Hoogstraten and Goeree in his analysis of what he called the 'classicist art criticism' in *Rembrandt en de regels van de kunst* (1964).[10]

Here I will follow another line. I will not look for similarities, nor place the authors on a progressive course. Instead, I will, quite deliberately and to some extent perhaps excessively, take them apart. I have two reasons to do so. The first is that, after reading the texts of both authors, I was struck by the differences rather than by the similarities. Goeree was only eight years younger than Van Hoogstraten and discusses comparable themes, but when reading his texts after those of Van Hoogstraten, one seems at times to enter another world. As I will argue, these differences are related to thoughts and views that went beyond the borders of art practice and theory and were informed by the author's general views on the world, philosophy, nature and science. The second reason is my unease with some of the recent interpretations of the authors. Both men are discussed in relation to the famous philosopher Benedictus de Spinoza (1632-1677). Goeree is typified as a radical, and Van Hoogstraten is also said to have had some relation to the philosopher.[11] We should be careful in identifying direct influences of philosophical schools, but, in comparative perspective, I do think it is possible to discern in Van Hoogstraten's writings traits that are, for a large part, rooted in the old humanistic and traditional Aristotelian-scholastic school, not only in his thoughts, but also in the way he transmits these thoughts, while Goeree's writings reveal at times traces of the New Philosophy. However, this was far removed from Spinoza.

It is true that both authors published their texts in turbulent times – a period when things were rapidly changing in the field of the arts as well as that of the sciences. It was the decade following Rembrandt's death and the same decade that Spinoza died. Traditional studies on the Dutch school of painting tell us that after the death of Rembrandt, a new style emerged that is usually – and somewhat inconveniently – indicated with the term 'Dutch classicism'. The main characteristics of this style were a stronger orientation towards antiquity; a predilection for the 'beautiful'; a general pursuit of clarity, order and harmony; as well as a growing preoccupation with the rules of art. New research over the last decennia made the term Dutch classicism more comprehensive and, as a result, more diffuse. Some studies pointed out that distinct classicist tendencies were present in Dutch art throughout the entire seventeenth century, while several scholars started to pay serious attention to neglected painters of the last decades of the Golden Age. Nevertheless, most art historians do agree that something changed in the art climate in the Republic in the last quarter of the seventeenth century.[12] Additionally, the changes in the field of science during this period in the Dutch Republic are much studied and debated. Rienk Vermij gives an outline of the basic tenors elsewhere in this volume.[13] In the middle of the century the ideas of René Descartes (1596-1650) had mostly replaced old scholastic and humanistic ways of acquiring knowledge. But the postulations of the French philosopher were discredited by radical and materialist interpretations, which eventually led to the subversive and (supposed) atheistic views of Spinoza. Spinoza's ideas about the relation between God and nature left no room for an omnipotent Creator and were seen as an abomination by many. Vermij is quite right to state that Spinozism was very present, but 'rather as a danger to be averted than a position that deserved serious consideration'. I think this could also be applied to the cases of Van Hoogstraten and Goeree.

My first goal is not to find direct influences of philosophers and scientists on the works of Van Hoogstraten and Goeree. Instead, I will take by comparison a more general approach. Various aspects will be discussed to highlight some significant differences between the two authors. One of the above-mentioned tendencies of the so-called Dutch classicism will play a central role: the growing preoccupation with rules. It is an aspect that is particularly interesting in light of the theme of this volume, since a correlation can be found between the arts and the sciences. In both fields a rational and mathematical rendering of procedures was increasingly emphasized. Traditionally, Dutch classicism is treated as an art-related movement, where alien influences from France and Italy were incorporated into the Dutch school of painting – and subsequently heralded its downfall.[14] Recent research emphasizes the broader cultural base of Dutch classicism by discussing it in the context of other artistic fields like architecture, literature and theatre.[15] Few attempts were made to link this shift with developments in the field of the natural sciences.[16] This might be surprising, because Emmens already mentioned the philosophy of Descartes and the rise of natural sciences as one of the constituting factors of some classicists' opinions.[17] He noticed a correlation between some of the main arguments of classicist criticism and Descartes's propagation of the use of a consequent methodology in mathematics.[18] Similarly to Descartes's views on gaining knowledge through clear and distinct ideas, writers on art advocated the use of clear and distinct rational (and if possible, mathematical) rules in the creation of art as an alternative for a more practice-based approach. This interesting parallel calls for further exploration, and together, Van Hoogstraten and Goeree form a good case to pursue this. It is also this inclination to mathematize and analyze that forms a distinct contrast between the two authors. Where Van Hoogstraten presents an almost unrestricted accumulation of data, we discern in Goeree's writings a stronger tendency towards the regulation and intellectualization of the arts. These differences become manifest on different levels: in the organization of their chapters, in the way they rendered their information, in their opinions on the intellectual development of the artist and in their general views on the relation between art and nature. These aspects will successively be discussed; however, I will not do justice to the full content of the writings of the authors. More systematic research is certainly needed, especially in the case of the prolific and fascinating polyhistor Willem Goeree. My goal is to reflect comparatively on selections from their writings as a reaction to recent interpretations. In addition, I intend to look at how insights within the history of science can help to illuminate some of the topics discussed by art historians.

Chapters

Van Hoogstraten and Goeree discuss similar subjects, but the manner in which they commit their thoughts to paper diverge, particularly in how they assemble, order and present their material. The distribution of topics among the chapters can illuminate this. Goeree planned to dedicate his chapters to the seven liberal arts, whereas Van Hoogstraten ordered his chapters according to the nine Muses, the classical goddesses of inspiration. Czech, who has meticulously analyzed *Inleyding tot de hooge schoole der schilderkonst*, states that the overlaying structure of Van Hoogstraten's text does not reveal itself easily.[19] The scheme he presents shows that the author's distribution of the subject matter throughout the chapters is not always coherent. Some chapters are dedicated to one general theme, others seem to alternate between two themes, and some are devoted to one general theme with interlacing subthemes.[20] Yet upon closer examination, Czech asserts, a distinct organization becomes manifest: the chapters represent hierarchically ordered stages of the educational training programme for the young painter. They evocate classes where each Muse represents a step in the learning process, which prompts the student to pass through different subjects. These are, successively, the premises of becoming a painter, and draughtsmanship (Euterpe); human proportions (Polymnia); choice of subject matter (Clio); depicting appropriate settings and surrounding for given subjects (Erato); inventing compositions (Thalia); colour (Terpsichore); the history of painting, and light, shadow and perspective (Melpomene); beauty and grace (Calliope); and finally, the usefulness, virtues and benefits of painting (Urania).[21] Czech links the educational disposition of the topics to Descartes's didactic appeal to divide difficult subjects into as many parts as possible and to begin with the simplest and ascend to the more complex.[22] With this he explains the incoherent, multi-themed chapters. These parts of the text appear only as supplementary subjects ('Neben-fächer') in the later chapters, after the student has obtained a basic foundation in the first four chapters. They can be considered as addenda to the main course.

However, taking the educational programme as clarification might explain some features, but still leaves inconsistencies. For instance, Van Hoogstraten discusses the human anatomy rather early in the second chapter, after a short introduction about draughtsmanship. Considering the didactic principle of advancing from the simple to the complex, this location is surprising because the human body was considered the most complex structure in nature to depict. Its place at the beginning is better explained by the importance of the subject.[23] Also, the reasoning behind the position of the exposé on the history of painting at the beginning of the chapter about light and perspective is not evident. Czech says it is 'only useful' when understanding of colour is accomplished, the most important medium of expression.[24] But as most stories of the history of painting start with the discovery of drawing an outline, we might expect to find this subject as well in the first chapter, which is partly devoted to draughtsmanship.[25] Reading Van Hoogstraten's book and studying Czech's illuminating scheme, it is not the lucidity of the organization of the topics that catches the eye, but rather the flexibility with which the author handles his subjects. Maybe Van Hoogstraten is less schematic and ordered than we would like him to be, or rather, he orders his material by premises that are lesser known to us. This will be discussed later, but first I will confront his arrangement with Goeree's chapter division.

As mentioned previously, Goeree planned a grander work in which all topics about painting would be discussed. These correlate to some extent to Van Hoogstraten's topics: the art of drawing, architecture, perspective, the human body, composition and colour. In the *Inleyding (...) schilder-konst* of 1670 Goeree proposes to structure his chapters according to the *artes liberales*, and he repeats this intention in the second edition from 1697. Moreover, in this later edition he reacts to Van Hoogstraten's chapter division, which was published in the meantime. Goeree speaks about it in a rather amusing, somewhat bantering way.[26] He states that the author probably borrowed his idea from Franciscus Junius (1589-1677) and that he embellished it with some 'nice adornments for pleasure' ('aardige dertelheden tot vermaak'). Alternatively, he suggests the *artes liberales* as a guiding principle and explains how each *ars* correlates with an aspect of painting. Grammar stands for general knowledge, dialectics for drawing, rhetoric for architecture, music for perspective, arithmetic for anatomy, geometry for ordering inventions, and finally, astronomy for colour.[27] From a modern viewpoint it is easier to follow the train of thought behind Goeree's order. After an exposé about the general knowledge an artist should posses, the student learns the basic principles of drawing and, subsequently – as a guiding principle – about the binding measurement systems of antique architecture, after which he can practise the rules of perspective. The prominence of the architectural orders in this scheme can be seen as a personal trait of Goeree, since he had a great passion for architecture.[28] In the next chapter the trainee learns about the human body. Upon accomplishing all of this, he can combine the two by practising inventing good compositions, which, in the end, must be expressed with colours.

Each author has a specific way of relating his themes to the different constituents of his adopted system. Van Hoogstraten's relations are often brought about by casual and creative association. For instance, he connects Thalia, the muse of comedy, to designing and testing compositions because one should guard against melancholy to order correctly.[29] Polymnia, the rhetorician, teaches about human proportions because she represents memory and a student must remember these things.[30] Terpsichore stands for colour because she is the most charming of the Muses and wears many coloured feathers.[31] And, to give a final example, Melpomene, the muse of tragedy, instructs about light and dark because in tragedy one finds comparable contrasts between happiness and sadness.[32] It is possible to give more examples, but we may be missing the point by trying to find the logic behind Van Hoogstraten's chapter division. He did not choose this scheme for its substantive suitability to the topic of painting, but for its versatility and fluid character. It was not his aim to create a crystal clear and self-evident scheme, but to find an elaborate mould into which to incorporate his subjects. The ancient and dignified scheme suggested a sense of completeness, coherence and universality, and it enabled him to order his extensive documentation and thoughts. At the same time, it could help as a mnemonic device for trainees.[33]

In contrast, Goeree did not connect his chapters to mythological figures, choosing instead the more cerebral scheme concerning knowledge and science. The scheme itself leaves less room for playful interpretation, as it does not involve anthropomorphic goddesses with specific features that serve as multifaceted models to articulate his text. Goeree reduces the themes on painting to their concise basic principles and subsequently connects them with the idiosyncratic characteristics of one of the *artes*. For instance, grammar represents general knowledge because it is needed to read literature; this is the premise for acquiring knowledge. Draughtsmanship equals dialectics, as both aim for truth: the first by arguing and the second by finding true forms. Music parallels perspective because both are based on correct systems of maxims and both offer diversion. And arithmetic equals anatomy because both are concerned with correct numerical relations.[34] Thus Goeree presents the act of painting in a context of other intellectual activity. By connecting them to these fields of knowledge he tries to show that the principles of painting are as fundamental and steadfast as those of the liberal arts. The arrangement of the chapters is a first indication of Goeree's more intellectual approach and the different analytical mode he applied when writing on the arts. This is also discernable in the organization of the text itself.

Text

Art historians have commented on the versatility and profusion found in the text of Van Hoogstraten, and sometimes on its lack of unity. Emmens characterized his book as a 'not unpleasant amalgam' of quotations of earlier authors.[35] Weststeijn points to the great variety of sources used by the author, stating that it is pointless to try to fit his theory 'into any specific school of thought'. His book, which contains 'references to both the mediaeval and the mechanistic world view', is better explained with the aid of the rhetorical aims of the author and the epideictic character of the text.[36] Ernst van de Wetering states in his recent discussion of Van Hoogstraten that 'it is important to realize that books in Rembrandt's time were not coherent, systematically underpinned arguments, but rather compendious accumulations of knowledge, which could contain chunks of information that were often incompatible'.[37] Perhaps these remarks can gain emphasis by making a detour in the history of sciences.

In the last decennia historians of science started to work to a greater extent from a broader notion of 'scientific culture' as an alternative to the grand narratives that dominated their field for so long, such as the Scientific Revolution or the mechanization of the worldview, as well as a fixation on the 'great names' of science. Next to an interest in the social context of science, they also focused on the different epistemological models and discursive strategies that were used by larger groups to produce and transmit knowledge. To give some examples, Antony Grafton studied the importance of textual transmittance of knowledge and bookish culture; James Bono explored the discursive practices of 'reading the Book of Nature' and noticed a shift from textual exegesis to hermeneutic strategies in the seventeenth century; and Eric Jorink paid attention to the same metaphor in a Dutch context, stressing its importance as a constituting element in the scientific culture of the Republic.[38] Of course, the epistemological and discursive modes described by these authors were not restricted to books on nature alone. Texts on other subjects – for instance, the arts – should bear traces of them. To illustrate this, and to help illuminate the difference between Van Hoogstraten and Goeree, I will use a classic article by the historian of science William Ashworth, 'Natural history and the emblematic world view'. Ashworth's goal is to describe some specific characteristics of natural history between 1550 and 1650, for which he introduces the term 'emblematic world view'.[39] He explains how the work of naturalists from this period differed fundamentally from the works of their successors. To make a specimen known, they did not restrict themselves to describing the visible features of the specimen and classifying these in a system. They collected all the information they could get about an animal or plant, drawing from many traditions – hieroglyphics, coins and medals, Aesopic fables, classical mythology, adages, epigrams, emblems and devices. This accumulation of data was considered the core of natural history, and only later were the specimens 'scraped clean' from 'their crust of analogies, resemblances and signs'.[40] In the second half of the seventeenth century, naturalists started to focus exclusively on the perceptible features of the specimens, which eventually resulted in an emphasis on classification.

Even though Ashworth's inciting exposé has some generalizing dangers, I do think his notions can be used to mark the contrast between Van Hoogstraten and Goeree more sharply.[41] Whereas the former was more imbedded in what we could call a 'bookish', 'emblematic', 'compiling' or 'accumulative' way of producing and transmitting knowledge, the latter shows signs of a more systematic and analytical approach to unfolding his thoughts. Van Hoogstraten's chapters are compilations of information that are based partly on his own experiences in the workshop and partly on what he read or heard elsewhere, for which he used an abundant quantity of classical and contemporary sources.[42] Loosely inspired by general themes, he threads chains of interesting data that might be useful for the young artist and interlards it with a multitude of quotations, stories and experiences. Compared to Goeree, his focus is more on the sequence of the narrative than on the systematic structure of the total text. When this style of research is taken for granted as a different mode to process and transmit knowledge, it becomes less necessary to find explanations for the perceived inconsistencies and the way he wanders sometimes from one subject to another.

Two examples will illustrate this. In the sixth chapter, Van Hoogstraten discusses colours. He stresses the importance of knowledge about this subject, but instead of practical information about the characteristics of specific pigments and the way they should be applied, his emphasis is on – to follow Ashworth's words – an 'emblematic' discussion of colours. He gives some practical remarks about which pigments are used by modern painters, but the greater part of the text concerns aspects such as to which planets the colours correspond and the virtues and vices they represent. So we read that white belongs to the Moon and blue to Jupiter, that the mixture of gold and blue stands for the lust of the world and that gold and grey stand for meticulousness.[43] In the following sections, Van Hoogstraten gives practical advice – for instance, that friendly colours mix well and hostile colours do not, and that one should follow the colours in nature – but specific information with regards to actual practice in the studio is relatively limited.[44] It would be illuminating to confront this with Goeree's treatment of colours, but sadly this part of the project was never written. Goeree did publish an adaptation of a treatise on watercolours by Gerard ter Brugge, two years before his *Inleyding (...) schilder-konst*. In it he gives detailed information about the practice of using colours: the preparation of paints, some characteristics of the pigments, the implements used, an overview of 32 pigments, how to prepare paints, how to apply them, how to keep them and what colours to use to depict skies, clouds, walls, rooms, flesh, hair, trees, wooden objects, cities, castles, rocks and all sorts of animals, fruits, vegetables, flowers and minerals. A one-to-one comparison is hazardous, but we can use this little book to show what Van Hoogstraten did *not* tell. Even though Van de Wetering convincingly showed that Van Hoogstraten's book should be taken for what it is, a manual for the young artist, many things the author recounts do not directly relate to the act of painting. This is not to say that his information was not useful for the painter. On the contrary, it shows that the field of knowledge that Van Hoogstraten considered suitable for the painter expanded in many divergent directions.

A second example will illustrate the author's accumulative manner of collecting data. In the fourth chapter, concerning the correct settings and details for certain scenes, he spends eighteen pages discussing hairstyles, clothing, weapons and armour. Van Hoogstraten sums up what he believes is appropriate and assembles fragments of information that can be useful. Thus the reader learns about the washed hair of Spartans, the beard of Charles V, the clothing of Persians, Egyptians, Romans and Russians, and about the Devil. Of course these things can be useful for the young artist, because he wants to know how to depict things, but sometimes the necessity of what the author does and does not tell is elusive. Revealing with regard to this manner of thinking is the concluding remark after his long digression: 'I avoid to speak of offerings, slaughter knives, axes, basins, cover cloths, and festoons, and only mention what pleases me and comes into mind'.[45] Again we see here an author who freely assembles information from a multitude of sources.

Goeree is just as long-winded as Van Hoogstraten, often even more so. His manner of recounting can also be, at times, associative.[46] But in contrast to Van Hoogstraten's train of thought, we notice in Goeree's text from time to time another mode of structuring his material, a more systematic way of unfolding his thoughts and instructing the trainee. (That is not to say 'better' or 'more 'scientific', but at the most 'more recognizable' for the modern reader.) This becomes clear in the way he discusses the human body in *Natuurlyk en schilderkonstig ontwerp der menschkunde*, from 1682. He treats each body part separately, beginning at the head and ending at the back. In almost every section, Goeree follows the same procedure. He discusses first the construction and function ('maaksel en dienst'), and then by the appearances and the physiognomic readings of specific traits. And so he discusses hair, countenance, forehead, brows, eyes, nose, ears, cheeks, mouth, lips, tongue, teeth, chin, neck, arms, hands, fingers, armpits, breasts, belly, navel, legs, feet, buttocks, back and skin.[47] Instead of the somewhat random variety presented by Van Hoogstraten, we see here an author striving systematically for completeness.

Goeree's aspiration to capture the diversity of the human body in all its appearances shows classifying tendencies. He even talks about different classes ('geslachten'), when he tries to map the diverse features of the countenance. He discusses the features in running text, but with little effort we can translate his divisions into a taxonomy:

1. Hair
    a. bristly
    b. curly
    c. straight

2. forehead
    a. flat and smooth
    b. hollow
    c. full and round
        I. full and round above
        II. full and round below
        III. full and round above and below

3. nose
    a. completely
        I. straight
        II. flat
        III. uniformly hollow
        IV. uniformly elevated (hooked)
    b. somewhat flat or elevated
    c. above straight, below hollow
    d. above straight, below elevated
    e. above hollow, below straight
    f. above hollow, below elevated
    g. above elevated, below straight
    h. above elevated, below hollow
4. mouth
    a. small lips
    b. big lips
    c. curled lips
5. chin
    a. short
    b. long
    c. thick
    d. hollow
    e. round[48]

This systematic method of description also appears in his discussion of the stances of the head. He distinguishes four classes of the upright stance, of which the first two classes have no variations (from the front and from the back), the third class has two variations (from the side: from the left and from the right) and the fourth class has many variations (receding, in many gradations; see fig. 1). Goeree calls these upright positions 'natural' and sets them apart from what he calls 'coincidental' positions. These are divided into two basic classes: the first is receding forwards or backwards, the second receding left or right. These two classes can be combined together in many variations, and the classes of the 'natural' stances can also be transposed on them. Later in the book, with the description of bodily movements, he follows the same procedure and divides them into seven classes.[49]

By modern standards Goeree's classifications still harbour many inconsistencies and are surely not very elaborate – sometimes even amusing. But compared to Van Hoogstraten, his way of informing the reader and instructing the trainee shows signs that are fundamentally different. It is based not on an alignment of data, but on a method of analysis where the material he wants to pass on is ordered and classified. The way Goeree does this is not compulsory. Just as Van Hoogstraten does, he likes to show off his knowledge and expose his erudition. He recounts stories, gives advice, and outlines things he has heard and read, but in between the stories and recollections we see a Goeree who is observing, analyzing and classifying. The desire to systemize is certainly present.

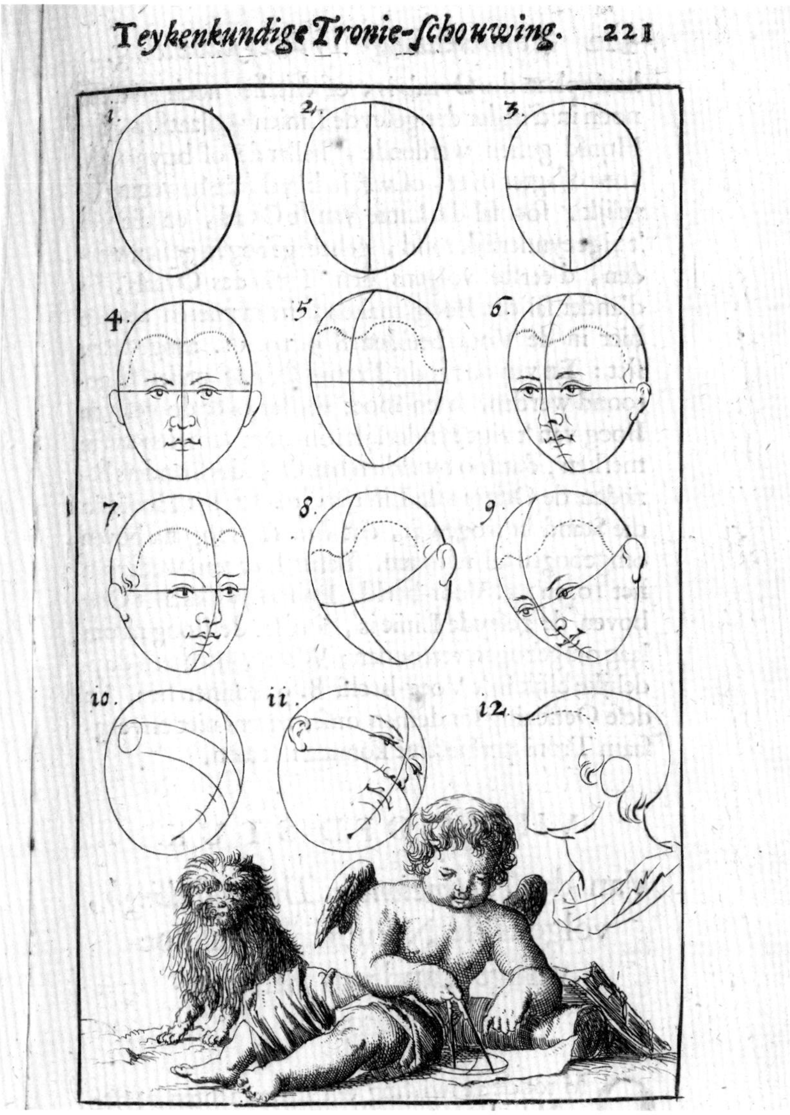

1

Artist unknown, *The human head in different positions,* engraving, Willem Goeree, *Natuurlyk en schilderkonstig ontwerp der menschkunde (...)* (Amsterdam 1682), Universiteitsbibliotheek Amsterdam (UvA), Bijzondere Collecties, O 61-6997.

Knowledge

Having explored some structural elements of the texts, I will now turn to the content. Earlier I mentioned a preoccupation with the rules of the arts as a characteristic of classicism. This coincided with a more knowledgeable approach to the arts. Both authors advocate a well-instructed and informed painter, but they hold different opinions about the extent and character of this knowledge. Whereas Van Hoogstraten states that both head and hand should be skilled at the same time, Goeree thinks the head should be trained first.[50] Goeree, like most so-called classicists, tried to relocate the art of painting from the domain of the hand to the domain of the brain. He fulminates against the practical attitude prevalent among artists in the Republic, about whom he states that the art lies not in their brains, but in their fingertips.[51] An artist should not begin painting without reflective knowledge ('beschouwige kennis') of many things. To underline this he gives a survey of nearly one hundred titles of recommended books concerning history, antiquity, mathematics, land surveying, architecture, natural philosophy, natural history and human anatomy.[52] Van Hoogstraten also emphasizes the importance of knowledge in many places, and even calls the art of painting – as it is often cited – a 'science' ('wetenschap').[53] He, too, acknowledges the benefits of natural philosophy, history, poetry and mathematics,[54] but in other instances he also expresses his reservations about too much learning. A premise seems to be that the acquired knowledge should be directly applicable to the act of painting itself. When discussing the arrangement of compositions, he states that knowledge ('wetenschap') alone will help little; sketching and practice will be of greater value.[55] Likewise, he utters his scepticism with regard to reading books on human anatomy. He wonders who has time and the desire to read Vesalius, Laurentius or Cabrolius. Even the work of Jacob van der Gracht (1593-1652) – an author who published an anatomy book especially for the artist – seems to him more applicable to surgeons than to painters. In comparison, Goeree numbers the works of Vesalius, Spiegelius, Vestingius, Van der Gracht and Bartholinus among his recommended literature. Both authors aim at artistic knowledge, but Van Hoogstraten presents many diverging and interesting data with the act of painting as a point of departure, while Goeree more often starts outside the act of painting and seeks knowledge that can feed the process.

The case of anatomy can serve as a good example to further illuminate the disparity. Again, a comparison might seem somewhat inappropriate, since Goeree wrote an entire book about the human body and Van Hoogstraten discusses it only in the second chapter of his treatise, but that the latter needed only 31 pages, whereas Goeree needed 432, may serve as an indicator of his more knowledgeable approach. According to Goeree, proportions change with every movement, which is why an artist should also know about the functions and the mechanics of the body parts.[56] He discusses these things extensively, giving an overview of the different measuring systems of the human body. In contrast, Van Hoogstraten keeps his information limited and recounts only what he considers useful for the painter. This consists mainly of reproductions of five prints depicting bodies of men, women and children with explanatory tables and measuring systems (fig. 2).[57] Van Hoogstraten confirms that copying these prints and learning the names on the tables by heart is sufficient to make a young artist 'fully qualified'. Even though he suggests the possibility of further study ('for those who want that'), he warns against a profusion of measurements.[58]

The case of perspective serves as a second example to emphasize the authors' distinct views on artistic scholarship. Several times, Van Hoogstraten stresses the importance of this science, stating that it is impossible to make good art without it, yet dedicates only three pages to the topic. He does not explain perspective, but gives – in his known accumulative way – a small collection of stories of persons who applied it

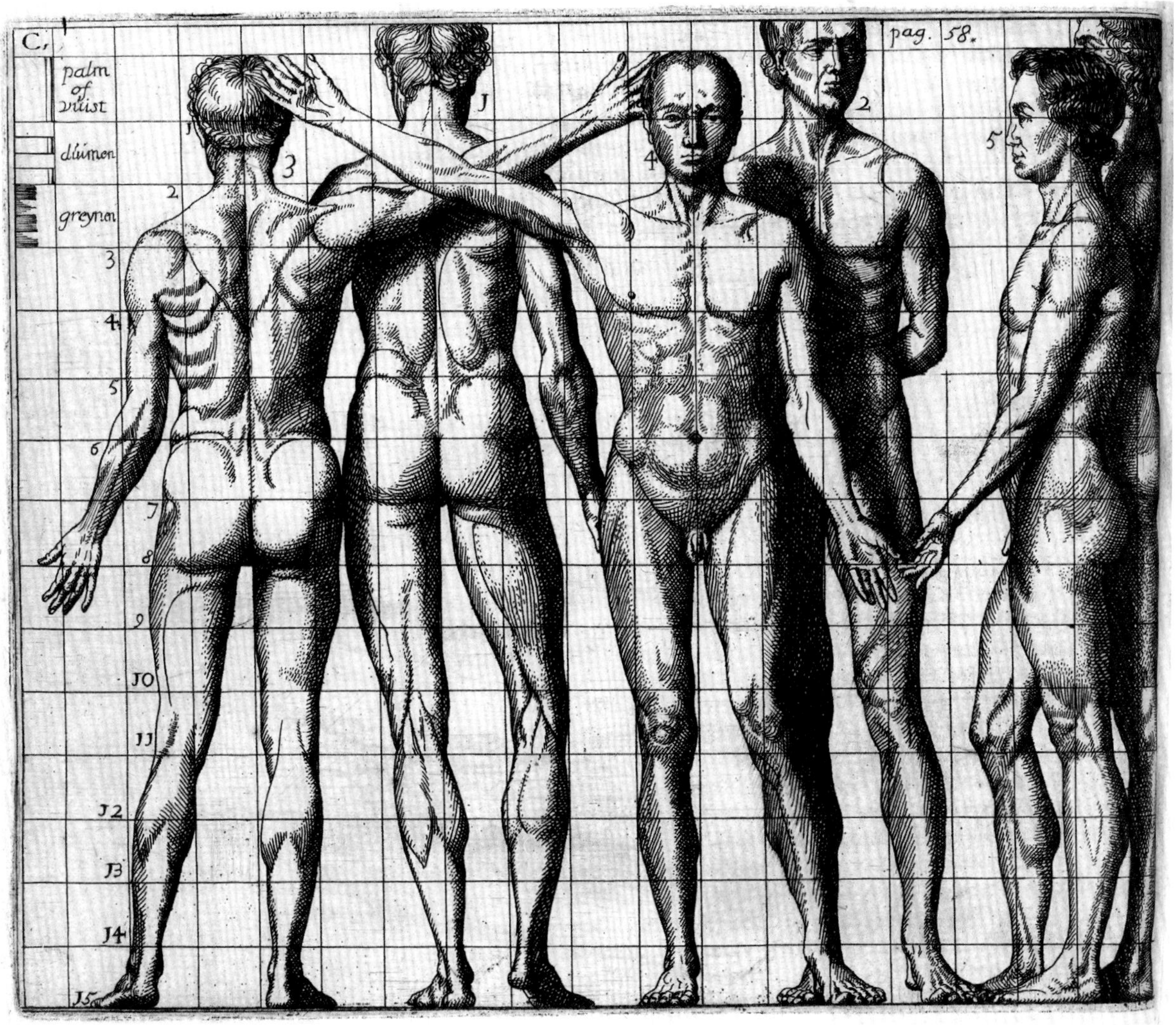

2

Samuel van Hoogstraten, *Six male bodies with different proportions,* etching, Samuel van Hoogstraten, *Inleyding tot de hooge schoole der schilderkonst. Anders de zichtbaere werelt* (Rotterdam 1678), Universiteitsbibliotheek Amsterdam (UvA), Bijzondere Collecties, O 62-6094.

in an effective way: Giulio Romana and Carel Fabritius used it in murals to make rooms look bigger, Giorgione used it with mirrors to make things simultaneously visible from different angles, Claudius Pulcher used it to make convincing theatre stages, and Raphael learned it from a monk. [59] Subsequently, the author declares he will not go into detail because many before him already did so and it would be a 'bore' to repeat this.[60] Instead, he refers his readers to a mixed assembly of authors, for instance, Albrecht Dürer, Hans Vredeman de Vries, Samuel Marolois and Girard Desargues.

Like Van Hoogstraten, Goeree stresses the importance of perspective, but instead of recounting stories, he chooses one distinct method. He declares himself an adherent to the finding of the French mathematician and engineer Girard Desargues (1591-1661) and considers this to be an understandable method based on 'steady foundations', which equates to mathematics. It can be applied perfectly to the arts, inspiring certainty and even enabling an artist to measure the distances of depicted objects in three-dimensional space.[61] Goeree never wrote his chapter on this topic, but a treatise titled *Perspectiva of Doorzichtkunde* was published in 1703 by his younger friend and protégé, Dirk Bosboom (c. 1648-c. 1707).[62] The content of this work corresponds remarkably to the plan Goeree uttered in 1670: a discussion of Desargues's method, preceded by a long digression on the nature of light, the construction of the human eye and its functioning. Bosboom's views on optics were, to a large extent, based on the insights of Descartes.[63] The reader learns in detail that light is caused by the straightforward movement of subtle matter and that it moves in straight lines into the eyes of the beholder. After this 30-page exposé, he turns to the method of Desargues. This stands in sharp contrast to Van Hoogstraten, who states in his discussion about light and perspective that he will bypass what philosophers have said about the nature of light and recounts only what is needed for the artist.[64]

The method Goeree and Bosboom propagated was known to be too intellectual and rational for the everyday artist. In *The science of art*, Martin Kemp describes how in the first half of the seventeenth century Desargues, who was influenced by Descartes's analytic geometry, developed a new means to create the illusion of depth on a flat plane. On the basis of projective geometry, he tried to replace the old Renaissance method of the 'checker-board floor' with a more accurate, mathematically grounded method. In France, after a brief proliferation, the method was considered too theoretical and unpractical for artists and was expelled from the French academy by Charles Le Brun (1619-1690).[65] Nevertheless, apparently Goeree and Bosboom found it fit to spread among Dutch artists.

Bosboom and Goeree also explain why painters should know these basic fundamentals that seem so far removed from the daily practice in the workshops. Bosboom emphasizes the importance by stating that an artist who has acquired knowledge of these matters will be able to rely on it in the future and will find solutions to all kinds of painterly problems.[66] Otherwise, one has to rely on untrustworthy chance, and, as Bosboom adds, 'Guessing is mistaking' ('Gissing maakt missing'). Goeree thinks likewise. In his *Inleyding (...) schilder-konst* he stresses the importance of fundamental knowledge and states that things such as draughtsmanship, architecture, perspective, anatomy, ordering and colours should be learned at an early age. One should not step over them too hastily, but an enlightened mind should explore them and learn to govern them. This has to precede the creation of art and is a premise for good art. Apart from the direct influence of Descartes, some distinct features of the analytical thinking of the New Philosophy can be found in Goeree's and Bosboom's approaches. On different levels they attempt to reduce the procedure of painting to some clear basic principles that would form the fundamentals on which an artist could rely. While Van Hoogstraten is presenting a rich tapestry consisting of diverse information about painting, Goeree is trying to unstitch the different threads of which it is made. It is clear that basic knowledge of human anatomy and physiology can help to depict characters in paintings and that mathematics and optics are needed for correct rendering of perspective, light and shade, but also the way Goeree classifies the countenances and gestures can serve as an example of bringing complex things back to some simple basics. He states that an artist who has imprinted the basic 'classes of the face' ('tronie-geslachten') in his mind will easily be able to depict someone he has seen only once. He must look for the general model in his head that corresponds with the impression of the observed person and can then fill in the basic form with further details.[67]

Rules and chance

The aforementioned characteristics of Goeree's work do not conform to Weststeijn's interpretation of the text. He uses the writer to suggest a relationship between Van Hoogstraten's concept of depicting the 'visible world' and Spinoza's *natura sive Deus*, in which the concept of the visible and invisible world overlap, and transposes the supposed influence of Spinoza on Goeree to Van Hoogstraten.[68] But Weststeijn's reading of Goeree is incomplete. He states: 'In Goeree's view it is not the universal fundamental principles that concern the artist'.[69] The writer was 'basing the significance of the depiction of nature not in the study of ideal beauty, but precisely in the fact that nature is "unfathomably rich in bringing forth multitudes of every species (...)"'.[70] Also, Goeree is supposed to have come to 'radical conclusions' that 'beauty is a relative phenomenon', which is further discussed in connection to Spinoza's views.[71] However, many of the views cited by Weststeijn are actually paraphrased by Goeree in order to refute them. This can be corroborated by citing quotes wherein Goeree airs his own opinion instead of those of others. In *Menschkunde* he reacts to 'philosophers' who suggest beauty is in the eye of the beholder, and he counters this by stating it

> (…) is in general not true of all true beauty, certainly not of that [beauty] which is mainly based on sound reasoning and on the equilateral proportional compositions of the parts to the whole.[72]

The author expresses here the traditional notion from classical architectural theory that defines 'true' beauty as a design in which the relations between all parts of a composition must agree with the harmonious whole, so that nothing could be changed without its disruption.[73] It is clear that Goeree presupposed a fundamental beauty that goes beyond the human experience and is inherent to things. He does not refer to cosmological analogies or heavenly harmonies, but relates it directly to truth found in mathematics. He admits that sometimes beauty might elude the ignorant beholder, but that it will remain true 'like a mathematical truth [that] can rely on itself'.

The excursions above already demonstrate that Goeree *did* target basic fundamentals in order to buttress the practice of painting. His texts on art – and also on world history, for that matter[74] – show a preoccupation with rules and regularity that is often expressed in a mathematical way. In six of his seven explanations of the chapters, discussed earlier, he uses phrases like 'distinct order', 'sure measurements' or 'according measure'. This wish to regulate and mathematize was certainly not new in art theory; but it is justified to assume that these old notions acquired a fresh radiance in this period due to new insights in the field of natural philosophy. This blended with his religious convictions. The following significant quotation shows this:

> And like God has laid in all parts of his Creation mathematical rules, according to which nature produces, retains and destroys her works, so the goodness of the same Creator bestowed such an ingenuity upon men that they found themselves infallible rules, to imitate nature.[75]

We should not too easily disregard these remarks as commonplace. Rather, I believe they formed the frame of reference from which Goeree wrote about the arts. Goeree's opinions on art found footing in a fluid merging of notions from classicist art theory, Protestant beliefs and the New Philosophy. These were the three constituents of his thinking. He was convinced that men should produce art according to the same regular procedures with which God governed his creation. For him the Biblical notion that God created everything according to measure, number and weight – a notion described in the Book of Wisdom, and to which he refers explicitly – was as directive as the new mathematical insights of Descartes and his followers.[76] He felt no opposition between these views, but regarded them as two truths that confirmed each other. It also informed his view on arts. As the quotation shows, in Goeree's thinking there is no sign of equating God with nature. On the contrary, he conceived nature as a precise and mathematical artwork crafted and ruled by the Supreme Artist. The reciprocal relation between God and nature was clear-cut, and because of this relation nature could function as a double-layered example for art. Goeree mentions imitation as an end goal, but the parallel between art and nature goes further than mere mimesis: nature not only supplies the examples that art should follow, but also reveals the processes by which this imitation should be accomplished. More explicitly stated, the regularity with which God moved matter in his creation serves as an example for men when utilizing artistic materials in order to create works of art.

There was hardly anything radical about this, though some orthodox thinkers may have frowned upon these words. Goeree did not need Spinoza's *natura sive Deus* to solve the paradox between the *res cogitans* and *res extensa* posed by Descartes.[77] If his thoughts coincided with a philosophical school at all, it would be wiser to turn to the ideas that were simultaneously developed outside the Dutch Republic, in England. There, thinkers associated with the Royal Society, like Robert Boyle (1627-1691) and John Ray (1627-1705), developed a research strategy where the mechanistic interpretation of nature was used as a confirmation of God's omnipotent power. A direct influence is not tenable, but the ideas these researchers developed and laid down in their writings coincided with the opinions of many writers and researchers in the Dutch Republic: a fertile combination of the New Philosophy and Calvinist beliefs, which could be used as an antidote for the atheistic tendencies within an overly materialist view on nature.[78]

An important feature of this view was the increasing consideration of nature as a work of art. This metaphor was not as clearly defined as that of the Book of Nature, and it had many guises: nature as painting, nature as clockwork, nature as automaton or nature as embroidery. But in the minds of researchers of the late seventeenth century it was certainly as vibrant as the Book of Nature analogy. Lorraine Daston and Katharine Park explain in *Wonders and the order of nature, 1150-1750* how, due to mechanistic philosophy, this metaphor of 'nature as art' gained vigour and replaced the older notion of 'nature as artist'. In the beginning of the seventeenth century, congruent with the scholastic Aristotelian views, Nature was seen as a maid or a worker of God. She was totally dependent upon her Creator, but was still allowed some autonomy. Nature harboured hidden powers and qualities, and this enabled her to err, to paint or to play. These perceptions eroded under the rise of mechanistic natural philosophy's influence. Nature's limited autonomy and her anthropomorphic character faded, and it – because she now became an 'it' – was seen eventually as a construct assembled from tiny particles that moved with mathematical regularity. This construct was incredible in its complexity, but in the end inert and a thing *made*. 'The ancient opposition of nature and art had been transformed into an opposition between the works of God and the works of man. Nature had become "the Art of God", no longer able to create art of her own'.[79]

Did these broad notions somehow affect the practice in the studios in the Dutch Republic, and can they help to delineate the difference between Goeree and Van Hoogstraten further? In looking for an answer it can be illuminating to, as a last excursion, pay some attention to those aspects of the artistic practice that eluded a rational or quantitative approach. For instance, the absence of a regularized system for mixing and applying colours was still felt as a deficiency by many. Goeree's explanation for connecting his chapter on colour to astronomy is telling in this respect. He binds the two because of their shared lack of certainty: just like astronomers had to guess the partly invisible orbits of the planets, artists had to guess the outcome of colour mixtures.[80] Others also noticed a gap in this knowledge. In his *Groot schilderboek*, Gerard de Lairesse (1640-1711) contrasts applying colour with anatomy and perspective, and finds it 'remarkable' that there are still no ground rules for this part of painting.[81] When autodidact Pieter Rabus (1660-1702) wrote in his *Boekzaal* a review on a little treatise on painting, he stated that the nature of basic paints belonged to philosophy, but that no person until now was able to demonstrate mathematically how paints were used and how colours were mixed.[82] By the absence of clear rules, the artist had to rely on practice and chance.

But chance itself was being challenged. Rabus's phrase 'until now' suggests that he thought this problem might be solved in the future. Also, Bosboom's 'guessing is missing' discredits chance as something to rely on in producing art. As mechanistic natural philosophy tried to deny the existence of chance in nature, writers on art that honoured the classicist principles tried to eliminate chance from the process of painting.

According to them, representatives of the older schools of painting relied too much on the fortuitous outcome of their manual skills.[83] Painters like Rembrandt and his last pupil, Arent de Gelder (1647-1727), served as negative examples. These artists deliberately used the uncertain outcome of brush strokes as a substantial part of their working process. One of the most famous examples of this bold manner of painting, with loose brushstrokes and dripping impasto, is the gold brocade sleeve of the male sitter in Rembrandt's *Jewish bride*. The painting and the sleeve were eloquently described by Ernst van de Wetering: '(...) it almost appears to be the outcome of a geological process; in some elusive way chance seems to have played an important part in this manner of painting'.[84] In contrast to what classicist criticizers might have thought, Van de Wetering attests in his discussion of two self-portraits of Rembrandt that this reliance on chance must have been a conscious choice in the act of painting.[85] An artist could allow the effects of chance to create convincing illusions by activating the beholder's imagination, a phenomenon coined by Gombrich as 'the beholder's share'.[86]

Compared to Goeree, Van Hoogstraten was still more open towards an accidental and less controlled way of working. For instance, he advocates a rough way of sketching (*in gros*) that the artist should apply to recognize shapes in the different, imprecise forms. He advises starting with indicating the bigger forms roughly and then working towards the details while searching for forms. This way of working is for him the basis of good drawing, which is again the basis of good painting.[87] For Van Hoogstraten, chance was still an option, even though he also pleads in many cases for more regularized procedures. Elucidating in this respect is the story he recounts about the painting contest between the painters François Knibbergen, Jan van Goyen and Jan Porcellis.[88] The goal was to finish a painting as quickly as possible. Each artist represented a distinct way of working. Knibbergen painted figures on his canvas as letters written with a pen (a painter relying on his habits and tricks). Van Goyen first filled the entire panel with paint like 'a many coloured agate or marbled paper' and found thus shapes and figures that he accentuated from the chaos (a painter relying on chance and his imagination). Porcellis first took his time to make an elaborate design in his head and then started to work (a painter relying on his intellect). Despite that all three paintings were fairly good, the outcome demonstrated that the method of the latter led to better results, because Porcellis acted 'certain and sure' ('zeker en gewis'). Here Van Hoogstraten manifests as a mouthpiece for a more intellectual approach, but, on the other hand, the example shows that 'chance' was still an option for painters. It is this accidental method of creation that was refuted by Goeree and Bosboom.

The tension between chance as a legitimate working procedure and a more controlled and regulated approach to the arts can pose some interesting questions for further investigation. In art history the scrutinizing research into the chemical formula of pigments, the complex recipes for paints and varnishes, the quantity of threads in canvases and the number of growth rings in wooden panels have not yet inspired questions about the basic premises with which early modern artists considered their materials. Did representatives of the old schools see nature as a productive force, which in itself could help or stimulate the artistic process? Could the qualities inhabiting the pigments in paints contribute to a greater effect of illusionism? Did painters who used chance in their technique paint not only 'after' nature, but also 'with' nature? Did the concept of natural products as a congregation of inert corpuscles that move according to mathematical laws contribute in any way to the diminishment of this way of working? Probably these questions were far removed from daily life in the studio, but in the end it must have made a difference whether an artist saw his working materials as substances preserving overt and hidden qualities, or as congregations of tiny corpuscles that move and act according to mathematical regularity. The wish to bind art to the same regular laws with which creation was governed became stronger in the first decades of the eighteenth century. In the Dutch Republic it found its most pronounced advocate in Lambert ten Kate (1674-1731), who is the focus of the essay by Fokko Jan Dijksterhuis elsewhere in this volume. Apart from precise calculations of human proportions, Ten Kate experimented with mathematical ratios for mixing colours and thus tried to fulfil an absence that was felt by many, as expressed before in the words of Rabus and De Lairesse. Hessel Miedema, who meticulously studied the correspondence between Ten Kate and the painter Hendrik van Limborch (1681-1759), describes their reflections. The two discussed anatomy and colour theory extensively, while Van Limborch functioned as an executor of Ten Kate's ideas. They devoted much effort to finding correct mathematical relations, but their findings were not very successful.[89] In the end, the gap between theory and everyday practice was too big. It seems that, with the perspective method of Desargues and the colour experiments of Ten Kate, the regulation and quantification of the arts had reached its limits. Art and science could no longer inform each other without discrediting their respective endeavours. Art was soon to find another direction. The irrefutable objective gradually made way for a more subjective approach to the arts, which eventually led to the new discipline of aesthetics. Science and art lost their shared, solid base of objective truths, and the bond between them was loosened. However, for men like Goeree, Bosboom and Ten Kate, this bond was still crystal clear and formed a premise for their reflections on art.

Notes

* I would like to thank Frans Grijzenhout, Ernst van de Wetering, Eric Jorink, Rienk Vermij, Fokko Jan Dijksterhuis and Jason Swinehammer for their valuable comments on my text.

1 Willen Goeree wrote *Joodsche Oudheden of voorbereidselen tot de Bybelsche Wysheid* (Amsterdam 1690); *Mosaize historie der Hebreeuwse Kerke* (Amsterdam 1700); *De Kerklyke en weereldlyke historien* (Amsterdam 1705).

2 Also, the manuscript did not survive. Van Hoogstraten mentions this work occasionally in his writings. Roscam Abbing 1993, 21; Czech 2002, 74-78; Weststeijn 2008, 38-39; Van Hoogstraten 1678, 86.

3 On Van Hoogstraten, see Brusati 1995, Czech 2002, Blanc 2008, and Weststeijn 2008.

4 On architecture, see Van den Heuvel 1997; on world history and antiquities, see Leemans 2004. Kwakkelstein presents a good overview of studies in which Goeree is mentioned. See Kwakkelstein 1998, 13-16.

5 *Inleydinge tot de al-ghemeene teyken-konst* (Middelburg 1668, ^{2}1670; Amsterdam ^{3}1697, ^{4}1705, ^{5}1739); *Inleyding tot de practijck der al-gemeene schilder-konst* (Middelburg 1670; Amsterdam ^{2}1697, ^{3}1704, ^{4}1770); *d'Algemeene Bouwkunde* (Amsterdam, 1681, 1705^{2}); *Verlichtery-konst*, (Middelburg 1668, ^{2}1670; Amsterdam ^{3}1697, ^{4}1705); *Natuurlyk en schilderkonstig ontwerp der menschkunde*, (Amsterdam, 1682, ^{2}1704, ^{3}1730, ^{4}1753).

6 This article will focus mainly on

Inleyding tot de practijck der al-gemeene schilder-konst and *Natuurlyk en schilderkonstig ontwerp der menschkunde. Verlichtery-konst* was an adaptation of another work. *Inleydinge tot de al-ghemeene teyken-konst* is already well studied by Kwakkelstein. The fifth work on architecture will be touched upon only briefly.

7 Czech 2002, 122. Czech has overlooked the fact that Goeree proposed his chapter division eight years prior to the publication of Hoogstraten's treatise. The chapter divisions will be discussed later.

8 De Vries 1998, 85.

9 Weststeijn 2008, 15. His remark also refers to the works of Karel van Mander and Philips Angel.

10 Emmens 1979.

11 In Jonathan Israel's index, Goeree is described as a 'Dutch radical writer'. See Israel 2001, 168, 363-364, 376, 428, 432, 449. Inger Leemans follows this lead in her discussion of the writings of Goeree. See Leemans 2004. For Goeree's rejection of Spinoza, see Goeree KWH 1705, fol. * 2 verso, fol. * 3 recto, 664, 671; and Goeree MK 1682, 359-360. Also, Weststeijn tries to link Goeree and Van Hoogstraten to Spinoza. See Weststeijn 2008, 329-351. I will return to this later.

12 Blankert 1999, 15; De Vries 1998, 26-27; Mai 2006; Mai & Paarlberg 2006.

13 See Vermij's essay elsewhere in this volume.

14 See, for example, Martin 1935-1936 and Haak 1984. Bob Haak, who authored the most recent and comprehensive compendium on Dutch art of the seventeenth century, mentions a 'lack of nuance' when speaking about the reception of art of the late seventeenth century, but he also uses words like 'stagnation' and 'weak extract' to describe it. Furthermore, he devotes a sparse five pages on the topic (from 503). Haak 1984, 499-503.

15 See the essays from Koen Ottenheym and Arie Jan Gelderblom in Blankert 1999 and De Vries 1998. De Vries's study on De Lairesse emphasizes the importance of the interaction between painting and theatre in the work of the painter.

16 Studies that consider the relation between science and Dutch art in the early modern period are Alpers 1983, Kasteleyn 1992 and Miedema 2006. These studies consider earlier or later periods.

17 Emmens 1979, 116, 244.

18 Ibid., 244-245

19 Czech 2002, 163.

20 Ibid., 164-167.

21 Ibid. See also Weststeijn 2008, 49-50.

22 Czech 2002, 173.

23 This explanation is given by Van de Wetering 2011, 21. Van Hoogstraten might have followed Karel van Mander with his placement of anatomy, who also discusses human proportions after an exhortation and a chapter on drawing.

24 Czech 2002, 175.

25 Plinius tells in his *Naturalis historia* how painting started with tracing lines round human shadows and how the daughter of Butades drew a line round the shadow of her beloved (see book 35.15 and 35.151).

26 Goeree TK 1670, 65.

27 Goeree SK 1670, 81-84; and 1697, 65-67.

28 About Goeree's writings on architecture, see Van den Heuvel 1997.

29 Van Hoogstraten 1678, 174.

30 Ibid., 38.

31 Ibid., 215.

32 Ibid., 244.

33 In this I follow Van de Wetering's recent interpretation of Van Hoogstraten's work. He reacts on Hessel Miedema's interpretation of Karel van Mander and Thijs Weststeijn's interpretation of Van Hoogstraten, which he finds both too textual and literary. They overlook the fact that the works were actually intended as instructive books for painters and trainees. See Van de Wetering 2011, 15-28.

34 Goeree SK 1670, 81-84; and 1697, 65-67.

35 Emmens 1979, 167.

36 Weststeijn 2008, 64.

37 Van de Wetering 2011, 14.

38 Grafton 1992; Grafton 2009; Bono 1995; Jorink 2010.

39 Ashworth 1990. The article was reprinted in D. Preziosi & C. Farago, *Grasping the World. The idea of the museum*, Aldershot 2004. See also Ashworth 1996.

40 Ashworth 1990, 317. Ashworth quotes here F. Jacob, *The logic of Live. A History of Heredity*, New York 1976.

41 For reactions on Ashworth's thesis, see Bono 1995, Bono 1999, Ogilvie 2006, 16, n. 61.

42 For an overview of all the sources used by Van Hoogstraten, see the useful appendix I.D in Czech 2002, *85-*149.

43 Van Hoogstraten 1678, 220-222.

44 For a discussion of the practice-orientated remarks on colour, see Van de Wetering 2011, 103-111.

45 Van Hoogstraten 1678, 161. Also, the three sections that are titled 'continuations of the previous' are exemplary of his accumulative way of working. See Van Hoogstraten 1678, 85, 210, 313.

46 Kwakkelstein 1998, 23. See, for example, Goeree MK 1682, 307-308.

47 Goeree MK 1682, chapter IV.

48 Ibid., 203-209.

49 Ibid., 282.

50 Van Hoogstraten 1678, 46; Goeree SK 1697, 96-97.

51 Goeree MK 1682, fol. *6 recto.

52 Goeree SK 1670, 46-49. The titles are mentioned only in the first edition.

53 Van Hoogstraten 1678, 24.

54 Ibid 1678, 88.

55 Ibid., 191.

56 Goeree MK 1682, chapter 4.

57 Van Hoogstraten 1678, 52. See the prints opposite pages 53, 55, 59, and 61-62. The prints show 1) a skeleton and man from the front, 2) a man from the side and back showing musculature, 3) six men with different proportions, 4) five women with different proportions, and 5) five toddlers with different proportions.

58 Ibid., 63.

59 Ibid., 273-276.

60 Emmens suspects he wanted to keep his knowledge about perspective secluded from his trainees. Emmens 1979, 124.

61 Goeree SK 1670, fol. (a) 7 recto; and 1704, 73.

62 For the connections between Goeree and Bosboom, see Van de Roemer 2005, 170-172.

63 Bosboom 1703, 8.

64 Van Hoogstraten 1678, 257.

65 Kemp 1990, 119-127.

66 Bosboom 1703, 23.

67 Goeree MK 1682, 204.

68 Weststeijn 2008, 350: 'Since we do not have the *Onzichtbare Werelt*, we are left with Goeree's remarks on the subject to reconstruct how Spinozism can be incorporated in a theory of painting'.

69 Ibid., 107.

70 Ibid., 347.

71 Ibid., 254, 344.

72 Goeree MK 1682, 1, 21.

73 Wittkower 1988.

74 Goeree wrote extensively about the measurements of Noah's ark, Solomon's temple and Ezekiel's vision of the temple.
75 Goeree SK 1670, 20.
76 Goeree MK 1682, 42. Book of Wisdom 11.21: '(...) thou hast ordered all things in measure, and number, and weight'. On this notion, see Bredekamp 1993, 41, and Folkerts *et al.* 1998. For Goeree's opinion about Descartes, see Goeree KWH 1705, 640.
77 Weststeijn 2008, 329-331, 349-351.
78 Deason 1986.
79 Daston & Park 1998, 301 and chapter 7.
80 Goeree SK 1697, 67.
81 De Lairesse 1707, 205.
82 Rabus, 1693. He reviews Willem Beurs's *De groote waereld in 't kleen geschildert* (1692).
83 Emmens 1979, 114.
84 Van de Wetering 1997, 157. See also Van de Wetering 1998. In Rembrandt's work it was connected to different aspects, like the concept of *sprezzatura* - the thorough skilfulness that enabled an artist to work with ease and supposed nonchalance - the challenge of the Titian's late style and specific pictorial means to suggest material structure and spatial depth.
85 Van de Wetering et al. 2005, 271-273, 303-311. He discusses the self-portraits in Kenwood House in London and the Kunsthistorisches Museum in Vienna.
86 Gombrich 1996, 153-244.
87 Van Hoogstraten 1678, 27.
88 Ibid., 237-238.
89 Miedema 2006, vol. 2, 3, 6, 36-38.

Bibliography

Alpers 1983
S. Alpers, *The art of describing. Dutch art in the seventeenth century*, Chicago 1983.

Ashworth 1990
W. Ashworth, 'Natural history and the emblematic world view', in: D. Lindberg & R. Westman (eds.), *Reappraisals of the Scientific Revolution*, Cambridge 1990, 303-332.

Ashworth 1996
W. Ashworth, 'Emblematic natural history of the Renaissance', in: N. Jardine, J.A. Secord & E.C. Spary, *Cultures of natural history*, Cambridge 1996, 17-37.

Beckers 1998
D. Beckers, 'Meetkunde als de korte en zekere weg naar kunst. Gérard de Lairesse (1640-1711) en zijn "Grondlegginge der Teekenkonst"', *Gewina* 21 (1998), 81-93.

Bijker & Dufais 1998
D. Bijker & N. Dufais (eds.), *Arent de Gelder (1645-1727), Rembrandts laatste leerling*, Gent 1998.

Blanc 2008
J. Blanc, *Peindre et penser la peinture au XVIIe siècle. La théorie de l'art de Samuel van Hoogstraten*, Bern 2008.

Blankert 1999
A. Blankert (ed.), *Hollands classicisme in de zeventiende-eeuwse schilderkunst*, Rotterdam & Frankfurt am Main 1999.

Bono 1999
J.J. Bono, 'From Paracelsus to Newton. The word of God, the Book of Nature, and the eclipse of the emblematic world view', in: J. Force & R.H. Popkin (eds.), *Newton and religion. Context, nature, and influence*, Dordrecht 1999, 45-76.

Bono 1995
J.J. Bono, *The word of God and the languages of man. Interpreting nature in early modern science and medicine. 1: Ficino to Descartes*, Madison, WI. 1995.

Bosboom 1703
D. Bosboom, *Perspectiva of doorzichtkunde, onderscheyden in voorbereytzelen, bespiegeling, en werkstellige oeffening (...)*, Amsterdam 1703.

Bredekamp 1993
H. Bredekamp, *Antikensehnsucht und Maschinenglauben. Die Geschichte der Kunstkammer und die Zukunft der Kunstgeschichte*, Berlin 1993.

Brusati 1995
C. Brusati, *Artifice and illusion. The art and writing of Samuel van Hoogstraten*, Chicago 1995.

Czech 2002
H. Czech, *Im Geleit der Musen. Studien zu Samuel van Hoogstratens Malereitraktat Inleyding tot de hooge schoole der schilderkonst. Anders de zichtbaere werelt (Rotterdam 1678)*, Münster 2002.

Daston & Park 1998
L. Daston & K. Park, *Wonders and the order of nature, 1150-1750*, New York 1998.

Deason 1986
G. Deason, 'Reformation theology and the mechanistic conception of nature', in: D.C. Lindberg & R.L. Numbers (eds.), *God and nature. Historical essays on the encounter between Christianity and science*, Berkeley 1986, 167-191.

De Lairesse 1707
G. de Lairesse, *Het groot schilderboek*, Amsterdam 1707.

Descartes 1657
R. Descartes, *Principia philosophiae. Of Beginselen der wysbegeerte (...)* (J.H. Glazenmaker, transl.), Amsterdam 1657.

De Vries 1998
L. de Vries, *Gerard de Lairesse. An artist between stage and studio*, Amsterdam 1998.

Dijksterhuis 1998
E.J. Dijksterhuis, *De mechanisering van het wereldbeeld*, Amsterdam 1998.

Dolders 1985
A. Dolders, 'Some remarks on Lairesse's "Groot schilderboek"', *Simiolus* 15 (1985), 197-220.

Emmens 1979
J.A. Emmens, *Rembrandt en de regels van de kunst*, Amsterdam 1979.

Field & Gray 1987
J.V. Field & J.J. Gray, *The geometrical work of Girard Desargues*, New York, Berlin & Heidelberg 1987.

Folkerts *et al.* 1998
M. Folkerts, E. Knobloch & K. Reich, *Mass, Zahl und Gewicht. Mathematik als Schlüssel zu Weltverständnis und Weltbeherrschung*, Weinheim 1989.

Gaukroger 2002
S. Gaukroger, *Descartes' system of natural philosophy*, Cambridge 2002.

Goeree BK 1681
W. Goeree, *d'Algemeene bouwkunde (...)*, Amsterdam 1681.

Goeree BK 1705
W. Goeree, *d'Algemeene bouwkunde (...)*, Amsterdam 1705.

Goeree KWH 1705
W. Goeree, *De kerkelijke en wereldlijke historiën*, Amsterdam 1705.

Goeree MK 1682
W. Goeree, *Natuurlyk en schilderkonstig ontwerp der menschkunde (...)*, Amsterdam 1682.

Goeree MK 1704
W. Goeree, *Natuurlyk en schilderkonstig ontwerp der menschkunde (...)*, Amsterdam 1704.

Goeree SK 1670
W. Goeree, *Inleyding tot de practijck der al-gemeene schilder-konst (...)*, Middelburg 1670.

Goeree SK 1697
W. Goeree, *Inleyding tot de praktijk der al-gemeene schilder-konst (...)*, Amsterdam 1697.

Goeree SK 1704
W. Goeree, *Inleyding tot de praktijk der al-gemeene schilder-konst (...)*, Amsterdam 1704.

Goeree TK 1668
W. Goeree, *Inleydinge tot de al-ghemeene teycken-konst (...)*, Middelburg 1668.

Goeree TK 1670
W. Goeree, *Inleydinge tot de al-ghemeene teycken-konst (...)*, Middelburg 1670.

Goeree TK 1697
W. Goeree, *Inleydinge tot de al-gemeene teyken-konst (...)*, Amsterdam 1697.

Goeree TK 1705
W. Goeree, *Inleydinge tot de al-gemeene teyken-konst (...)*, Amsterdam 1705.

Goeree VK 1668
W. Goeree, *Verlichtery-konst (...)*, Middelburg 1668.

Goeree VK 1670
W. Goeree, *Verlichterie-kunde (...)*, Middelburg 1670.

Goeree VK 1697
W. Goeree, *Verligterie-kunde (...)*, Amsterdam 1697.

Goeree VK 1705
W. Goeree, *Verligterie-kunde (...)*, Amsterdam 1705.

Gombrich 1996
E.H. Gombrich, *Art and illusion. A study in the psychology of pictorial representation*, London 1996.

Grafton 1992
A. Grafton, *New worlds, ancient texts. The power of tradition and the shock of discovery*, Cambridge, MA & London 1992.

Grafton 2009
A. Grafton, *Worlds made by words. Scholarship and community in the modern West*, Cambridge, MA 2009.

Grijzenhout & Veen 1992
F. Grijzenhout & H.T. Veen (eds.), *De Gouden Eeuw in perspectief. Het beeld van de Nederlandse zeventiende-eeuwse schilderkunst in later tijd*, Nijmegen 1992.

Haak 1984
B. Haak, *Hollandse schilders in de Gouden Eeuw*, Amsterdam 1984.

Hooykaas 1973
R. Hooykaas, *Religion and the rise of modern science*, Edinburgh 1973.

Israel 1995
J.I. Israel, *The Dutch Republic. Its rise, greatness, and fall, 1477-1806*, Oxford 1995.

Israel 2001
J.I. Israel, *Radical Enlightenment. Philosophy and the making of modernity 1650-1750*, Oxford 2001.

Jorink 1999
E. Jorink, *Wetenschap en wereldbeeld in de Gouden Eeuw*, Hilversum 1999.

Jorink 2010
E. Jorink, *Reading the Book of Nature in the Dutch Golden Age, 1575-1715*, Leiden 2010.

Kasteleyn 1992
L.P. Kasteleyn, '"Op de waarheid klaar in uwe tekeninge spele". De topografische tekenkunst in de Republiek gedurende het tweede kwart van de 18de eeuw en de theorieën van Newton (II)', *Delineavit et Sculpsit* 7 (1992), 16-35.

Kemp 1990
M. Kemp, *The science of art. Optical themes in the art from Brunelleschi to Seurat*, New Haven & London 1990.

Kwakkelstein 1998
M. Kwakkelstein, *Willem Goeree. Inleydinge tot de al-ghemeene teycken-konst. Een kritische geannoteerde editie*, Leiden 1998.

Leemans 2004
I. Leemans, 'De weg naar de hel is geplaveid met boeken over de bijbel. Vrijgeest en veelschrijver Willem Goeree (1635-1711)', *Nederlandse letterkunde* 9 (2004), 255-273.

E. Mai (ed.), *Holland nach Rembrandt. Zur niederländischen Kunst zwischen 1670 und 1750*, Cologne 2006.

Mai & Paarlberg 2006
E. Mai & S. Paarlberg (eds.), *Vom Adel der Malerei. Holland um 1700*, Cologne 2006.

Martin 1935-1936
W. Martin, *De Hollandsche schilderkunst in de zeventiende eeuw*, Amsterdam 1935-1936.

Miedema 1969
H. Miedema, 'Rembrandt en de regels van de kunst (book review)', *Oud-Holland*, 84 (1969), 249-256.

Miedema 2006
H. Miedema, *Denkbeeldig schoon. Lambert ten Kates opvattingen over beeldende kunst. Tekstuitgave met commentaar*, Leiden 2006.

Nieuwentijt 1715
B. Nieuwentijt, *Het regt gebruik der wereltbeschouwingen*, Amsterdam 1715.

Ogilvie 2006
B.W. Ogilvie, *The science of describing. Natural history in Renaissance Europe*, Chicago & London 2006.

Rabus 1693
P. Rabus, *De boekzaal van Europe, ontsloten met de maanden mei en juni*, Rotterdam 1693.

Roscam Abbing 1993
M. Roscam Abbing, *De schilder & schr ver Samuel van Hoogstraten, 1627-1678. Eigent dse bronnen & oeuvre van gesigneerde schilder en*, Leiden 1993.

Van den Heuvel 1997
C. van den Heuvel, 'Willem Goeree (1635-1711) en de ontwikkeling van een algemene architectuurtheorie in de Nederlanden', *Bulletin Koninklijke Nederlandse Oudheidkundige Bond* 96 (1997), 154-176.

Van de Roemer 2005
G.M. van de Roemer, *De geschikte natuur. Theorieën over natuur en kunst in de verzameling van zeldzaamheden van Simon Schijnvoet (1652-1727)*, Amsterdam 2005 (unpub. diss. Amsterdam).

Van de Wetering 1997
E. van de Wetering, *Rembrandt. The painter at work*, Amsterdam 1997.

Van de Wetering 1998
E. van de Wetering, 'Opmerkingen over de relatie tussen techniek, stijl en toeval bij Arent de Gelder; een vergelijking met Rembrandt', in: D. Bijker, N. Dufais & F. de Graaf (eds.), *Arent de Gelder (1645-1722), Rembrandts laatste leerling*, Gent 1998, 19-35.

Van de Wetering 2011
E. van de Wetering, 'Towards a reconstruction of Rembrandt's art theory', in: E. van de Wetering *et al.*, *A corpus of Rembrandt paintings V. Small-scale history paintings*, Dordrecht 2011, 3-140.

Van de Wetering *et al.* 2005
E. van de Wetering, K. Groen, P. Klein, *et al.*, *A corpus of Rembrandt paintings IV. The self-portraits*, Dordrecht 2005.

Van Hoogstraten 1678
S. van Hoogstraten, *Inleyding tot de hooge schoole der schilderkonst. Anders de zichtbaere werelt*, Rotterdam 1678.

Vermij 1991
R. Vermij, *Secularisering en natuurwetenschap in de zeventiende en achttiende eeuw. Bernard Nieuwentijt*, Amsterdam 1991.

Weststeijn 2008
T. Weststeijn, *The visible world. Samuel van Hoogstraten's art theory and the legitimation of painting in the Dutch Golden Age*, Amsterdam 2008.

Wittkower 1988
R. Wittkower, *Architectural principles in the Age of Humanism*, London & New York 1988.

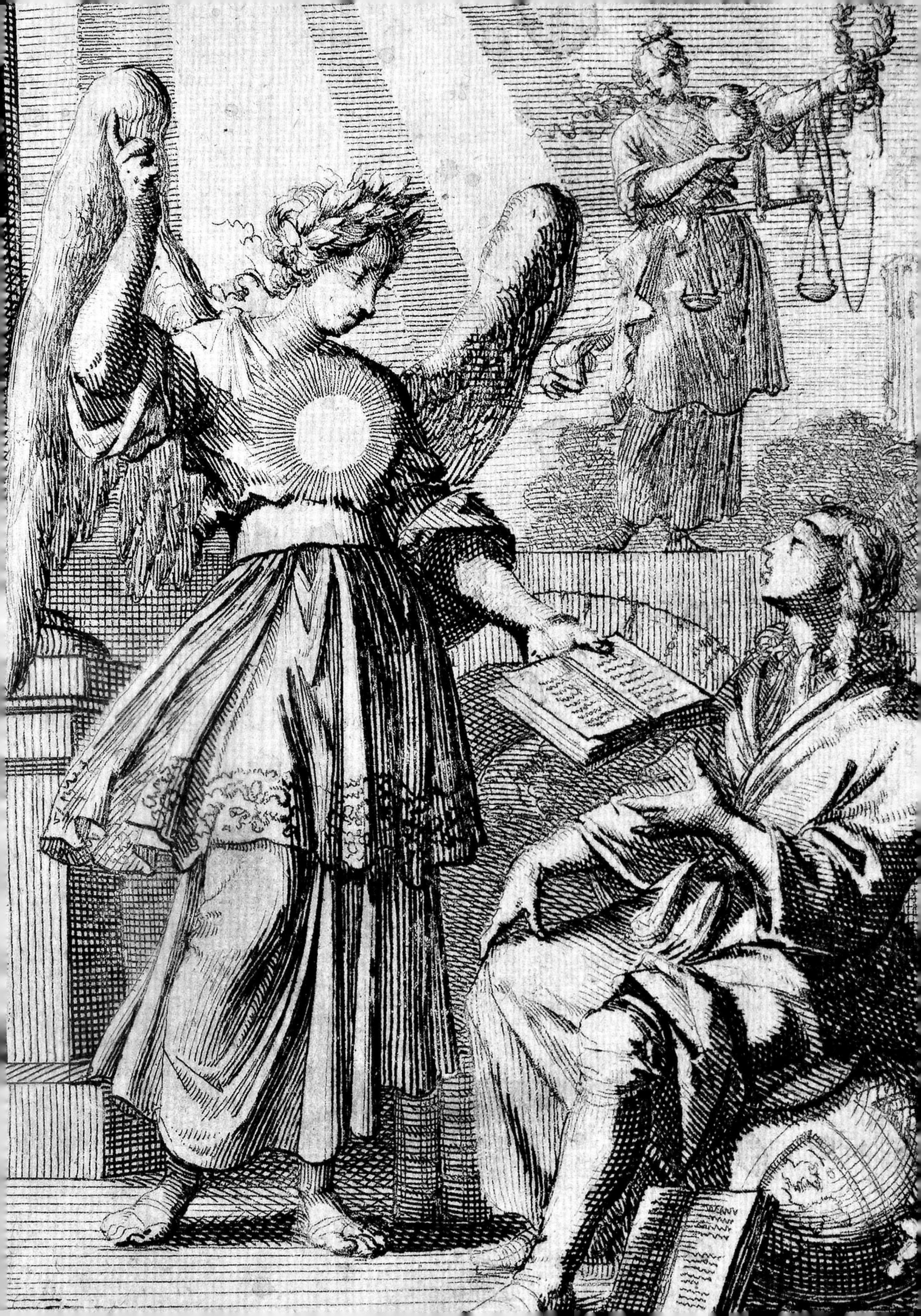

The light of nature and the allegorisation of science on Dutch frontispieces around 1700*

Rienk Vermij

Introduction

Scientific knowledge as it emerged in the seventeenth century was supposed to be exact and factual. Scientists preferred plain speech, unadorned reports and bare factual accounts. But that is not to say that they completely repudiated contemporary artistic culture. When it came to presenting their position to a wider audience, they unhesitatingly made use of all the rhetorical and emblematic tricks of the trade. When the Royal Society of London was founded in 1660, for example, the choice of a coat of arms was a serious business. The members discussed various designs before settling upon the present one.[1] It might be assumed that emblems and allegories belonged more to the magical and emblematic worldview of the sixteenth century, but the proponents of the new science of the second half of the seventeenth century used them too. Such allegorical representations can in particular be found in the frontispieces of scientific or philosophical books.

At a time when books did not have dust jackets or descriptive, laudatory 'blurbs', the title page was one of the few places where the content could be publicised. A frontispiece could be used to give potential readers some foretaste of the contents of the work; in some cases, it had a programmatic character.[2] In studying the philosophical and scientific debates of the period, it might therefore be worthwhile to look not only at the various views as they were expressed in books, pamphlets and other printed texts, but also at the way they were expressed in allegories and images. For this reason, frontispieces are increasingly drawing the attention not just of historians of art, but also of historians of ideas. Systematic studies have been undertaken on the frontispieces for specific types of literature, such as historical, mathematical and astronomical works.[3] Other authors have analysed them to throw light on specific historical issues. For example, Jim Bennett has analysed a frontispiece in his discussion of the pretentions of surveyors around 1600, and Werner Schneiders has attempted to sketch the self-image of the German Enlightenment as it was reflected in frontispieces of philosophical works.[4]

Rather than tracing the history of a general theme, in this article I want to concentrate on a specific place and period: the Dutch Republic of the late seventeenth and early eighteenth centuries. Philosophical debates,

Detail fig. 8
Jan Luiken, frontispiece to Arnold Geulincx, *Ethica of zeden-konst.*

even if they addressed more general problems, were in the first place very often local. Authors reacted to books that were talked about in their communities and networks and that were often by authors they knew. The language of images too, although in principle international, drew upon local traditions. The United Provinces, in particular Holland, had an important artistic tradition, and it may be assumed that the makers of frontispieces were also working within local conventions. In this article, I will try to establish how frontispieces were part of the philosophical debates of their time. That is, I will mainly be concerned with their iconography.

Moreover, I will concentrate on one particular element: light. Some years ago, William Ashworth devoted a study to the image of light in frontispieces of scientific works.[5] However, he ignored the Dutch examples, which are actually quite distinct. Light came to have a specific meaning in Dutch religious and philosophical debates of the end of the century. It stood for rational or natural knowledge, in connection with rather far-reaching epistemological and religious claims. The metaphor was used in philosophical writings and was prominent in several frontispieces. I will try to show how the use of this image varied in response to changing philosophical views.

I first will give a short introduction on the intellectual issues that these works addressed. Second, using specific examples, I will discuss some factors that contributed to the production and design of the frontispieces, in order to help establish how 'meaningful' a frontispiece was actually intended to be. Third, the meaning of light in these frontispieces can become clear only in the context of their general message. It is important, therefore, to understand some main iconographical conventions of the period. I will try to map these conventions by comparing various frontispieces (not all of them depicting light – but then, the absence of light may also be meaningful). Finally, I will discuss light – the significance attributed to it and the way its representation changed. I will first discuss how frontispieces to Cartesian works came to use representations of light in a specific philosophical sense, and then I will investigate how, partly in reaction to these earlier representations, one of the key works of experimental philosophy addressed the same element.

Intellectual background

The period around 1700 represents a critical juncture in the history of science and philosophy. By the middle of the seventeenth century, the many discoveries and new ideas of the preceding century had completely discredited the Aristotelian worldview that had dominated European thinking in earlier periods. Several alternatives were proposed, but the first to come up with a viable new system was René Descartes (1596-1650). Modern textbooks on the history of philosophy tend to emphasise those contributions of Descartes that are still considered philosophically relevant, especially in metaphysics and epistemology. It is important to realise that Descartes was also, or even foremost, a natural philosopher who proposed a new physical worldview. Although, after a few decades, most of his specific theories proved to be untenable, his general idea that material reality could be described by geometrical properties and was governed by universal laws of nature was eminently important and showed the way for later research.

Descartes's philosophy soon was attacked in its turn. Concern arose especially about the religious consequences. Even though his metaphysics were predicated on the existence of God, Descartes's mechanical world, governed by fixed and unalterable laws, seemed to leave no room for divine intervention or special providence. These consequences were pointed out with relentless logic by Benedictus Spinoza (1632-1677). Spinoza's rejection of the biblical miracles and of any transcendent divine power was deemed outrageous by most people. Although Spinozism was very present in the philosophical debates of the seventeenth century, it was so rather as a danger to be averted than as a position that deserved serious consideration.

By the end of the seventeenth century, a promising alternative to both Aristotelian and Cartesian philosophy was the 'experimental philosophy' promulgated especially by the English scientist Robert Boyle (1627-1691). Experimental philosophers accepted most basic tenets of the Cartesian worldview, but they rejected the claim to mathematical certainty. Taking their departure from a voluntarist theology, they denied that the laws of nature were necessarily fixed and immutable. Only experience can tell us what really happens in nature. In this way, they succeeded in avoiding the religious pitfalls of the new philosophy and in reconciling the new scientific worldview with the demands of traditional theological doctrine: rejection of a traditional interpretation of a biblical sentence based on mere philosophical presuppositions was simply not warranted.[6]

These debates were followed closely in the Netherlands. The Dutch Republic was an important intellectual centre as well as a major centre of book production, and Dutch authors contributed both in Latin and in the vernacular to the debates of their time. Descartes had lived for twenty years in the Dutch Republic and still had an important following at the end of the century. A Dutch version of his complete works was republished in Amsterdam in the 1690s (with a frontispiece, which will be discussed below). There were also many religious groups and self-proclaimed spiritual leaders that tried to incorporate the new philosophical ideas. Their views could range from quite orthodox to very libertine, sometimes even bordering on Spinozism. The new experimental ideas too had their Dutch proponents, such as Bernard Nieuwentijt (1654-1718), whose book *Het regt gebruik der wereltbeschouwingen* (The right way of contemplating the works of the Creator, 1715) attempted to demonstrate that science and orthodox religion are very compatible.

These philosophical debates are well known, but so far, nobody has looked at the way they were articulated in the imagery of the time, in particular in the frontispieces of the various books. It is not that these have remained completely unknown; the frontispiece of the book by Nieuwentijt is actually quite often reproduced in historical studies on the Dutch Enlightenment. However, it has not been studied in relation to the earlier material.

Author, publisher and artist

In order to interpret the frontispieces, one should ideally know by whom they were ordered or made, and with what purposes in mind. Unfortunately, only in rare cases is there any direct information on their production. It seems fair to assume that in most instances, the frontispiece was commissioned by the publisher, with the purpose of giving the book a more attractive appearance and enhancing sales. The actual design was normally the work of a professional artist. He was probably the real expert in imagery and must have had some impact, but, as a paid hand, he may have had to take hints or even directions from the publisher, the author, an editor or a patron.[7]

That the artist could have considerable impact can be shown in the case of Jan Luiken (1649-1712), one of the most famous engravers of the period. His designs are generally original and powerful and appear to have had an important influence on allegorical imagery in Dutch frontispieces. Luiken not only made many frontispieces and illustrations for other people's works, but also published several emblematic works of his own. He was an acknowledged master of allegorical imagery; it would have been strange for a publisher to order an allegorical frontispiece from him without drawing upon his expertise.

Luiken came from a family with strong connections to libertine religious groups. As a young man he had experienced a religious 'conversion', and after that time, he emphasised the transitory nature of worldly things and the all-importance of religious faith. He was close to the mystical and spiritualist tendencies of his time, but without distancing himself from orthodox Christianity. These convictions clearly informed his artistic program, not just in works to which he took the initiative himself, where they are ubiquitous, but also in works commissioned by others. It is unlikely to have been accidental that Luiken illustrated many religious works, like John Bunyan's. He even refused work that went against his convictions. In 1689, he undertook to make a frontispiece for a religious book by Barend Joosten Stol (1631-1713). However, when he discovered that the book contained an attack on the German mystic Jacob Böhme (1575-1624), Luiken changed his mind and refused to have anything further to do with it.[8] Such an attitude was definitely not considered normal for an artist (Stol considered it rather crazy), but the impact of exceptional artists like Luiken should not be neglected.

A case where the author appears to have been in control is the well-known frontispiece of Nieuwentijt's *Het regt gebruik der wereltbeschouwingen* (fig. 1). This book was published in Amsterdam in 1715 by Joannes Pauli and the widow of Joannes Wolters, who appear to have continued in that very year the business of the deceased Joannes Wolters. The latter had published between 1687 and 1714 over a hundred mostly scholarly books. *Het regt gebruik* was an apologetic work, written with the aim of convincing atheists and unbelievers of God's wisdom, power and goodness. Nieuwentijt regarded the atheism of his day as a real danger. He was therefore very much engaged with the book and clearly wanted to get his message across. He probably chose the publisher; in the 1690s, Wolters

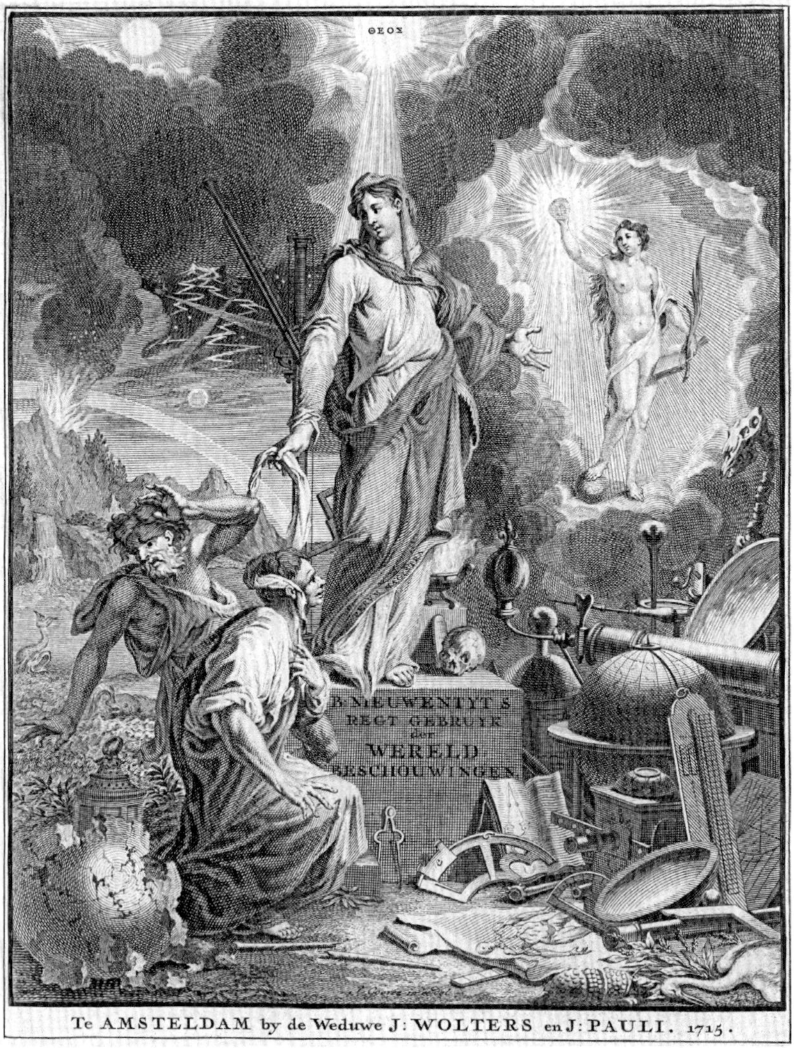

1

Jan Goeree, frontispiece to Bernard Nieuwentijt, *Het regt gebruik der wereltbeschouwingen*

engraving (Amsterdam 1715), Utrecht Universiteitsbibliotheek, P. qu. 578. For accompanying explanation, see appendix 1.

had published some treatises by Nieuwentijt on infinitesimal calculus, so that author and publisher appear to have had an established relationship.

Het regt gebruik was a comprehensive and richly illustrated work, and its publication must have demanded a heavy investment, but Nieuwentijt was a rich man and could afford to have things his way. It would seem plausible that he also kept a close eye on the frontispiece. The frontispiece is strikingly handsome; it would seem that more care (and money) was spent on it than would have been justifiable for purely commercial reasons. It was designed and executed by Jan Goeree (1670-1731), an artist who had some reputation at the time. Not many frontispieces by him are known. This one is amazingly rich in detail; the images are quite original and are clearly designed to convey the book's message. It is hard to miss the message, as the frontispiece is accompanied by an explanation in verse (probably also by Goeree, who is known to have written verse at other occasions as well). *Het regt gebruik* was a book with a message, and

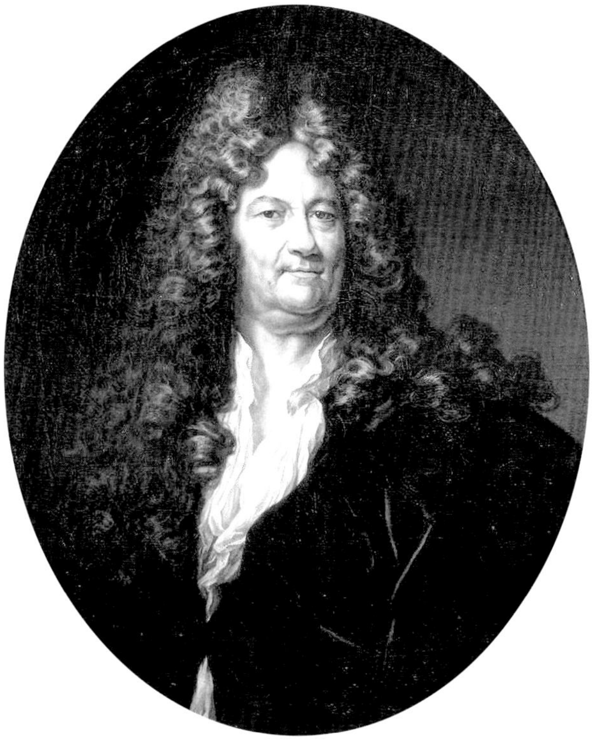

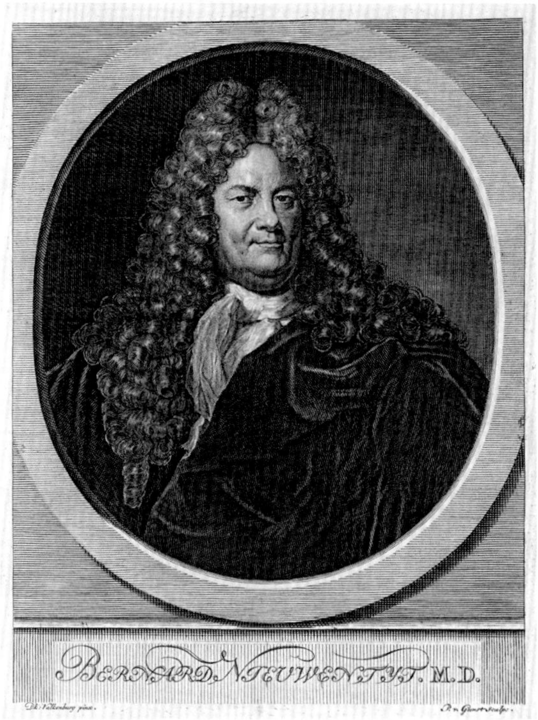

2
Dirk Valkenburg, *Portrait of Bernard Nieuwentijt,* undated
painting, 87.5 x 70 cm,
Universiteitsbibliotheek Amsterdam.

3
P. Gunst (after Dirk Valkenburg), *Portrait of Bernard Nieuwentijt*
engraving, Bernard Nieuwentijt, *Het regt gebruik der wereltbeschouwingen* (Amsterdam 1715), Utrecht Universiteitsbibliotheek, P. qu. 578.

everything was done to get the message across. The frontispiece should be regarded as programmatic and reflecting the author's intentions.

The book also includes a portrait of the author, a practice not uncommon in the period. The engraving mentions that it was done after a painting by Dirk Valkenburg (1675-1721), which has also been preserved (fig. 2). When one compares the engraved portrait to the painting, it appears that the former has been considerably modified. In the engraving, the rather wild or even frivolous open collar of the painting has been decently covered up (fig. 3). Instead of a worldly regent, we encounter a grave and serious man – well in line with the way Nieuwentijt presents himself as an author. The 'public face' of the book, and the message it was supposed to convey, was apparently given considerable attention.

The case is quite different with the frontispiece of what would appear as another high-profile philosophical work of the period, Descartes's complete works, published in 1690 in Dutch translation (fig. 4). For one thing, the author had been dead for many years and cannot have been involved in the production. Responsibility must have fallen completely on Jan ten Hoorn, an active Amsterdam publisher who aimed at a popular audience and published nearly exclusively in the vernacular. He must have been a clever businessman, making sure that his books were attractive to his customers. They were often illustrated and many had engraved frontispieces.[9]

However, the fact that Ten Hoorn was guided by commercial considerations does not preclude him from caring about the content or the meaning of the frontispieces. For either ideological or business reasons, Ten Hoorn specialised in radical philosophical and religious works. He published works of Cartesian philosophy, many by local Amsterdam physicians, and works of religious freethinkers, also often local men. In several instances he was fined for selling forbidden books. His shop was a meeting place for local freethinkers. For his frontispieces, he often collaborated with like-minded artists, and it seems reasonable to assume that he wanted them to appeal to the intended audience. He had an intensive collaboration with Jan Luiken. Between 1679 and 1711, Luiken made dozens of frontispieces for Ten Hoorn, more than for any other publisher.[10]

The frontispiece to Descartes's works is by an anonymous artist and is less meticulously done than that in Nieuwentijt's work. There is no accompanying explanation, and the meaning of the image is rather unclear. Moreover, the frontispiece appears to be rather ill-suited to the work itself. The book was in fact a re-edition of the Dutch translation by

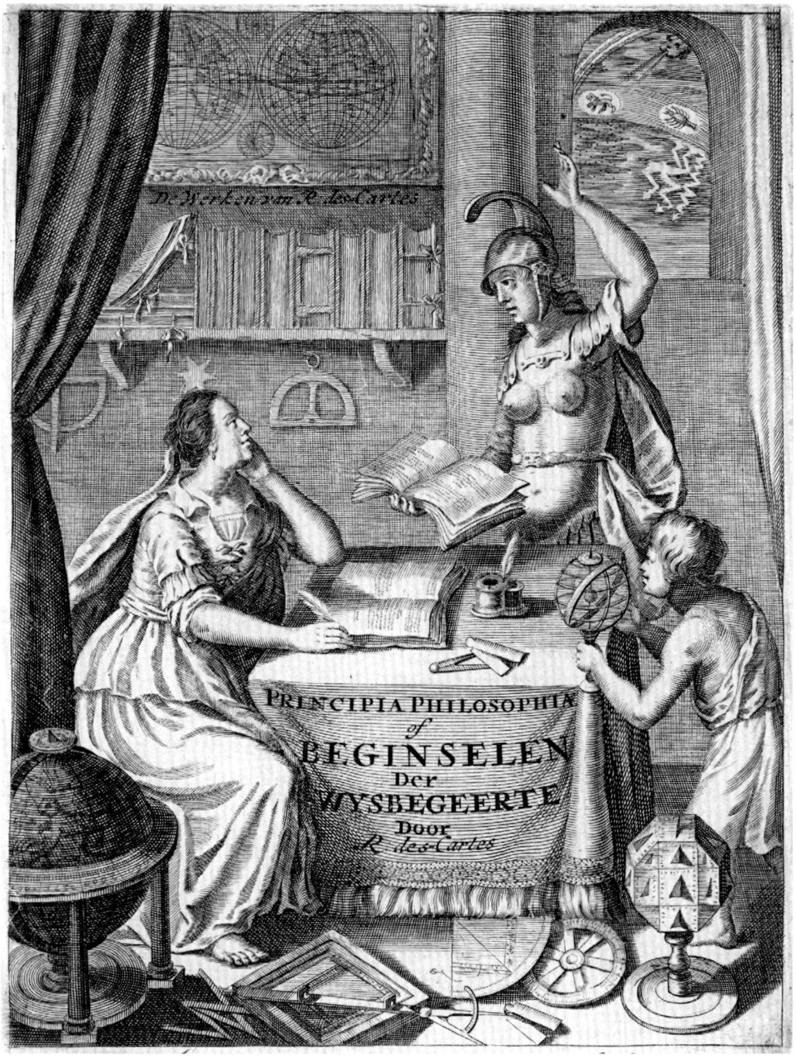

4
Unknown artist, frontispiece to René Descartes, *Beginselen der wysbegeerte,* engraving (Amsterdam 1690)
Utrecht, Universiteitsbibliotheek, Y. qu. 229.

Jan Hendrik Glazemaker (1620-1682), which had originally been published in 1656-1657 by Jan Rieuwerts (1617-c. 1685). The first volume, which has the frontispiece, comprehends Descartes's *Principles of philosophy* along with two smaller works: *Meteorology* and *The world* (titled here *Treatise of light, and the foremost of the senses*). What appears as a second print-run of this edition included a justification by Jan ten Hoorn of the total project, dated 8 December 1691, and an index of all four volumes. (These other volumes were published in 1690-1692.) The frontispiece, on the other hand, mentions prominently the title *Principles of philosophy* (in Dutch, *Beginselen der wysbegeerte*) but does not mention the other works contained in the volume, nor is there any hint that this is the first of four volumes. There is a caption 'The works of Renatus Descartes', but this appears to refer to the books that are depicted in the frontispiece.

The frontispiece omits the publisher's name and the year of publication. These are normally to be found at the bottom of the print, regardless of the fact that all of these books also have a typographical title page. Actually, there are some traces of lettering at the very bottom, and so it would seem that these data have been cut from the original copper plate.[11] In other words, it appears as if the frontispiece had originally been made for a publication by another publisher, for a slightly different work, and was adopted for the present volume afterwards. A pamphlet published in the same year, 1690, indicates that Jan ten Hoorn had a fight with his brother Timotheus, also a publisher, over the right to publish the translation of 'the principles of Descartes' first philosophy' [*sic*]. In the pamphlet, the former is staged as claiming a privilege for publishing the work (as Jan actually had), and the latter retorting that he had initiated and first printed it.[12] The pamphlet is polemical and hardly unbiased, but it might well be true that there was a fight. In that case, the frontispiece was probably originally commissioned by Timotheus and later appropriated by Jan.

As a publisher, Timotheus ten Hoorn was far less important than his brother Jan. He was above all a publisher of novels of an often facetious and even pornographic character. While he also published some medical works, partly by the same Cartesian physicians who also published with his brother, these mainly concerned sexual topics like venereal diseases and obstetrics. They seem to have served as a respectable facade for his main merchandise.[13] If the frontispiece was really commissioned by him, one might well wonder who was responsible for the design. Jan ten Hoorn apparently felt that it was good enough – at least for business purposes. The only thing that really needed change was the name of the publisher.

The above considerations indicate that the intentions behind the allegories on the frontispieces depend on the specific circumstances under which they were produced, and therefore they give us some preliminary indications as to how they should be approached. The frontispiece to Nieuwentijt's book is highly programmatic and carefully conceived, both in the overall design and in the many details. The Descartes frontispiece appears rather an obligatory embellishment, and it would be rash to look for any specific message of the artist or the publisher. Both images are clearly allegorical and must have some meaning. But in the case of the Descartes frontispiece, we would expect the artist to have followed some standard conventions, whereas in the case of Nieuwentijt, we should not be surprised to find he used allegorical elements in a new way.

The image of philosophy

Confronted with a frontispiece like that in the 1690 Descartes edition, one is tempted first of all to look up the various figures in a standard iconographical handbook of the period. The most authoritative work was Cesare Ripa's *Iconologia*. This book was certainly known in the Netherlands. There were several Dutch editions, one of them reprinted in 1699 by Timotheus ten Hoorn. However, there is in general very little correspondence between Ripa's images and the frontispieces. For the most part, the frontispieces appear to draw on local iconographic idioms, established by popular artists like Jan Luiken.

The appropriate way to investigate the meaning of the frontispieces is to work from the known to the unknown – from frontispieces (or other allegories) whose meaning is clear because of contemporary explanations, to frontispieces without these. Of course, artists were not held to fixed meanings and might playfully subvert or modify established conventions, but even so, they had to start from common expectations. Establishing some 'canon' of established meanings is the only way to avoid arbitrariness. I will limit myself to frontispieces, largely ignoring other allegorical images, partly for practical reasons and partly because this seems methodologically more proper, considering our intention to study the way in which frontispieces influenced other frontispieces.

Let us start with the frontispiece, designed and engraved by Jan Luiken, of the Dutch translation of Robert Boyle's *Disquisition about the final causes of natural things* (1688). The book was published in Amsterdam by the widow of Steven Swart, who specialised in translations from the English and works on English affairs. The frontispiece shows a seated woman, absorbed in thought, with an oversized quill in her hand and an open notebook on her lap (fig. 5). This engraving is accompanied by an explanation in verse, probably by Luiken himself, who was also an accomplished poet. The poem identifies the woman in the foreground as a personification of philosophy: 'Here, philosophy is depicted, absorbed in contemplating the final causes of all things arranged in the universe'.[14]

The woman is shown leaning on a terrestrial globe; a much larger celestial globe is placed behind her. Books, a pair of compasses and a pair of scales are at her feet. The text explains that it is on earth and in the heavens that Philosophy finds the subject of her studies of final causes. The pair of globes probably refers to the common (Ciceronian) idea that philosophy entailed the contemplation of both terrestrial and celestial things. A similar symbolism is to be found in Luiken's frontispiece for the Dutch translation of John Ray's *Three physico-theological treatises* (1693). Here, the terrestrial globe is combined with an armillary sphere.[15] One might wonder whether the books depicted in the frontispiece serve as

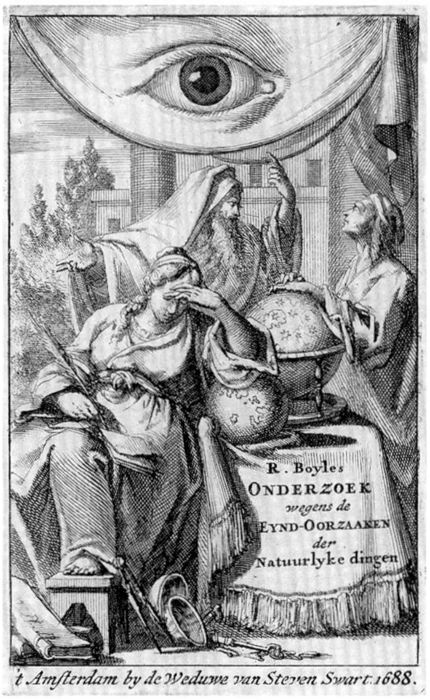

5

Jan Luiken, frontispiece to Robert Boyle, *Onderzoek wegens de eynd-oorzaaken der natuurlyke dingen*

engraving (Amsterdam 1688)

For accompanying explanation, see appendix 2.

attributes of philosophy or, as they are put carelessly on the ground, rather stand for rejected, bookish knowledge. However, the fact that instruments lie around in the same careless way in this print and in some subsequent prints makes the latter explanation improbable. In a 1684 frontispiece by Luiken, to be discussed below, scales serve as an attribute of Truth.

In European iconography, allegorical figures are generally characterised by their attributes. We may assume that if we find, in an appropriate context, another woman with similar attributes, she probably has the same meaning as the one in the Boyle frontispiece. Indeed, Luiken's image of Philosophy was taken up in other frontispieces by other artists. An interesting example is offered by a frontispiece to a volume of Cartesian treatises published in Leiden in 1688 under the title *Metaphysica*.[16] The book contains treatises on metaphysics, on motion and on animal economy (roughly, physiology) by the prominent Cartesian physician Cornelis Bontekoe (1647-1685), as well as a treatise on 'true physics' by the famous Cartesian philosopher Arnold Geulincx (1624-1669). In the frontispiece, by an unknown artist, we again see a seated woman with a quill and an open book. A (presumably terrestrial) globe is at her feet, and in her hand she holds an armillary sphere as a representation of the heavens. Some books are also at her feet (fig. 6). There is no explanation accompanying the print, but the common elements, and the fact that this print is from the same year as Luiken's frontispiece to Boyle's book, strongly suggest that this woman too represents Philosophy.

A very similar woman also appears on the abovementioned frontispiece of Ten Hoorn's 1690 edition of Descartes's works. She is seated at a table in a contemplative pose, again with a quill in her hand and a book in which to write in front of her. There is a celestial globe placed beside her, but the earth now is represented by a world map on the wall. Mathematical instruments are lying around or hanging from the wall, and books are placed on a shelf (fig. 4). In spite of the lack of an explanation, we may safely assume that she too represents Philosophy. That the books are not at her feet but are placed on a shelf may be in imitation of an engraved portrait of Descartes, made by Cornelis Hellemans, probably around 1690, in which Descartes is seated at a table, writing a book, in a similar posture.[17] Descartes's foot is on a volume of Aristotle's work, which clearly shows his rejection of Aristotle's ideas. In this scene, the floor clearly is not the appropriate place for books as philosophical attributes, which may be the reason that Hellemans drew a bookshelf above Descartes's head.

A somewhat enigmatic element is the six-pointed star above the woman's head. It can hardly be there by accident. At first sight, it looks like some kind of instrument hanging from the wall. Instruments of this shape, however, are not known. Maybe the engraver himself was not quite sure what to think of it. Stars are a familiar attribute of Urania, the muse or personification of astronomy, most often in the form of a crown of stars but sometimes as a single star on her head. An example of the latter can be found in the frontispiece of *Theatrum cometicum* by Stanislaus

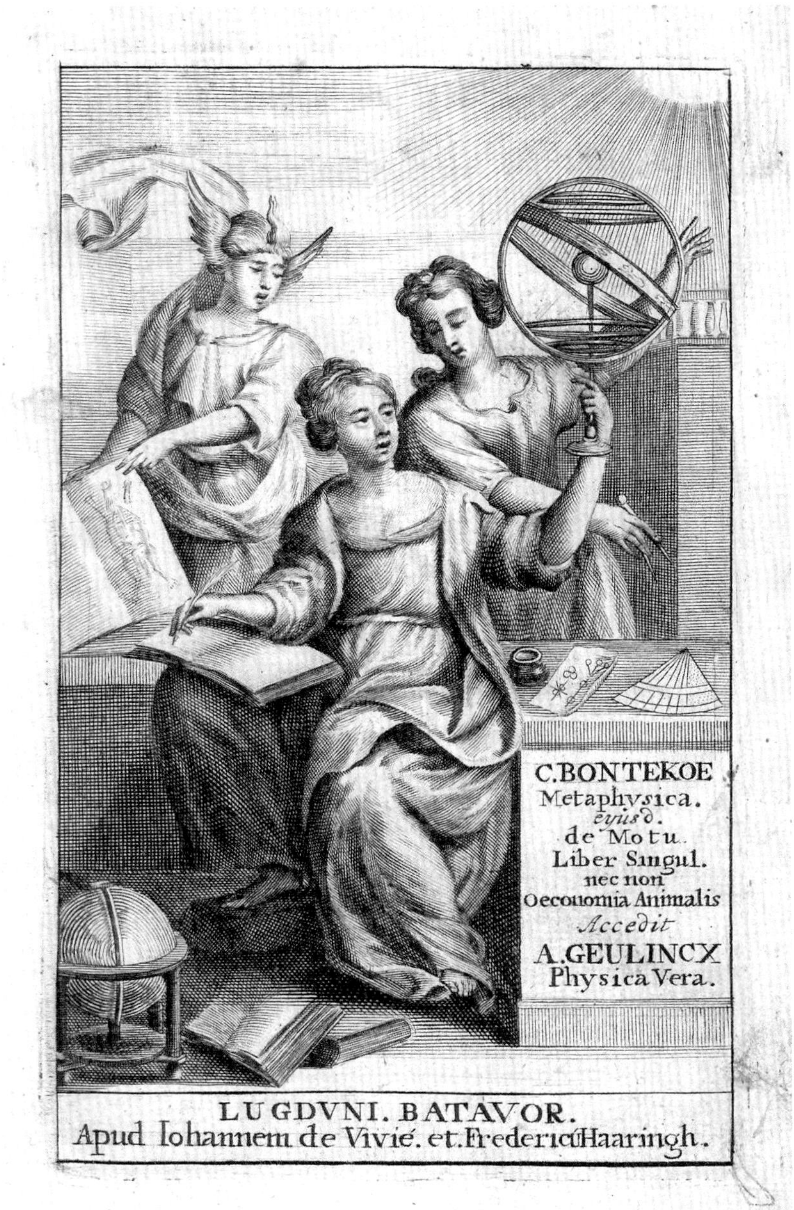

6
Unknown artist, frontispiece to Cornelis Bontekoe, *Metaphysica*
engraving (Leiden 1688), Universiteitsbibliotheek Amsterdam, Special Collections, OTM: O 62-7777.

Lubieniecki (1623-1675), published in Amsterdam in 1667 and reprinted in Leiden in 1681.[18] An interesting case from about the same time as the edition of Descartes's works is the frontispiece of the mathematical dictionary by Jacques Ozanam (1640-1718), published in Amsterdam and Paris in 1691. The frontispiece depicts personifications of both geometry and astronomy. Astronomy is depicted with the usual crown of stars, and Geometry bears a large pentagram on top of her head.[19] It seems improbable that the woman in the Descartes frontispiece represents astronomy, as all other attributes are lacking, but apparently, the designer found it appropriate to allot Philosophy with some astronomical or mathematical attributes. (See also the mathematical instruments lying around.) This would make some sense. In Aristotelianism, there was a strict separation between philosophy and mathematics (which included astronomy), but Descartes tended to conflate the two. Representing philosophy in this particular way might be seen as an anti-Aristotelian statement.

This personification of philosophy was not a common allegory of the time. Standard personifications of philosophy most often drew on Boethius's description, in his *Consolation of philosophy*, of philosophy as a stately woman, with a stairway. The figures we have seen represent a local convention that probably was initiated by Luiken. The prints are from a limited time-span. The books, of course, would circulate for a longer period, and the images and the iconographic conventions were probably remembered for some time. Indeed, a somewhat similar woman is depicted in the frontispiece of Wyer Willem Muys's *Elementa physices* from 1711. She is using a pair of compasses, books are at her feet and she is surrounded by instruments (only one globe, however). In a few cases, elements were picked up abroad. An interesting case is a portrait of Descartes (his likeness clearly copied from the portrait by Frans van Schooten Jr. [1615-1660]) that was published in an edition of his works in Frankfurt, Germany, in 1692 (fig. 7). Descartes is here depicted in the same way as Philosophy in the frontispieces we have seen: seated at a table, writing in a book, a pair of globes at his feet, some other instruments lying around. It would seem that the artist was inspired by the Dutch frontispieces of a few years earlier.

Three frontispieces about teaching wisdom

Several of these frontispieces show a scene of teaching or instruction. Although this is certainly not the only way philosophical books could be illustrated, it was apparently a well-known iconographical convention. In the Boyle frontispiece (fig. 5), Philosophy is pondering all by herself, whereas the instruction is taking place in the background – an old, experienced man teaches his pupil the wonderful construction of the eye (as the poem specifies). In the other cases, it is Philosophy herself who is being taught. Three different frontispieces from roughly the same period share common imagery. Their meanings are probably similar, or they refer to common concepts. These are the frontispieces to the 1690 edition of Descartes's own work and to two works by prominent Cartesians: Cornelis Bontekoe's volume of 1688, mentioned above; and Arnold Geulincx's *Ethica* from 1690. The frontispiece of the latter was again by Jan Luiken.

We have already discussed the 1690 Descartes volume at some length, but the other two works still need some introduction. The 'Bontekoe' volume consists largely of works by Geulincx: even the three treatises attributed to Bontekoe himself are based on notes that he had taken at

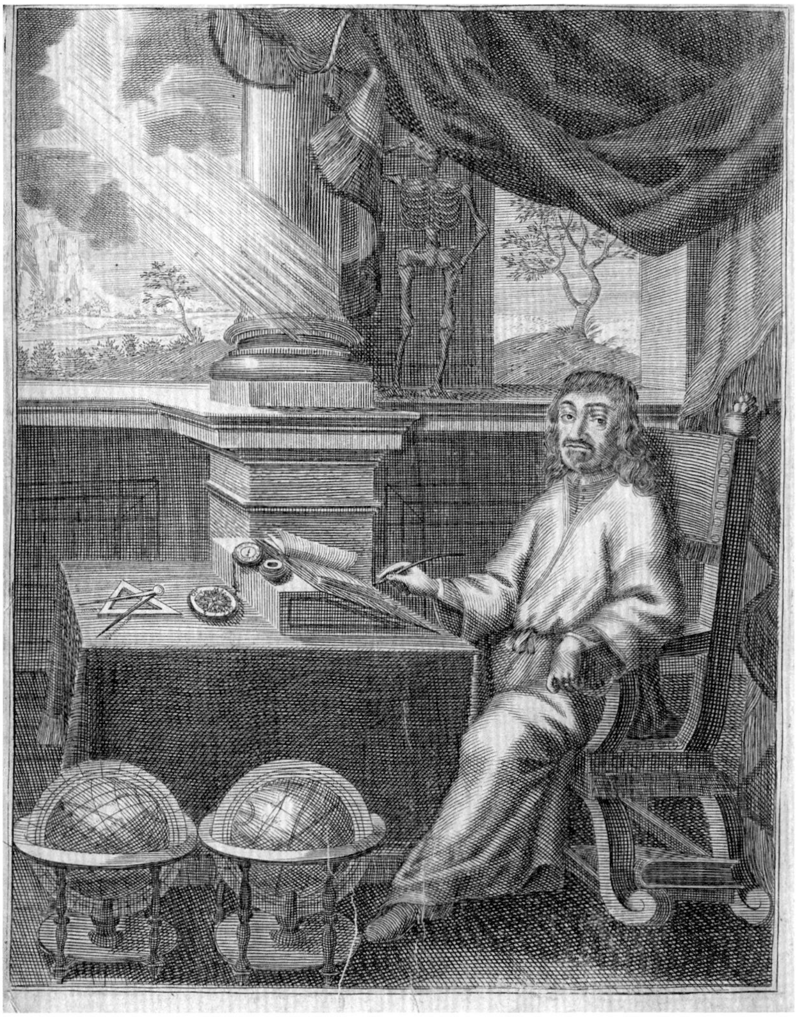

7
Unknown artist, *Portrait of Descartes*
engraving, René Descartes, *Opera omnia* (Frankfurt 1692)
Wolfenbüttel, Herzog August Bibliothek, H: O 13a. 4to Helmst.

Geulincx's Leiden lectures. It is not clear who edited the volume or was responsible for the design of the frontispiece. Both Geulincx and Bontekoe were dead by 1688 – Geulincx from the plague in 1669, Bontekoe in 1685 in Berlin, where he had been invited by the Elector of Brandenburg. The book was published in Leiden by Frederik Haaring and Johannes du Vivié. Du Vivié had published, among other things, various theological works, whereas the Bontekoe volume was Haaring's first undertaking as a publisher. In the following years, he published many other works: medicine, travel reports, theology, both in Latin and in the vernacular.

As for Geulincx's *Ethica*, the original Latin version had been edited in 1675, after the author's death, by the same Cornelis Bontekoe, using the pseudonym Philaretus. (A second edition was published in 1683 by the above-mentioned Du Vivié.) The translation, for which Luiken made the frontispiece, was published by the Dordrecht publisher Dirk Goris.[20] Although not a very prominent figure, Goris published several works on the new philosophy and its effects on religion. The frontispiece (fig. 8)

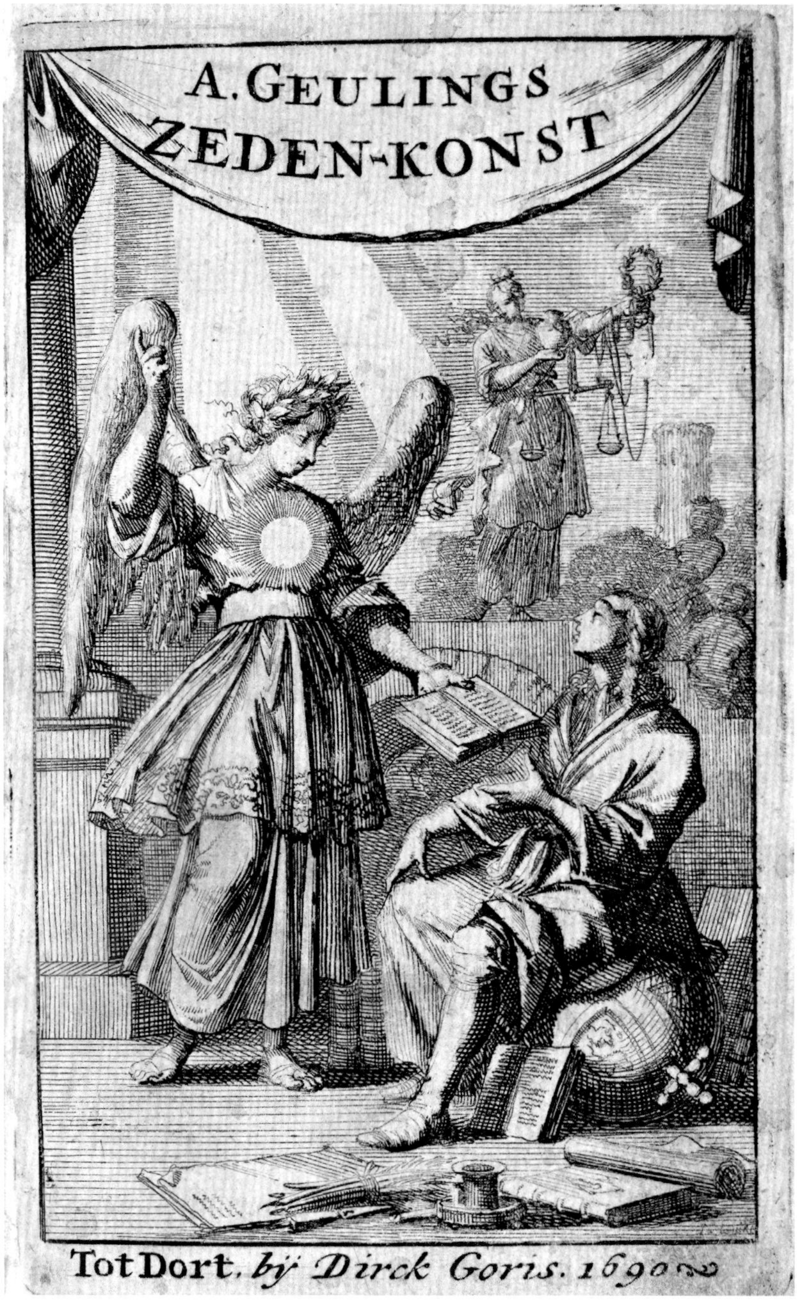

8
Jan Luiken, frontispiece to Arnold Geulincx, *Ethica of zeden-konst*
engraving (Dordrecht 1690), Den Haag Koninklijke Bibliotheek, 943 F 97.

was apparently well received. Goris reused it for a re-edition of the book in 1697, and a copy on a smaller scale was used by the Amsterdam publisher Gillis Janssonius van Waesberge for his re-edition of the original Latin work in 1696.[21] It seems reasonable to assume that it also influenced the Descartes frontispiece of 1690, especially as this print probably was not really programmatic, and so the author more likely simply followed existing examples.

In the Bontekoe frontispiece, Philosophy is taught by two other women, one with a pair of compasses and gesturing towards the sun, the other showing an open book with an illustration of a dissected animal (fig. 6). In the Descartes frontispiece, Philosophy is clearly instructed by the Minerva-like figure standing in front of her, who also shows an open book, and gestures towards a window (behind her), through which several natural phenomena are visible: wind, lightning, a rainbow, a part of the Zodiac (fig. 4). The frontispiece to Geulincx's *Ethica* shows a winged female figure, who is again pointing with one hand to the sun above, and with the other giving an open book to a male person, seated on a symbol of the world (fig. 8).[22] One notices the similar diagonal composition in all three frontispieces. In the Bontekoe example, the light of the sun reaches Philosophy via the armillary sphere and the figure representing geometry. In the Descartes frontispiece, there is a similar line from the window with its Zodiac (suggesting the heavens), via Minerva and the open book, to Philosophy. In the Geulincx frontispiece, the direction is top left to bottom right, instead of top right to bottom left, as in the other two, but the idea is still the same: the light of the sun reaches the seated figure via the figure of Virtue and the open book.

None of these frontispieces is accompanied by an explanation. In the Bontekoe and Geulincx frontispieces, the design appears inspired by the contents of the respective books, especially the preface and dedication, and the meaning can be grasped reasonably well from these. The Descartes frontispiece is harder to interpret. The text of the book offers no real clue. However, as said before, given the resemblances of these three images, made at about the same place and time, we may assume that common elements bear similar meanings. As for the Bontekoe volume, the preface has as its main theme the importance of mathematics to the understanding of metaphysics, motion and physics. Geulincx's 'True physics', it is explained, is based not on useless logical terms but on the method of mathematical demonstrations. By this method, one ascends from the most basic truths to the higher sciences. In metaphysics (according to this preface), the same method is followed, as is shown in the first treatise by Bontekoe. If we apply this to the frontispiece, the woman with the pair of compasses probably represents mathematics or geometry.[23] Note that the armillary sphere is held in such a way that it might well serve as one of her attributes. Lying on the table next to the inkstand is a leaf with mathematical schemata. This figure is actually found in Geulincx's treatise on true physics. So this represents physics, the fundamentals of which Geometry is teaching (according to the preface). The sun, to which Geometry is pointing, may then stand for metaphysics, the other discipline to which the preface says mathematics leads.

For the other figure, with winged head, no clue is given in the foreword. The book she is showing to Philosophy has an illustration of a dissected animal. This must be referring to the treatise on animal economy. (This treatise contains no illustrations, by the way.) Physiology is clearly not taught by geometry. An educated guess would be that the frontispiece shows how philosophy is taught by both reason, in the form of mathematics, and experience. However, the preface is silent about the foundations of physiological knowledge. Also, the meaning of the open book is not clear. It might be an allusion to the 'book of nature', a common metaphor at the time for the created world as a second revelation of God. However, this would not be very apt in the case of Bontekoe, who (in another work) ridiculed the very notion of the book of nature, calling it the 'Boek van turelure', *turelure* being a kind of nonsense word.[24]

Let us postpone judgment until we have looked at the other two frontispieces. The wings and the sun on the breast of the female figure in Luiken's frontispiece (fig. 8) are, according to Ripa, attributes of Virtue. Other elements of the description are absent in the print, but, as the book is on ethics, this identification makes perfect sense.[25] This figure's gesture appears to express a central tenet of the book. In his preface, Geulincx emphasised that he derived knowledge of virtue from 'the inborn light of reason' alone, without having recourse to revelation. So, it seems plausible that the sun to which the woman is pointing represents this light of reason. The book she is offering, then, is probably Geulincx's *Ethica* itself. Her gesture indicates that the book is based on, and in agreement with, the light of reason.

How far would the same explanation hold for the books on the other two frontispieces? In the case of the Bontekoe volume (fig. 6), this appears somewhat problematic. The book is held up to show its contents, and the figure with the winged head is pointing to it.[26] But it is not in the diagonal line from the sun to Philosophy, and there is no gesture to indicate its agreement with reason or truth. The latter arrangement seems to have been an innovation by Luiken, who felt that this expressed the message in a stronger way. In the case of the Descartes frontispiece (fig. 4), the parallels would appear more convincing. The artist seems to have simply followed Luiken's composition. The book Minerva is showing is therefore probably likewise the very book for which the frontispiece was made, Descartes's *Principles of philosophy*. Her gesture towards the natural phenomena seen through the window then indicates that this book is in full accordance with nature and gives a true description of it. Actually, the frontispiece contains another self-reference – 'the works of Renatus Descartes' are standing prominently on a shelf in Philosophy's study.

Jan Luiken and the light of nature

The prefaces to Geulincx's and Bontekoe's volumes emphasise still another point: that reason and Scripture do not contradict each other and that, actually, a proper use of reason supports religion. In his dedication to Geulincx's *Ethica*, Bontekoe made clear that this book was in full agreement not just with the principles of Descartes, but also with the prescriptions of Holy Scripture. According to him, Geulincx is showing us virtue 'by a clear and pure oracle of reason, as a reproach to paganism and to the glory and fame of the Holy Pages'.[27] Bontekoe's dedication has a strongly apologetic tenor and refutes those philosophers who derive materialistic ideas from Descartes. The very title of the book, in its Dutch translation, claims that it is directed 'against the ungodliness of this century, as well as some God-renunciating philosophers'.

Thus, an alternative and at first sight plausible interpretation of the Geulincx frontispiece would be that the woman representing virtue is making explicit this very agreement between reason, or the 'natural light', here represented by the light from above at which she points, and the Bible, which then would be represented by the book she holds in her other hand. For several reasons, I feel this explanation should be rejected. There is not even a hint of Scripture in Descartes's *Principles*, so it is hard to see how the book in the Descartes frontispiece could represent the Bible. However, it seems unlikely that these two very similar allegories would have completely different meanings. Moreover, it is striking that Nieuwentijt, roughly half of whose book is about the authority of the Bible, does not have a book depicted in his frontispiece, even if the accompanying explanation expressly states that it shows how God can be known from both his Word and his Creation. The omission can hardly have been an oversight. If there was a common expectation that a book figuring in a frontispiece would represent the work itself, it is more understandable that he did not use this motif.

If there is a reference to divine revelation in the frontispieces, it is not in the depiction of a Bible, but in the depiction of the sun or its light. Light is a very old symbol that traditionally had strong religious connotations. In the sixteenth and seventeenth centuries, it came to be used in philosophical and scientific contexts as well, but without completely losing its religious meaning. William Ashworth, in a detailed iconographical study, concluded that there were two main groups of scientists using light as a metaphor on their frontispieces: Jesuits, who referred to the light symbolism of the Church fathers and the Catholic tradition; and alchemists, who followed Paracelsus's idea of a light of nature (*lumen naturae*), developed in the Protestant tradition. These two groups 'form the sum total of all seventeenth century scientists whose title pages depicted the Lights of Reason and Nature in some form'.[28]

Rather than looking at Jesuits or alchemists, we might try to find examples closer at hand. Jan Luiken used the symbol of light not only in the Geulincx frontispiece but in other cases as well, and in some cases he provided an explanation. An interesting case concerns the books by the Amsterdam sectarian leader and amateur philosopher Willem Deurhoff (1650-1717). Deurhoff was a well-known figure in his time, with a considerable following and influence. Within the Reformed Church, he was considered a dangerous heretic, contaminated with Spinozistic ideas. He was certainly much influenced by the new philosophy, notably by Descartes, Spinoza and Geulincx. His interpretation of Christian doctrine was often unorthodox, but he emphatically denied that his teachings were contrary to Holy Scripture. Still, in many cases he preferred philosophical reason as a way to religious truth.

Deurhoff wrote several works, most of them published by Jan ten Hoorn and with frontispieces by Jan Luiken. These frontispieces regularly have explanations in verse. An example is *Beginzelen van waarheid en deuchd* (Principles of truth and virtue, 1684; reprinted, including

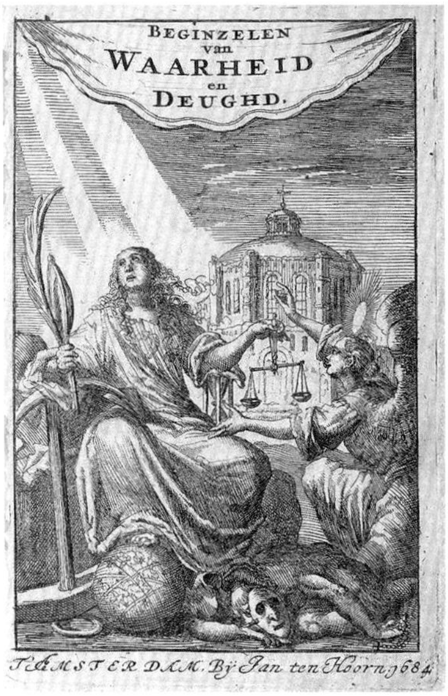

9

Jan Luiken, frontispiece to Willem Deurhoff, *Beginzelen van waarheid en deuchd*
engraving (Amsterdam 1684), Universiteitsbibliotheek Amsterdam, Special Collections OTM: O 62-5065. For accompanying explanation, see appendix 3.

frontispiece, in 1693).[29] In this work, Deurhoff claimed to deduce the principles of truth and virtue from reason alone, explaining the mysteries of faith in a philosophical way. Probably by coincidence, Luiken's frontispiece to this book is, in the arrangement of figures, reminiscent of the one he made for Geulincx's *Ethica*. We see two female figures with a church in the background (fig. 9). (In the Geulincx frontispiece, instead of the church, we see a personification of justice.) According to the accompanying explanation, the figure on the left represents Truth, treading upon Deceit and Hypocrisy, and the winged figure with the sun on her head on the right, who approaches Truth with open arms, is Virtue. In the frontispiece, Truth is illuminated from above by rays of light. In other frontispieces engraved by Luiken for books by Deurhoff, from 1697 and 1687 (reprinted 1707), the same light beams from above, so Luiken appears to have regarded this as an essential element in Deurhoff's work.[30] Moreover, the light in the Geulincx frontispiece is depicted in much the same way, suggesting a parallel. Both Geulincx and Deurhoff claim that the contents of their respective books are derived from reason alone. The light would appear to indicate this illumination by natural reason.

However, as the poem accompanying the Deurhoff frontispiece explains, Truth is expecting to receive the essence of true wisdom directly from God. The beams of light thus represent divine illumination. In other prints by Luiken too, light is clearly used as a representation of the divine. This concerns not only religious books (or what we would label as such), like the one by Deurhoff, but also works on natural philosophy. In 1697, Jan ten Hoorn, in cooperation with the Leeuwarden publisher Rintjes, published a Dutch translation of some works by Kenelm Digby (1603-1665), Nicholas Papin (ca. 1610-1653) and Athanasius Kircher (1602-1680) under the title *Theatrum sympateticum*. The book had earlier been published by Timotheus ten Hoorn and Jacob van Royen with a different frontispiece by an unknown artist. Luiken's frontispiece for the new edition, as far as the main scene is concerned, imitates this earlier picture. However, the earlier engraving depicted in its top corners the sun and moon, and in between, at the top of the page, an eye as the symbol of the divine (fig. 10). In Luiken's frontispiece, this symbol has been replaced by one he felt apparently more appropriate: a set of light beams shining from above (notice that the sun and the moon are still there). This light is again very similar to the light in his frontispiece for Geulincx's *Ethica* and other works (fig. 11).[31]

So, we are left with a question: Why would Luiken state that Truth is expecting everything directly from God in the frontispiece of a book that claims to be based on reason alone? More generally, why is light, as symbol of divine revelation, used especially for authors who emphatically refer to reason, even if they emphasise at the same time that their results are in perfect harmony with the Bible? In order to answer this question, we shall have to have a closer look at the meaning of light in the philosophy of this period.

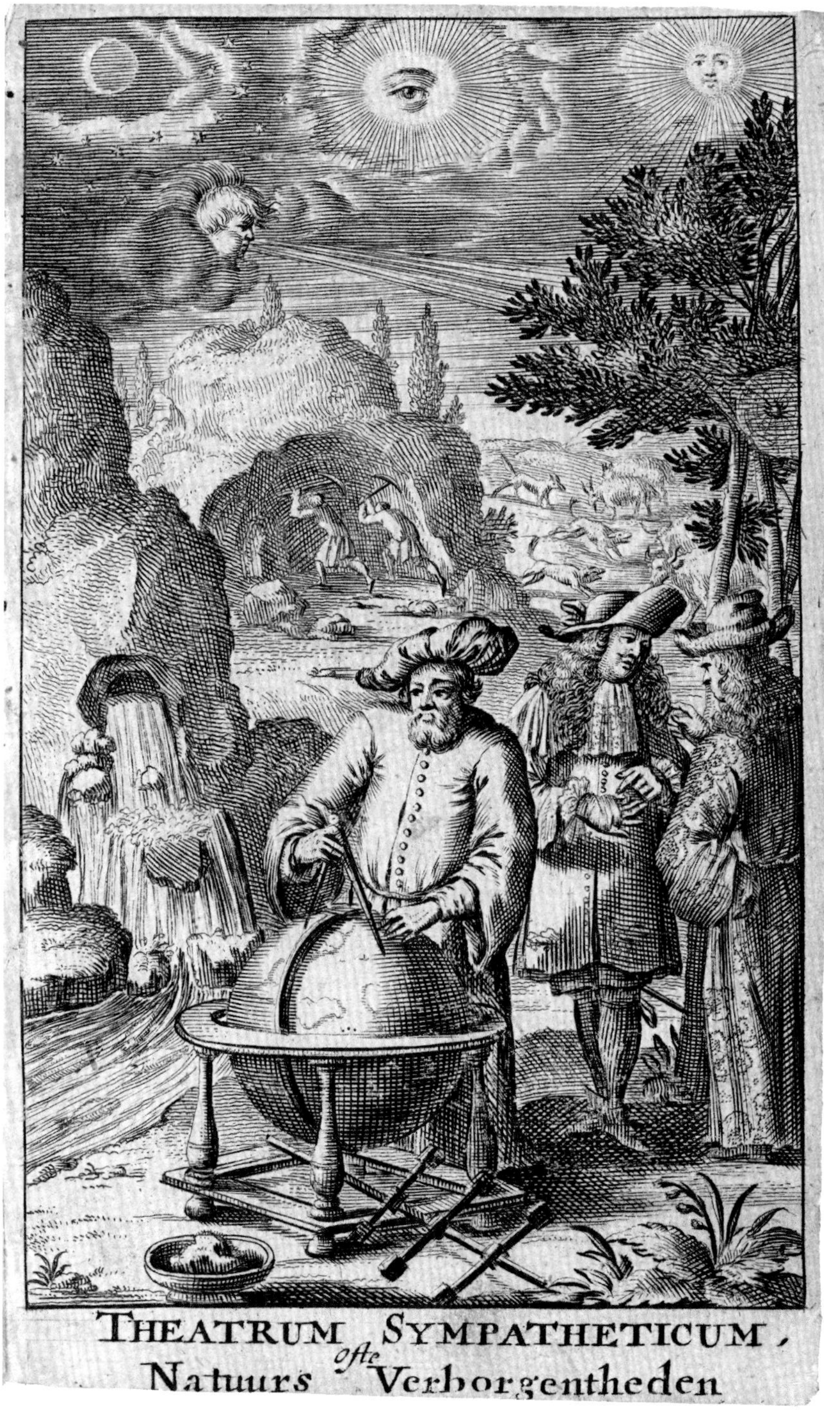

10

Unknown artist, frontispiece to Kenelm Digby and others, *Theatrum sympateticum*
engraving (Amsterdam 1681), Den Haag Koninklijke Bibliotheek, 28 E 25.

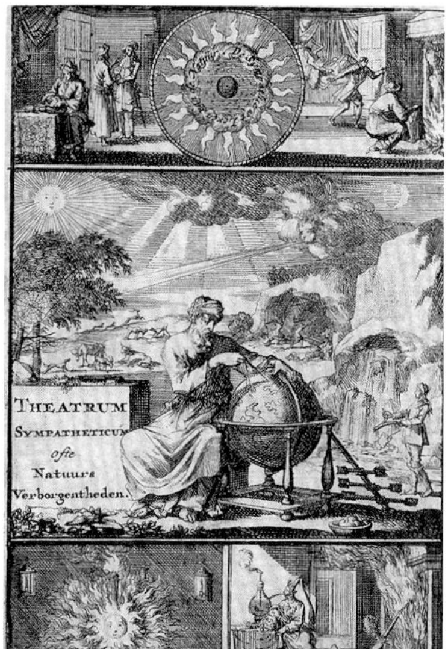

11

Jan Luiken, frontispiece to Kenelm Digby and others, *Theatrum sympateticum*
(second edition, Amsterdam and Leeuwarden 1697).

Wisdom by illumination: the light of nature

The metaphor of the light of nature was used not just by alchemists and Jesuits, but also by Descartes as a foundation for his philosophy. To Descartes, natural light became identified with one's natural reason. It was not by the ratiocinations of Aristotelian logic that people could discern truth from falseness, but by an inborn capacity, which for Descartes was largely identical with mathematical insight. This was 'the source of detailed, unquestioned knowledge of the causal principles, as they apply to God and our ideas'.[32] This light of nature, or the power to discern truth from falsehood, was an essential principle of Descartes's system. In this respect, it made perfect sense to use light as a symbol of reason, as happened on the Geulincx and Bontekoe frontispieces. Other artistic images associated philosophy, or truth, with light, as in the already mentioned portrait of Descartes, probably inspired by Dutch examples, published in the Frankfurt 1692 edition of his works (fig. 7).

However, the relation of philosophical, rational truth to religion remained a point of debate. Seventeenth-century philosophers spent much time in discussing the religious basis of philosophical ideas. Science and religion had to be harmonised in some way. As Amos Funkenstein has demonstrated, the new philosophy occasioned the rise of a new 'secular theology'.[33] Such concerns make it understandable that the originally religious metaphor of the light of nature became so popular in a philosophical context. It allowed one to see philosophical truth in a religious way. That in Cartesian philosophy the symbol of light was detached from its religious origins cannot be stated without qualification.

Reichardt, looking mainly to France, has argued that Cartesian light symbolism was mainly a transformation of the language of Roman Catholicism.[34] This would not really apply to the way it was used by Descartes's Dutch followers. They used the language of light as they knew it from the native spiritualist tradition. Spiritualism, mainly in the form of Anabaptism and later Mennonism, had gained a strong following in Holland in the sixteenth century and since that time never quite faded away. Their ideas on light probably have some connections with Paracelsus's *lumen naturae*, but they appear distinct from the alchemical tradition.

As an example of the way Cartesian natural light was understood religiously, one might refer to the merchant Pieter Balling, who belonged to Spinoza's circle. Balling wrote a treatise *Het licht op den kandelaar* (The light on the candlestick, 1662), on the inner light as a principle of both genuine knowledge and religion: 'the light (...) is a clear and distinct knowledge of truth, in the understanding of each man, by which he is so much convinced of the being and quality of things, that it is impossible for him to doubt it'. This is still very much Cartesian, but he further explained that, although he preferred to call it 'light', he felt it indifferent 'whether one calls it *Christ*, the *Ghost*, the *Word*, etc., as all these come down to one and the same'.[35] Here, the spiritualist idea of direct illumination by the Holy Ghost and the Cartesian idea of natural reason as an infallible gateway to truth become virtually indistinguishable.

The light in Luiken's frontispieces should be understood in the same vein. It represents both man's natural faculties and direct divine illumination. In other words, natural reason (rightly used) is the way God speaks to us. The double meaning of light is also exactly what we see in the use of the sun in the Bontekoe frontispiece of 1688. On the one hand, the sun is the light of nature by which we see the world: the sun's light is radiating upon Geometry and, further down, Philosophy (fig. 6). On the other hand, true mathematical insight leads us to an understanding of the source of this light: Geometry points up towards the sun. The sun is both a source of light and the subject of man's highest hopes.

Descartes and many of his followers felt that mathematical knowledge was the only absolutely certain knowledge, whereas other forms of knowledge could all be doubted. Therefore, Descartes had made mathematics the foundation of his new philosophy, and his light of nature was first of all mathematical insight. Indeed, the Bontekoe frontispiece hints at this close association between mathematics and light: it is via Geometry that the light of the sun reaches Philosophy. The position of the figure who seems to represent empirical knowledge is much less prominent. At the same time, we should acknowledge that this figure too is illuminated by the light of the sun. The frontispiece seems to underline that there is no contradiction between mathematical insight and experience. Indeed, Descartes did not repudiate experience. He rather felt that natural light was a precondition for other forms of true knowledge. That does not mean that knowledge of natural phenomena can simply be deduced from metaphysical insight, a position Geulincx in *Physica vera* explicitly rejected. But it is only the natural light that allows us to understand them.[36]

Still, it is with the geometrical method that the light of nature is most closely associated. Luiken did not use the allegory of light when he made a frontispiece for the Dutch translation of Robert Boyle's *Disquisition about the final causes of natural things*. In this book, Boyle argued that natural investigation showed that nature was not governed by blind necessity, but that all things were made for a purpose. His main example was the eye. A close, empirical study of the eye would show the existence of final causes and therewith the wisdom of the creator. Boyle was a religious author who emphasised that true science and religion were in harmony. No doubt, Luiken sympathised with his position. But Boyle was wary of the rationalism of Descartes. His 'true science' was based on experiments, not geometrical reasoning.

Luiken summarised Boyle's view by depicting a large eye in the frontispiece, drawn in a naturalistic way (fig. 5). As the accompanying verses explain, the 'experienced old man' draws the attention of his student to it: 'how neatly all characteristics of the eye are fitted by nature, how artfully every part is formed, and that it shows clear proof of the great and miraculous powers of the Creator on the whole.' At the same time, his gesture can be seen as pointing upwards to the heavens, very much in the same way as Virtue is pointing to the sun in the Geulincx frontispiece (fig. 8). That God reveals himself in empirical objects is a recurrent theme in Luiken's work, as is apparent from a short glance at any of his emblem books. The symbol of light was thus not a general sign of agreement with revelation, but it corresponded with specific philosophical and epistemological positions.

There were many different interpretations of Cartesian philosophy and many ways to bring the new science into line with religious thought. It deserves notice that we find the symbol of natural light especially in frontispieces of works by a small group of like-minded people, such as Geulincx, his pupil Bontekoe, and Deurhoff. Advocates of their ideas would be found in the circle around Jan ten Hoorn, who published Bontekoe's collected works in 1689,[37] and probably a few other publishers. Dirk Goris, the Dordrecht publisher who in 1690 published the translation of Geulincx's *Ethica*, also published a work by Willem van Blyenberg, a local magistrate who corresponded with both Spinoza and Deurhoff and who may well have served as a local patron of philosophical studies.

These strongly rationalist ideas certainly did not appeal to everybody. As noticed, the Descartes 1690 frontispiece, which otherwise appears to largely draw upon Luiken's frontispiece for Geulincx's *Ethica*, replaces the shining sun with an open window (fig. 4). This replacement may well be due to the spiritualist connotations of light. The publisher probably hoped to sell this book also to people not in sympathy with the ideas on the inner light.

Nieuwentijt and the light of truth

Let us try now to apply all this to the frontispiece of the book by Nieuwentijt. Nieuwentijt was a fierce critic of Spinoza. Spinoza in particular identified the light of nature with the geometrical method. He regarded geometrical knowledge as the way to gain knowledge of the highest things, including God. Nieuwentijt argued that Spinoza's 'geometrical way of reasoning' could not produce knowledge of the real world. In contrast, he propagated the experimental methods of Robert Boyle. His work would become very influential in the Dutch Republic.[38]

From Nieuwentijt's works, it is not always easy to derive his position regarding other religious thinkers. Though clearly critical of Cartesianism, Nieuwentijt was wary to engage in polemics. He hoped that the experimental method would end the useless polemics among Christians. In his books, therefore, he concentrated on 'facts' that (he felt) directly countered atheism, but he avoided theological controversies. He vehemently attacked Spinoza and Spinozism, but only rarely did he mention any followers of Spinoza by name. A rare instance happened in a later book in 1720. He then vented his anger about some sectarians with Spinozist leanings, notably Pontiaan van Hattem (1641-1706) and Jacob Brill (1639-1700). The reason for his indignation was not their physics, but their biblical exegesis. But it should be noted that these people, commonly classified as 'antinomians', stood clearly in the spiritualist tradition.[39] His work seems to have contributed to discredit ideas on natural light. In 1722, an anonymous author used Nieuwentijt's arguments to refute rationalist philosophers, 'who want to understand everything by what they call the inborn light'.[40]

Nieuwentijt was thus a decidedly anti-rationalist thinker. It is therefore somewhat surprising that the symbol of light figures very prominently in the frontispiece of his book (fig. 1). Looking more closely, however, it is used in a decidedly new way, which distances it from the way it was used in the earlier books we have discussed. Whereas in the earlier frontispieces there is only one source of light, in line with the idea that natural and supernatural light are indeed one, in Nieuwentijt's frontispiece there are three. In the middle, there is the divine light, radiating directly from God (indicated by the Greek word ΘΕΟΣ). To the right, we see the light of truth, held by a common iconological personification of truth and indicated as such in the accompanying poem. Top left, we see the sun, which may represent the light of nature.[41]

Let us first interpret the image following the accompanying description. As in the earlier prints, a central role is given to a female figure who appears to have some instructional role. As explained in the accompanying poem, this woman, standing on an altar, represents physics (*Natuurkunde*), 'unabashed because of the incontestable truth of experience'. She lifts the blindfold from the eyes of a kneeling figure, who in the poem is identified as 'the philosopher'. With her left arm, the woman is gesturing, not upwards to the divine light or the light of nature, as was the case in the earlier frontispieces, but to the light of truth in the distance. Notice that this light is placed amidst dark clouds, and one has to make some effort to discern it. This effort required to see the truth is further emphasised by the many instruments lying around. We notice a globe, a thermometer, a *laterna magica*, an air pump (a somewhat outdated model for 1715), a distillation apparatus, a burning mirror, a telescope, some animals and plants and some books, although these are not prominent.

These instruments may well be a reference to the earlier Descartes frontispiece, in which several instruments are also depicted (fig. 4). These, however, on closer inspection appear to be nearly all mathematical instruments: a pair of compasses, a surveyor's quadrant, an armillary sphere (held by a small boy) and a sundial. They hint that mathematics is the key to interpreting the phenomena of nature seen outside the window. Goeree's engraving, on the other hand, shows mainly physical instruments, making the point that insight into reality is gained not by mathematics, but by experience. The explanation to the Nieuwentijt frontispiece underlines this by stating: 'The manifold instruments which surround them, show how right investigation of things penetrates further into real wonders of nature's secrets, than the flattering mind, mere ideas, or reason'. Disabused by Physics, the philosopher can see the light of truth, whereas earlier, he had been blindfolded because he, 'deceived by himself, relies on his imagination'.

One should note, however, that mathematics is not completely repudiated. Although there are only a few mathematical instruments depicted, a pair of compasses is given a very prominent place, against the altar in the very centre. Earlier, as stated before, Nieuwentijt had published some work on infinitesimal calculus. As is especially clear from a later book, *Gronden van zekerheid* (Foundations of certitude), which appeared posthumously in 1720, Nieuwentijt regarded pure mathematics as an important field in itself but emphasised that it could not be used to draw conclusions about the real world.

The light of truth, which Physics reveals to the philosopher, is not welcomed by all. Next to the philosopher is another man, who appears frightened or angered by the light. Instead, he is turning to the little mole and other symbols of blindness and dumbness depicted in the bottom left corner (to which the real philosopher turns his back), diagonally opposed to the light of truth. Or, as the explanation has it, on seeing the brightness of truth, he withdraws 'into the darkness of his night'. He represents 'the *esprit fort*, who despises investigation'. As Nieuwentijt explains in the book, these *esprits forts* try to elevate themselves above the common people by laying aside all fear and respect for religion. As such, the *esprit fort* represents one of the highest degrees of atheism. Because of their misguided ambition, these people refuse to acknowledge the light even when they see it.[42]

Krifka has pointed to an opposition between the left-hand and the right-hand sides of the frontispiece: light and order on the right, chaos and darkness on the left. She reads this as an announcement of the Enlightenment program of progress: the right-hand side would represent the results of modern, experience-based science; the chaos on the left would represent the old world picture. The suggestion is interesting but contradicts the overall message of the book. An opposition between light and darkness may well be intended, but this would rather be an opposition between belief and unbelief than between old and new. Moreover, there is a rainbow amidst the 'chaos'. Krifka's interpretation also suffers because she knows the frontispiece only from a German translation of the book and therefore is not aware of the accompanying verse explanation. For example, she interprets the central figure as Veritas and the nude figure on the right as Nature.[43]

In the upper left corner we see the sun. Unlike the sun in the Bontekoe frontispiece or the various engravings by Luiken, however, it is not radiating upon the scene. It is surrounded by dark clouds that prevent it from illuminating the philosophers below. This seems to be in deliberate opposition to the earlier prints, indicating that, in the present fallen state of the world, people do not have any direct knowledge of things. That does not imply, however, that people cannot gain any knowledge at all. It can hardly be an accident that the telescope, placed behind the altar, is pointed exactly at the sun. Natural principles do not appear to the mind naturally, but they can be discovered by careful empirical investigation.

This natural light of the sun is again quite distinct from the divine light. From the top centre, a beam of light is descending directly from the divine. This light, however, does not illuminate the scene or the philosopher, but falls exclusively upon the figure of Physics. Unlike the light portrayed in the earlier frontispieces, the radiation from the divine is clearly one-way. Physics is not pointing or gesturing towards this divine light. She allows humans to obtain truth of some kind, but certainly not an insight into the divine itself. Rather, the illumination of Physics by the divine light indicates that true (that is, experimental) physics is divinely sanctioned and that its results are in agreement with revelation. As the poem says, 'A beam of the divine accompanies this knowledge, in spite of the boldest sacrilege, and shows in the universe the undeniable mark of the Speaker in his Word, and of the Maker in his Work'.

The light in Nieuwentijt's frontispiece clearly has religious overtones and must have immediately made clear to any potential reader that this book would maintain and defend religion. But at the same time, the artist is careful to avoid any hint at spiritualism, with which light commonly would be associated. The complicated way light and darkness are represented appears carefully chosen with an eye to both the religious and epistemological connotations – connotations that are not necessarily fully expressed in the text. Nowhere in his work does Nieuwentijt refute the works of Deurhoff or Geulincx. Still, if the frontispiece is to be believed, he wrote his work in conscious opposition to their ideas of an 'inner light'.

A study of the frontispiece to Nieuwentijt's book demonstrates therefore the importance of looking at the imagery against the background of local iconographic traditions and the use of similar imagery in other books. The direct message of the frontispiece is obvious from the content of the book and the accompanying poem. But its full visual impact on contemporaries can be appreciated only when we realise that it refers to a local iconographical tradition regarding light. The symbol of light had been used in earlier frontispieces in a specific sense, which was refuted by Nieuwentijt. His frontispiece therefore is part of a philosophical debate, waged with images rather than words.

Notes

*The author thanks Eric Jorink, Jo Spaans, Inga Elmqvist, Bert van de Roemer, Thijs Weststeijn, Marrigje Rikken and Joanna Woodall for comments on an earlier version of this article

1 Hunter 1989, xiv, 41-42. On the use of rhetorics in early science, see Vickers 1985.
2 Harms 1978; Elmqvist Söderlund 2010, 102-117.
3 Remmert 2005; Elmqvist Söderlund 2010; Ashworth 1989; Kintzinger 1995; Haitsma Mulier 1994.
4 Bennett 1991, 349-352. Schneiders 1992, 8-10 (explanations) and 16-25 (illustrations). For another example of the use of iconography in a historical study, see Shapin and Schaffer 1985, 30-35.
5 Ashworth 1989.
6 For an introduction to the intellectual background, see Israel 2001. See also Vermij 1991.
7 On the role of the artist in designing frontispieces, see Elmqvist Söderlund 2010, 89-102, and Haitsma Mulier 1994, 276-277.
8 Van 't Veld 2000, 67. The episode is known from a correspondence between Stol and Luiken, which Stol included in his book.
9 On Jan ten Hoorn, see Kleerkooper and Van Stockum 1914-1916, vol. 1, 273-279, and Cordes 2008, 622-625.
10 Van Eeghen 1905-1906, vol. 1, xxxiv; Van 't Veld 2000, 87-90.
11 In at least one 1690 volume (Utrecht, university library, call number Y qu 229), the frontispiece is a loose leaf, pasted on a blank page at the correct place in the book.
12 Anonymous 1690, 6. On the pamphlet, see Leemans 2002, 175.
13 On Timotheus ten Hoorn, see Leemans 2002, 175-180, 278-281, and Van Eeghen 1960-1978, vol. 3, 164-166.
14 Boyle 1688. Klaversma and Hannema 1999, Catalogus Luyken, no. 290; Van Eeghen 1960-1978, vol. 2, 157-158 (no. 139).
15 Ray 1694. Klaversma and Hannema 1999, no. 1232; Van Eeghen 1960-1978, vol. 1, 269-270 (no. 249), see also vol. 2, 800 (no. 1465).
16 Bontekoe 1688.
17 According to Van Someren 1888-1891, vol. 2, 223, the engraving was published by the Leiden publisher Johannes Tangena. (Information kindly provided by P.E. van Kooij of the Netherlands Institute for Art History.) The STCN mentions book publications by Tangena between 1687 and 1692.
18 Elmqvist Söderlund 2010, 80, 180. Another example is the frontispiece to Johannes Hevelius, *Machina coelestis* (Danzig 1673); see Elmqvist Söderlund 2010, 303. For examples of Urania with a crown of stars, see ibid., 221, 222.
19 Remmert 2005, 114-117. The frontispiece was made by Johannes van den Aveele.
20 Geulincx 1690.
21 Geulincx 1696. An earlier edition from 1691 did not contain this frontispiece. The frontispiece was used again in the re-edition of 1709.
22 Klaversma and Hannema 1999, Catalogus Luyken, no. 553. Van Eeghen 1960-1978, vol. 2, 189, no. 179.

23 Compasses were a familiar geometrical attribute. In Ripa 1644, 328, geometry is represented by a woman with either a pair of compasses and a plummet, or a pair of compasses and a triangle.
24 Jorink 2006, 95; see also Jorink 2010, 87-88.
25 Ripa 1644, 83, describes Virtue as 'Een schoone en aengenaeme Ionghe Maeghd, met vleugels aen de schouderen, houdende in de rechter hand een spiesse, en met den slincker hand een Lauwerkrans, hebbende een Sonne op de borst'.
27 In the prints of Romein de Hooghe, a winged head indicates the swift intellect (information from Jo Spaans). This would well correspond to the Minerva, as godess of wisdom, on the other print. Minerva was found often on frontispieces. A French example from 1679 is in Shapin and Schaffer 1985, 35.
27 Geulincx 1690, 6, ****4v.
28 Ashworth 1989, 104.
29 Deurhoff 1684. See Klaversma and Hannema 1999, 136, and Van Eeghen 1905-1906, 108-109, no. 83.
30 Klaversma and Hannema 1999, 437, 438. Van Eeghen 1905-1906, 139-140 (no. 118), 308 (no. 297).
31 Klaversma and Hannema 1999, 441. Van Eeghen 1905-1906, 309 (no. 298). The STCN has a link to the frontispieces of both editions of Digby.
32 Morris 1973, 184. For general overviews of the use of the light metaphor, see Reichardt 1988, and Mortier 1969 (for Descartes, 16).
33 Funkenstein 1986.
34 Reichardt 1988.
35 Balling's text, originally from 1662, has been republished in Klever 1988, 65-74. The quoted sentences, 68. Van Bunge 2003 states that Balling was especially concerned about maintaining his interpretation of the inner light against that of the Quakers.
36 Ashworth notes that in the Protestant tradition, the intellectual light is not uncommonly identified with the light of experience (Ashworth 1989, 100).
37 This edition does not have a frontispiece, but the two volumes do have special emblems on the title pages (accessible via STCN).
38 On Nieuwentijt, see Vermij 1991.
39 Vermij 1991, 86-87.
40 G.B. 1722, 59: 'die alles door het *ingeschapen ligt* (zo ze zeggen) verstaan en begrypen willen'.
41 Remarkably, the use of various sources of light is what Ashworth (1989, 96) notices among some representatives of the Jesuit tradition, although there appears to be no close parallel to the way this is done in the Nieuwentijt frontispiece.
42 Nieuwentijt 1715, 4-5.
43 Krifka 2000, 725.
44 A note in the book explains this word as 'Philosophie'.

Bibliography

Anonymous 1690
Anonymous, *Relaas van de beroertens op Parnassus. Ontstaan over het drukken van de Beginselen der eerste wijsbegeerte van de heer Renatus Descartes*, Amsterdam 1690.

Ashworth 1989
W. Ashworth, 'Light of reason, light of nature. Catholic and Protestant metaphors of scientific knowledge', *Science in Context* 3 (1989), 89-103.

Bennett 1991
J.A. Bennett, 'Geometry and surveying in early-seventeenth-century England', *Annals of Science* 48 (1991), 345-354.

Bontekoe 1688
Cornelis Bontekoe, *Metaphysica, et liber singularis de motu*, Leiden 1688.

Bots 1972
J. Bots, *Tussen Descartes en Darwin. Geloof en natuurwetenschap in de achttiende eeuw in Nederland*, Assen 1972.

Boyle 1688
R. Boyle, *Onderzoek wegens de eynd-oorzaaken der natuurlyke dingen*, Amsterdam 1688.

Cordes 2008
R. Cordes, *Jan Zoet, Amsterdammer 1609-1674. Leven en werk van een kleurrijk schrijver*, Hilversum 2008.

Descartes 1690
R. Descartes, *Principia philosophiae, of beginselen der wysbegeerte (...) Als mede een nette verhandeling vande verhevelingen, neffens dat van 't licht: en 't voornaamste der zinnen* (Jan Hendrik Glazemaker, transl.), Amsterdam 1690.

Deurhoff 1684
[W. Deurhoff], *Beginzelen van waarheid en deuchd, door zuivere reedeneringen van 't ingeschaapen denkbeeld afgeleid*, Amsterdam 1684.

Elmqvist Söderlund 2010
I. Elmqvist Söderlund, *Taking possession of astronomy. Frontispieces and illustrated title pages in 17th-century books on astronomy*, Stockholm 2010.

Funkenstein 1986
A. Funkenstein, *Theology and the scientific imagination from the Middle Ages to the seventeenth century*, Princeton 1986.

G.B. 1722
G.B., *Het denkbeeldig Spinosismus in 't kort vertoont. En volgens de gronden van de Heer Bernard Nieuwentyt wederlegt*, Harderwijk 1722.

Geulincx 1690
A. Geulincx, *Ethica of zeden-konst, zynde een kennisse syns selvs* (posthumously edited by Filaretus [=Cornelis Bontekoe], translated into Dutch by Ant. de Reus), Dordrecht 1690.

Geulincx 1696
A. Geulincx, *Gnothi seauton, sive Ethica*, Amsterdam 1696.

Harms 1978
W. Harms, 'Programmatisches auf Titelblättern naturkundlicher Werke der Barockzeit', *Frühmittelalterliche Studien* 12 (1978), 326-355.

Haitsma Mulier 1994
E.O.G. Haitsma Mulier, 'Woord en beeld. Titelprenten van enkele Nederlandse historische werken uit de 17de en 18de eeuw', *Holland* 26 (1994), 274-291.

Hunter 1989
M. Hunter, *Establishing the new science. The experience of the early Royal Society*, Woodbridge 1989.

Israel 2001
J. Israel, *Radical Enlightenment. Philosophy and the making of modernity 1650-1750*, Oxford 2001.

Jorink 2006
E. Jorink, *Het boeck der natuere. Nederlandse geleerden en de wonderen van Gods schepping 1575-1715*, Leiden 2006.

Jorink 2010
E. Jorink, *Reading the Book of Nature in the Dutch Golden Age, 1575-1715*, Leiden & Boston 2010.

Kintzinger 1995
M. Kintzinger, *Chronos und Historia. Studien zur Titelblattikonographie historiographischer Werke vom 16. bis zum 18. Jahrhundert*, Wiesbaden 1995.

Klaversma & Hannema 1999
N. Klaversma and K. Hannema, *Jan en Casper Luyken te boek gesteld. Catalogus van de boekencollecie Van Eeghen in het Amsterdams Historisch Museum*, Hilversum 1999.

Kleerkooper & Van Stockum 1914-1916
M.M. Kleerkooper & W.P. van Stockum, *De boekhandel te Amsterdam voornamelijk in de 17de eeuw. Biografische en geschiedkundige aanteekeningen*, The Hague 1914-1916.

Klever 1988
W.N.A. Klever, 'De Spinozistische prediking van Pieter Balling. Uitgave van "Het licht op den kandelaar" met biografische inleiding en commentaar', *Doopsgezinde Bijdragen*, N.S. 14 (1988), 55-85.

Krifka 2000
S. Krifka, 'Zur Konstruktion der Natur in wissenschaftlichen Experimenten', in: H. Holländer (ed.), *Erkenntnis Erfindung Konstruktion. Studien zur Bildgeschichte von Naturwissenschaften und Technik vom 16. bis zum 19. Jahrhundert*, Berlin 2000, 725-753.

Leemans 2002
I. Leemans, *Het woord is aan de onderkant. Radicale ideeën in Nederlandse pornografische romans 1670-1700*, Nijmegen 2002.

Morris 1973
J. Morris, 'Descartes' natural light', *Journal of the History of Philosophy* 11 (1973), 169-188.

Mortier 1969
R. Mortier, '"Lumière" et lumières". Histoire d'une image et d'une idée au XVIIe et au XVIIIe siècle', in: idem, *Clarté et ombres du siècle des lumières. Etudes sur le XVIIIe siècle littéraire*, Genève 1969, 13-59.

Muys 1711
W.M. Muys, *Elementa physices methodo mathematica demonstrata*, Amsterdam 1711.

Nieuwentijt 1715
B. Nieuwentijt, *Het regt gebruik der wereltbeschouwingen, ter overtuiginge van ongodisten en ongelovigen*, Amsterdam 1715.

Ray 1694
J. Ray, *De werelt van haar begin tot haar eynde*, Rotterdam 1694.

Reichardt 1988
R. Reichardt, 'Light against darkness. The visual representations of a central Enlightenment concept', *Representations* 61 (winter 1998), 95-148.

Remmert 2005
V. Remmert, *Widmung, Welterklärung und Wissenschaftslegitimierung. Titelbilder und ihre Funktionen in der Wissenschaftlichen Revolution*, Wiesbaden 2005.

Ripa 1644
C. Ripa, *Iconologia, of uytbeeldingen des verstands* (D.P. Pers, transl.) Amsterdam 1644 (reprint Soest 1971).

Schneiders 1992
W. Schneiders, 'Aufklärungsphilosophien', in: S. Jüttner and J. Schlobach (eds.), *Europäische Aufklärung(en). Einheit und nationale Vielfalt*, Hamburg 1992, 1-25.

Shapin & Schaffer 1985
S. Shapin & S. Schaffer, *Leviathan and the air pump. Hobbes, Boyle and the experimental life*, Princeton 1985.

STCN
Short Title Catalogue Netherlands, http://www.kb.nl/stcn/.

Thijssen-Schoute 1989
C.L. Thijssen-Schoute, *Nederlands cartesianisme* (2nd ed.), Utrecht 1989.

Van Bunge 2003
W. van Bunge, 'Balling', in: idem *et al.* (eds.), *Dictionary of seventeenth and eighteenth-century Dutch philosophers*, Bristol 2003, 45-47.

Van Eeghen 1960-1978
I.H. van Eeghen, *De Amsterdamse boekhandel 1680-1725*, 5 vols., Amsterdam 1960-1978,

Van Eeghen 1905-1906
P. van Eeghen (in cooperation with J. Ph. van der Kellen), *Het werk van Jan en Casper Luyken*, Amsterdam 1905-1906, 2 vols.

Van Someren 1888-1891
J.F. van Someren, *Beschrijvende catalogus van gegraveerde portretten van Nederlanders*, vol. 2, Amsterdam 1888-1891.

Van 't Veld 2000
H. van 't Veld, *Beminde broeder die ik vand op 's werelds pelgrims wegen. Jan Luyken (1649-1712) als illustrator en medereiziger van John*

Bunyan (1628-1688), Amsterdam 2000 (diss. Free University Amsterdam).

Vermij 1991
R. Vermij, *Secularisering en natuurwetenschap in de zeventiende en achttiende eeuw. Bernard Nieuwentijt*, Amsterdam 1991.

Vermij 2003
R. Vermij, 'The formation of the Newtonian philosophy. The case of the Amsterdam mathematical amateurs', *British Journal for the History of Science* 36 (2003), 183-200.

Vickers 1985
B. Vickers, 'The Royal Society and English prose style. A reassessment', in: B. Vickers & N. Struever, *Rhetoric and the pursuit of truth. Language change in the seventeenth and eighteenth centuries*, Los Angeles 1985, 1-76.

Appendix

Explanation of the frontispiece of Nieuwentijt, *Het regt gebruik der wereltbeschouwingen* (fig. 1).

Verklaring der tytel-prent
Natuur-kunde onbeschroomt door 't onbetwist'lyk Waar
Der Ondervinding, pronkt hier veiligh op't Altaar.
Sy ligt den Philosooph, die door sigh selfs bedrogen
Op syn Verbeelding rust, den blind-doek van syn oogen
't Veelvuldigh konst-tuigh, dat beneden haar omringt,
Leert, hoe 't regt ondersoek der dingen dieper dringt
In ware wond'ren van Natuurs verborgentheden;
Dan 't vleyende Verstant, 't bloot Denkbeelt of de Reden:
Die als Ervarentheit ontbreekt, hoe trots, alleen
Ryk in gedagten syn en arm in saak'lykheên.
Sy wyst met d'andre hand, waar ymand met vertrouwen
En eerbied 't heerlyk ligt der Waarheit kan beschouwen.
Terwyl een Sterke Geest, die 't ondersoek veragt,
Op desen glans vergrimt, in 't duist're van syn nagt.
Een straal der Godheit vergeselschapt dese kennis
Der schepselen, in spyt der stoutste Heilig-schennis;
En toont in 't groot Heel-al 't onloochenbaar merk
Des Sprekers in syn Woort, des Makers in syn Werk

2. Explanation of the frontispiece of Boyle, *Onderzoek wegens de eynd-oorzaaken der natuurlyke dingen* (fig. 5).

Op de tytel-print:
Hier zit de Wysgeerte[44] afgemaald,
Geheel aandachtig in 't doordringen
Der Eynd-oorzaaken aller dingen,
Die binnen 't Heel-al zyn bepaald.
In de Aardkloot vindt zy overvloed
Van stoffe, om klaarlyk te bemerken
Dat Godes hand in zyne werken
Zich kenlyk aan den mensch voordoet:
De Hemel die zo heerlyk pronkt
Met haar Groot-Licht en kleyner' Starren,
Heeft haaren yver, om te ontwarren
Der zelver Eynden, ook ontvonkt.
Terwyl de eervaaren Gryzaard vast
Aan zynen Leerling gaat ontdekken,
Hoe overaardig all' de trekken
Van 't Oog zyn door Natuur gepast,
Hoe konstiglyk een yder deel
Gevórmd is; dat het klaare blyken
Geeft van des Scheppers wonderlyk en
Groot vermoogen op 't Geheel.

3. Explanation of the frontispiece of Deurhoff, *Beginzelen van waarheid en deuchd* (fig. 9).

tot verklaaring van de tytel plaat:
Des Schryvers toeleg kan men leezen
In deeze Prent, daar Waarheid zit
Vol hoop, met een aandagtich weezen,
Om van God zelf, het innich pit
Der waare Wijsheid af te wachten;
Zy spiegeld zich in 't goed verstand,
Weegd alles naauw, door tijds betrachten,
Dat valsch is, wijstze van de hand;
En trapt, met onverschrokke treeden,
De loogen en schijnheiligheid:
Wijl Deuchd met uitgestrekte leeden
Zich zelven na de Waarheid vleid,
En poogd deez schoonheid aan te kleeven,
Om als twee zusters, hand aan hand,
Met God in eeuwigheid te leeven,
Daar 't vuur van zijne liefde brand.

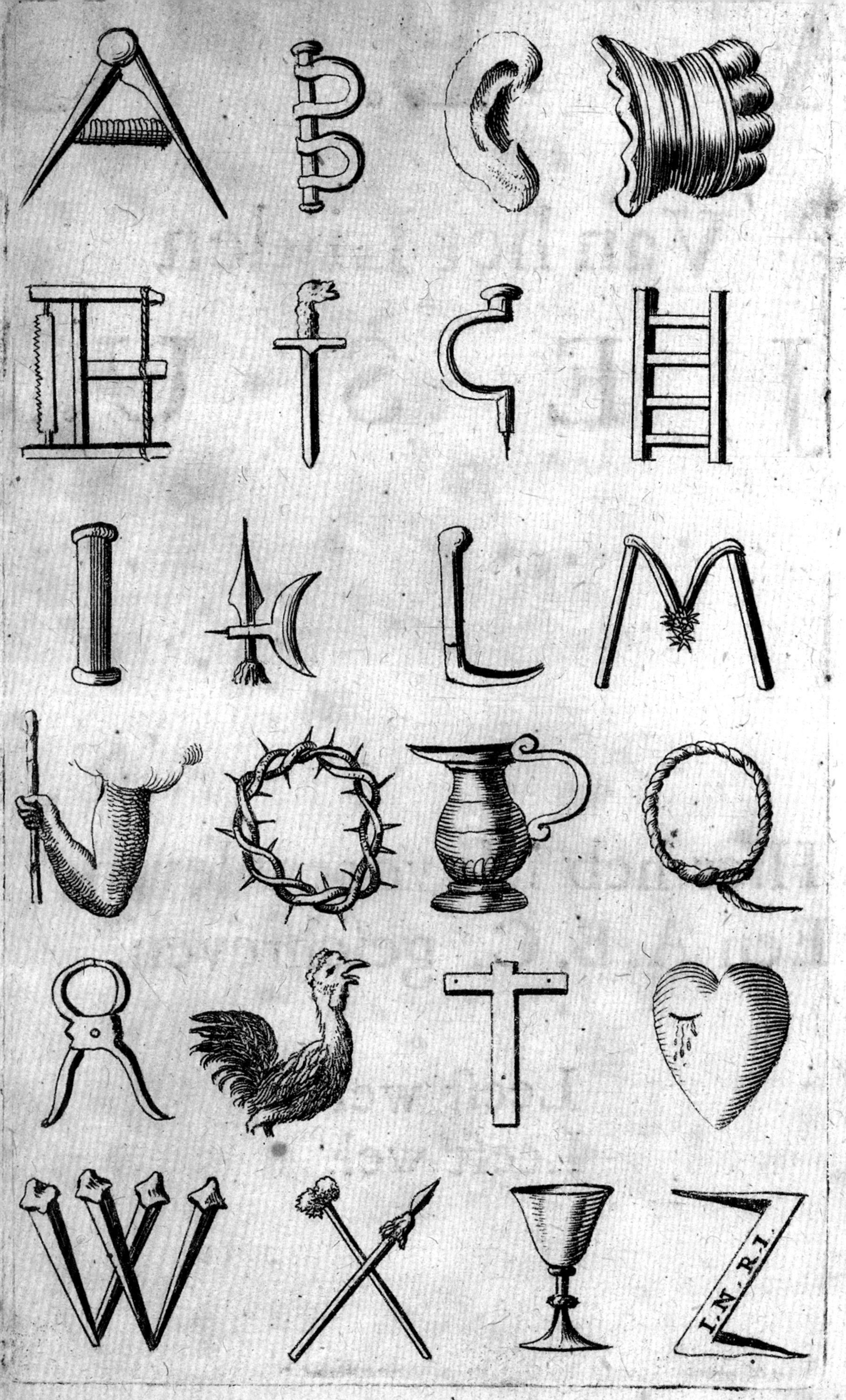
I.N.R.I.

From hieroglyphs to universal characters
Pictography in the early modern Netherlands*

Thijs Weststeijn

In his 1663 book *Het gheestelyck jubilee* (The spiritual celebration), the Antwerp Jesuit Joannes van Sambeeck (1601-1666) included an image of the letters of the alphabet constructed from the instruments of Christ's Passion (fig. 1).[1] This witty conceit had a serious undertone that combined artistic and scientific interests: it illustrated the seventeenth-century concern with how to adjust words to things, or *verba* to *res*. It implicitly harked back to a century of Jesuit scholarship concerned with pictography, or writing by means of images rather than alphabetic script. The Jesuits not only thought that meditation on the Passion benefited more from images than texts, but they also had great expectations for pictography related to history, geography, and science. Besides aiding their global proselytisation and mnemotechnical training, it would lay bare the Christian wisdom expressed in the most ancient history – via the Egyptian hieroglyphs – and the geographically most remote civilisations – via the Chinese and American scripts – and provide a universal language reflecting the order of nature itself.

Echoes of this debate reverberated throughout the Netherlands, which is not surprising in the light of the ubiquity of the figurative arts as well as the rise of linguistic scholarship at Dutch universities. Thus, Rembrandt's pupil Samuel van Hoogstraten (1627-1678) stated that a central commonplace in art theory – namely that painting is 'mute poetry' – was wrong, since Pictura 'speaks abundantly, in a hieroglyphic manner'. In his 1678 treatise on painting, he explained that the Chinese, Egyptians, and Mexicans had 'written their books with meaningful pictures, instead of letters; and their way of expressing themselves has reached us now in the art of painting'.[2] The issue also interested scientists. When Christiaan Huygens discussed the possibility of communicating with intelligent life on other planets in his *Cosmotheoros* (The Hague 1698), he mentioned 'painted scenes in use among the barbarous people of Mexico and Peru'.[3]

This article examines the Netherlandish chapter of the Renaissance fascination with pictograms, or *beeldletteren*, as a key aspect of the scientific enterprise to decode, represent, and master the Book of Nature.[4] What did Van Hoogstraten mean when he wrote that painting speaks 'hieroglyphically'? Clearly, he should not be interpreted according to the twentieth-century trend of highlighting the 'textual' nature of artistic utterances, but rather an opposite approach: the idea that communication

1
Image alphabet
engraving, Joannes van Sambeeck, *Het gheestelyck jubilee* (Antwerp 1663, p. 248) (photo: Universiteitsbibliotheek Amsterdam, Special Collections).

through images is more fundamental than that through alphabetic signs. When Renaissance writers on the visual arts used the formula *ut pictura poesis*, they elaborated upon Aristotle's assumption that 'the soul never thinks without images'.[5] Pictography was just one expression of a more general desire to sidestep the contingency of alphabetic signs and arrive at pure communication of knowledge. Although Plato had rejected the idea of a natural language 'in which the shapes and sounds of letters indicated the essential nature of things', Dutch scholars forgot his conclusions and envisaged an ideal language that directly mirrored reality.[6] To quote Descartes, they aimed at the identification of 'simple ideas that (...) form the basis of all the things that man can think (...) [these] would help the judgement to represent things so distinctly that it would be impossible to be mistaken about them, whereas the words that we actually use have mostly confused meanings'.[7] Thus, just like Dutch painters saw their art as a 'mirror of nature',[8] so Descartes envisioned a specular doctrine of knowledge and language denoting individual things 'similar to a picture or mirror that represents with absolute precision the image of the universe and its parts'.[9]

As we shall see, ideas of pictography touched on a wide range of speculations about the essence and origin of language, remote civilisations, memory, logic, and finally the study of the natural world and how it could be 'read' by scientists. Combining different strands of scholarship with the skills of painters and engravers, pictography was a consummate effort to arrive at encyclopaedic knowledge joining art, language, history, geography, and philosophy, which, like many of the century's other encyclopaedic projects, did not fulfil its scientific promises but had striking spin-offs especially from an artistic point of view. Studying this phenomenon by zooming in on the Netherlands is somewhat problematic, as it had a broad European scope linking scholars and artists in the Low Countries to their counterparts in Rome, the German states, and England. Nevertheless, the Netherlands were fertile ground not only for a sophisticated visual culture but also for innovative linguistics and alternative interpretations of divine revelation, which were interwoven with the revolution of knowledge that pictography promised.

Without aspiring to be comprehensive or incontrovertible, I will focus on four interrelated aspects. The first is the *hieroglyphic* argument. Archaeological remains and documents suggested that pictograms were revelatory of lost, ancient, or geographically remote 'hidden knowledge'. Second, when joined to ideas about non-European languages, this historical research inspired a *linguistic* argument. Ideas about a primitive vocabulary from which the modern languages were spurious offshoots were widely popular in the Netherlands; in a related context, the notion of Real Characters – signs that stand in direct relation to the things they represent – explored the art-theoretical commonplace that drawing and writing were fundamentally identical activities. The third argument is the *encyclopaedic* one. Ideas about the 'alphabet of nature', or the visible world as a text in which God addresses humankind, inspired the organisation of encyclopaedic collections according to a similar alphabetic structure. The fourth argument, finally, is *logical*. Crucial here is the importance of images in the tradition of mnemonic logic, as well as the idea that the creative process resides in the combination of primitive concepts.

The hieroglyphic point of view

Our discussion begins with the reception of Egyptian hieroglyphs. Various artists in the Netherlands designed emblematic pictograms; the Antwerp patrician Jan van der Noot (c. 1539-1595) stands out for combining them in a script for 'riddle obelisks' (fig. 2).[10] To welcome Charles V in 1540, Utrecht's Town Hall displayed a frieze with

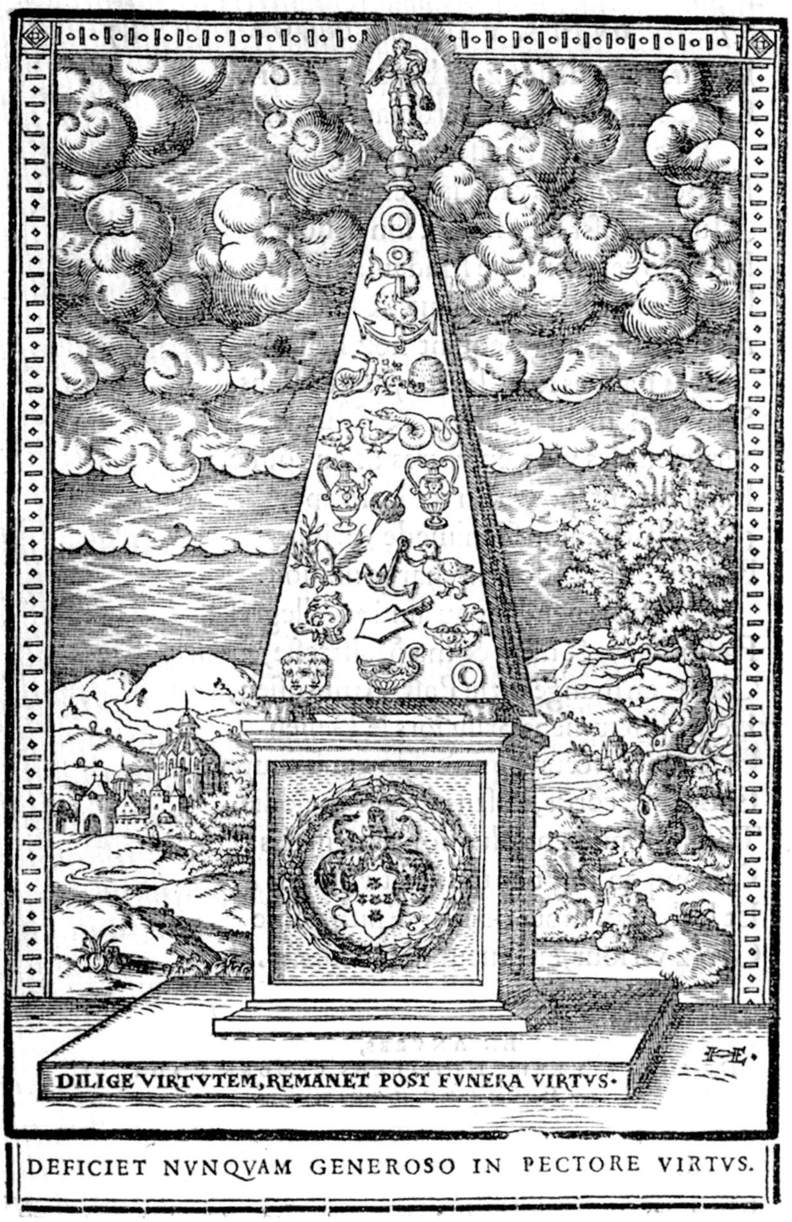

2
Monogrammist HE, *Obelisk with hieroglyphs*
woodcut, 225 x 145 mm, Jan van der Noot, *Een cort begryp der XII. Boecken Olympiados,* (Antwerp 1579)
(photo: Universiteitsbibliotheek Amsterdam, Special Collections).

'hieragliphie' [*sic*], or 'Egyptische letteren veel significatyf', which the painter Lieven van der Schelden (active 1550s-1590s) imitated for Alessandro Farnese's entry in Ghent in 1584.[11] Of a more lasting nature were the inscriptions on Hubert Mielemans's (d.1558) grave in Liège and the obelisk that Maarten van Heemskerck (1498-1574) erected for his father, representing the latter's motto pictographically (a painter's hand above a turtle) (fig. 3).[12]

Implicitly, these artists harked back to Leon Battista Alberti's (1404-1472) treatise on architecture that advised avoiding letters on monuments made for posterity and replacing them with 'Egyptian' pictograms that could always and under all circumstances be read by scholars.[13] Alberti thus turned to the inscriptions on obelisks in Rome that he thought could be deciphered even without knowing the language spoken in Egypt.[14] Artists in the Netherlands had as little specific knowledge about hieroglyphs as Alberti; Van der Schelden and Van Heemskerck's creations were largely products of the imagination.[15] Moreover, the few extant historical sources were unreliable. A book containing spurious

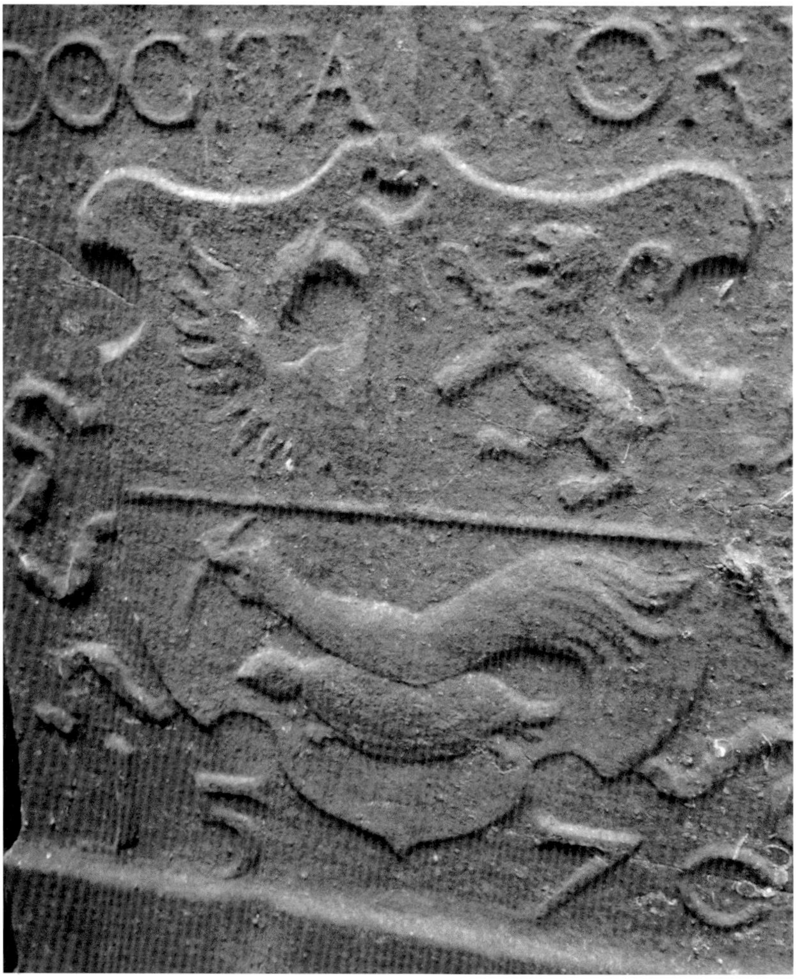

3
Maarten van Heemskerck, Memorial to Jacob Willemsz. van Veen, 1535 (detail)
224 cm x 61 cm (bottom), 24 cm (top). Formerly in the cemetery of the Dutch Reformed Church,Heemskerk; presently kept inside the church (photo: Maartje de Boer).

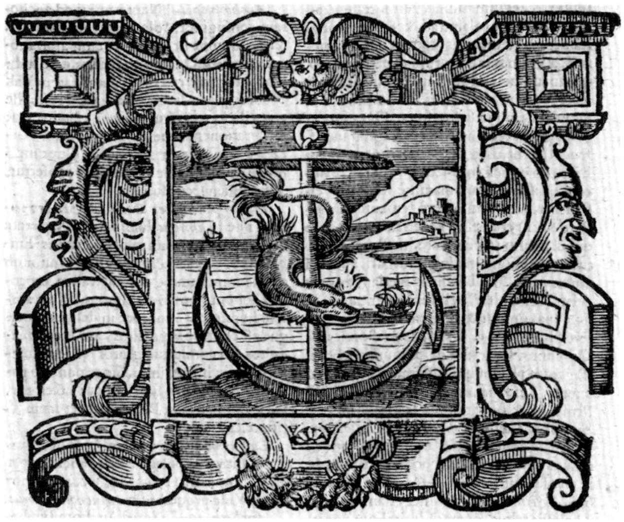

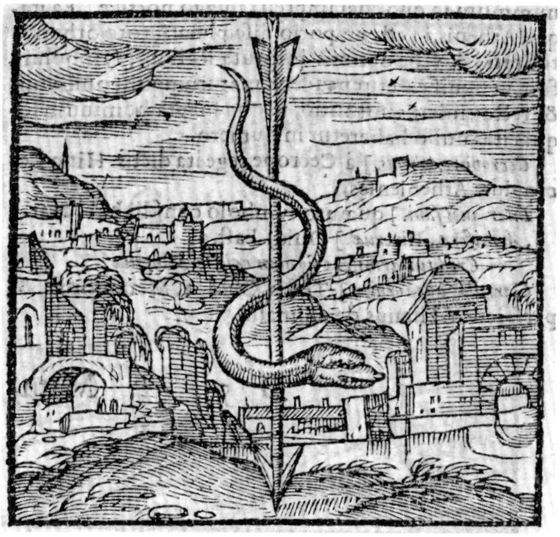

4
Princeps
engraving, Andrea Alciati, *Emblemata (...) cum (...) notis Laurentii Pignorii Patavini* (Padua 1661, p. 615)
(photo: Universiteitsbibliotheek Amsterdam, Special Collections).

5
Maturandum
engraving, Andrea Alciati, *Emblemata,* (Leiden 1584, p. 58)
(photo: Universiteitsbibliotheek Amsterdam, Special Collections)

hieroglyphic 'explanations' attributed to Horapollo, a priest of late antiquity, resurfaced in the fifteenth century. It was widely reprinted and catalysed the popularity of emblem books, which borrowed its imagery.[16] A metal tablet with hieroglyphs found in Rome in 1525 also attracted artists and scholars, such as the Paduan humanist Lorenzo Pignoria (1571-1631) who published a lavish edition of this 'Isiac table'. More recent insights have revealed that it was actually of Roman origin and merely mimicked Egyptian letters. Pignoria made an edition of Andrea Alciati's *Emblematica* (first edition 1505), one of the fashionable illustrated books that derived some inspiration from the hieroglyphs.[17]

Erasmus seems to have had Alciati's text in mind when he treated 'hieroglyphic' questions in a small excursus in his *Chiliades adagiorum* (Venice 1508) dealing with the motto *festina lente.*[18] Alciati's emblem illustrating this motto combines a dolphin, symbolising speed, with an anchor, denoting slowness (fig. 4). A visually similar image (combining a snake and an arrow) was given a different meaning, which suggests that 'hieroglyphic' symbolism was by no means unambiguous (fig. 5).[19] Its polysemic challenge was confronted in a linguistic context only a century later, when the Amsterdam professor Gerardus Vossius (1577-1649) analysed the feasibility of a pictographic script. His statements in *De arte grammatica* (Amsterdam 1635) were partly based on discussions with his 'best friend' Pignoria, revealing how the Dutch debate benefited from Northern Italy.[20]

Vossius's analysis reflects the incipient scientific interest in Egyptian picture writing in the early seventeenth century. For a *Thesaurus hieroglyphicorum* (Munich c. 1607), the Brussels artist and printer Nicolaas van Aalst (c. 1526-c. 1612) made engravings – which one historian characterised as 'remarkably accurate' – of inscriptions on obelisks in Rome.[21] Collectors such as brothers Gerard Reynst (c. 1570-1650) and Jan Reynst (1601-1646) in Amsterdam and Ernst Brinck (1582-1649) in Harderwijk acquired authentic Egyptian objects; Bernardus Paludanus's (1550-1633) Enkhuizen collection, which included three mummies, even attracted internationally renowned philologists such as Grotius and Scaliger.[22] Dutchmen would have had the opportunity of seeing real hieroglyphs on a monumental scale had the Earl of Arundel – whose household welcomed many artists, including Rubens and Van Dyck, and Dutch linguists such as Johannes de Laet (1581-1649) – succeeded in buying the obelisk found on the Appian Way in Rome.[23] The Earl's advisors admired its 'figures' (i.e., the hieroglyphs), 'the best [they had] seen'.[24] However, the outbreak of the English Civil War prevented this large-scale logistical operation that would have altered London's skyline. Eventually, the Jesuit Athanasius Kircher (1602-1680) placed the obelisk on top of the Fountain of the Four Rivers in Rome. This scholar, known as the 'father of Egyptology', published what he pretended to be the translation.

Kircher remained the leading authority on hieroglyphic matters for the remainder of the century. A truly integrated approach to *res* and *verba* was a *leitmotif* in this polymath's writings, which were aimed at winning readers for the True Faith. His quest for a 'universal language' that was also the reconstruction of humankind's primeval vocabulary, spoken by Adam and first written down in hieroglyphs, recurs in his writings on Egypt and China, the Tower of Babel, Noah's Ark, cryptography, and symbolic logic. Moreover, he drove home his point through lavishly illustrated folio-sized books. When he was first publishing in Rome, he depended on many of the Northern artists active in that city. From 1661 onwards, he had his works printed directly by Johannes Janssonius in Amsterdam, probably inspired by the greater pool of illustrators available there. As one historian notes, such a long-term commercial publishing arrangement was still a rarity in the seventeenth century;[25] around a dozen engravers were involved, including some of the most accomplished of their day, such as Cornelis Bloemaert (1603-1692), Gérard de Lairesse (1640-1711), and Romeyn de Hooghe (1645-1708).[26] The priest and painter Lieven Cruyl (1634-1720), who was based in Rome for some time, seems to have been one of Kircher's favourites.[27] Not only artists,[28] but scientists too became familiar with the Jesuit's ideas, many of which were even published in Dutch.[29] Christiaan Huygens acquired Kircher's works (which inspired his *Cosmotheoros*) and no one less than Spinoza came to consult them.[30] Some of the Jesuit's earlier writings were also reissued in Amsterdam. Ultimately, to quote one historian, 'the team of Jesuit author and Protestant printer symbolised the cosmopolitan appeal of Kircher's books'.[31] It is therefore not surprising that Kircher's ideas and images depended in part on earlier Dutch scholarship.

In this regard, Kircher's predilection for Hermes Trismegistus – or the Egyptian deity Thot, the 'thrice-great' – stood out. This sage was credited with inventing writing and figurative art in order to preach proto-Christianity before Moses. He played a key role in Kircher's 'translations' of Egyptian inscriptions, which responded to a general trend: as Yates argues, 'in order to think on Renaissance lines about the hieroglyphs it was absolutely necessary to keep up belief in Hermes Trismegistus'.[32] Even though its attribution to Trismegistus was disputed, the rediscovery of the *Corpus Hermeticum* in the fifteenth century nevertheless sparked a lively interest in pictography. We must bear this in mind because before the findings of nineteenth-century Egyptology, these ideas were common coinage for artists and scholars. In treatises on painting, the Hermetic myth presented an integrated approach to words and things, joining the revelation of the divine *logos* to the first works of art. Alberti's *Della pittura* (1435) stated that 'Trismegistus, an ancient writer, judged painting and sculpture to be born at the same time as religion'.[33] As the Dutch art theorist Franciscus Junius (1591-1677) echoed, 'Egyptian Hermes', a contemporary of Osiris, Noah, and Dionysus, 'began the veneration of the gods (...) and embellished his life with all arts'.[34] The most explicit, however, was the Spanish painter Vicente Carducho (1576-1638), who added that the Egyptian sages 'made themselves understood in an elegant manner with symbols of men and animals'.[35]

In the early seventeenth century, the most ardent defender of the 'thrice-great' sage was Otto Heurnius (1577-1652), keeper of the collection of Leiden University's anatomical theatre, the most important of its kind in all of Europe. Frantically seeking Egyptian objects through a contact in Aleppo, he managed to acquire several stones with inscriptions and images of Isis 'that are full of hieroglyphs'.[36] He published his observations in *Barbaricae philosophiae antiquitatum libri duo* (Leiden 1600). This book sought the primeval wisdom as God had revealed it to Adam: Trismegistus played an important role as he had been taught directly by Noah and one of his sons.

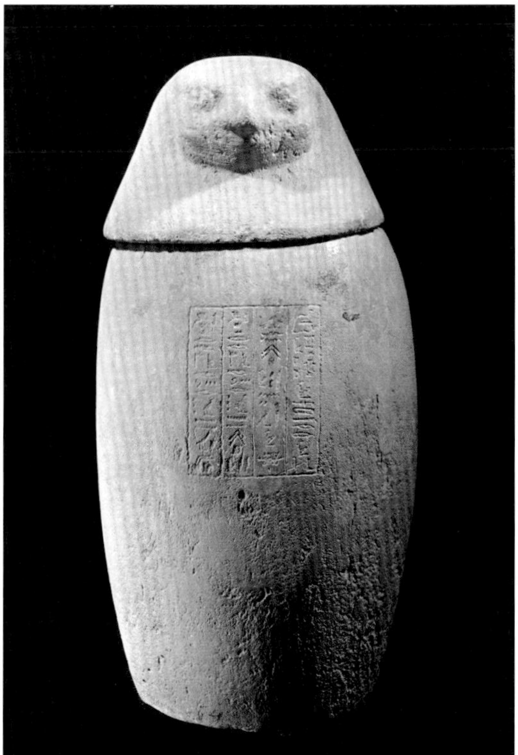

6

Egyptian burial vase 'Leu Leu'
Leiden, Rijksmuseum voor Oudheden
(photo: Rijksmuseum voor Oudheden, Leiden).

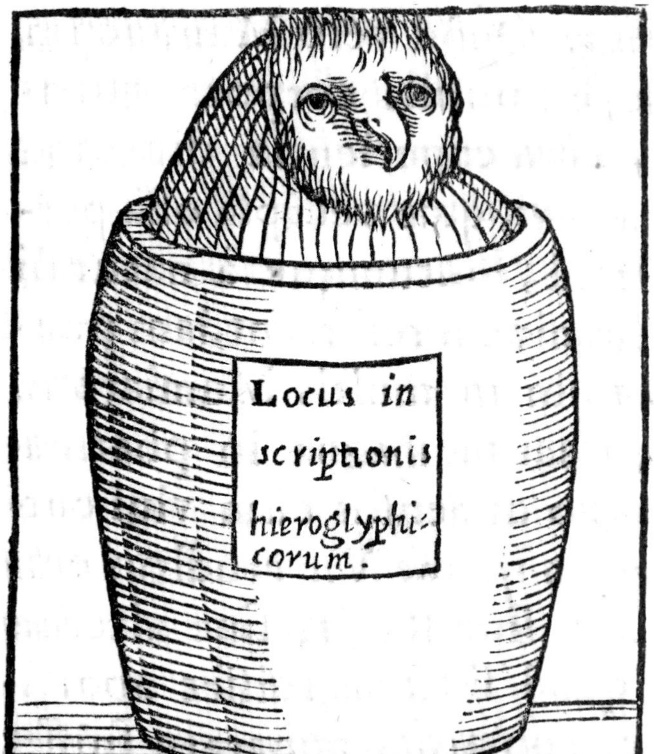

7

Image of Egyptian burial vase (see fig. 6)
Athanasius Kircher, *Oedipus Aegyptiacus*,
Rome 1652-1654
(photo: Universiteitsbibliotheek, Leiden).

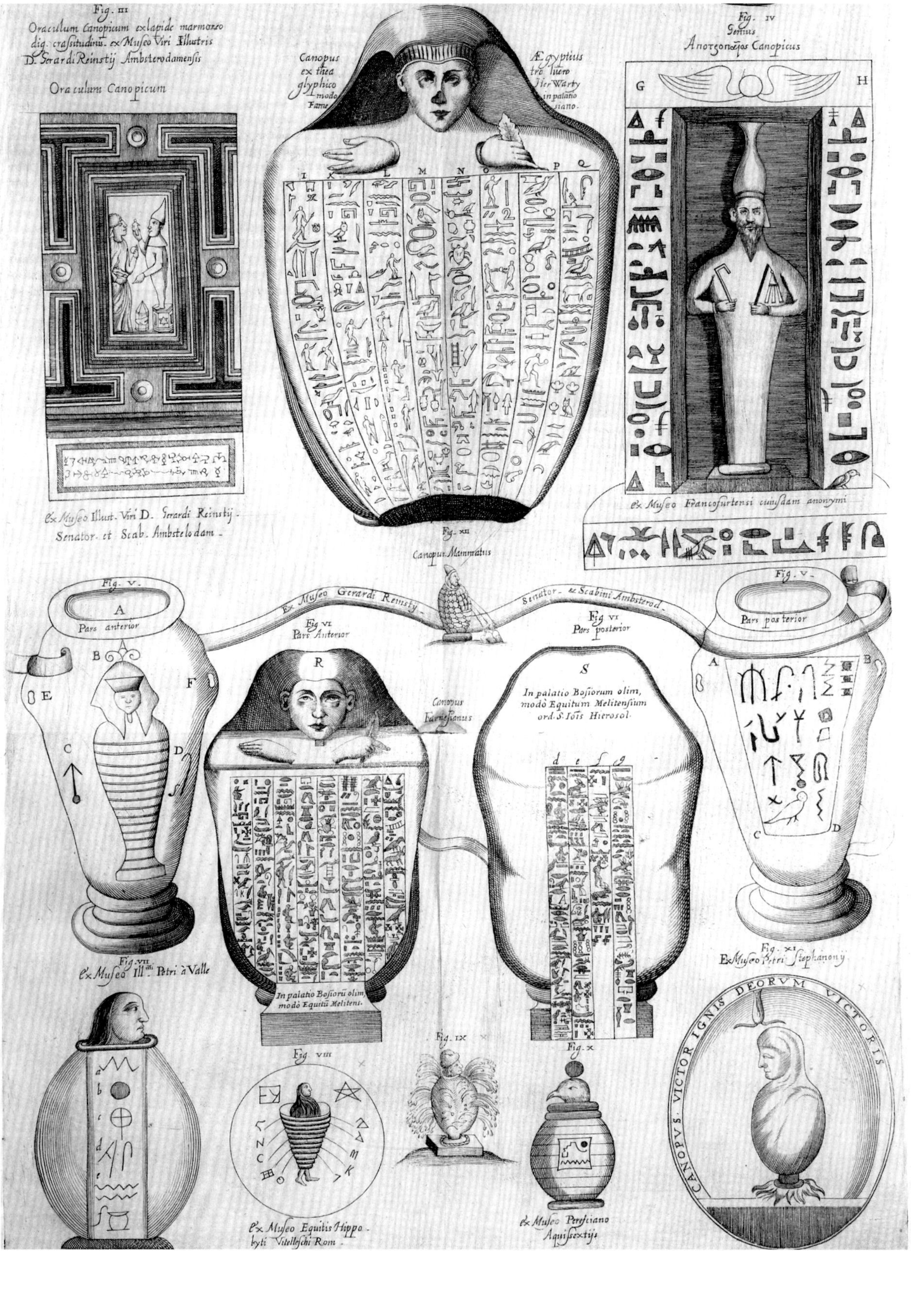
Fig. III
Oraculum Canopicum ex lapide marmoreo dig. crassitudinis ex Museo Viri Illustris D. Gerardi Reinstij Ambsterodamensis
Oraculum Canopicum
Canopus ex thea glyphico modo Farne
Ægyptius tre luero Her Warty in palatio siano.
Fig. IV
Genius
Canopicus
G
H
I K L M N O P Q
Ex Museo Illust. Viri D. Gerardi Reinstij Senator. et Scab. Ambstelodam.
ex Museo Francofurtensi cuiusdam anonymi
Fig. XII
Canopus Mammatus
Fig. V.
A
Pars anterior
Ex Museo Gerardi Reinstij Senator. et Scabini Ambsterod.
Fig VI
Pars Anterior
Fig VI
Pars posterior
Fig. V.
Pars posterior
B
E
F
C
D
R
S
Canopus Farnesianus
In palatio Bosiorum olim, modò Equitum Melitensium ord. S. Iöis Hierosol.
d e f g
In palatio Bosioru olim modò Equitu Meliten.
Fig. VII
ex Museo Ill.mi Petri à Valle
Fig. XI
Ex Museo Petri Stephanony
Fig. VIII
Fig. IX
Fig. X
CANOPVS VICTOR IGNIS DEORVM VICTORIS
a
b
c
d
e
ex Museo Equitis Hippolyti Vitelleschi Rom.
ex Museo Perosciano Aquisextijs

Kircher came to depend greatly on Heurnius, who placed the invention of hieroglyphs much earlier in history than was usual. To the Dutch scholar, they were direct remnants of the lost Adamic vocabulary, the 'primitive language' in which words and things were identical. This idea became the basis of Kircher's ideas on not only pictography but also the history of the world. He argued that all civilisations were offshoots from the 'one primeval religion' that had originated in Egypt (with modern Catholicism as its purest representative). The main 'proof' for this thesis was the allegedly pictographic nature of the Egyptian hieroglyphs, Chinese characters, and Meso-American writing. Pictography thus became essential to Kircher's account of sacred history, in which he contended that all languages could be traced back to a common source, the primitive tongue spoken before the Flood: 'Just as the senses give the same report to all men at all times, so too the first language (...) had perfectly understood reality'.[37]

To substantiate this far-reaching view, Kircher published his correspondence with Heurnius in the book in which he claimed to have translated the hieroglyphs, *Oedipus Aegyptiacus* (Rome 1652-1654). For the accompanying illustrations, Kircher's envoy in the Netherlands, Barthold Nihusius (1589-1657), ordered drawings of one of Heurnius's objects, an Egyptian urn still in the collection of Leiden's Rijksmuseum voor Oudheden (figs. 6, 7).[38] Some of Brinck's Egyptian works were also depicted in the book; nineteen illustrations were based on objects in the Reynst brothers' collection (figs. 8-10).[39]

Heurnius's belief that the rediscovery of the *Corpus Hermeticum* heralded a new era of Egyptian wisdom in the Netherlands may have

8

'Statua constans marmore albo ex Cimeliarchio Domini Gerardi Reinst, Senatoris et Scabini Amstelaedamensis, ab eruditissimo Bertholdo Nihusio mihi transmissa'
burial vase from the Van Reynst collection, reproduced recto and verso (Fig. V, center, lateral sides) and *'Oraculum Canopicum ex lapide marmoreo dig. crassitudinis ex Museo viri illustris D. Gerardi Reinstij Amsterodamensis'* (Fig. III, top left), in: Athanasius Kircher, *Oedipus Aegyptiacus*, Rome 1652-1654, vol. III, 1654, facing p. 30 (photo: Universiteitsbiliotheek, Leiden).

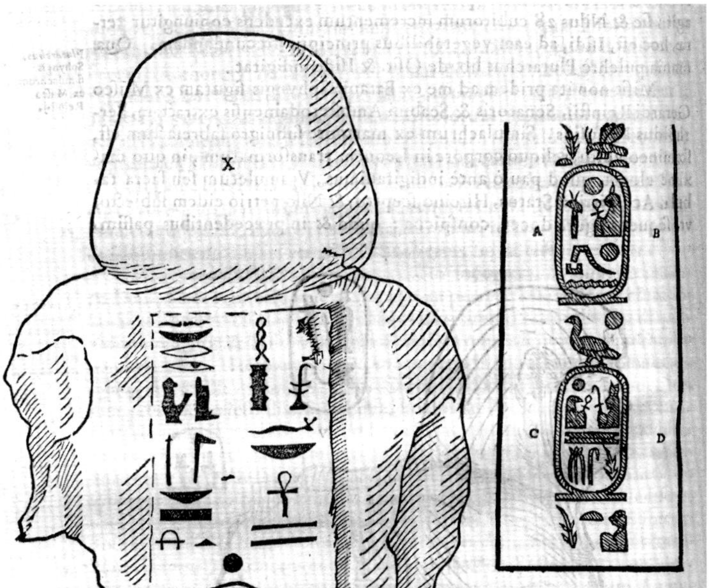

9

'Sphyngis simulachrum ex Museo Reinstij'
Athanasius Kircher, *Oedipus Aegyptiacus* (Rome 1652-1654, vol. 3, p. 458)
(photo: Universiteitsbibliotheek, Leiden).

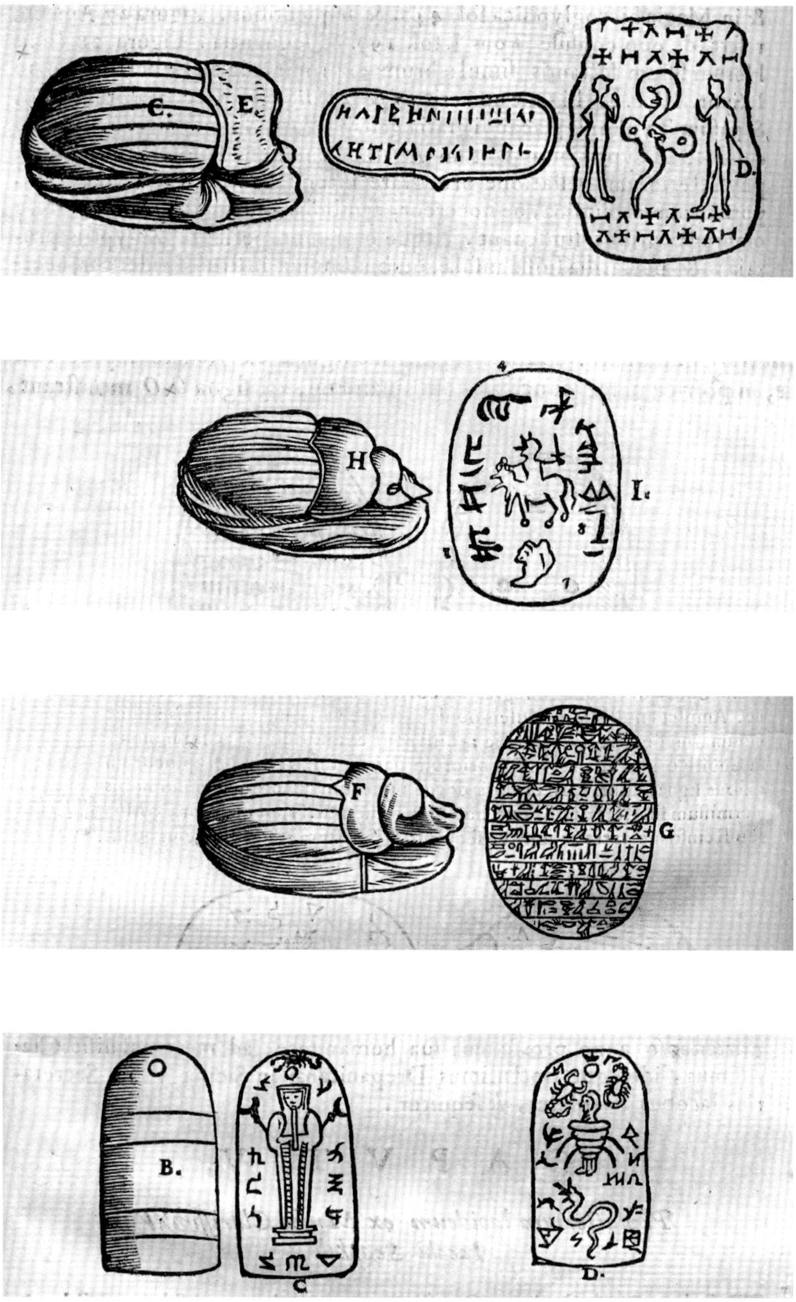

10

'Figurae ex Reinstiano museo'
Athanasius Kircher, *Oedipus Aegyptiacus* (Rome 1652-1654, vol. 3, p. 524-525) (photo: Universiteitsbibliotheek, Leiden).

raised the eyebrows of modern historians of science, but it was certainly not an isolated case.[40] The *Corpus* was translated into Dutch no less than three times: in the circles of the Antwerp printer Christopher Plantin (c. 1580), the Alkmaar physicist Cornelis Drebbel (1607), and the merchant-mystic Abraham van Beyerland (1643), respectively.[41] As we shall see, the discussion of the Adamic origin of pictography had a particular Dutch chapter.

The linguistic point of view

Besides the archaeological view engendered by the hieroglyphs in the Netherlands, there was a particularly lively discussion about the reconstruction of the 'primitive language' from before the Confusion of Tongues, and a somewhat less ambitious search for a universal script that could restore some of the lost linguistic unity. Here, theories about humankind's first speech looked to Dutch rather than Hebrew as a *lingua sancta*. Authors pleading for the emancipation of the Dutch vernacular, such as Simon Stevin (1548-1620) and Dirck Volckertsz. Coornhert (1522-1590), developed the theory that Dutch was just as 'primeval' as Hebrew: in contrast to the ornate and mannered modern Romance languages, Dutch exemplified stoic brevity and thus stood in a more direct relation to the things themselves. The most explicit writing on the subject was *Origines Antwerpianae* (Antwerp 1569) by Joannes Goropius Becanus (1519-1572), who interpreted compound words, which are so frequent in Dutch, as combinations of the basic monosyllabic signs.[42]

As a matter of course, since Dutch was the language spoken by Adam, Becanus connected it to the hieroglyphs and their invention by Hermes Trismegistus, referring to the Isiac table in particular.[43] In his *Hieroglyphica* (1580), he disagreed with those who credited Moses with the invention of writing: the prophet's Egyptian education itself testified to the earlier use of hieroglyphs. Becanus thus derived Thot from the Dutch ''t Hoot' or 'het Hoot', meaning 'the head' and clearly referring to God as the head of the universe. Adding scholarly clout to his etymologies, the author connected them to works of art, noting that many ancient Hermae – statues of Hermes – were simple squared columns lacking all detail save a head: 'Do you see how the vocal symbol coincides with the statue and how aptly each sign expresses the same notion?'[44]

Adamic language gained urgency towards the year 1656. Those who took a literal view of biblical revelation had lofty expectations: the renewal of Noah's covenant, putatively instituted 1,656 years after the creation of Adam, was supposed to take place and divine knowledge, lost after Babel, would be revealed once more. In the Netherlands, this belief surfaced in the controversy surrounding the ideas of the Antwerp-based scholar Isaac la Peyrère (1596-1676). In his *Prae-adamitae*, printed in Amsterdam in 1655 and promptly translated into Dutch, La Peyrère held that there had been people before Adam, which explained why the earth was extensively inhabited so soon after Adam's creation.[45] While still in manuscript form, these ideas sparked a Dutch discussion about how humankind had populated the entire world:[46] in 1642, Hugo Grotius penned a refutation arguing that all people descended from Adam, based on his observation that the American languages developed from the Germanic ones.[47] In response, Johannes de Laet (who had met La Peyrère in Leiden in 1646) severely criticised Grotius's work, demonstrating that the Amerindian languages had no affinity to Hebrew or the European vernaculars.[48] For our discussion, it is relevant to note that Amerindian writing was generally seen as pictographic. Reacting to Grotius, the Harderwijk professor Georg Hornius (1620-1670; no relation to

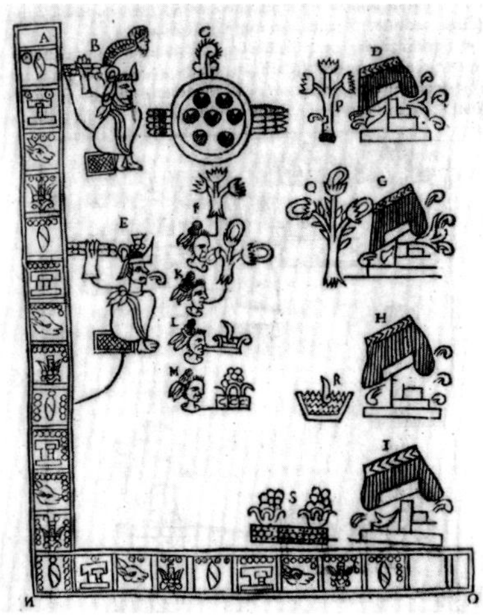

11

Sample of Mexican writing

Athanasius Kircher, *Oedipus Aegyptiacus* (Rome 1652-1654, cap. IV, p. 33) (photo: Universiteitsbibliotheek, Leiden).

Heurnius) discussed the role of pictures in Maya and Aztec writing in order to derive American letters from Chinese.[49] He thus echoed Kircher, who printed an image of Aztec signs in *Oedipus Aegyptiacus* and argued that Egyptians ended up in the Americas via China (fig. 11).[50] Using pictography as a central argument, Hornius posited in his *De originibus Americanis* (1652) that the difficulty in learning the great number of Chinese characters (in his view at least 120,000) had led the Amerindians to use simplified characters. Whereas Chinese signs could express abstract notions, American pictograms were merely images of visible objects:[51] 'I confess that the way of writing among the ancient Mexicans and Chinese was not the same, but it was also not very different (...). The Chinese write with a paintbrush and one character comprises many letters and makes an entire word'.[52] In the grand debate about 'world history' and how to connect the civilisations of the New World to those of East Asia and the *terra australis nondum cognita*, the supposed hieroglyphic origin of the Chinese and Amerindian writing systems supported the catholic argument that all writing, language, and culture derived from a common Middle Eastern source.

Hornius's main opponent was another scholar inspired by La Peyrère, namely Isaac Vossius (1618-1689).[53] Dubbed a pocket-sized Kircher, Vossius was the most influential Dutchman to revise notions of sacred history and he elaborated many of the Jesuit's ideas.[54] In 1642 he even went to Rome, attracted by the 'Museo Kircheriano' with its ancient Egyptian and Chinese objects.[55] Vossius's text 'De artibus et scientiis Sinarum' (1685) was an implicit reaction to Kircher's monumental book on China.[56] Praising the clarity of Chinese draughtsmanship in particular, the Dutch scholar became the first European to rank Asian visual art above that of the West.[57] He also claimed that Chinese writing represented an uninterrupted cultural tradition of five thousand years: the letters might be 'very laborious, but the benefit reaped from them is in accordance with the labour, because even if one had learned only [Chinese] characters, one would seem to have accomplished more than if one had acquired knowledge of all languages that exist or have existed'.[58] Here, Vossius appears to have adopted Kircher's idea that Chinese letters, like pictograms with universal meaning, were independent of spoken language. The characters or *beeldletteren* had developed from images: 'The ancient Chinese based their letters on all things that present themselves to the eye, and revealed and expressed the thoughts of their mind solely through the varied combinations and compositions of these things'.[59] Kircher, who had the Haarlem artist Willem van der Laegh (1614-1674) engrave his characters, observed that the Chinese did not write with pens, but with 'paintbrushes'; their writing should rather be termed 'painting'.[60] His illustration of the Chinese way of writing therefore included a monkey, symbolising visual art as the 'ape of nature' (fig. 12).[61] In the earliest stage of Chinese civilisation, these images would have been painted by the mythical first Emperor, 'Fohi', obviously at the instruction of someone from the Near East, perhaps Noah's son Cham. In the course of the centuries, his figurative forms had become increasingly abstract:

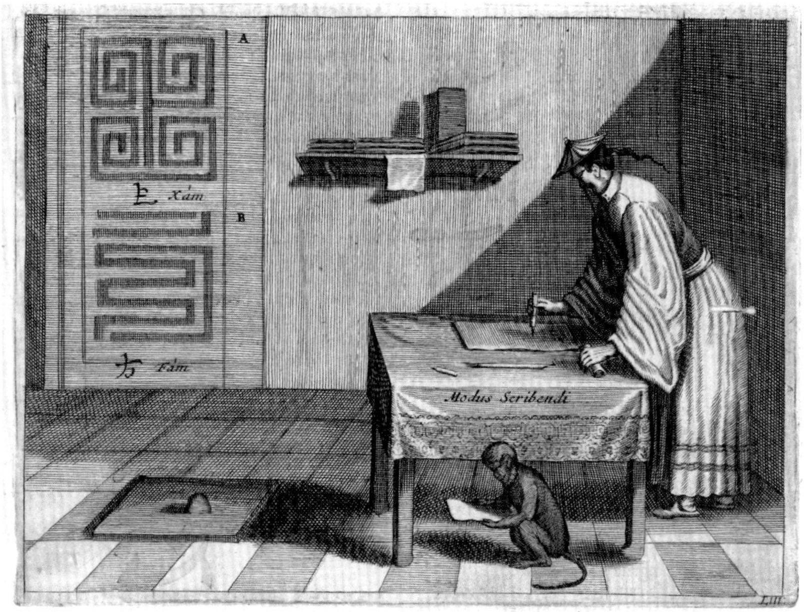

12
'Modus scribendi'
Anathasius Kircher, *China illustrata* (Amsterdam 1667) (photo: Universiteitsbibliotheek Amsterdam, Special Collections).

> The origins of their writing (...) the [Chinese] learned from Father Cham, and from Mercurius Trismegistus, counsellor to his son Nefraim and the first instigator of pictograms. And surely, those ancient characters of the Chinese, which are equal to pictograms in all aspects, are an important proof for me to believe in this [Egyptian origin]'.[62]

Upon concluding his 'Grand Tour' and returning to the Netherlands, Vossius tried to rival Kircher's network in various ways, seeking contact with other Jesuits and with the governor of the Dutch East India Company to obtain manuscripts and printed texts on China.[63] However, he failed to consult the East Asians in the Netherlands, such as the Chinese Michael Shen Fuzong (c. 1658-1691), who visited the playwright Vondel in Amsterdam;[64] the two students at Leiden University, sons of a Dutch official and a Japanese woman, who were born and raised in Japan;[65] or the Japan-born mathematician of Harderwijk, Pieter Hartsinck (1637-1680). In contrast, Vossius's views on Chinese were informed chiefly by his vision of sacred history, which departed from Kircher's pious account. The Dutchman's controversial claims in 'De artibus et scientiis Sinarum' implied that Chinese writing had survived uncorrupted since before the Confusion of Tongues and even the Flood (which supposedly took place in 2349 BCE). This interpretation fuelled his doubts about the status of Hebrew as *lingua sancta*: since the Hebrew Bible was incompatible with Chinese chronology, he came to the radical conclusion that the text was unreliable.[66]

Vossius suggested replacing the accepted biblical chronology with an alternative one, but refrained from drawing the ultimate conclusion: if the arts and letters of China were as old as he assumed, Chinese could well have been the language God spoke to Adam when he named the animals and had perfect and complete knowledge. On the basis of Vossius's writings, however, the British architect John Webb (1611-1672) claimed that all languages are essentially derived from Chinese in his *Essay endeavouring the probability that the language of the Empire of China is the primitive language* (1669).[67] Webb – the only seventeenth-century architect whose designs were inspired by Chinese calligraphy – referred extensively to Vossius to argue that the Chinese had letters before the Flood, and that they lived too far away to be guilty of Babel.[68] Suggesting that their script might be older than the hieroglyphs, he ultimately concluded that the Chinese valued their written characters above works of art:[69]

> The Chinois give willingly great sums of money for a copy of their antient characters well-formed, and they value a good writing of their now [*sic*] letters far more than a good painting, whereby from being thus esteemed, they come to be reverenced. Insomuch that they cannot endure to see a written paper lying on the ground.[70]

In the Netherlands, ideas on Chinese art and writing, and their consequences for sacred history, became common and met with popular ridicule. In a 1698 satirical magazine, a mock inventory of East Asian imports listed a drawing 'in which 12 mandarins were sketched with a single stroke of the brush', echoing Vossius's praise of Chinese draughtsmanship, as well as 'the genealogical register of the Preadamites in Chinese characters'.[71]

Besides this linguistic approach to pictography – which, at the risk of being anachronistic, may be called 'historical' or 'comparative' – there was a complementary and more strictly philosophical approach. Gombrich has addressed how, according to Renaissance scholarship, vision conveyed a specific kind of pre-predicative knowledge, more fundamental than alphabetic language. Some images were regarded as direct reflections of Platonic ideas, something that made seeing the most truthful kind of perception.[72] The pictographic project entailed this great ambition of Renaissance scholarship: a language in which words and things stood in a direct one-on-one relationship.

It should first be noted that the tradition of art theory offered several arguments in support of this idea. In his treatise on painting of 1585, Romano Alberti mentioned Amerindian writing in order to call to mind the original designation of painting as ζωγραφία – writing not based on alphabetic letters, but on living things.[73] In the Netherlands, Gerardus Vossius's treatise on painting, 'De graphice' (1650), resorted to the same argument from etymology, explaining how the Greek verb γράφειν meant 'painting' as well as 'writing', to prove that the two activities were closely interrelated.[74] Karel van Mander saw training in draughtsmanship as even more fundamental than literacy (or 'grammar', as he called it), stating that 'the noble and learned art of grammar has been reared and nurtured swiftly by [draughtsmanship], which taught her where to obtain those letters and characters through which people in various languages adequately understand each other's meanings'.[75]

Echoing these artistic sentiments, some scholars developed the idea that because images represent nature more closely than language, they could be used as signs with universal meaning or Real Characters (expressing *res* rather than *verba*). This concept was introduced by Francis Bacon (1561-1626), who differentiated these characters from 'vocal characters', such as letters of the alphabet: 'it is the use of China (...) to write in Characters Real, which express neither letters nor words in gross, but things or notions'.[76] In the Netherlands, the polyglot and emblematist Hermannus Hugo (1558-1629) elaborated this idea in *De prima scribendi origine* (Antwerp 1617), the earliest systematic analysis of script from its origin to the printed book.[77] Although Hugo was indebted to Becanus's hieroglyphic theories, his true interest was the Real Character and he demonstrated his knowledge of the missionaries' observations about Chinese writing (a Jesuit himself, he translated East Asian mission letters into Latin). Noting that the Chinese wrote with brushes identical to those used for painting in Europe, Hugo aspired to a universal script:[78] 'When individual letters are qualified to denote not words, but the things themselves, and when all these [letters] are common to all people, then everyone would understand the writing of the various peoples even though each one would call those things by very different names'.[79] The ideal of universal communication through pictures may have inspired his emblem book *Pia desideria* (1624) illustrated by Boëtius à Bolswert (c.1580-1633), which became popular throughout the Netherlands.[80] In any case, another Antwerp missionary, Johannes Steegius, eventually tried to 'explain the Christian doctrine more accessibly through a picture language' (in *De Christelycke leeringhe verstaenelycker uytgeleyt door eene beelden-sprake,* 1647). Although aimed at the conversion of illiterate Flemish peasants, this project demonstrated the impact of the Jesuits' overseas experiences.[81]

Vossius's *De arte grammatica* explored further how pictography could aid cross-cultural communication. To explain the possibility of Real Characters the author discussed hieroglyphs, the numbers in arithmetic, and astrological symbols representing planets.[82] Vossius's highest hopes, however, were for Chinese writing. From the fact that a single script was used for unrelated languages, he concluded that each sign was a miniature picture of the object it referred to:

> (...) for the Chinese, there are no fewer letters than there are words: however, they can be combined together, so that their total number does not exceed 70,000 or 80,000 (...). The Chinese and Japanese, although their languages differ just as much as Hebrew and Dutch, still understand one another if they write in this manner. For even if some might pronounce other words when reading, the concepts would nevertheless be the same. Indeed, now as people of different languages who see the same thing, understand the same thing, likewise, those who see the sign of a thing would have the same understanding of it.[83]

Since Chinese characters express concepts without the medium of language, they are similar to paintings: 'In the same sense, all people obtain the same concept from the painting of a human figure, a horse, a tree, or a house'.[84] Because Chinese signs expressed things directly, Vossius concluded that the number of letters equalled that of the categories of objects: it was possible to learn 'as many characters as there are things', if done from an early age onwards.[85] This idea must have seemed plausible to his contemporaries as the art of memory itself was based on the visualisation of concepts. After all, the mind's capacity to store images had been trained in Western scholarship since antiquity, and the Chinese, who seemed to be superior in other arts too, might have brought the artificial memory to perfection.[86]

Although the Jesuits in Rome and Paris dominated scholarship on China, linguistic speculation was a decidedly Northern European occupation. Together with Bacon, Hugo and Vossius were the first European scholars to propose a universal language based on Chinese.[87] Their ideas were particularly influential in England, where the philosophers John Wilkins (1614-1672) and George Dalgarno (1626-1687) assiduously developed a complete Real Language. Dalgarno, who also worked on sign language, addressed the 'far-seeing Chinese': 'do not, we beseech you, render blind us one-eyed ones, anxious as we are to look more intently at your affairs, by displaying enchanting images in place of letters'.[88] For this author, the validity of pictograms lay in their antiquity: 'As evidence that *real* characters were in use before *vocal* characters, I may cite the most ancient Chinese and Egyptian peoples'.[89] Probably mindful of Vossius's search for the roots of language in principles that were universally understood, Wilkins praised *De arte grammatica* for exploring the possibility of a 'natural grammar'.[90] For the convenience of Englishmen, a 1657 polyglot Bible likewise advocated Vossius's proposal for a universal language.[91]

The encyclopaedic point of view

In this quest for a historically primitive language and a philosophically universal one, the two strands of research coincided in scholarly projects devoted to mapping out human knowledge – historical, natural, and artificial – in an ordered system. One example was Jan Amos Comenius's (1592-1670) ambition to classify and communicate knowledge of the world through simple images as outlined and illustrated in his *Orbis sensualium pictus* (1658) (fig. 13).[92] From 1656 onwards, he developed his educational project in Holland, possibly attracted by the writings of Gerard Vossius, whom he had hoped to visit.[93] As Slaughter points out, this project was closely related to a universal language scheme.[94] Comenius expected Real Characters to be the basis for the reconciliation of humankind. Speaking of the 'pleasing light' granted by the 'knowledge we received about the symbolic characters used by the Chinese, namely that these characters help men speaking different languages to understand one another', he asked: 'If this is a good thing, why should we not devote ourselves to the discovery of a Real Language,

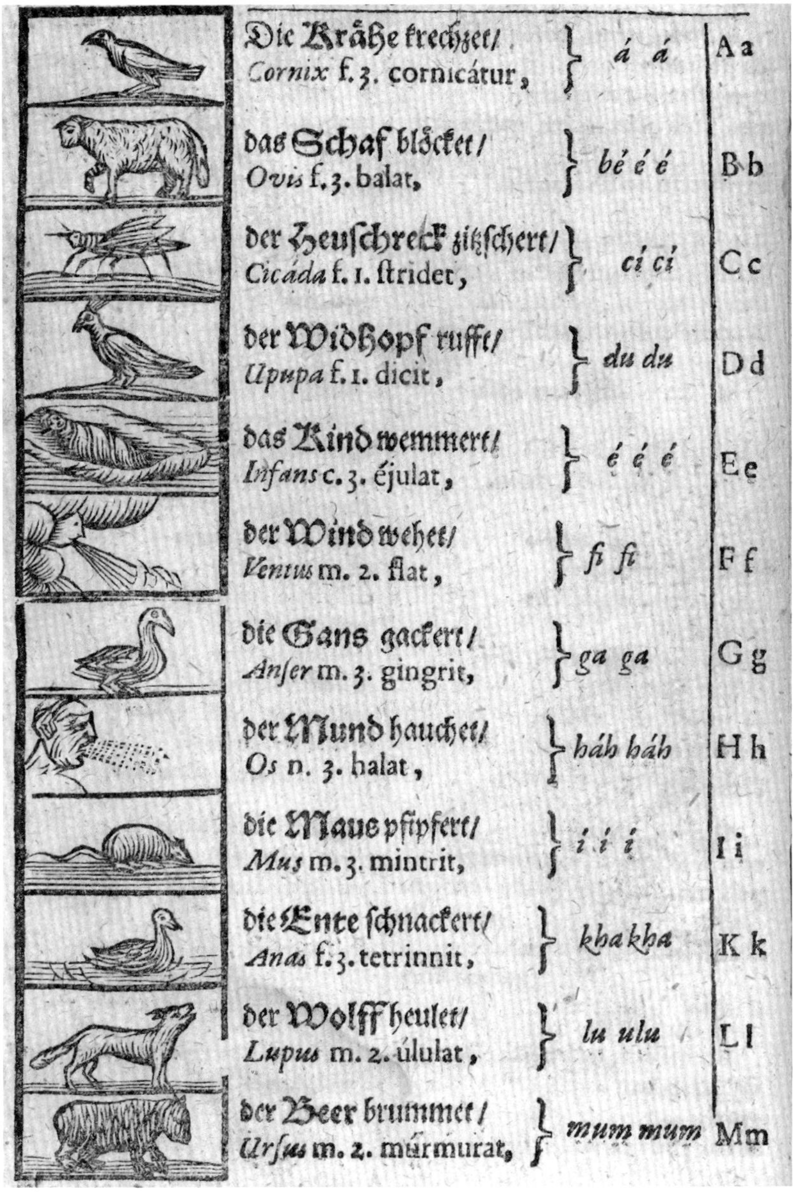

13
'Ein lebendiges und stimmbares Alfabeth'
J.A. Comenius, *Orbis sensualium pictus quadrilinguis* (Nuremberg 1679) (photo: Universiteitsbibliotheek Amsterdam, Special Collections).

which all men equally should understand?'[95] Ultimately, Comenius wanted to replace education based on alphabetic letters with one based on pictures. *Orbis pictus* thus begins with an admonishment to the reader:

> Before all other things, you must learn the simple voices that constitute human speech: those to be produced by animals, and imitated by your tongue, and painted by your hand. That is what we shall seek in the world, and we will observe every object: here you have a living and speaking alphabet.[96]

In relation to his ideal of *pansophia*, or universal knowledge that was communicated through universal language, Comenius envisaged *pantaxia*, an ordered system of all things.[97] This brings us to the relations between pictography and encyclopaedism. His educational treatise appeared in German as *Die sichtbare Welt* (1662),[98] and thus is linked to other books with encyclopaedic pretensions in the Netherlands. We are first reminded of Van Hoogstraten's treatise on painting subtitled *De zichtbaere werelt* (The visible world). Its encyclopaedic scope surfaces in the organisation of the chapters, dedicated to the nine Muses and each associated with one of the planets; the elaborate title pages provide visual résumés of the contents of each chapter and probably served a mnemonic function. Emblem books often shared such a format following a physical or philosophical structure, such as the planets, the ages of man, the hierarchy of creation, or simply the days of the week.[99] One example is Jacob Harrewyn's *'t Jaar d'xii maanden, vii dagen en iv getyden* (The year in twelve months, seven days and four seasons; Amsterdam 1698); this author also published *De xxv letteren van het ABC* (The twenty-five letters of the ABC; Amsterdam 1694), with 25 chapters devoted to the letters of the alphabet.

These emblematic works suggested that a sequence (alphabetic or otherwise) of images would be able to represent the variety of Creation in a structured manner. Emblem books titled *Cabinet*, *Schatzkammer*, or *Musaeum* were similar to taxonomic projects such as cabinets of precious rarities and 'paper museums' reproducing antiquities – to quote a recent historian, 'There were also more exotic creatures and delicate monsters than could ever fill any princely Wunderkammer'.[100] One such book in which word and image supported each other in an encyclopaedic fashion was Kircher's *Arca Noë* (Amsterdam 1675), which presented the Ark as the first museum. The images by Lieven Cruyl show the precise location in the boat of more than a hundred species (fig. 14).[101] It is not difficult to situate this project within the larger context of Kircher's linguistic interests. In the Ark, the animals still had the names given by Adam when he first coupled words to things.

Slaughter's analysis of the cross-pollination of scientific taxonomy and universal language schemes raises the question of whether the organisation of encyclopaedic collections in the Netherlands was related to the pictographic ideas formulated by Hugo and Vossius.[102] The activities of some collectors suggest it was, at least in a general sense. Responding in part to the division of the planetary spheres, the Antwerp physician Samuel Quiccheberg (1529-1567) initially envisaged a comprehensive taxonomy for 'naturalia, mirabilia, artefacta, scientifica, antiquitates, and exotica'.[103] Other collectors soon drew a parallel between similar taxonomic views and their interest in language: probing for the meaning of the objects in their possession, for instance, they too tried to decipher the hieroglyphs. Trained in a Bolognese collection based on a fascination with Egypt and the primitive language, the Dutchman Aelius Vorstius (1565-1624) may have initiated the enduring Northern European interest in an integrated approach to words and things.[104] Brinck's cabinet of antiquities thus came to include texts from East Asia. De Laet, who collected hieroglyphs and Chinese writing, even examined Amerindian script: as director of the Dutch West India Company he contacted the governor of Dutch Brazil.[105] Various Dutchmen eventually gravitated towards the museum of the Dane Olaus Wormius (1588-1655), who tried to decipher the runes by comparing them to hieroglyphs.[106] In this connection, it is worth remembering Jorink's observation that Dutch collections were usually housed in libraries and thus expressed the interplay of *res* and *verba*.[107] He points out Jan Woudanus's (1570-1615) image of the Theatrum Anatomicum in Leiden, which foregrounds the skeletons of Adam and Eve, possibly referencing the age in which man gave names to all things (fig. 15). Jorink even interprets a rarity in the theatre, the philologist Isaac Casaubon's (1559-1614) bladder, as testimony to the integrated approach of language and *naturalia*. In any event, the collection contained books written in Chinese and Egyptian artifacts. We should not be surprised to find that its keeper was Heurnius, whose influential hieroglyphic work has been mentioned above.[108]

By now it should be obvious that more was at stake than scholarly exoticism. The desire to adjust things to words responded directly to didactic theories. Historians have pointed to the connection between cabinets of curiosities and the reform of pedagogy in the seventeenth century:[109] the didacticism of authors such as Comenius was based on the vernacular and what German scholarship calls *Realienkunde*, to teach by demonstration using images and objects (also useful, for instance, in educating the Amerindian heathens).[110] As 'the sciences find easier entrance through the eyes than through the ears', the 1666 catalogue of a Schleswig collection compared the Book of Nature to a learning book for young people.[111] In England, similar statements inspired the Royal Society, with its motto *Nullius in verba* (on the basis of nobody's words), to compile a scientific cabinet. Most eloquent in this respect was Robert Hooke's (1635-1703) statement that material objects should be classified according to the order of nature, just as words are organised by grammar. He aspired to 'as full and compleat a collection of all varieties of natural bodies as could be obtain'd, where an inquirer (...) might peruse, and turn over, and spell, and read the book of nature, and observe the orthography, etymologia, syntaxis, and prosodia of natures grammar'.[112]

In the Netherlands, the concern with education through a grammar of nature found expression in the metaphor *schoole der werelt* (school of the world), to use the title of an emblem book with edifying exegeses of objects from everyday reality.[113] The metaphor harked back to Comenius's statement on the living and speaking alphabet of nature: to this scholar, 'the physical world is a visible showplace of God'.[114] Such a view was welcomed by Calvinist readers, who looked for divine revelation not in the stories of saints or devotional imagery but in God's Creation. After all, had not Calvin himself called upon 'the speechless teaching of heaven and earth'?[115] As Jorink demonstrates, the metaphor of the Book of Nature inspired scientific culture throughout the Netherlands,[116] and Bakker observes analogous developments in seventeenth-century landscape painting.[117] Treatises of art theory support their conclusions. The British draughtsman Henry Peacham (1576-1643), who often crossed the North

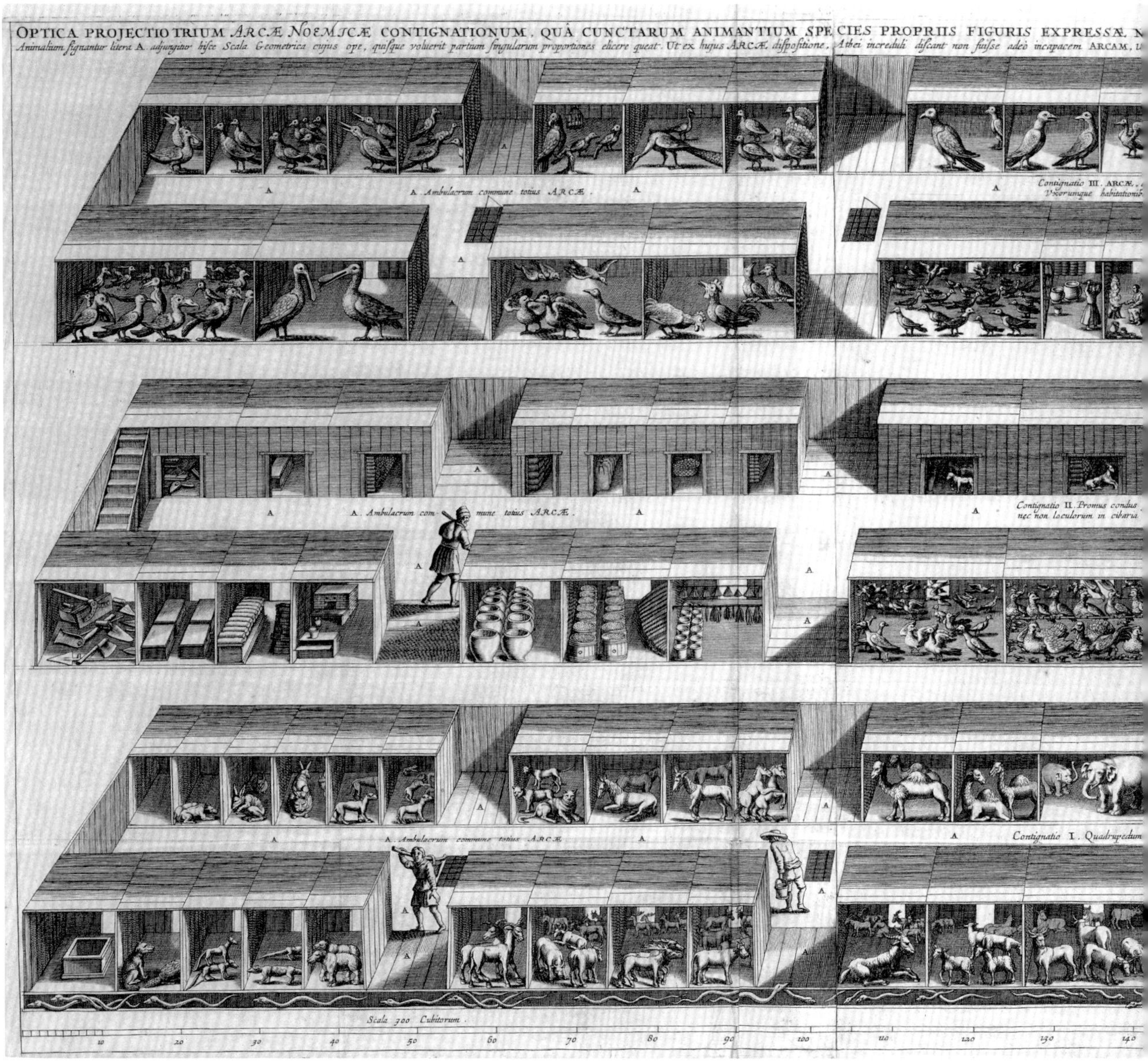

14
Lieven Cruyl, *Noah's Ark*
Athanasius Kicher, *Arca Noë* (Amsterdam 1675)
(photo: Koninklijke Bibliotheek, The Hague).

Sea and counted Dutch artists among the best he knew, attested that 'since [painting] is onely the imitation of the surface of nature, by it as in a booke of golden and rarelimned letters (...) wee reade a continuall lecture of the wisedome of the almightie Creator'.[118] He shared his outlook with Catholics such as Cardinal Gabriele Paleotti (1522-1597), who observed that the Bible was written so late (i.e., 2,370 years after the Creation) because the Creator first wished to teach mankind through the things themselves 'that can be universally understood by all (...) the silent voices that come from the things created by God'.[119] The link to pictograms and emblems, then, was considered obvious: a 1638 English emblem book stated that 'before the knowledge of letters, God was

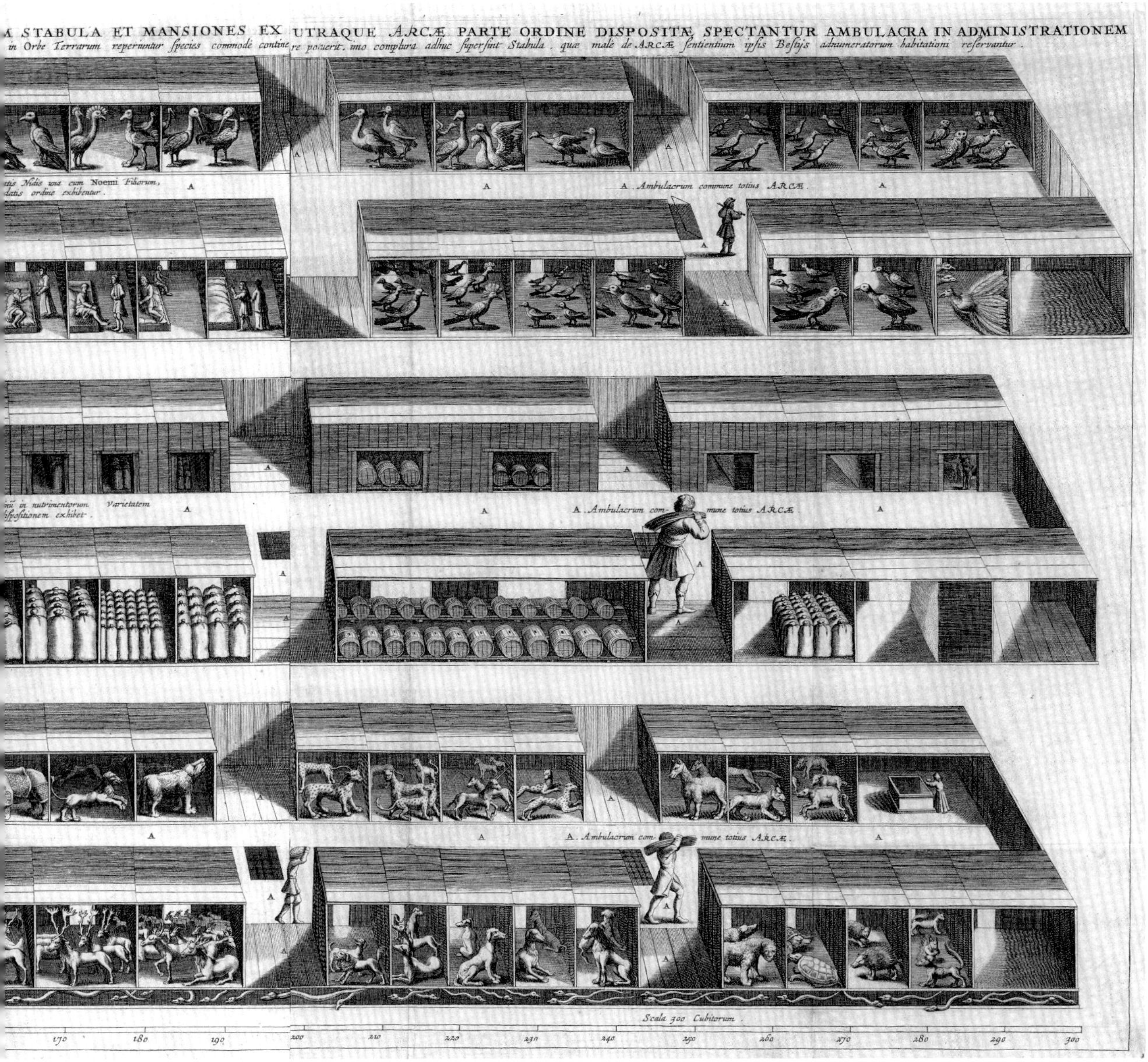

known by hieroglyphicks; and, indeed, what are the heavens, the earth, nay every creature but hieroglyphicks and emblemes of His glory?'[120]

The 'reading' of the Book of Nature thus inspired an essential metaphor for our discussion, namely the equation of the elements of the visible world, these 'rarelimned letters', with an alphabet. Already in 1598, an explanation of the catechism of the Dutch Reformed Church stated that 'all visible creatures are like letters in a beautiful book that refer us to their Creator'.[121] According to the Leiden professor Friederich Spanheim (1632-1701), the existence of God is revealed in 'this book of nature, in which just such a beautiful combination of creatures can be found as one made by so many letters that constitute a wonderful meaning'.[122] He clarified the metaphor of

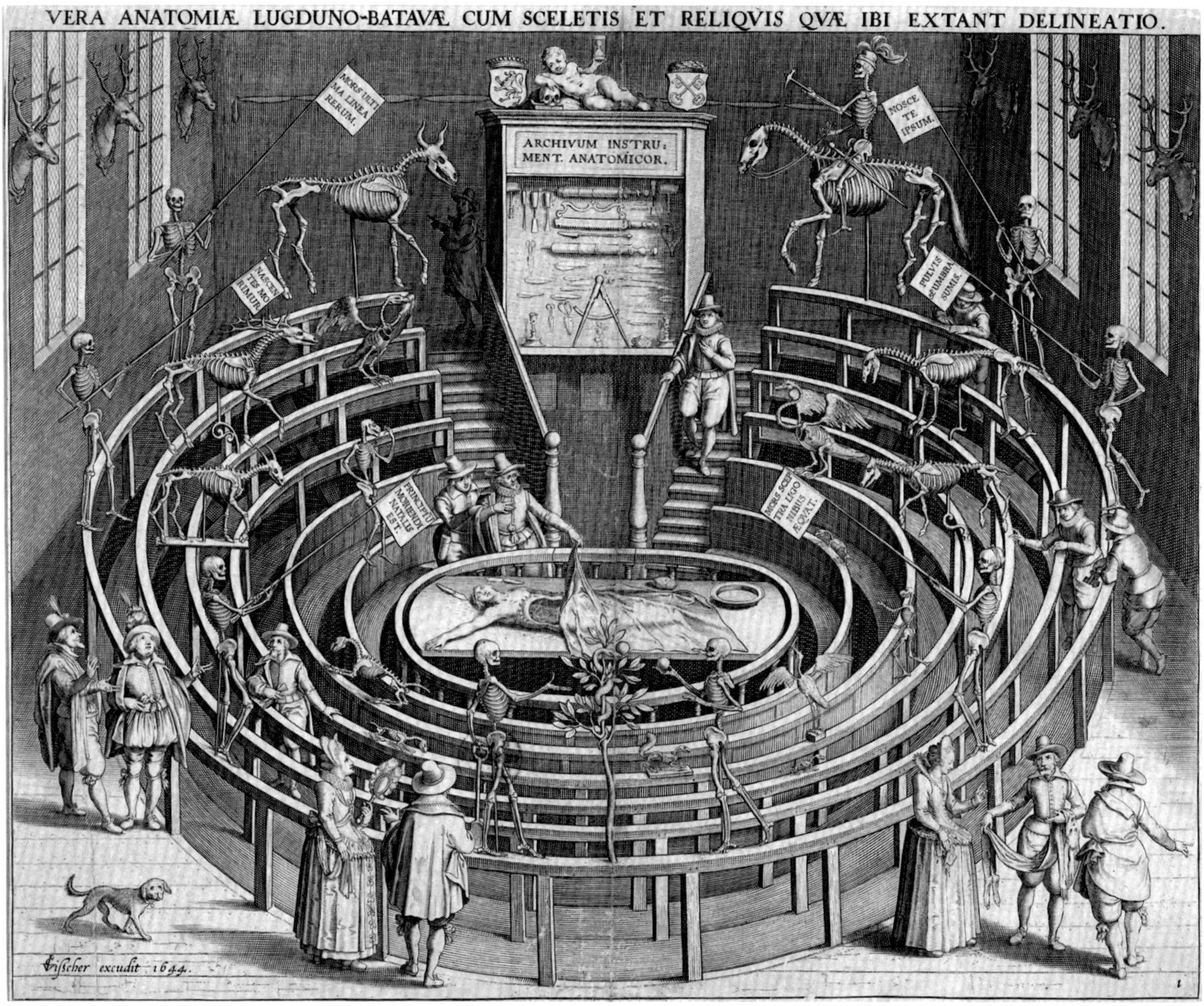

15

Claes Jansz. Visscher, after Joannes Woudanus, *Leiden's anatomical theatre,* 1644

engraving, 38 x 32 cm

(photo: Universiteitsbibliotheek, Leiden).

seeing visible things as letters, stating that they only obtained their meaning through a certain order. Atheists who maintain that the order in nature does not prove God's presence are as witless as those who think that a 'heap of characters', or the letters in a printer's shop, 'thrown in the air and there confused and mixed together, could combine themselves so evenly and with such order that one could see those letters and characters assemble into an edifying book or a well-made poem'.[123]

The logical point of view

The didactic importance allotted to vision that seems to inspire the 'alphabet of the world' metaphor points to a fundamental aspect of education: the art of memory and the theories of invention and argumentation that were based on it. Here, as we shall see, notions about 'nature's grammar' converge with ideas about series of images that reflect principles of logical organisation.

As, according to Aristotle, pictures leave a more lasting impression than words in the mind, the central procedure in early modern mnemotechnics was to transform written or oral information into mental images of every kind and organise them in an imaginary spatial architecture, responding to Cicero's statement that 'one could use the places [of memory] instead of the wax [of the writing tablet] and the images instead of the letters'.[124] The formation of mental images was interpreted as the condensation of inchoate sense impressions into clear-cut elements that could serve as the basis for new ideas. One of the masters of this technique, Giordano Bruno (1548-1600), described artificial memory as the presentation of thoughts, singled out for attention 'in the likeness of painting', in 'a distilled and developed order of conceivable species, arranged as statues, or a microcosm, or some other kind of architecture (...) by focusing the chaos of the imagination'.[125]

In the Netherlands, the importance of images for memory was stressed particularly in the education of children. Erasmus, for instance, stated that

> a youth will learn [texts] more eagerly and retain them better in his mind when one shows him the themes well illustrated before his eyes (...). Most children enjoy painted scenes of hunting: what a multitude of trees, animals, birds, and quadrupeds there is for them to learn while they play![126]

Later, Vossius, outlining how the teaching of Latin vocabulary should be enlivened by painted scenes from the Bible, referred to an image 'that represents the message of creatures (...) the sun, the moon, the stars, a man, a woman, a tree, a snake (...). When the child is asked to identify every individual thing in this image, he will also learn to name them in Latin'.[127] The author who came closest to putting these ideas into practice in an educational framework was Comenius. He pleaded for displaying pictorial summaries of textbooks in the schoolroom: 'numerous texts can be represented concisely by means of illustrations (...), which will daily exercise the students' senses and memory'.[128]

The relevance of mnemonics for the pictographic project is apparent in the notion that an artificial language based on totally arbitrary symbols would be extremely difficult to memorise. Pictograms, therefore, were the most reliable Real Characters. In the words of Gottfried Wilhelm Leibniz (1646-1716), the script of a universal language should be composed of 'geometrical figures and pictures of the type used in the time of the Egyptians, and today by the Chinese; pictures which cannot be reduced to a fixed alphabet, or to letters, which are a terrible impediment to memory'.[129] For our discussion, it is interesting to note that within the context of education, Leibniz referred to one of the teachers of artificial memory in the Netherlands, Lambert Schenkel (1547-after 1624).[130] This wandering scholar from Den Bosch, a Dutch rival to Bruno, offered a ten-day course in mnemotechnics, extracting an oath of secrecy from all the participants.[131] He devoted attention to the subject in *De memoria libri duo* (Douai and Antwerp 1593) and *Gazophylacium artis memoriae*

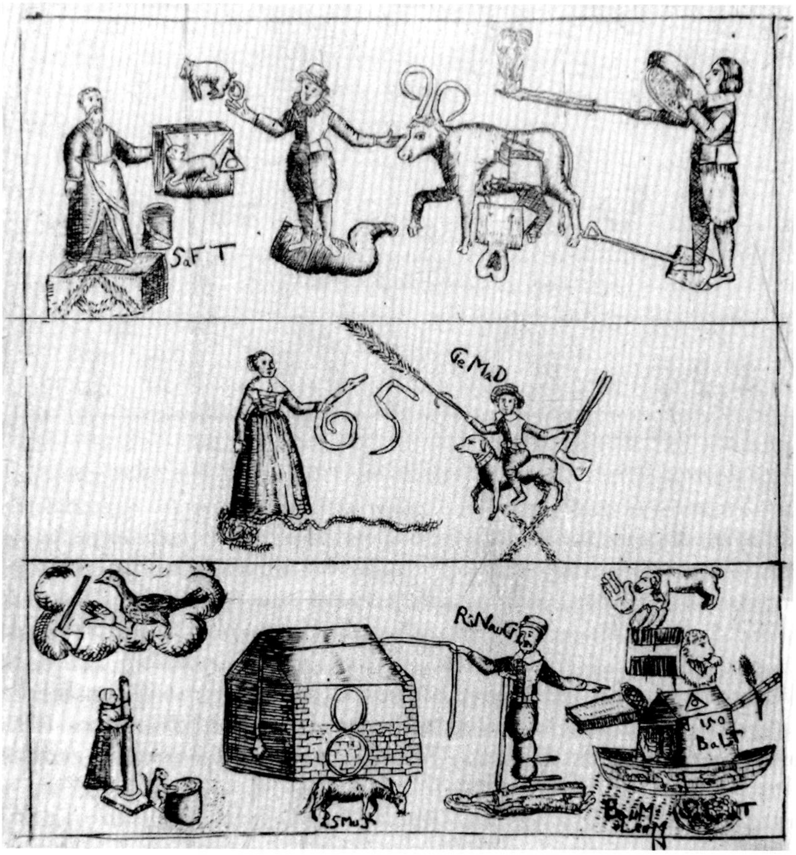

16
Mnemotechnical images of the patriarchs Adam, Seth, and Enos
Johannes Buno, *Tabularum mnemonicarum (...) simulacris et hieroglyphicis figuris delineata exhibetur* (Lüneburg 1662-1664).

(Strasbourg 1610).[132] Later, however, his works and reputation were deemed to have discredited the art of memory as being devoid of scientific value. Descartes even dubbed him 'that muddle head'.[133] Schenkel may have used imaginative rebus-like signs that were popular among artists from Leonardo onwards.[134] Still, it is important to point out that precisely the fantastic nature of the artificial memory's imagery was one of its central tenets: it assumed that bizarre images stick best in the mind.[135] This idea led to remarkable concoctions: the German educationalist Johannes Buno (1616-1697; not to be confused with Bruno) wanted to teach history by literally combining primitive images. To give just one example, his image representing the story of the patriarch Enos involved a dog and a large hand above a comb growing out of a figure's back, all on top of a house inscribed with numbers and letters (fig. 16).[136] Regardless of the question of whether these images fulfilled their mnemotechnical scope, they shed light on the *raison d'être* of some of the more abstruse imagery in seventeenth-century book illustrations and propaganda prints.[137]

Especially in German-speaking Europe, the presumed importance of images for mental processes inspired 'hieroglyphic bibles', 'image catechisms', rebuses, and figure alphabets basing devotional practices on techniques from the art of memory – even card games were used as mnemonic aids in this sense.[138] Van Sambeeck's figure alphabet, mentioned at the beginning of this article, was surpassed by far by Melchior Mattsperger's *Geistliche Herzenseinbildungen* (Augsburg 1685, translated into Dutch in 1720), which consistently replaces words in biblical verses with pictures. Here, the images should no longer be interpreted in an emblematic fashion by looking for multiple layers of meaning: they are little more than visually perspicacious aids for memorising Scripture (figs. 17-18).

The idea that thought processes are the reordering and combining of mental images made extrapolation from memory to logic and obvious step. Thus, the most exalted notions connected to pictography not only entailed the view that an image script makes a more direct imprint on the mind than alphabetic writing, but also suggested that a complete notation system could be the perfect model for thought. In regard to logical theory in the Netherlands, Rudolf Agricola's (1444-1484) writings, in which the visual arts play an important role, provided a foundation.[139] For the seventeenth century, however, the Dutch contribution can be valued only in an international context. Leibniz was the author who most comprehensively explored the themes discussed above in a coherent whole, including pictography, Chinese characters, primitive language, logic, and collecting. (Not surprisingly, he also became involved in the debate about sacred history sparked by La Peyrère and Vossius.[140])

A great deal of Leibniz's scholarly efforts was aimed at constructing a universal language, or *lingua characteristica*, to act as a system of notation for the 'alphabet of human thoughts' – for primitive concepts.[141] Leibniz's basic premise was that only by combining primitive concepts could one

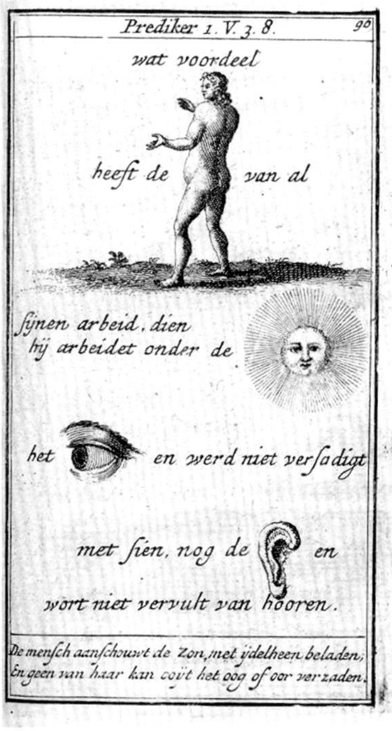

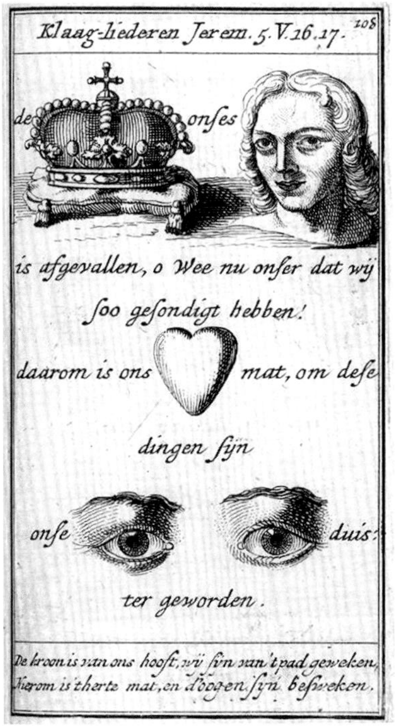

17 and 18

Two images

Gerard Puppius Hondius, *De kleine printbybel. Waar in door verscheide afbeeldingen een meenigte van bybelsche spreuken verklaart werden. Tot vermaak der jeugd, en om (...) de spreuken der H. Schrift by na zonder moeite in de geheugenis te brengen* (Amsterdam 1720 [orig. 1685]) (photo: Universiteitsbibliotheek Amsterdam, Special Collections).

produce more complex ones and thus gain new knowledge.[142] Studying these combinations according to what Leibniz called the 'combinatorial art' would reveal the workings of the mind, which he concluded from his belief that 'all our thinking is nothing but the connection or substitution of signs, whether they be words, or marks, or images'.[143] The signs for the primitive concepts, being the building blocks of all knowledge, should represent them directly and unambiguously. Dimly echoing Becanus's search for basic monosyllabic elements, Leibniz therefore asked for *figures significantes par elles mêmes* that could be understood immediately: a script with 'signs as natural as possible' (*notas quam maxime naturales*) that could be read without a dictionary.[144]

In Leibniz's work we again discern the influence of Kircher, who was one of his correspondents. Both authors looked to China as the utopian state in which the structure of language directly reflected the structure of reality. Ideas about the Chinese writing system were therefore an important source of inspiration for Leibniz's logical theories.[145] He must have been drawn to his Jesuit predecessor's assumption that the Chinese classified their pictograms according to the four physical elements: 'When they discussed of a fiery nature, they used serpents, asps, and dragons, which, arranged in this or that pattern, stood for this or that thing. To describe airy subjects, arrangements of birds were needed; for watery topics, fish; for the description of vegetable nature, flowers, leaves, and branches' (fig. 19).[146] Such a view of pictograms as classification symbols explains why Leibniz expected Chinese script to reflect the way the Chinese organised knowledge. Deciphering the characters would be a great contribution to epistemology.[147]

Taking into account the scholarly background of Leibniz's logical theory involving Chinese and Egyptian writing, contrary to what is commonly thought he did not invent symbolic logic: Kircher's *Ars magna sciendi* (Amsterdam 1669) appeared earlier. In its attempts at designing symbolic logic this book makes conspicuous use of pictograms. The title page features an image of Kircher's *alphabeta artis*, listing various symbols that index argumentative combinations, such as a heart for 'concordance' and an omega for 'the end' (fig. 20). The list concludes with small images of a man, an animal, a tree, and a stone to denote categories that respond to the hierarchy of creation; an angel's head refers to the angelic aspects of

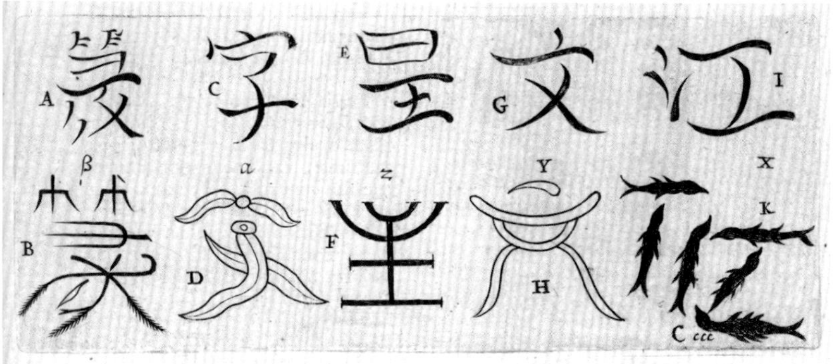

19
Willem van der Laegh, *Chinese characters*
Athanasius Kircher, *China illustrata* (Amsterdam 1667)
(photo: Universiteitsbibliotheek Amsterdam, Special Collections).

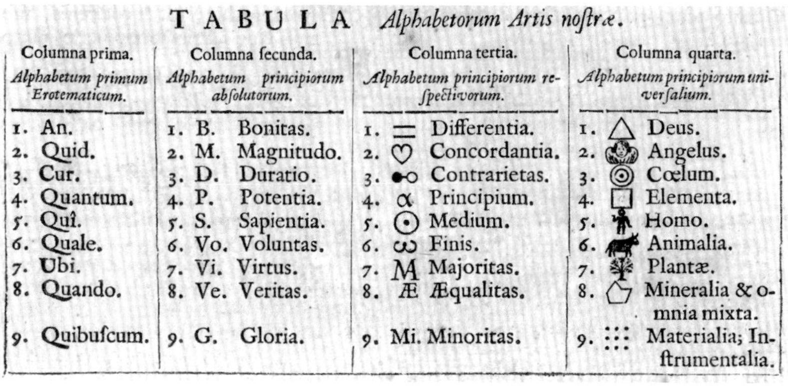

TABULA *Alphabetorum Artis noſtræ.*

Columna prima. *Alphabetum primum Erotematicum.*	Columna ſecunda. *Alphabetum principiorum abſolutorum.*	Columna tertia. *Alphabetum principiorum reſpectivorum.*	Columna quarta. *Alphabetum principiorum univerſalium.*
1. An.	1. B. Bonitas.	1. = Differentia.	1. △ Deus.
2. Quid.	2. M. Magnitudo.	2. ♡ Concordantia.	2. Angelus.
3. Cur.	3. D. Duratio.	3. Contrarietas.	3. ◎ Cœlum.
4. Quantum.	4. P. Potentia.	4. α Principium.	4. □ Elementa.
5. Quî.	5. S. Sapientia.	5. ⊙ Medium.	5. Homo.
6. Quale.	6. Vo. Voluntas.	6. Finis.	6. Animalia.
7. Ubi.	7. Vi. Virtus.	7. M Majoritas.	7. Plantæ.
8. Quando.	8. Ve. Veritas.	8. Æ Æqualitas.	8. Mineralia & omnia mixta.
9. Quibuſcum.	9. G. Gloria.	9. Mi. Minoritas.	9. ::: Materialia; Inſtrumentalia.

20
'Alphabeta artis'
Athanasius Kircher, *Ars magna sciendi* (Amsterdam 1669)
(photo: Universiteitsbibliotheek Amsterdam, Special Collections).

intelligence. Kircher developed an even more complete pictographic system in the manuscript *Novum hoc inventum quo omnia mundi idiomata ad unum reducuntur* (now in the Vatican Library). This includes other pictographic symbols for the verbs drinking (chalice), giving (hand), and seeing (eye), and for nouns such as hour (hourglass), city (skyline), law (scales), faith (cross), and the Holy See (chair) (fig. 21).[148]

Any utterance written in the alphabet of *ars magna sciendi* would lay bare the structure of its argument in an unambiguous way. This assumption illustrates the appeal of the supposedly pictographic Chinese characters to philosophers in Europe. Westerhoff calls attention to Leibniz's assumption of the 'structural isomorphism' between matter, thought, and language.[149] To Leibniz, just as reality is the combination of primitive forms, thought processes are the combination of primitive mental images, and they are to be expressed most unequivocally in combinations of universal characters. This idea harks back to one of Comenius's expectations about pictography; if Chinese characters aided cross-cultural communication, he wondered whether they would also lead to 'the discovery not only of a language but of thought, and what is more, of the truth of things themselves at the same time?'[150]

Thus, at the end of the seventeenth century, the different expectations about pictography merged into what Rossi calls the 'logico-encyclopaedic tendency'. Leibniz apparently derived from Comenius the thesis that 'the universal language and the encyclopaedia [i.e., a system for organising all knowledge] were coincidental'.[151] Leibniz's statement on his own *lingua characteristica* confirms this interrelationship: 'One who learns this [universal] language, simultaneously learns the encyclopaedia which is the true entrance to the sciences'.[152] Discussing logico-encyclopaedism further would take us too far afield from Netherlandish art. Yet, one aspect is relevant for our discussion, namely the link between memory, logic, language, and collecting.

We can only briefly point out that early modern culture assumed memory and collections to be similar, primarily because both were considered to be spatially extended.[153] As already noted, memory was taken to be an array of different *loci* where the objects to be remembered could be deposited. Quiccheberg's taxonomic text thus referred

21

'Pasigraphia'

Athanasius Kircher, *Novum hoc inventum quo omnia mundi idiomata ad unum reducuntur*, (Rome, Vatican Library).

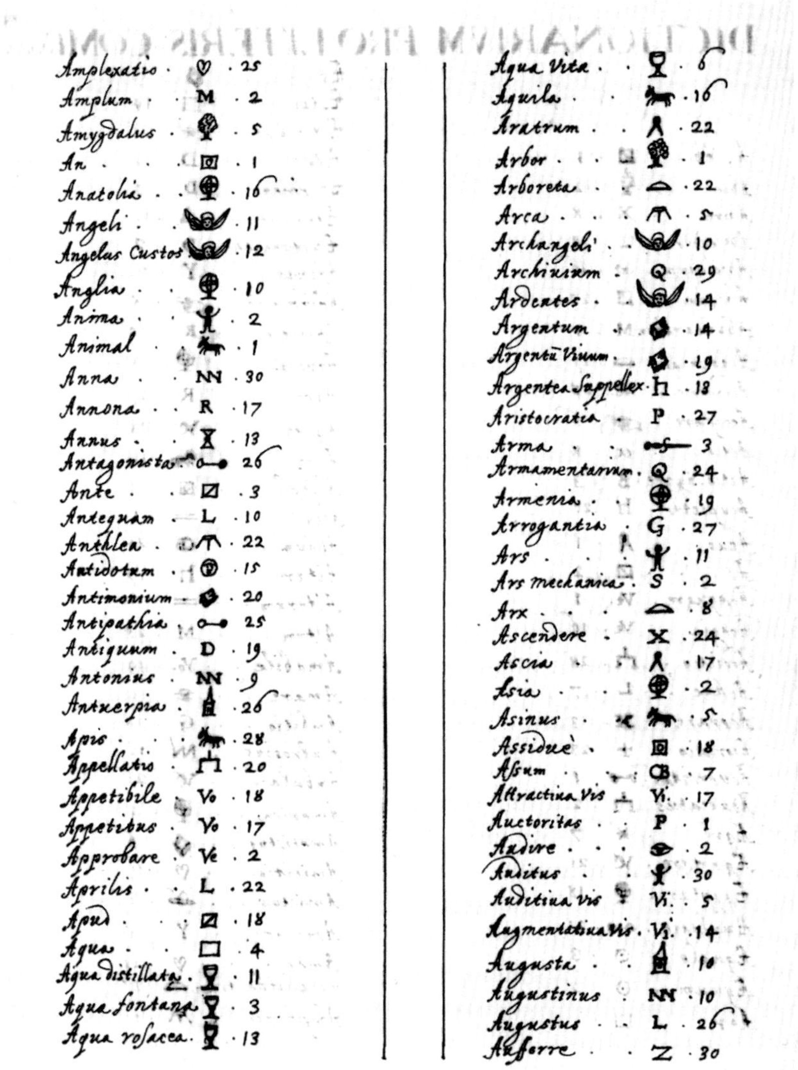

repeatedly to Giulio Camillo's mnemotechnical treatise *Teatro della memoria* (1550), which he called a 'semicircular museum'.[154] The supposed link between the structure of nature and the structure of thought explained the philosophical importance of *naturalia* collections: they would reveal how the 'grammar of things' overlapped with the 'grammar of the mind'. One is reminded of Hooke's metaphor that the Royal Society's display reflected nature's grammar. Collections that adequately organised natural objects would disclose how God had written his Creation. Accordingly, the poet and playwright Jan Zoet (c. 1608-1674) described an Amsterdam scientific collection as 'A book in which God has described himself splendidly. / The mortal mind is dumbfounded in the face of such letters'.[155]

The two-pronged concern with philosophical language and taxonomy explains why Leibniz was interested not only in pictography but also in the display of *naturalia* and *artificialia*. In 1675, he devised an outline for a museum, or rather a centre for universal scholarship comparable to Kircher's, including a library, theatre, garden, and laboratory to accommodate conferences and performances. These would extend, to name just a few, to demonstrations of the camera obscura, shadow projections, calculators, Chinese arts, and the taxonomy of insects established by the Dutch physician Johannes Swammerdam (1637-1680).[156] In tandem with the 'logico-encyclopaedic tendency', the end of the seventeenth century witnessed the proliferation of new ways of organising collections obeying the laws of physics or natural philosophy in general. In Hamburg, Johann Daniel Major's *Unvorgreiffliches Bedencken von Kunst- und Naturalienkammer insgemein* (1671) replaced the common taxonomy based on cosmological principles, such as the four elements, with a more properly scientific one based on materials. This effort seems to run parallel to a development Jorink detected in the Netherlands: he singles out Swammerdam, who looked for scientific constants in nature such as the arrangement of animals according to genus and the stages in their development.[157] Swammerdam's ideas were widely influential. For instance, Levinus Vincent (1658-1727), author of *Wondertooneel der nature* (Theatre of nature's marvels, 1706), structured his collection according to similar categories (fig. 22).[158]

The most literal equation between the principles of universal language and the principles of collecting transpired in the Royal Society. Its motto *Nullius in verba* expressed its scepticism of scholarly Latin and alphabetic language in general. The Society's secretary, Henry Oldenburg (c. 1617-1677), condemned the earlier Scholastic framework as follows: 'Let us emancipate ourselves from that servitude, and let us be eager henceforth to scrutinise nature and things as they are'.[159] And so when it desired to arrange its collection 'according to the exact method of the ranks of all the species of nature', the Society turned to John Wilkins's project for a universal language discussed above.[160] We may quote from his main philosophical work, *Essay towards a Real Character and a philosophical language* (1668), to stress the importance of pictographic theories. He was inspired by 'the men of China, who have for many ages used such a general Character, by which the inhabitants of that large kingdom (...) communicate with one another, everyone (...) reading it in his own language'.[161] In his own alphabet of Real Characters, Wilkins used symbols that are not truly pictograms but what may be called ideograms, signs representing the categories of reality. Still, as Slaughter points out, his project was based on the idea that the highest form of knowledge is observational rather than mathematical, and that the faculties of sense provide the essential concepts for thought (fig. 23).[162] Wilkins assigned specific symbols to the elements, such as stone, metal, and vegetation, in a manner not dissimilar to how Kircher thought that Chinese pictograms expressed classifications in nature. Wilkins's writings thus testify most literally to the link between the pictographic project and scientific taxonomy. It seems no coincidence that he built on Vossius and Hugo, two authors from the Low Countries, a region where writing with images was such a topical subject and where the logical aspect was complemented by hieroglyphic, linguistic, and encyclopaedic theories.

Conclusion

In *Gulliver's travels* (1726), Jonathan Swift parodied pictographic ambitions in his account of the school of languages at the Lagado Academy of Projectors. Here, words are simply abolished and replaced in conversation by the things they represent, bundles of which could then be assembled in advance as required for each day's business. Another fictitious travel account, *Le Nouveau Gulliver* (1730), returned to this idea describing a utopian island off the coast of China, called 'Babilary', which is ruled by women and uses a script based on pictures.[163] It seems that in the early eighteenth century, pictography was banned to the realm of imaginary voyages, thus validating Descartes's scepticism: 'Don't expect ever to see it in use, for that would presuppose huge changes in the order of things, and the whole world would have to be nothing other than a terrestrial paradise, which is fine only in the land of novels.'[164]

22
A. van Buijsen
engraving, 215 x 305 mm, title page of Levinus Vincent, *Het tweede deel of vervolg van het wondertooneel der natuur* (Amsterdam 1715) (photo: Universiteitsbibliotheek, Leiden).

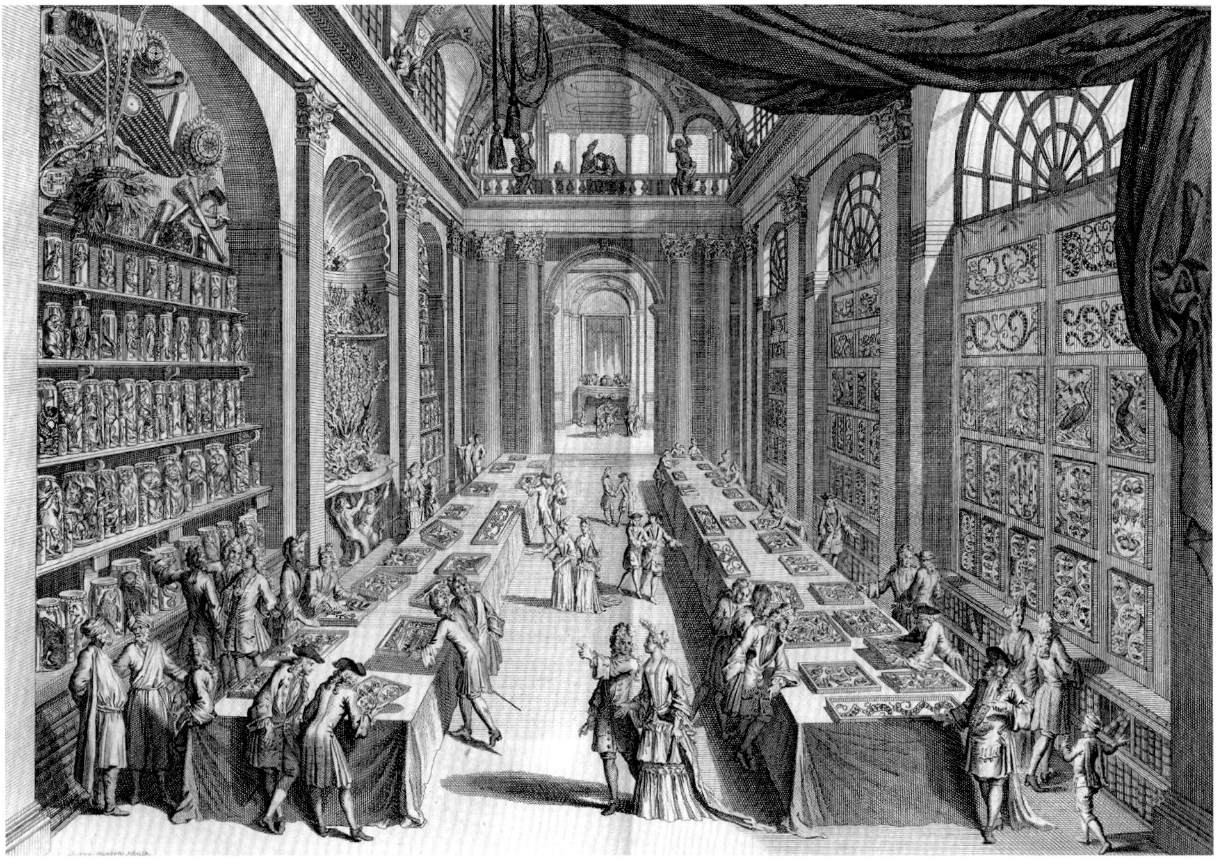

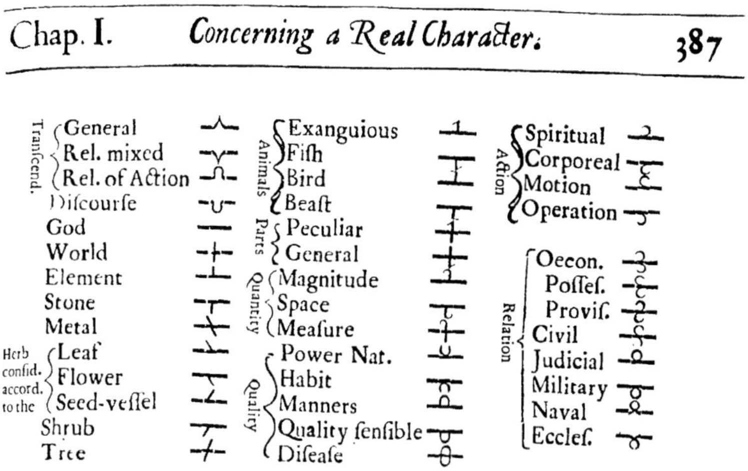

23
Ideograms
John Wilkins, *Essay towards a Real Character and a philosophical language* (London 1668).

This overview, focused on the Netherlands, has revealed that the possibility of pictography chiefly remained a promise, a conclusion that, from the perspective of a century after the 'linguistic turn' in the humanities, does not surprise us.[165] The urgency of the pictographic project for seventeenth-century Dutchmen can be understood only by looking at the political context – religious wars in Europe and the clash between revealed wisdom and new discoveries. The popularity of images of Babel from the second half of the sixteenth century onwards suggests that the attempt to write with pictures was just one chapter in a grander debate, questioning the status of Hebrew as *lingua sancta*. Fuelled by the rise of the vernaculars and the decline of Latin, it went hand in hand with a growing layman's interest in natural philosophy and the inventorying and categorising of the visible world.[166] Not only did Jesuits such as Van Sambeeck, Hugo, and Kircher think about pictograms as a means for the militant Church to compensate, by global proselytisation, for the souls lost to the Reformation, but Comenius too had a clear utopian program. Having arrived in Holland after travelling through war-torn Europe, he proposed a world council to institute a universal language whose fixed definitions would eliminate the ambiguous use of words.[167] In a millenarian context, pictography promised to undo the curse of Babel and regain Adamic dominion over nature. Such expectations explain how scholarly discussions about issues such as Chinese writing could spark theological controversy. According to the Frankfurt professor of logic Elias Grebnitz (1627-1689), every time the name of God was written in Chinese, a sin was committed against the Second Commandment which prohibits the depiction of the divinity.[168] When another Protestant, the Leiden-educated orientalist Andreas Müller (c. 1630-1694), finally found his key to Chinese, which was simultaneously a key to the hieroglyphs, he was accused of heresy and, in the hour of his death, threw his materials into the fire.[169] Isaac Vossius's Chinese utopia was also perceived as dangerous to accepted authority. One critic even deemed it 'spoiled by a false metaphysics' that would lead to 'nothing less than a total subversion of all religion'.[170]

It soon transpired that this criticism was unnecessary. From the scientific point of view, as far as Real Characters were concerned, according to the elder Vossius and Heurnius, the tens of thousands of signs of Chinese and Amerindian writing discouraged linguists and philosophers. Dalgarno, for instance, complained about the 'near infinitely burdensome' number of letters making them unfit for philosophical language.[171] Nor did the ambitions for a pictographic language that perfectly reflected the order of nature come true. But there was one, largely unintended, scientific spin-off. When Leibniz studied the signs used in the *Yi Jing* (Book of changes), his misguided ideas on Chinese writing inspired him to interpret them as Fohi's 'hieroglyphs' representing the Infinite and Chaos from which God had rescued mankind. He also thought, however, that he had revealed a system of binary relations, which became a source of inspiration for his ideal language: 'a new *characteristica* that will appear to be a continuation of Fohi's, and provide the beginning of the analysis of ideas and of that marvellous calculus of reason that I am planning'.[172] This *characteristica univeralis* catalysed the development of modern symbolic logic and eventually of computer languages, thus fulfilling the seventeenth-century wish for a language whose structure would be identical to the structure of thought itself.[173]

From the artistic point of view, one can hardly say that Samuel van Hoogstraten's reference to Egyptian, Chinese, and Meso-American pictography was inspired by an insight more profound than his ambition to raise the intellectual status of painting. Yet, as an echo of the many-sided debate sketched above, it also suggests that on a conceptual level art and scholarship were deemed to be closely interwoven – perhaps more so than historians often realise. In the Dutch Golden Age, discussions about the figurative arts were embedded in wide-ranging assumptions about the visual foundation of mental processes. On the one hand, the faculty to process sense input and make argumentative combinations was understood as the spatial arrangement of images stored in memory. On the other hand, ideas about pictographic signs highlighted the role of primitive concepts in language and in thought, as the basis for creativity and complex knowledge. In contrast to modern notions, pictographic ambitions also underscored the idea that the structure of language could directly reflect the structure of the physical world.

Finally, we have seen that in the Netherlands, the discussion on pictography occupied scholarly disciplines involving the most remote places in history and geography, as well as the farthest recesses of the human mind. This extended to the animals in Noah's ark, Hermetic hieroglyphs, runes, emblems, Amerindian writing, rebuses, and early attempts at philosophical language and computing. The quest for the grammar in which God speaks through his Creation involved *naturalia* and *artificialia* in a comprehensive manner. It is safe to say that the pictographic project could flower only in the scholarly hothouses provided by encyclopaedic collections, equipped to cultivate a crossbreed of words and things. It was doomed to wither.

Notes

*I am grateful to Jennifer Kilian and Katy Kist for correcting my English. Of course, any and all mistakes are mine alone.

1 Van Sambeeck 1663, 248.

2 'd'Egyptenaeren, Chinezen (...) hebben hare boeken met Zinnebeelden, in plaets van letteren, geschreven; en deeze wijze van uitbeelden is met de schilderkonst ook tot ons gekomen', Van Hoogstraten 1678, 90.

3 '[G]eschilderde verbeeldsels, welke by de woeste barbaren van Mexiko en Peru in gebruik waren', Huygens 1754, 79.

4 The Dutch term *beeldletter* was used in Goeree 1697 (ed. princ. 1670), 25, and Kircher 1668, 274.

5 Aristotle, *De anima* III, 3 (Bekker 431a).

6 Plato quoted by Coudert 1978, 65.

7 Rossi 2000, 173.

8 Van Hoogstraten 1678, 25.

9 Descartes to Mersenne (1639), quoted from Rossi 2000, 114.

10 Lambert Lombard (1506-1566) based hieroglyphs on Italian examples. See Cat. Liège 2006, 87-91, 409-410. He included some in *The refusal of Joachim's sacrifice*, oil on panel, 112 x 80 cm, Liège, Musée de l'Art Wallon. Zacharias Heyns (c.1566-before 1638) drew an obelisk with hieroglyphic symbols that constitute an easy rebus (*Album amicorum* for Abraham Ortelius, f. 104, Cambridge University, Pembroke College Library); see Van Dorsten & Hamilton 1980. Joris Hoefnagel (1542-1601), who defined himself as 'exornator hieroglyphicus', combined images of natural history with emblems and image alphabets; cf. Wilberg Vignau-Schuurman 1969.

11 These examples are discussed in Waterschoot 1978, 59. See also Iversen 1961, 159.

12 *Funerary monument of Canon Hubert Mielemans,* c. 1558, pierre de Theux, 285 x 130 cm, Liège, Église Sainte-Croix. On Van Heemskerck see Veldman 1977, 144-145. Joachim Beuckelaer painted hieroglyphs in his *Market scene with Ecce Homo,* 1570, Stockholm, Nationalmuseum.

13 Iversen 1961, 65; Alberti, *De re aedificatoria* X, VIII, 4.

14 Iversen 1961, 65; and more recently Curran 2007.

15 Van Heemskerck drew an authentic obelisk, then on the Capitoline Hill in Rome in 1532-1536; the drawing is now in the Staatliche Museen, Berlin.

16 The *editio princeps* in Greek was printed by Aldus in Venice (1505). Also Kircher (on whom more below) used the interpretations attributed to Horapollo. On the book's Renaissance reception see Iversen 1961, 47-49, 65, 72-75.

17 On Alciati and Pignoria, see Manning 2002, 116. In Antwerp, Christopher Plantin printed new editions of Alciati's book.

18 Iversen 1961, 75.

19 The second image refers to the idea that things need time to mature. The literature on emblems is vast; see the overview in Manning 2002, 13-36.

20 Vossius calls Pignoria 'amicissimus' in G. Vossius 1635, 140. On the two scholars' correspondence, see Van der Lem & Rademaker 1993, 468, Rademaker 1981, 218-219 and notes.

21 The book was issued by the Bavarian scholar Johannes Georgius Herwarth von Hohenburg (1553-1626); see Iversen 1961, 86. North of the Alps, the first description of the hieroglyphs in the vernacular was Herold 1554.

22 Jorink 2006, 276; Jorink 2010, 270-271; Jorink 2011.

23 All of the artists working for the Arundels either came from the Low Countries or were trained there, just like their librarian, the linguist Franciscus Junius. The artistic and scholarly world of the Arundel circle is discussed in my *Art and antiquity in the seventeenth-century Netherlands,* Amsterdam & Philadelphia, forthcoming (esp. chapter 1).

24 Arundel's agents were John Greaves and William Petty. See Howarth 1985, 138. Arundel also visited Kircher's collection in Rome; Reilly 1974, 160.

25 Van Waesberghe offered 2 200 scudi for the publishing rights in the Holy Roman Empire, England, and the Low Countries; Stolzenberg 2001, 10.

26 Others were Coenraet Decker (1651-1685), Willem van der Laegh (1614-1674), Theodoor Matham (1606-1676), Jean van Munnichuysen (1654/55-after 1701), Crispijn van de Passe the Younger (1597/8-1670 or after), and Athonie Heeres Sioertsma (born 1626/27).

27 Kircher's illustrators are discussed in Godwin 2009, 47-58; for his iconography, see Tongiorgi Tomasi 1986, 165-175. The honour of drawing Kircher's portrait in Rome on 2 May 1655 fell to Bloemaert; it became the basis for the scholar's later effigies in print and painting.

28 Among the artists, the amateur draughtsman Goeree (Goeree 1697, 26) refers to Kircher's take on the hieroglyphs formulated in Kircher 1652-1654; Goeree 1682, 43, refers to Kircher's *d'Onderaardse weereld* of 1682. Drawing on the *Oedipus*, the Leiden-based artist and humanist Johannes Schefferus (1621-1679) concluded that the Egyptian pictograms derived from classical ekphraseis, such as Apelles' *Calumny*; see Ellenius 1960, 131.

29 Kircher's works in Dutch were Kircher 1662 (other editions 1681, 1697, 1709), Kircher 1668, and Kircher 1682. The international impact of the book on China is discussed in Shen-Ching Chang 2003.

30 Huygens referred to Kircher's *Herssenschilderagtige reis* (*Iter extaticum*); cf. Huygens 1754, 123. Spinoza was interested in the *Onderaardse weereld,* as his correspondence with Henry Oldenburg testifies (cited in a letter of 7 October 1665), reprinted in Wolf 1935, 202.

31 Stolzenberg 2001, 10.

32 Yates 1964, 417. This idea drew on various sources, including Plato's reference in Philebus to the Egyptian deity Thot as the inventor of writing; later authors such as Philo and Cicero identified Thot with Hermes; Iversen 1961.

33 Alberti 1966, 63-65.

34 Junius 1991, vol. 2, 202, no. 613. Other references to Hermes Trismegistus as the inventor of painting are in Biondo 1549, fol. 8v; R. Alberti, *Trattato della nobiltà della pittura* (Rome 1585), quoted from Barocchi 1960, vol 3, 230; Paleotti 1960, 191.

35 'Trismegisto antiguo escriptor, quiere que la Pintura naciese con la Religion (...) Los Egipcios es cierto que se davan a entender por modo elegante, con figuras de hombres, y animales, como lo dize Cornelio Tacito: "Primi AEgyptii per figuras animalium sensus mentem effingebant"', Carducho 1979, 109-110.

36 The inventory is published in Barge 1934. On Heurnius's collection, see also Jorink 2011, and Huisman 2009, 46-76.

37 Cornelius 1965, 7.

38 Kircher 1652-1654, vol. 3, 514-517; see Jorink 2006, 303.

39 Jorink 2006, 300; Jorink 2010, 291-292; cf. Logan 1979.

40 According to Tim Huisman, 'the degree of idiosyncrasy in Heurnius' intellectual activities and convictions appears limited', Huisman 2009, 68.

41 Hermes Trismegistus s.a., Hermes Trismegistus 1607, and Hermes Trismegistus 1643.

42 In the 17th century, the Amsterdam-based Franciscus Mercurius van Helmont (1614-1698) developed an alternative view, which again took Hebrew as its source; see Coudert 1999. On Becanus's monosyllabic primeval language, see Van Hal 2008, 83-136; Eco 1995, 95-103. On Adamic language, see Aarsleff 1999 and Krop 1996.

43 Becanus, 'Origines Antwerpianae', Liber IX, in: *Opera* (Antwerp 1580); discussed in Iversen 1961, 88, 159.

44 'Viden' quam vocale symbolum cum statua quadret, quamque apte eandem utrumque signum exprimat notionem?', Becanus, 'Hieroglyphica', in: *Opera* (Antwerp 1580), quoted from Purnell 2002, 115.

45 Schnapp 2006, 399-412; Jorink 2011. La Peyrère's book was translated into Dutch as *Preadamiten,* s.l. 1661.

46 La Peyrère 1655, 318, 420.

47 Grotius 1642.

48 De Laet 1643.

49 For the Hornius-Vossius debate on Chinese history, see Weststeijn 2007.

50 Kircher 1652-1654, chapter 4, 33; Hornius 1652, 229, 236. Few historians have addressed 17th-century European knowledge of Amerindian writing; see Perri 1995. Alciati had some interest in the Amerindian pantheon; see McCormack 1995, 122.

51 'Existimo autem Sinensium scribendi morem apud Mexicanos et Iaponenses paulatim exolevisse propter summam difficultatem. Habent enim notarum CXX millia, ex quibus mediocriter doctum ad LXXX millia nosse opportet. Quid mirum notis illis abolitis alias et faciliores figuras surrogatas? Mexicani quae picturis exprimi poterant, nolebant longo et difficili labore siglis notare', Hornius 1652, 271.

52 'Fateor non unam scribendi penitus apud Mexicanos et Sinenses rationem, nec tamen penitus diversa fuit (...) Cataini scribunt penicillo pictorio et una figura multas literas complectitur ac verbum facit', Hornius 1652, 270-271.

53 Vossius refers to La Peyrère's thesis in I. Vossius 1659a, 47.

54 The term is Karel Davids's. Although sceptical towards Kircher's work on the hieroglyphs, Vossius expressed regard for the Jesuit's 'ingenuity and erudition', I. Vossius 1659b, 41.

55 On the museum, see Mastroianni, 2001. For Vossius's contacts with Kircher in Rome, see his remarks on 'Kircherus (...) mihi olim Romae amicus', I. Vossius 1659b, 41. On his visit to Rome in 1645, Coenraad Burgh gave Vossius's regards to Kircher; see Blok 2000, 154.

56 I. Vossius 1685.

57 See Weststeijn forthcoming.

58 'Soli in hoc nostro mundo sunt Seres, qui iam a quinque fere annorum millibus perennem & nunquam interruptam conservavere literaturam, illam quidem operosam, sed cujus fructus tanto respondeat labori, cum si quis vel solos ipsorum perdidicerit characteres, plus possit videri profecisse, quam si quis omnium quot sunt quosve [*sic, presumably for* quotve] fuere linguarum assecutus fuerit peritiam', I. Vossius 1685, 69-70.

59 'd'[E]erste Sinezen hun merkletteren uit alle de dingen, die zich voor't gezicht vertonen, genomen, en alleenlijk naar de verscheide schikking en ordening dezer samengezette dingen de bevattingen van hun geest geopenbaart en bekent gemaakt hebben', Kircher 1668, 275.

60 Kircher 1668, 275, 281. On Van der Laegh, see Godwin 2009, 56, 238.

61 On ape symbolism see Janson 1952.

62 'd'[E]erste beginselen der letteren, die zy van Vader Cham, en van Merkurius Trismegistus, raat van zijn zoon Nefraim, en d'eerste insteller van de beeltletteren (...) geleert hadden. En zeker, deze oude merkletteren der Sinezen, in alles met de beeltletters gelijk, zijn een gewichtig bewijs, daar door ik bewogen word om dit te geloven', Kircher 1668, 274; cf. Szczesniak 1952.

63 Peters 2008, 34-35.

64 Peters 2008, 159. English scholars rose to the occasion in 1687, when the Oxford librarian Thomas Hyde asked Shen Fuzong to catalogue his Chinese books. Some of Shen's Latin letters survive, partially published in Hyde 1796.

65 It can be assumed they had some knowledge of Chinese characters. The family had left Japan in 1641 and the sons had been born in 1620 and 1634. The younger François was listed in the matriculation register as *japonensis*; see Kornicki 1993, 505.

66 The Septuagint states that the world was created in the year 5200 BCE, the Vulgate in the year 4004 BCE. The Deluge was calculated to have taken place in the years 2957 and 2349 respectively; see Finnegan 1964, 191, 184, and Weststeijn 2007, 544.

67 Webb 1669, references to Vossius on 154-157, 175-176, 188. See Harbsmeier 1995. On the 'primitive' status allotted to Chinese, see Marrone 2004, 85-92.

68 Bold 1981. References to Vossius in Webb 1669, 154-157, 175-176, 188.

69 Webb 1669, 169, 187.

70 Ibid., 177.

71 *Haegse Mercurius* no. 97 (9 July 1698), [3] and no. 103 (30 July 1698), [1-2]; quoted from Van Kley 2003, 230.

72 Gombrich 1948.

73 '[N]on senza cagione fu dalli Greci la pittura detta ζωγραφία, cioè viva scrittura (...) Come narra Alfonse de Castro a proposito che la pittura sia viva scrittura a ciascheduno in universale, ancorché idiota, navigando li Spagnoli sotto Carlo Quinto nelle nuove parti occidentali del mondo, ritrovarono che gli uomini di quel paese in luogo di lettere e caratteri talmente dipingevano varie imagini', R. Alberti, 'Trattato della nobiltà della pittura (1585)', quoted from Barocchi 1971-1977, vol. 1, 367.

74 'Ubi *scribere* dixit [Statius Papinius] pro *pingere,* more Graeco, quibus utrumque notat; unde pingendi ars iis graphice vocatur. Ac sic quoque non solum scripta, sed etiam picta, legi dicimus', G. Vossius 1650, 84. The indefinite boundary between words and images in early modern scientific explanation is addressed in Lüthy & Smets 2009.

75 'Teycken-const omhelsend' alle dinghen, / Houdt alle Consten in matighe stringhen. / Sy is een Voedster aller Consten goedich (...) Iae oock d'edel Grammatica bevroedich, / Is door haer ghehooght en ghewassen spoedich, / Leerend' haer letters en caracten halen, / Waer door de Menschen in verscheyden talen / Malcanders meeninghe verstaen accoordich', Van Mander 1604, fol. 8V, vol. 2, par. 2. Goeree likewise argued that painting was just as fundamental as theology for human civilisation, since literacy was based on 'hieroglyphic image-letters', Goeree 1697, 25.

76 *The advancement of learning*, II.xvi.2.

77 Hugo 1617. See Strasser 1988, 90-95.

78 Hugo 1617, 59-62, 88.

79 'Si singulae literae impositae essent, non vocibus, sed rebus ipsis significandis, eaeque essent hominibus omnibus communes; omnes omninio homines, etiamsi gentes singulae res singulas

diversis nominibus appelent, singularum gentium scriptionem intelligerent', Hugo 1617, 60.

80 There were 40 editions and many translations (a Dutch one by Justus de Harduyn).

81 Steegius 1647. This *beelden-sprake* consists in three dimensions: narrative or figural representations, metaphors or parables, and symbols. Muller 2006 discusses it in relation to the Chinese mission, Dutch almanacs, and the Lima-based missionary Diego Martinez, who burned his writings and entered the jungle, in order to teach through images painted on panels.

82 G. Vossius 1635, vol. 1, 140: 'quomodo item gentes plurimae conveniunt in notis Arithmeticis, quas cifras vocant: quomodo etiam variarum linguarum Astrologi conspirant in notis planetarum', G. Vossius 1635, vol. 1, 142.

83 '[P]rodidit Nicolaus Trigaultius (...) non pauciores Sinensibus literas esse, quam voces numerantur: eas tamen iter inter se componere, ut LXX, aut LXXX milia non excedant. Imo idem refert, Sinenses & Japanenses, etsi lingua aeque differant, ac Hebraei, & Belgae; tamen, quae sic scribuntur, intelligere (...) Utcumque enim in legendo alii alia verba pronunciassent: tamen iidem fuissent conceptus. Nempe uti nunc variarum linguarum homines rem eandem conspicientes eandem rem concipiunt: ita idem [*sic*] rei signum intuentes, eundem haberent conceptum', G. Vossius 1635, vol. 1, 143. Vossius discusses Chinese writing on pages 140-143 and 122.

84 'Quomodo ex pictura hominis, equi, arboris, domus, homines omnes eundem habent conceptum', G. Vossius 1635, vol. 1, 142.

85 '[T]ot prope characteribus fore, quot res sunt', G. Vossius 1635, vol. 1, 142. The remark is all the more striking as Vossius's estimate of the total number of characters was too high: the most complete Chinese dictionary of the 17th century, Mei Yingzuo's *Zihui* of 1615, listed 33,179 characters.

86 On early modern mnemotechnics, see Yates 1974, also below, note 124 ff.

87 Cornelius 1965, 29.

88 G. Dalgarno, 'Omnibus quo omnino hominibus (1660)', in: Cram & Maat 2001, 121.

89 G. Dalgarno, 'Ars signorum, vulgo character universalis et lingua philosophica (1661)', in: Cram & Maat 2001, 157.

90 Cornelius 1965, 32.

91 Walton 1657, 'prologomena', 10.

92 Comenius 1658; on the ideal of eduction through images, see Sadler 2007. See also Forster 1961.

93 Rademaker 1981, 280.

94 '[O]f all of Comenius' ideals of universal language, the *Orbis pictus* came closest', Slaughter 1982, 101.

95 Comenius 1938, 186-189.

96 'Vor allen Dingen, must du lernen die schlechten Stimmen, in welchn die menschliche Rede bestehet: welche die Thiere wissen zu bilden, und deine Zunge weiss nachzumachen, und deine Hand tan [*sic*] mahlen. Darnach wollen wir gehen in die Welt, und wollen beschauen alle Ding. Hier hast du een lebendiges und stimmbares Alfabeth', Comenius 1679, 4.

97 Geissler 1959, 151, 141.

98 Comenius 1662.

99 Manning 2002, 110-132, esp. 121.

100 Ibid., 115.

101 See Kircher 1675, 60-61, 66-68, 116-117, 122-123; for the individual animals, 54, 71-94.

102 Slaughter 1982.

103 S. Quiccheberg, *Inscriptiones vel tituli theatri amplissimi* (1565): its program is reprinted in Hüllen 1989, 121-130.

104 The collector was Ulisse Aldrovandi (1522-1605) whose *Opera omnia* were published in 1599-1668; see Jorink 2006, 274; Jorink 2010, 264-265, and Bono 1995, 123-124.

105 Jorink 2006, 308; Jorink 2010, 297.

106 Jorink 2006, 287, 310; Jorink 2010, 277, 299. Wormius echoed his fellow Scandinavians Olaus Magnus (1490-1557) and Johannes Bureus (1568-1652). The latter's mystical view on pictograms fused Hermetic and cabbalistic lines of thought. See Magnus 1655, lib. I, cap. XXXVI, 58; Ellenius 1960, 273, points out that similar ideas were conceived by the French Jesuit Nicolaus Caussinus (in *De symbolica Aegyptiorum sapientia,* 1618).

107 Jorink 2006, 305.

108 Jorink 2006, 295; Jorink 2010, 295.

109 One example is the Gotha Collection; its catalogue is reprinted as an appendix in Collet 2007.

110 Cornelius 1965, 33.

111 The first statement is from Johann Valentin Andreae's *Theatrum physicum* (1619); quoted in Collet 2007, 44. Adam Olearius's catalogue is quoted by Hüllen 1989, 131.

112 Hooke 1705, 338.

113 Van Hoogstraten 1682.

114 '[M]undus corporeus est Theatrum Dei visibile', Comenius 1966, col. 431.

115 Jorink 2006, 298.

116 This is the central thesis of Jorink 2006; Jorink 2010.

117 Bakker 2004; see also De Klijn 1982.

118 *The compleat gentleman* (London 1622) quoted in Ellenius 1960, 228-229. Peacham travelled to the Netherlands in 1613; as exemplary artists he mentioned Hendrik Goltzius, Johannes Stradanus, Jan Wierix, Chrispijn van de Passe, and Michiel van Mierevelt; Peacham 1622, 108-109.

119 '[N]e' primi tempi volse Iddio ammaestrare gli uomini per l'istesse opere sue e cose create, che potessero essere universalmente apprese da tutti, allegando il detto del salmo: "Caelo enarrant gloriam Dei et opera manuum eius annunciat firmamentum." (Psalm 18,2) (...) Non ci è gente o lingua o condizione di persone, che non possa intendere bene quelle voci tacite ch' escono dall' opere create d' Iddio', Paleotti 1960, 144, 148.

120 Cited in Holtgen 1988, 183-207.

121 '[A]lle sienlijcke creaturen zijn als letteren in eenen schoonen Boeck om ons den werckman derselve aen te wijsen', Corstens 1598, fol. 50r; cf. Jorink 2006, 63 and 353; Jorink 2010, 52.

122 '[D]it boeck der nature, daer soo een fraeye samenvoeging van schepselen in gevonden wordt, als soo veele letteren die een verwonderlijcke sin maken', Spanheim 1677, 56, quoted from Jorink 2006, 426; see also Jorink 2010, 404-405.

123 '[D]e gene, de welcke meynen soude, dat de letteren in een druckerye sonder keur genomen zijnde, of dat een hoop Characteren (...) in de locht opgesmeten zijnde, en daer verwerdelijck dooreen beroert, sich soo gelijckigh te samen souden konnen voegen, en met soo veel ordre by een vergaderen, dat men van die letters, en Charachters [*sic*] een boeck vol van leeringen of een welgemaeckt gedicht soude sien voortkomen', Spanheim 1677, 110, quoted from Jorink 2006, 426; see also Jorink 2010, 404.

124 Cicero, *De oratore,* II, 86; see Yates 1974.

125 Bruno, *De umbris idearum* (1582), paragraph 101; quoted by Ingrid Rowland, who paraphrases Bruno's idea that 'the art of memory isolated individual sense perceptions from the stream of consciousness and endowed these perceptions with special characteristics that transformed them into thoughts', Rowland 2009, 123.

126 'Fabulas et apologos hoc discet [puer] libentius ac memineret melius, si horum argumenta scite depicta pueri oculis subjiciantur, et quidquid oratione narrantur, in tabula demonstretur. Idem aeque valebit ad ediscenda arborum, herbarum et animantium nomina, simul et naturas (...) Plerique gaudent pictis venationibus, hic quot species arborum, herbarum, avium, quadrupedum, per lusum disci possunt?' Erasmus 1966, 447, 571-573.

127 For additional examples, he pointed to the imagery in fables and card games. G. Vossius 1699, fols. 77r-80v.

128 Comenius, *Philosophia prodromus et conatuum pansophicorum dilucidatio* (Leiden 1644), quoted from Rossi 2000, 134.

129 Gerhardt 1875-1890, vol. 4, 73; Rossi 2000, 178-9; cf. Kneale 1966.

130 Rossi 2000, 190.

131 Strasser 2000, 22.

132 Ibid., 32-33.

133 Descartes 1964-1972, vol. 10, 230; quoted from Rossi 2000, 112.

134 On this topic see Margolin 1986, vol. 2, 53-91.

135 Strasser 2000, 71.

136 On Buno, see Strasser 2000.

137 Compare the images of many-headed, many-limbed monsters from the Arminian-Gomarist debate, *Den teghenwoordighen arminiaen* (1623), *Warminiaen* (1618), and *Den nieuwen Barnevelschen handel* (1619), images in Cat. Utrecht 1994, 55, 71. Of course, these were not mnemotechnical images; however, we should realise that what strikes us as abstruse and far-fetched in the images was expected to make a more unequivocal and lasting impression than the accompanying text.

138 Strasser 2000, 54-55; see Clouston 1894.

139 See Baxandall 2003.

140 Wenchao Li 2008.

141 Gerhardt 1875-1890, vol. 7, 198-99; Rossi 2000, 185.

142 Heinekamp 1972.

143 'Omnis Ratiocinatio nostra nihil aliud est quam characterum connexio et substitutio, sive ille characteres sint verba, sive notae, sive denique imagines', Leibniz quoted from Gerhardt 1875-1890, vol. 3, 605, cf. vol. 7, 31.

144 Leibniz followed Becanus's view that German would come closest to the Adamic language, but dismissed his fanciful etymologies as mere 'Goropiser'. Leibniz, quoted from Gerhardt 1875-1890, vol. 5, 379; Leibniz, *Dissertatio de arte combinatoria* (ed. princ. 1666), quoted from Gerhardt 1875-1890, vol. 4, 27-104, esp. 73.

145 Perkins 2004, 42. The misguided view that Chinese letters represented basic concepts may have been inspired by one of the striking contingencies determining the Jesuits' success among the Chinese literati. Some missionaries were able to memorise Chinese poetry and recite it backwards, associating the characters with images in their artificial 'memory palace'; see Spence 1984. The presumed interrelations between Chinese 'pictograms', the layout of these images in the memory theatre, and the logical principles determining this layout seem a promising topic for further research.

146 Kircher 1667, 227. Kircher also lists other natural elements identified by the Chinese as the basis for characters.

147 On Leibniz's ideal of an *ars combinatoria* reflecting the structure of reality, see Westerhoff 1999.

148 Marrone 1986. For the manuscript in the Vatican library, see Marrone 2002, 125. The *Novum hoc inventum* is not a philosophical language but more comparable to sign language ('Pasigraphia') making communication possible between people speaking different languages.

149 Westerhoff 1999, 456.

150 Comenius 1938, 186-189. For his influence on Wilkins, see DeMott 1955.

151 Rossi 2000, 184.

152 Rossi 2000, 181; Gerhardt 1875-1890, vol. 7, 26.

153 See Bolzoni 1994. Some authors extend the comparison from the memory to the imagination. Cf. De la Puente 1619, 194: 'my imaginative facultie is, like a hall painted with many images and figures'.

154 See Westerhoff 2001, 645.

155 'Een boek, daar in zig God zelfs heerlik heeft beschreven. / Het vleeschelik vernuft staat, voor die lett'ren stom', Zoet 1675, 213. The statement refers to Jan Volckertsz.'s collection; see Jorink 2006, 323; Jorink 2010, 310.

156 Leibniz 1994.

157 Jorink 2006, 330.

158 Ibid., 344.

159 Letter to Jerónimo Lobo (13 April 1668), in Hall & Hall 1965-1986, vol. 4, 316-317.

160 The words are Robert Hooke's, reprinted in Hooke 1705 and quoted by Slaughter 1982, 159. Although the catalogue of the collection was compiled only after Wilkins's death, it was finished by one of his protégés, Nehemia Grew (1641-1712); see Collet 2007, 274, 280. Cf. Lewis 2007, 200; Pearce & Arnold 2000.

161 Wilkins 1668, 13, 451. At the Society, Robert Hooke tried to talk to some visiting Chinese merchants in 1693.

162 Slaughter 1982, 193.

163 Guyot Desfontaines 1730.

164 Descartes 1970, 9-12.

165 In the 18th century, Giambattista Vico continued to see pictography, or 'a language with natural significations' that was putatively developed in a primeval age of mankind, as the basis of all human culture. See Vico 1984, 138-146. In this context, he harked back to Becanus (p. 139), De Laet (p. 143) and Heurnius (pp. 29, 48, 51, 287).

166 Borst 1957-1963, vol. 3, 1048-1261; Wegener 1995.

167 Sadler 2007, Geissler 1959.

168 Lach 1940, 573.

169 Peters 2008, 205; Lach 1940, 573.

170 '[G]astés par une fausse métaphysique (...). qui ne vont pas à moins qu'à un renversement général de toute Religion', Renaudot 1718, 28-29.

171 G. Dalgarno, 'Character universalis (1657)', in: Cram & Maat 2001, 85.

172 Leibniz to Bouvet, quoted in Walker 1994, 447. See also Perkins 2004, 42, Sun Xiaoli 1995, and Porter 2001, 55-57.

173 Cf. Davis 2000. For a modern analysis of the feasibility of expressing logical relations in pictures, see Westerhoff 2005.

Bibliography

Aarsleff 1999
H. Aarssleff, 'The rise and decline of Adam and his Ursprache', in: A. Coudert (ed.), *The language of Adam*, Wolffenbüttel 1999, 277-295.

Alberti 1966
L.B. Alberti, *On painting* (J.R. Spencer, transl.), New Haven & London 1966 (ed. princ. 1435).

Bakker 2004
B. Bakker, *Landschap en wereldbeeld. Van Van Eyck tot Rembrandt,* Bussum 2004.

Barge 1934
J. Barge, *De oudste inventaris der oudste academische anatomie in Nederland*, Leiden 1934.

Barocchi 1960
P. Barocchi (ed.), *Trattati d'arte del cinquecento*, Bari 1960, 3 vols.

Barocchi 1971-1977
P. Barocchi (ed.), *Scritti d'arte del cinquecento*, Milan & Naples 1971-1977, 3 vols.

Baxandall 2003
M. Baxandall, 'Rudolph Agricola on art and on patrons', in: *Words for pictures,* New Haven & London 2003, 69-82.

Becanus 1580
J.G. Becanus, *Opera*, Antwerp 1580.

Biondo 1549
M. Biondo, *Della nobilissima pittura,* Venice 1549.

Blok 2000
F.F. Blok, *Isaac Vossius and his circle. His life until his farewell to Queen Christina of Sweden,* Groningen 2000.

Bold 1981
J. Bold, 'John Webb. Composite capitals and the Chinese language', *The Oxford Art Journal* 4 (1981), 9-17.

Bolzoni 1994
L. Bolzoni, 'Das Sammeln und die ars memoriae', in: Andreas Grote (ed.), *Macrocosmos in microcosmo. Die Welt in der Stube. Zur Geschichte des Sammelns 1450-1800*, Opladen 1994, 129-168.

Bono 1995
J.J. Bono, *The word of God and the languages of man. Interpreting nature in early modern science and medicine,* Madison 1995.

Borst 1957-1963
A. Borst, *Der Turmbau von Babel. Geschichte der Meinungen über Ursprung und Vielfalt der Sprachen und Völker,* Stuttgart 1957-1963, 4 vols.

Carducho 1979
V. Carducho, *Diálogos de la pintura* (F.C. Serraller, ed.), Madrid 1979 (ed. princ. 1633), 109-110.

Clouston 1894
W.A. Clouston, *Hieroglyphic Bibles, their origin and history*, Glasgow 1894.

Collet 2007
D. Collet, *Die Welt in der Stube. Begegnungen mit Außereuropa in Kunstkammern der Frühen Neuzeit,* Göttingen 2007.

Comenius 1658
J.A. Comenius, *Orbis sensualium pictus,* Nuremberg 1658.

Comenius 1662
J.A. Comenius, *Die sichtbare Welt, das ist aller vornehmen Weltdinge, und Lebensverrichtungen, Vorbildung und Benamung,* Nuremberg 1662.

Comenius 1679
J.A. Comenius, *Orbis sensualium pictus quadrilinguis,* Nuremberg 1679.

Comenius 1938
J.A. Comenius, *The way of light of Comenius* (E.T. Campagnac, transl.), Liverpool & London 1938.

Comenius 1966
J.A. Comenius, *De rerum humanarum emendatione consultatio catholica* (J. Cervenka *et al.*, eds.), Prague 1966.

Cornelius 1965
P. Cornelius, *Languages in seventeenth- and early eighteenth-century imaginary voyages,* Geneva 1965.

Corstens 1598
C. Corstens, *Wtlegghinge des catechismi der gereformeerde Christelicke Kercke in Neerduytsland*, Leiden 1598.

Coudert 1978
A. Coudert, 'Some theories of a natural language from the Renaissance to the seventeenth century', in: A. Heinekamp (ed.), *Magia Naturalis und die Entstehung der modernen Naturwissenschaften*, Wiesbaden 1978, 56-119.

Coudert 1999
A. Coudert, *The impact of the Kabbalah in the seventeenth century. The life and thought of Francis Mercury Van Helmont (1614-1698)*, Leiden 1999.

Cram & Maat 2001
D. Cram & J. Maat (eds.), *George Dalgarno on universal language,* Oxford 2001.

Curran 2007
B. Curran, *The Egyptian Renaissance. The afterlife of Egypt in Renaissance Italy*, Chicago 2007.

Davis 2000
M. Davis, *The universal computer. The road from Leibniz to Turing,* New York 2000.

De Klijn 1982
M. de Klijn, *De invloed van het Calvinisme op de Noord-Nederlandse landschapsschilderkunst 1570-1630,* Apeldoorn 1982.

De Laet 1643
J. de Laet, *Notae ad dissertationem Hugonis Grotii De origine gentium americanarum, et observationes aliquot ad melior indaginem difficillimae illius quaestionis,* Leiden 1643.

De la Puente 1619
L. de la Puente, *Meditations upon the mysteries of our Holie Faith* (J. Heigham, transl.), St. Omer 1619, 2 vols.

DeMott 1955
B. DeMott, 'Comenius and the Real Character in England', *Periodical of the Modern Language Association* 70/5 (1955), 1068-1081.

Descartes 1964-1972
R. Descartes, *Oeuvres* (C. Adam & P. Tannery, eds.) Paris 1964-1972.

Descartes 1970
R. Descartes, *Philosophical letters* (A. Kenny, ed.), Oxford 1970.

Eco 1995
U. Eco, *The search for the perfect language,* Oxford 1995.

Ellenius 1960
A. Ellenius, *De arte pingendi. Latin art literature in seventeenth-century Sweden and its international background*, Uppsala & Stockholm 1960.

Erasmus 1966
D. Erasmus, *Declamatio de pueris statim ac liberaliter instituendis* (J.-C. Margolin, ed.), Geneva 1966 (ed. princ. 1529).

Finnegan 1964
J. Finnegan, *Handbook of Biblical chronology,* Princeton 1964.

Forster 1961
L.W. Forster, 'Comenius und die Emblematik. Ein Hinweis', *Zeitschrift für slavische Philologie* 29 (1961), 247-250.

Geissler 1959
H. Geissler, *Comenius und die Sprache,* Heidelberg 1959.

Gerhardt 1875-1890
C.I. Gerhardt (ed.), *Die philosophischen Schriften von Gottfried Wilhelm Leibniz,* Berlin 1875-1890.

Godwin 2009
J. Godwin, *Athanasius Kircher's theatre of the world,* London 2009.

Goeree 1682
W. Goeree, *Natuurlyk en schilderkonstig ontwerp der menschkunde,* Amsterdam 1682.

Goeree 1697
W. Goeree, *Inleyding tot de practijck der algemeene schilderkonst,* Middelburg 1697 (ed. princ. 1670).

Gombrich 1948
E. Gombrich, '"Icones symbolicae". The visual image in Neo-Platonic thought', *Journal of the Warburg and Courtauld Institutes* 11 (1948), 163-192.

Grotius 1642
H. Grotius, *De origine gentium Americanarum dissertatio,* Paris & Amsterdam 1642.

Guyot Desfontaines 1730
P.-F. Guyot Desfontaines, *Le nouveau Gulliver ou voyages de Jean Gulliver fils du capitain Gulliver,* Paris 1730.

Haegse Mercurius 1698
Haegse Mercurius 97 (9 July 1698) and 103 (30 July 1698).

Hall & Hall 1965-1968
A.R. Hall & M.B. Hall (eds.), *The correspondence of Henry Oldenburg,* Madison & London 1965-1986, 8 vols.

Harbsmeier 1995
C. Harbsmeier, 'John Webb and the early history of the study of the classical Chinese language in the West', in: M. Wilson & J. Cayley (eds.), *Europe studies China. Papers from an international conference on the history of European Sinology,* London 1995, 297-338.

Heinekamp 1972
A. Heinekamp, 'Ars Characteristica und natürliche Sprache bei Leibniz', *Tijdschrift voor filosofie* 34 (1972), 469-63.

Hermes Trismegistus s.a.
Hermes Trismegistus, *Pimandre,* manuscript on paper, after 1574, Antwerp, Museum Plantin-Moretus, Ms M 40.

Hermes Trismegistus 1607
Hermes Trismegistus, *Wondervondt van de eeuwighe bewegingh, die den Alckmaersche philosooph Cornelis Drebbel (...) ooc mede by gevoeght een boeck Pymander, beschreven van Mercurius driemael de grootste, die oock een philosooph en priester, en coningh soude in Egypten gheweest hebben, in de tijdt van Moses,* Alkmaar 1607.

Hermes Trismegistus 1643
Hermes Trismegistus, *Sesthien boecken (...) met groote naarstigheyt, uyt het Grieckx ghebracht in ons Nederduytsch,* Amsterdam 1643.

Herold 1554
J.B. Herold, *Heydenweldt und irer Götter anfängcklicher ursprung,* Basel 1554.

Holtgen 1988
K.-J. Holtgen, 'Francis Quarles' second emblem book "Hieroglyphikes of the life of man"', in: K.-J. Holtgen *et al.* (eds.), *Word and visual imagination. Studies in the interaction of English literature and the visual arts,* Erlangen 1988, 183-207.

Hooke 1705
R. Hooke, *The posthumous works of Robert Hooke* (R. Waller, ed.), London 1705.

Hornius 1652
G. Hornius, *De originibus Americanis libri quator,* Leiden 1652.

Howarth 1985
D. Howarth, *Lord Arundel and his circle,* New Haven 1985.

Hugo 1617
H. Hugo, *De prima scribendi origine et universa rei literariae antiquitate,* Antwerp 1617.

Huisman 2009
T. Huisman, *The finger of God. Anatomical practice in seventeenth-century Leiden,* Leiden 2009.

Hüllen 1989
W. Hüllen, '*Their manner of discourse*'. *Nachdenken über Sprache im Umkreis der Royal Society*, Tübingen 1989.

Huygens 1754
C. Huygens, *De wereldbeschouwer, of gissingen over de hemelsche aardkloten, en derzelver cieraad,* Rotterdam 1754.

Hyde 1796
T. Hyde, *Syntagma dissertationum*, Oxford 1796.

Iversen 1961
E. Iversen, *The myth of Egypt and its hieroglyphs*, Copenhagen 1961.

Janson 1952
H.W. Janson, *Apes and ape lore in the Middle Ages and the Renaissance,* London 1952.

Jorink 2006
E. Jorink, *Het 'Boeck der Natuere'. Nederlandse geleerden en de wonderen van Gods schepping 1575-1715*, Leiden 2006.

Jorink 2010
E. Jorink, *Reading the Book of Nature in the Dutch Golden Age*, Leiden & Boston 2010.

Jorink 2011
E. Jorink, 'Noah's Ark restored (and wrecked). Dutch collectors, natural history and the problem of Biblical exegesis' in: S. Dupré and C. Lüthy (eds), *Silent messengers. The circulation of material objects of knowledge in the early modern Low Countries*, Berlin 2011, 153-184

Junius 1991
F. Junius, *The literature of classical art,* (K. Aldrich, P. Fehl & R. Fehl, eds.), Berkeley, Los Angeles & Oxford 1991 (ed. princ. 1694), 2 vols.

Kircher 1652-1654
A. Kircher, *Oedipus Aegyptiacus, hoc est, universalis hieroglyphicæ veterum doctrinæ temporum iniuria abolitæ instauratio*, Rome 1652-1654.

Kircher 1662
A. Kircher, *Het eerste deel van het groote licht der natuurkunde of tweede vervolgh op de eminente oratie van (...) Digby* (J. Casteleyn, transl.), Haarlem 1662.

Kircher 1667
A. Kircher, *China illustrata*, Amsterdam 1667.

Kircher 1668
A. Kircher, *Toonneel van China* (J.H. Glazemaker, transl.), Amsterdam 1668.

Kircher 1669
A. Kircher, *Naturelyke en geneeskonstige navorsching der peste* (Z. van de Graaf, transl.), Rotterdam 1669.

Kircher 1675
A. Kircher, *Arca Noë,* Amsterdam 1675.

Kircher 1682
A. Kircher, *d'Onderaardse weereld,* Amsterdam 1682.

Kneale 1966
W. Kneale, 'Leibniz and the picture theory of language', *Revue internationale de philosophie* (1966), 204-215.

Kornicki 1993
P.F. Kornicki, 'European Japanology at the end of the seventeenth century', *Bulletin of the School of Oriental and African Studies* 56/3 (1993), 502-524.

Krop 1996
H.A. Krop, 'The antiquity of the Dutch language. Renaissance theories on the language of Paradise', in: J. Fenoulhet & L. Gilbert (eds.), *Presenting the past. History, art, language, literature,* London 1996, 203-226.

Lach 1940
D.F. Lach, 'The Chinese studies of Andreas Müller', *Journal of the American Oriental Society* 60/4 (1940), 564-575.

La Peyrère 1655
I. la Peyrère, *Praeadamitae,* Amsterdam 1655.

Leibniz 1994
G.W. Leibniz, 'Drôle de pensée, touchant une nouvelle sorte de représentation', in: cat. Bonn, Kunst- und Ausstellungshalle der Bundesrepublik Deutschland, *Wunderkammer des Abendlandes. Museum und Sammlung im Spiegel der Zeit,* Bonn 1994, 122-124.

Lewis 2007
R. Lewis, *Language, mind and nature. Artificial languages in England from Bacon to Locke,* Cambridge & New York 2007.

Cat. Liège 2006
Cat. Liège, Musée de l'Art Wallon, *Lambert Lombard, Renaissanceschilder. Luik 1505/06-1566. Interdisciplinaire essays en tentoonstellingscatalogus* (G. Denhaene), Liège 2006.

Logan 1979
A.-M.S. Logan, *The 'cabinet' of the brothers Gerard and Jan Reynst,* Amsterdam 1979.

Lüthy & Smets 2009
C. Lüthy & A. Smets, 'Words, lines, diagrams, images. Towards a history of scientific imagery', *Early Science and Medicine* 14 (2009), 398-439.

Magnus 1655
O. Magnus, *Historia de gentibus septentrionalibus,* Rome 1655.

Manning 2002
J. Manning, *The emblem,* London 2002.

Margolin 1986
J.-C. Margolin, 'Rébus et hieroglyphs, pictogrammes et idéogrammes', in: *Histoire du Rébus,* Paris 1986, vol. 2, 53-91.

Marrone 1986
C. Marrone, 'Lingua universale e scrittura secreta nell'opera di Kircher', in: M. Casciato *et al.* (eds.), *Enciclopedismo in Roma barocca. Athanasius Kircher e il museo del Collegio Romano tra Wunderkammer e museo scientifico*, Venice 1986, 78-86.

Marrone 2002
C. Marrone, *I geroglifici fantastici di Athanasius Kircher*, Rome 2002.

Marrone 2004
C. Marrone, *Le lingue utopiche,* Viterbo 2004.

Mastroianni 2001
A. Mastroianni, 'Kircher e l'Oriente nel Museo del Collegio Romano', in: Cat. Rome, Palazzo Venezia, *Athanasius Kicher. Il museo del mondo* (E. lo Sardo, ed.), Rome 2001, 65-78.

McCormack 1995
S. McCormack, 'Limits of understanding. Perceptions of Greco-Roman and Amerindian paganism in early modern Europe', in: K.O. Kupperman (ed.), *America in European consciousness, 1493-1750*, Chapel Hill 1995, 79-129.

Muller 2006
J. Muller, 'Jesuit uses of art in the Province of Flanders', in: J.W. O'Malley & G.A. Bailey (eds.), *The Jesuits II. Cultures, sciences, and the arts,* Toronto, London & Buffalo 2006, 113-156.

Paleotti 1960
G. Paleotti, 'Discorso intorno alle imagini sacre e profane' (Bologna 1582), in: P. Barocchi (ed.), *Trattati d'arte del cinquecento*, Bari 1960, 3 vols., vol. 2, 117-509.

Peacham 1622
H. Peacham, *The compleat gentleman,* London 1622.

Pearce & Arnold 2000
S. Pearce & K. Arnold, 'John Wilkins experiments with a philosophical museum language', in: *The collector's voice. Critical readings in the practice of collecting, Vol. 2. Early voices,* Aldershot 2000, 63-66.

Perkins 2004
F. Perkins, *Leibniz and China. A commerce of light*, Cambridge 2004.

Perri 1995
A. Perri, 'Un'ipotesi di Utopia. La scrittura azteca e le lingue perfette', *Il Mondo 3*, 2/3 (1995), 306-326.

Peters 2008
M. Peters, *'Mercator sapiens' (de wijze koopman). Het wereldwijde onderzoek van Nicolaas Witsen (1641-1717), burgemeester en VOC-bewindhebber van Amsterdam.* Groningen 2008 (diss. Groningen).

Porter 2001
D. Porter, *Ideographia. The Chinese cipher in early modern Europe,* Stanford 2001.

Purnell 2002
F. Purnell, 'Francesco Patrizi and the critics of Hermes Trismegistus', in: M. Mulsow (ed.), *Das Ende des Hermetismus,* Tübingen 2002, 105-160.

Rademaker 1981
C.S.M. Rademaker, *Life and work of Gerardus Johannes Vossius 1577-1649*, Assen 1981.

Reilly 1974
C. Reilly, *Athanasius Kircher. A master of a hundred arts, 1602-1680*, Wiesbaden 1974.

Renaudot 1718
E. Renaudot, *Anciennes relations des Indes et de la Chine, de deux voyageurs mahométans, qui y allèrent dans le neuvième siècle, traduites d'arabe en francois*, Paris 1718.

Rossi 2000
P. Rossi, *Logic and the art of memory. The quest for a universal language*, Chicago 2000.

Rowland 2009
I. Rowland, *Giordano Bruno. Philosopher/heretic,* Chicago 2009.

Sadler 2007
J.E. Sadler, *J.A. Comenius and the concept of universal education*, New York & London 2007 (ed. princ. 1966).

Schnapp 2006
A. Schnapp, 'The Pre-Adamites. An abortive attempt to invent pre-history in the seventeenth century?', in: R. Ligota & J.-L. Quantin (eds.), *History of scholarship. A selection of papers from the seminar on the history of scholarship, held annually at the Warburg Institute,* Oxford 2006, 399-412.

Shen-Ching Chang 2003
Shen-Ching Chang, *Natur und Landschaft. Der Einfluß von Athanasius Kirchers 'China Illustrata' auf die europäische Kunst,* Berlin 2003.

Slaughter 1982
M.M. Slaughter, *Universal languages and scientific taxonomy in the seventeenth century*, Cambridge 1982.

Spanheim 1677
F. Spanheim, *Den atheist of Godverloochenaar overtuycht in vier predikatien over de Woorden Psalm 14 V.1*, Amsterdam 1677.

Spence 1984
J.D. Spence, *The memory palace of Matteo Ricci,* New York 1984.

Steegius 1647
J.G. Steegius, *De Christelycke leeringhe verstaenelycker uytgeleyt door eene beelden-sprake noodigh soo voor kinders als groote die niet en kunnen lesen, gherievigh voor een ieder om beter te vatten ende t'onthouden,* Antwerp 1647.

Stolzenberg 2001
D. Stolzenberg, *The great art of knowing. The Baroque encyclopedia of Athanasius Kircher,* Stanford 2001.

Strasser 1988
G.F. Strasser, *Lingua universalis. Kryptologie und Theorie der Universalsprachen im 16. und 17. Jahrhundert,* Wiesbaden 1988.

Strasser 2000
G.F. Strasser, *Emblematik und Mnemonik der frühen Neuzeit im Zusammenspiel. Johannes Buno und Johann Justus Winckelmann,* Wiesbaden 2000.

Sun Xiaoli 1995
Sun Xiaoli, 'Leibniz, les symboles I Ching et les idéogrammes', in: Y. Daiyun & A. le Pichon (eds.), *La Licorne et le dragon. Les malentendus dans la recherche de l'universel,* Beijing & Paris 1995, 109-122.

Szczesniak 1952
B. Szczesniak, 'The origin of the Chinese language according to A. Kircher's Theory', *Journal of the American Oriental Society* 72 (1952), 21-29.

Tongiorgi Tomasi 1986
L. Tongiorgi Tomasi, 'Il simbolismo delle immagini. I frontespizi delle opere di Kircher', in: M. Casciato *et al.* (eds.), *Enciclopedismo in Roma barocca. Athanasius Kircher e il Museo del Collegio romano tra Wunderkammer e museo scientifico,* Venice 1986, 165-175.

Cat. Utrecht 1994
Cat. Utrecht, Catharijneconvent, *Rekkelijk of precies. Remonstranten en Contraremonstranten ten tijde van Maurits en Oldenbarnevelt* (T.G. Kootte), Utrecht 1994.

Van der Lem & Rademaker 1993
G.A.C. van der Lem & C.S.M. Rademaker, *Inventory of the correspondence of Gerardus Johannes Vossius (1577-1649),* Assen & Maastricht 1993.

Van Dorsten & Hamilton 1980
J. van Dorsten & A. Hamilton, 'Two puzzling pages in Ortelius' Album Amicorum', in: C.C. Barfoot *et al.* (eds.), *Times and tide. Writings offered to A.G.H. Bachrach,* Leiden 1980, 45-53.

Van Hal 2008
T. van Hal, *'Moedertalen en taalmoeders'. Methodologie, epistemologie en ideologie van het taalvergelijkend onderzoek in de renaissance, met bijzondere aandacht voor de bijdrage van de humanisten uit de Lage Landen,* Leuven 2008 (diss. Leuven).

Van Hoogstraten 1678
S. van Hoogstraten, *Inleyding tot de hooge schoole der schilderkonst. Anders de zichtbare werelt,* Rotterdam 1678.

Van Hoogstraten 1682
F. van Hoogstraten, *De schoole der wereld,* Dordrecht 1682.

Van Kley 2003
E. van Kley, 'Qing Dynasty China in seventeenth-century Dutch literature 1644-1700', in: W.F. Vande Walle (ed.), *The history of the relations between the Low Countries and China in the Qing Era (1644-1911),* Leuven 2003, 217-234.

Van Mander 1604
K. van Mander, 'Grondt der edel-vry schilderconst', in: *Het schilder-boeck,* Haarlem 1604.

Van Sambeeck 1663
J. van Sambeeck, *Het gheestelyck jubilee,* Antwerp 1663.

Veldman 1977
I. Veldman, *Maerten van Heemskerck and Dutch humanism in the sixteenth century,* Maarssen 1977.

Vico 1984
G. Vico, *The new science of Giambattista Vico* (T.G. Bergin & M.H. Fisch, transl.), Ithaca & London 1984.

G. Vossius 1635
G.J. Vossius, *De arte grammatica libri VII,* Amsterdam 1635.

G. Vossius 1650
G.J. Vossius, 'De graphice', in: *De quatuor artibus popularibus,* Amsterdam 1650, 61-92.

G. Vossius 1699
G.J. Vossius, 'Dissertationis generalis bipartitae pars (...) altera, quae ostendit, quomodo variae adeo disciplinae optime addiscantur', in: *Opera,* Amsterdam 1695-1701, vol. 4 (1699), fols. 77r-80v.

I. Vossius 1659a
I. Vossius, *Dissertatio de vera aetate mundi, qua ostenditur natale mundi tempus annis minimum 1440 vulgarem aeram anticipare,* The Hague 1659.

I. Vossius 1659b
I. Vossius, *Castigationes ad scriptum Georgii Hornii de Aetate Mundi,* The Hague 1659.

I. Vossius 1685
I. Vossius, 'De artibus et scientiis Sinarum', in: *Variarum observationum liber,* London 1685, 69-85.

Walker 1994
D.P. Walker, 'Leibniz and language', in: R.S. Woolhouse (ed.), *Gottfried Wilhelm Leibniz. Critical assessments,* New York & London 1994, 436-451.

Walton 1657
B. Walton, *Biblia sacra polyglotta,* London 1657.

Waterschoot 1978
W. Waterschoot, 'Hieroglyphica te Gent in 1584', *Verslagen en mededelingen van de Koninklijke Academie voor Nederlandse Taal- en Letterkunde* 1 (1978), 47-85 (reprinted in: *Schouwende fantasye. Opstellen van Werner Waterschoot,* Ghent 2002, 227-262).

Webb 1669
J. Webb, *An historical essay endeavouring the probability that the language of the Empire of China is the primitive language,* London 1669.

Wegener 1995
U.B. Wegener, *Die Fascination des Maßlosen. Der Turmbau zu Babel von Pieter Bruegel bis Athanasius Kircher*, Hildesheim 1995.

Wenchao Li 2008
Wenchao Li, 'Leibniz, der Chronologiestreit und die Juden in China', in: D.J. Cook *et al.* (eds.), *Leibniz und das Judentum*, Stuttgart 2008, 183-208.

Westerhoff 1999
J.C. Westerhoff, 'Poeta calculans. Harsdorffer, Leibniz, and the "Mathesis Universalis"', *Journal of the History of Ideas* 60/3 (1999), 449-467.

Westerhoff 2001
J.C. Westerhoff, 'A world of signs. Baroque pansemioticism, the polyhistor and the early modern Wunderkammer', *Journal of the History of Ideas* 62/4 (2001), 633-650.

Westerhoff 2005
J.C. Westerhoff, 'Logical relations between pictures', *Journal of Philosophy* 102/12 (2005), 603-623.

Weststeijn 2007
T. Weststeijn, '"Spinoza sinicus". An Asian paragraph in the history of the Radical Enlightenment', *Journal of the History of Ideas* 68/4 (2007), 537-561.

Weststeijn forthcoming
T. Weststeijn, 'Vossius' Chinese utopia', in: E. Jorink & D. van Miert (eds.), *The challenger. Isaac Vossius (1618-1689) and the European world of learning,* Leiden & Boston forthcoming.

Wilberg Vignau-Schuurman 1969
T.A.G. Wilberg Vignau-Schuurman, *Die emblematischen Elemente im Werke Joris Hoefnagels,* Leiden 1969.

Wilkins 1668
J. Wilkins, *Essay towards a Real Character and a philosophical language,* London 1668.

Wolf 1935
A. Wolf, 'An addition to the correspondence of Spinoza', *Philosophy* 10/38 (1935), 200-204.

Yates 1964
F.A. Yates, *Giordano Bruno and the Hermetic tradition,* Chicago 1964.

Yates 1974
F.A. Yates, *The art of memory,* Chicago/London 1974 (ed. princ. 1966).

Zoet 1675
J. Zoet, *d'Uitsteekenste digtkunstige werken,* Amsterdam 1675.

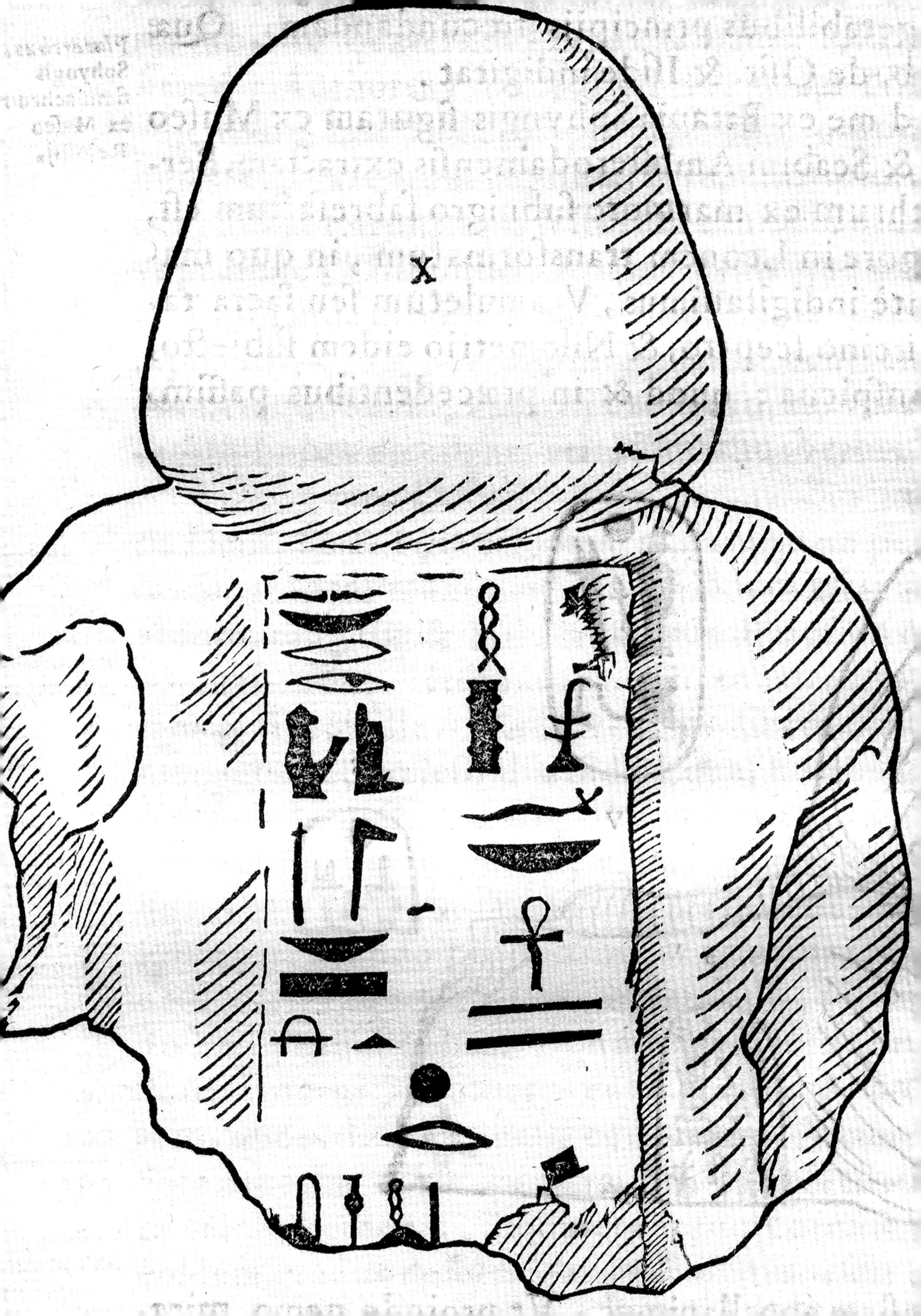
x

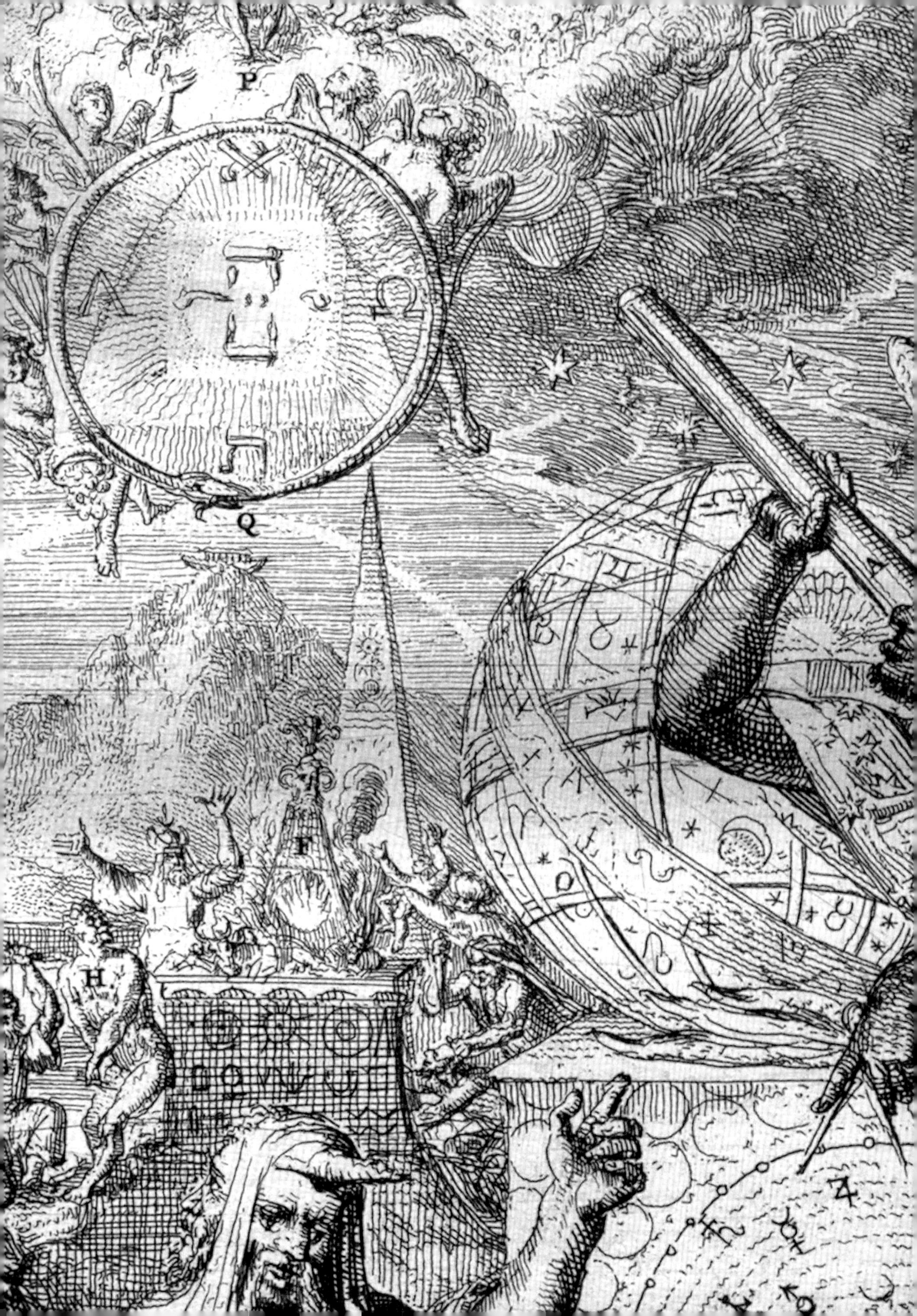
P
Λ
Ω
Q
F
H.

Art, science and religion in Romeyn de Hooghe's *Hieroglyphica*

Joke Spaans

Romeyn de Hooghe (1645-1708) is best known as the prolific producer of high-quality loose-leaf prints, book illustrations and frontispieces. He left an astounding volume of etchings and engravings, in a wide variety of genres, but all bearing his distinctive style. Near the end of his life, however, he also wrote two substantial books that so far have hardly attracted scholarly attention. The first one, his *Spiegel van Staat des Vereenigde Nederlands*, has recently been noted for its contribution to the history of republicanism.[1] His *Hieroglyphica of merkbeelden der oude volkeren, namentlyk Egyptenaren, Chaldeeuwen, Feniciers, Joden, Grieken, Romeynen, enz. Nevens een omstandig Bericht van het Verval en voortkruypende Verbastering der Godsdiensten door verscheyde eeuwen; en eyndelyk de Hervorming, tot op deze tyden toe vervolgt,*[2] published posthumously, reflects De Hooghe's mature views on religion. Among its more than 400 pages of text, De Hooghe placed 63 full-page etchings. Rather than being simply illustrations accompanying the text of his book, the etchings present his argument, whereas the text 'illustrates' – in its literal meaning, 'sheds light on' – the plates. Each plate and chapter is dedicated to a specific theme in the genealogy of the Christian religion.

As the title promises, De Hooghe uses many of the conventions of a genre that enjoyed some popularity during his time. Owing to the inspiration of the impenetrable Egyptian hieroglyphs, the sixteenth and seventeenth centuries saw the production of several books in which images and objects were treated as references to multiple layers of meaning. Besides the obvious 'literal' connection between an image and the concrete object it depicted, 'hieroglyphical' authors identified each and every *possible* meaning of an image (or word), with a predilection for metaphorical and moral connotations and hidden clues to ancient, occult wisdom.[3] In Chapter 2, De Hooghe claims to have expanded and modernised this genre in his *Hieroglyphica*. He argues that from antiquity, the production of hieroglyphs has always continued. And, as many of the symbols of the ancients have been lost – and, moreover, in many subjects modern scientists have superior knowledge – modern symbols, like those devised for use in navigation, distillation techniques or telescopy, are of even more use to his European readers than those derived from antiquity.[4] Scattered through the *Hieroglyphica* we find many references to the natural sciences. Usually they support De Hooghe's critique of the forms

Detail fig. 9
De Hooghe, *Toepassing,* 1690-1708 (detail: left margin).

of organised religion he saw in his own time and his proposals for further reformation, which are the mainstay of the book as a whole.

De Hooghe and the scientific knowledge of his time

Romeyn de Hooghe's claims to scholarship are shaky at best. In his youth he reputedly visited the Latin school run by Franciscus van den Enden (1602-1674), where the young Baruch de Spinoza (1632-1677) also went.[5] Later in life he gained – or, as rumour had it, bought – a doctorate in law at the University of Harderwijk.[6] His real education, however, came from a wide network of contacts among the intellectual elite and access to their conversations, libraries and collections. Despite his lack of formal education, it can be said that, in a very real sense, the prints in *Hieroglyphica* were the result of a lifetime of study.

In the *Hieroglyphica*, De Hooghe tells an anecdote, which perfectly illustrates how he, as an ambitious artist, participated in a network of regents and scholars. Considerable difference in social rank and level of education did not preclude forms of conviviality providing someone like De Hooghe with the latest news on discoveries in the natural sciences and on the debates they provoked in the field of theology. Fun was to be had here too, as in the cartoon that was the conversation piece at the occasion described by De Hooghe:

> Once, at the home of the late Lord Exchequer Burgh, I saw Atlas and Hercules struggling to keep the earth in its orbit, now that Copernicus had upset the system of the cosmos. Tycho Brahe had his shoulders set under the globe, and Ptolemy came to his help, screaming in distress, to prevent the world from shattering in its fall. Meanwhile Copernicus kicked in the crystal spheres surrounding the earth and under his foot extinguished their twinkling lights. Gassendi came running and with huge effort pushed the sphere from its mountain top, so that, once fallen from the shoulders of Atlas, the world attained uniform motion. The design was by the Lord of Zelhem and the Oxford Society. It also contained several wellknown stargazers and moonrummagers of our times coming to blows over the question whether the earth is stationary or moving.[7]

Here we find De Hooghe at the house of a prominent Dutch regent. The 'late Lord Exchequer Burgh' is Coenraad van den Burgh (1623-1669). He was a very wealthy man, son of an Amsterdam burgomaster, himself an alderman of that great city. As befitted a man of his stature he lived in a large house on the Heerengracht. He also possessed a manor and was entitled to call himself Lord of Kortenhoef. Since 1666 he was head of the Treasury of the States General. His wife was Cornelia Hooft, a daughter of the famous poet Pieter Cornelis Hooft (1581-1647), who was also a high government official in the Gooi area, east of Amsterdam.[8] The artist and the regent had enjoyed a satirical drawing or print on the lively debates on heliocentrism in the second half of the seventeenth century.

Perhaps the designer of the image, Christiaan Huygens (1629-1695),

Lord of Zelhem, had also been present. Huygens was the scion of a family that had been in the service of the Stadholders for three generations, and a renowned scientist. He may have brought the piece of artwork that engaged the company at the house of Van den Burgh with him.

Christiaan Huygens had a life-long passion for astronomy. He designed and used telescopes to observe the moon, the planets and the stars. He was the first to describe the rings of Saturn, and he discovered one of its moons, Titan. In his views of the world system, he held with the heliocentrism of Nicolaus Copernicus (1473-1543), and he worked to strengthen the arguments in favour of this theory, which during his time was not yet universally accepted.[9] He calculated the orbits, distances and sizes of the planets. He published several tracts on cosmology and corresponded on the subject with other scholars through the Royal Society in London and the *Académie royale* in Paris.

Aside from his serious work, Huygens also appears to have produced satire. The above-mentioned print or drawing – unknown so far, and probably lost – ridicules Ptolemy, the architect of the geocentric cosmos and its defender Tycho Brahe (1546-1601). The latter had devised an improved system of a geocentric cosmos, one that predicted the positions of the heavenly bodies as well as the Copernican system did. At the same time it salvaged the authority of Aristotelian philosophy and the Bible. Reformed theologians objected to Copernicanism because of its contradiction of these traditional authorities. The most adamant of them was Gisbertus Voetius (1589-1676), professor of theology and rector magnificus at Utrecht University. Voetius was no astronomer. His position in this debate was determined by his rejection of the philosophy of René Descartes (1596-1650). The latter's hermeneutics of systematic doubt struck many theologians as godless. Descartes's cosmology, in which he followed Copernicus, was just one of the dangerous innovations that, in the eyes of Voetius, threatened theology and religion.[10] Huygens despised these theological objections. Among the skirmishing defenders and opponents of geocentrism in the above-mentioned cartoon, Voetius may well have been included.

Besides the elusive cartoon, Huygens's *Cosmotheoros* may have a satirical meaning as well. The treatise – originally written in Latin but quickly translated into all main European languages – is written in the form of two consecutive letters to Christiaan's brother Constantijn Jr. (1628-1697). It is obviously firmly based on Huygens's own observations and computations. It describes the system of the world as it would appear not only to Earth-based observers, but also to the presumed inhabitants of the other planets. It does so with an impressive array of calculations and estimations of orbits and relative planetary masses.[11] The largest part of the first letter, however, is taken up with elaborate speculations about inhabitants of other worlds, the so-called Planetarians. Huygens argues that, as other planets have geological and meteorological characteristics – mountains, seas, and clouds – similar to those of Earth, it stands to reason that they also have life, and most probably sentient life. From there it naturally follows that these hypothetical Planetarians also have economies and cultures, providing them with everything from food and shelter to music, and even pocket watches like those Huygens himself had invented. Imperceptibly, Huygens's argument segues from bona fide scientific theory into literary pastiche. This part reads like a satire on the absurdities resulting from attempts to describe the natural world without proper recourse to observation and experiment. It seems directed at the anti-Copernican theologians. Their rejection of a plurality of worlds was based on theological claims that Man is the centre and final cause of Creation, and that it is sinful to inquire beyond that which God has revealed in Scripture.

The seventeenth century saw a small crop of literary fantasies about extraterrestrial life. All are utopian satires.[12] Like no other scholarly pursuits at that time, astronomy and cosmology demonstrated the need for a new natural philosophy – and for serious reflection on the underpinnings of theology and religion. We find this awareness explicitly formulated by the Amsterdam minister Balthasar Bekker (1634-1698) in his controversial book *De betoverde weereld*. He deplored the relative neglect of mathematics and astronomy – studies that did not give access to lucrative jobs, as did the traditional studies of law, medicine or theology – because nowhere was it more obvious that the orthodox worldview depended on 'prejudice', that is, upon blindly following traditional authority instead of open-minded rational inquiry.[13] Bekker's alternative readings of Biblical texts on devils, angels and demons, rejecting – and occasionally ridiculing – traditional authorities, came up against the same theological objections that also riddled the reception of Copernicanism.[14] In *Cosmotheoros*, Huygens likewise pokes fun at the unwillingness or inability of scholars to acknowledge the insufficiency of current theories to explain the more baffling observations of nature and to think outside their conceptual boxes.[15] It is probably no simple oversight that Huygens does not credit his watch-toting Planetarians with having a religion.[16]

Participation in the seventeenth-century debates on the compatibility of science and religion, and more generally on Cartesian philosophy, was not restricted to the learned world of the academy. Often the more original contributions were ventilated in socially mixed networks of intellectuals, meeting in bookshops and coffee houses, sharing their insights in correspondence networks, private libraries and cabinets of curiosities. Here we find leisured regents, technical professionals like engineers and navigators, and a growing class of men of letters who served as diplomats, agents, spies and newsagents. The meeting between Van den Burgh and De Hooghe is a good example. Some of these cultured amateurs rejected much of traditional scholarship and Christian doctrine in favour of a more rationalist worldview and/or new political philosophies. When they published their views, they usually did so in the vernacular, instead of academic Latin.[17] In these proposals for radical reform, ministers were often painted in the blackest of colours, as demagogues and rabble-rousers, who undermined the authority of the state and ridiculed modern scholarship before the ignorant masses.[18]

Attacks upon the established authorities in philosophy and theology were not without risk, however. From the middle of the seventeenth century the States of Holland repeatedly issued resolutions that prohibited any philosophical activity that might undermine the doctrine of the Reformed Church. Hefty fines and prison sentences discouraged straightforward publication of the more radical ideas.[19] Censorship, however, could be evaded. A literary underground of radical philosophical texts developed in the latter decades of the century.[20] Literature and the graphic arts offered an alternative way to represent controversial ideas covertly – that is, in such a way that they would be noticed only by a discriminating intellectual audience. The established poets kept aloof, but the satire of *Cosmotheoros* is an example of such a literary disguise. References to the new cosmology could also be shown in the ornamental borders of expensive decorative wall-maps and the plates of atlases.[21]

Perhaps the most spectacular example of a radical-philosophical argument presented in the pictorial language of allegorical print is Romeyn de Hooghe's *Hieroglyphica.* In this work, De Hooghe develops a critique of early modern religion, based, among other things, on the new cosmology. Here De Hooghe proves that he was confidently at home in the post-Cartesian intellectual climate (fig. 1), where new discoveries in the natural sciences and reports about exotic cultures were evaluated in the light of the ongoing study of antiquity and the Bible, transforming the traditional understanding of the world based on Aristotelian principles. He was, moreover, very well informed on current political and theological discussions. Religion had itself become a focal point in early Enlightenment thought, and De Hooghe's *Hieroglyphica* offers one of the more radical contributions to this field.[22] Placed within the context of De Hooghe's 'religious' *oeuvre*, it becomes an eloquent blueprint for an enlightened spirituality.

From Catholic devotion to enlightened critique of religion

De Hooghe does not have a reputation for religiosity: quite the contrary. In the early 1690s he was the target of an incredibly vile smear campaign, originating with anti-Orangist Amsterdam regents. As an avowed partisan of William III (1650-1702), De Hooghe had refused a commission to lend them his pen and etching needle. They retaliated with a flurry of defamatory pamphlets, portraying De Hooghe as a kleptomaniac, a sexual pervert and an atheist. None of these accusations could be substantiated, but De Hooghe, who gave as good as he got in the exchange, was never officially rehabilitated. By that time he had left Amsterdam to settle in Haarlem, where he opened a drawing academy. There he lived as an honourable member of the urban elite and a minor regent. Not only did he receive several commissions from the magistracy, but also he was considered fit to serve as deacon in the Reformed Church and as judge in one of the lower courts of justice. Of course the juicier pieces of slander could never be unsaid, however, and have blackened his reputation ever since.[23]

Although clearly a libertine, De Hooghe was probably not the stark atheist later historiography has made him. Much of his work is concerned

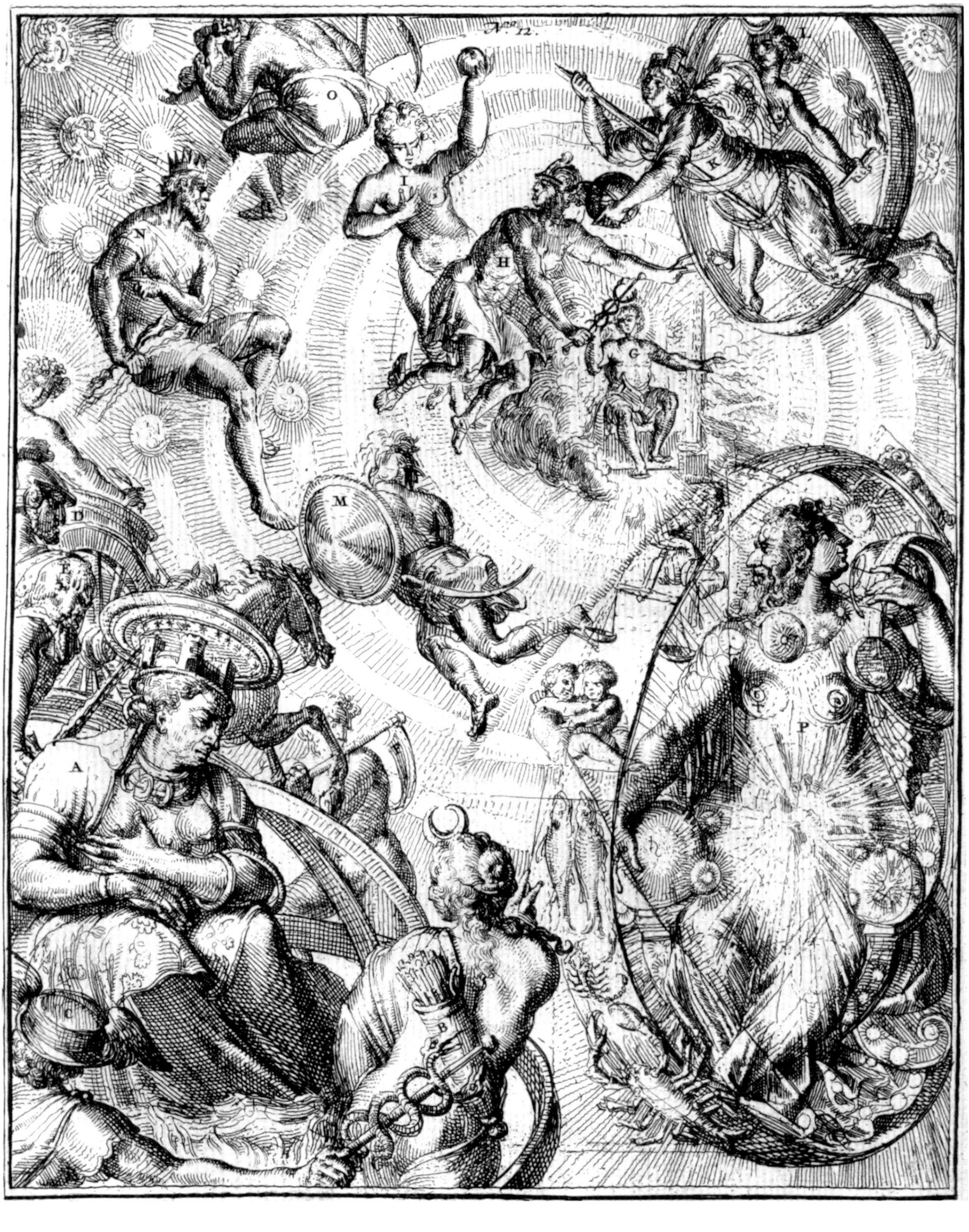
O
L
I
K
N
H
G
M
D
E
A
F
P
C
B

with religion. From the very beginning of his artistic career he acquired commissions for book illustrations, many of them for works of religious edification. In itself this is not surprising: a large proportion of early modern book production was made up of theology and devotion.[24] He illustrated religious works by authors of a wide variety of confessional backgrounds. He also contributed title pages and illustrations to Lutheran and Reformed Bibles. This ecumenicity was not unusual at the time.[25] In the case of De Hooghe, however, it clearly fed into a personal view of True Religion.

Many of his book illustrations are quite literal depictions of the texts they accompanied. However, in the allegorical title pages, De Hooghe often presents a personal interpretation. Here he shows a consistent concern for True Faith, or the invisible True Church, that does not coincide with any of the formal confessions or religious groups of his time. The title print of *Het Voorhof der Ziele*, a volume of religious emblems by the Catholic poet Frans van Hoogstraten (1632-1696),[26] is fairly typical. It shows the figure Pious Soul, approaching his heavenly destination. He wears a tunic printed all over with religious symbols. In his hand Pious Soul carries a burning heart, symbol of love, and he goes barefoot as a penitent. His head is crowned with laurels, and an angel guides him towards Christ who meets him with wide-open arms. In the background a steep stairway leads to Heaven. Behind the figure of Christ we see an allegory of the Pious Soul: a naked man, carrying a cross towards Heaven.

Cross-bearers occur often in prints by De Hooghe. They refer to the theme of the Imitation of Christ and seem to be derived from earlier Jesuit devotional works. Boëthius à Bolswert (c. 1580-1633) put them in the illustrations he made for Antoni Sucquet's *Via vitae aeternae* (1620), and we also see them on the engraved title page of Herbert Rosweyde's *De Naervolginghe* (1631). One of the illustrations in *Via vitae aeternae* even has hovering angels, or probably the winged souls of the saved, carrying crosses. We find a very similar image on a later title page by De Hooghe, where Christ himself is shown soaring among the clouds, carrying his cross.[27]

Among the authors whose work De Hooghe illustrated were prominent politicians like Constantijn Huygens and Nicolaes Witsen (1641-1717). In the course of his career De Hooghe drifted more and more towards collaboration with authors with known enlightened views. Apparently he was in a position to be discriminating. He contributed to quite a few antiquarian and historical works critical of religion in general and of the official, State-protected Churches of the time in particular. The allegories he designed for these books are often much more complex. Examples are his illustrations for *Verhandeling van de oude Orakelen der Heydenen* by the Haarlem Mennonite physician Anthoni van Dale (1638-1708)[28] and *Historie der Kerken en Ketteren,* the Dutch translation of the well-known critical church history by Gottfried Arnold (1666-1714), Lutheran minister in Giessen, Germany.[29]

Van Dale gives a very erudite, but also very sceptical, description of the various forms of divination in classical antiquity. He systematically

1

De Hooghe, *Van de loop van Hemel en Aarde,* 1690-1708

etching, *Hieroglyphica*, plate 12

Three cosmological models: in the lower left corner, the Ptolemaic system, with Earth as an old but fertile woman (multiple breasts, flowered skirt) surrounded by a complicated array of circles. In the lower right corner, an equally contrived system found among the peoples of Siam. In the upper half, the Copernican view of the solar system, with the Sun (G) in the centre; orbited by Mercury (H), Venus (I), Earth (K) and Moon (L), Mars (M), Jupiter (N) and Saturn (O). The text mentions Saturn's five moons. Note that the solar system is surrounded by infinite space, filled with other stars, each with their own planetary systems, reminiscent of the Cartesian theory of 'vortices'.

compares the oracles of the ancient world with claims of exceptional spiritual gifts or demonic powers by his contemporaries. The latter had unfailingly been proven false, and even worse: as fraudulent ploys to part gullible fools and their money.[30] De Hooghe's illustrations provide impressions of what the sanctuaries housing the oracles might have looked like. One print, however, resumes the argument of Van Dale's entire work. False Religion is here depicted as an idol in the form of a (golden) calf accompanied by two female figures. One is veiled and masked and stands on cat's feet: symbols of falseness and deceit. She holds the consciences captive, depicted as two prisoners in chains, carrying the attributes of inconstancy and blindness. The second female figure has five breasts – a reference to the ancient cult of Diana in her embodiment of fertility. Her head has three faces and is crowned by a mousetrap. She represents the temptation of superfluous knowledge.

Only one figure in this crowded print can escape the lures of False Religion: a donkey. This humble animal has its eyes fixed upon a crucifix, which is firmly rooted in a rock next to an open Bible and a skull, denoting the pious contemplation of mortality and salvation. Patiently, the donkey suffers the weight of a fat monk marked with the donkey's ears of stupidity – a devotee of False Religion and an icon of priestcraft. In the background we see a temple court surrounding a high mountain. It is inspired by the popular image of the Tabula Cebetis, a Renaissance allegory on the itinerary of the human soul with its ethical choices. Here sacrifices are being offered on altars decorated with crosses. Small human figures carry crosses along a steep road to the top of the mountain. Again, self-sacrifice and imitation of Christ are presented as the marks of True Religion. Emphatically, the clergy of the established Churches is shown as lacking knowledge of True Religion, in contrast to the simple souls, represented in the humble beast of burden.[31]

For the Dutch translation of Arnold's *Unpartheyische Kirchen- und Ketzer-historie*, De Hooghe made the title print. It shows Naked Truth: a nude woman, crowned with a wreath of laurels and carrying the standard of Christ in her right hand. Like her Lord, she rises from an open grave, triumphing over False Religion, which is embodied in four allegorical figures. Two of these symbolise priestly tyranny over the consciences of the faithful, carrying as attributes, among other things, the many-breasted Diana, a mask and cat's claws. One of them has fallen on his face. Two others represent the collusion of religion with political power: a Moor with a drawn sabre and a helmeted bishop carrying a world orb, bolts of lightning and a bunch of keys.

These emblems of oppression are contrasted with figures representing True Faith: one of them a simply dressed, venerable old man with his hands joined in prayer. Elsewhere in the work of De Hooghe we find a similar figure standing for the persecuted Church 'under the Cross', the very subject of Arnold's work.[32] The old man rests his foot on the prone priest figure, sharing in the triumph of Naked Truth. The other represents the method for gaining truth. It is a sitting man, with his back to the viewer and his gaze fixed upon the central figure of Naked Truth. His tunic is covered with wide-open eyes. The wings of quick intellect adorn his head. In his lap he has an open book. In a balance he weighs a sword and a pillar: symbols of attack and defence in debate. Condescendingly, he wards off fame and fortune, represented in a trumpet, a cauldron full of coins and the coiled snake of envy. Like Arnold's text, the print calls for unfettered and unbiased inquiry into the mysteries of religion. Such an inquiry requires clear vision, sharp intellect and a careful weighing of arguments, without considerations for worldly gain.

Many elements from these title pages – with their emphasis on inner piety, free inquiry into religious truth and the rejection of ecclesiastical power – return in De Hooghe's posthumous *Hieroglyphica*. Here, however, we also find the reflection of an interest in the natural sciences and in the 'cultural anthropology' inspired by the voyages of exploration and the enormously popular travel literature. As with True Religion and the enlightened critique of religion, we can trace De Hooghe's interest in these fields through the books to which he contributed his artwork. The impressive list of these books contains microscopical studies of Antoni van Leeuwenhoek (1632-1723); medical books by De Hooghe's good friend Govert Bidloo (1649-1713); Francesco Redi's (1626-1697) work on insects; and descriptions of travels to China, the Indies and Muscovy. In the same way, books on antiquity, law and history for which he produced plates and title pages probably informed De Hooghe's *Spiegel van Staat*.[33]

An artistic critique on religion

It is not an easy task to give a good impression of *Hieroglyphica*, and plainly impossible to do justice to its complexity in only a few pages. Although book-sized, the work does not offer a continuous text. For each of the plates, De Hooghe wrote an extensive legend, providing an audience less erudite than he was himself with the information necessary to grasp the significance of its various elements. Like the plates themselves, the text is often open to various readings. It is left to the reader to construe the overall argument of the book from these disjointed and, probably deliberately, ambiguous explanations. The 63 etched plates are crowded with allegorical figures and abound in symbolism and learned references.[34] Despite De Hooghe's declaration of intent in the introduction, it is not quite an iconology in the sense of Cesare Ripa's *Iconologia* (1593) or Johannes Pierius's *Hieroglyphica* (1566).[35] De Hooghe refers to the difficulties artists often encountered when they had to depict the ancient gods correctly and gallantly offers to share his deep erudition in this field with fellow artists lacking in formal education, but the plates in his *Hieroglyphica* are absolutely unsuited as models to be copied for drawing.

Many emblem books claim usefulness as a treasure trove for later artists, and De Hooghe's introduction follows this convention of the genre.[36] In doing so, however, De Hooghe also self-consciously presented himself as a master of his art. At the time the *Hieroglyphica* was made, De Hooghe was the director of a drawing academy in Haarlem.[37] As such, he was, in a way, the successor of Carel van Mander (1548-1606), who had a similar school there a century earlier. In fact the *Hieroglyphica* may have been inspired by Van Mander's *Schilderboeck*. The *Schilderboeck* was

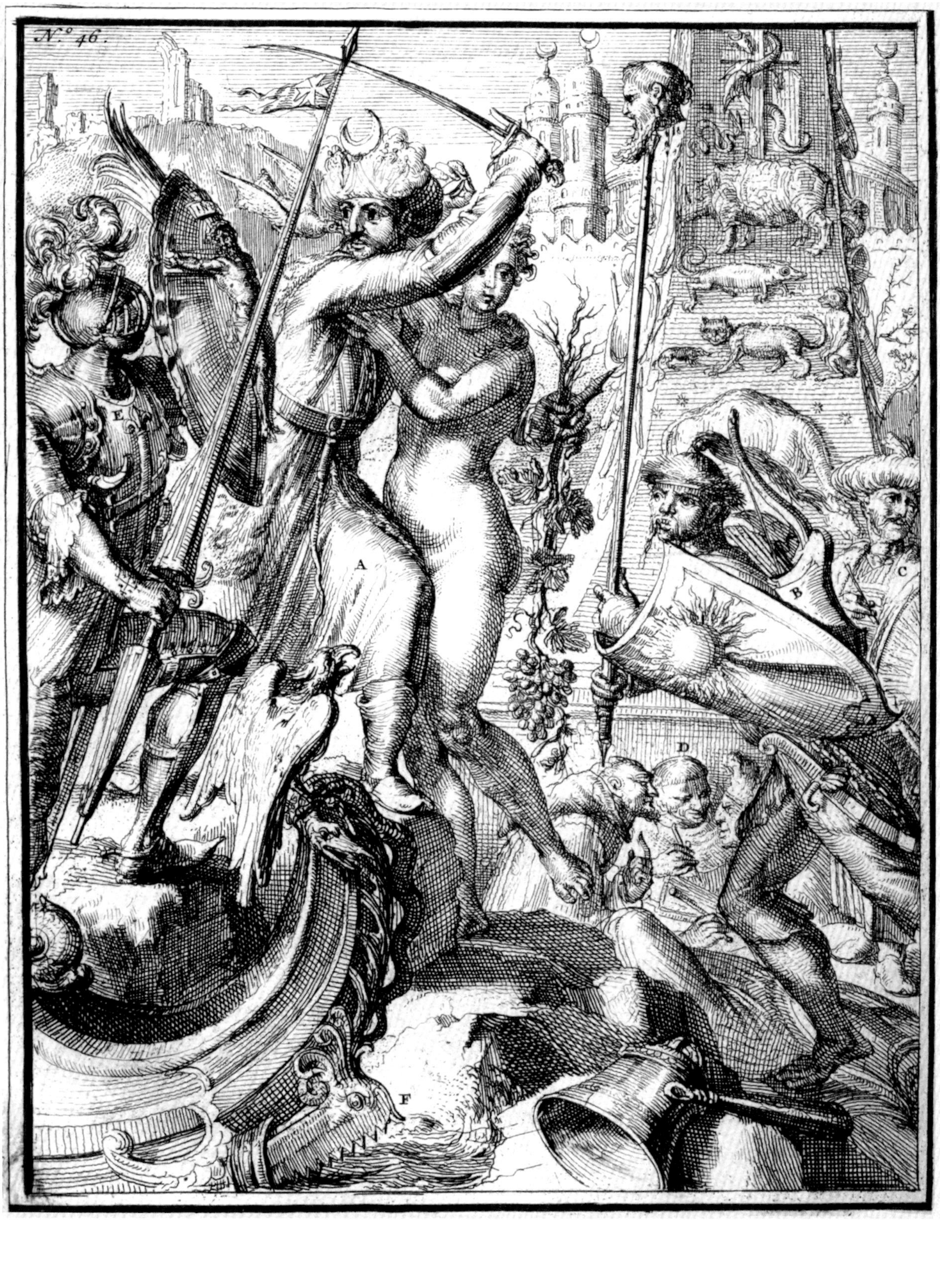
N.º 46.
A
B
C
D
E
F

published in 1604, and besides a compendium of famous painters, it contains a eulogy to the art of painting and its ability to edify and entertain at the same time.[38] The first chapter of the *Hieroglyphica* eloquently discusses the similar, or even superior, powers of the emblem.

The *Hieroglyphica* may even have been intended as a homage to the work of Van Mander and to the Haarlem magistrates who had offered De Hooghe citizenship when he moved to their city in 1679 after his conflicts with the regents of Amsterdam. On one of its plates he included a ship with a sawtoothed keel, the invention claimed by Haarlemmers as the key to success in the capture of Damietta during the Fifth Crusade, and the central icon of Haarlem's civic pride (fig. 2).[39] In the accompanying text the glorious deeds of the medieval Haarlemmers are explicitly compared to the valour of the Dutch navy that used their technique in capturing the English flagship at Chatham (1668) and a Spanish silver transport in the bay of Vigos (1702).[40] The penultimate chapter contains a veiled reference to the controversial work *Den Hemel op Aarden* (1703) by Frederik Leenhof (1647-1712), a Reformed minister of Zwolle and admirer of Spinoza (fig. 3).[41] De Hooghe must have completed the *Hieroglyphica* between 1703 and 1708, the year he died, a full century after Van Mander's

2

De Hooghe, *Van de Mahometaansche beginselen,* 1690-1708

etching, *Hieroglyphica*, plate 46.

This allegory on Islamic culture contains the image of a Crusader (left margin) disembarking from a ship with a sawtoothed keel, the emblem of the city of Haarlem. Contemporary views on Islam often emphasised its violent character; De Hooghe shows here that violence is anything but an exclusively Islamitic vice.

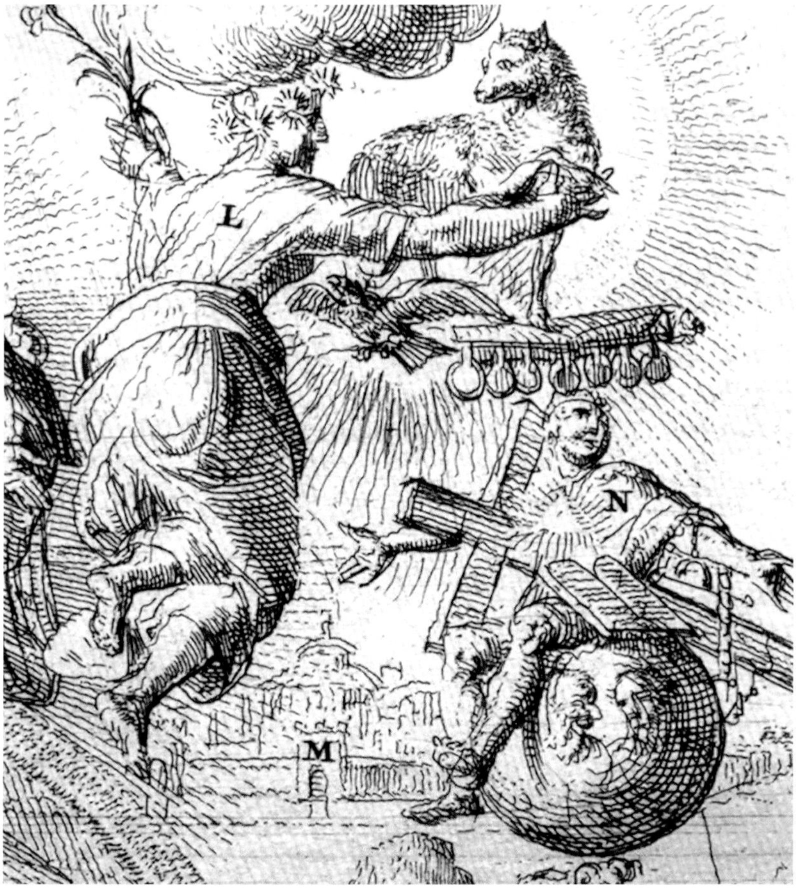

3

De Hooghe, *Van verscheyde gewaande en den waaren Hemel,* 1690-1708 (detail: centre, top third)

etching, *Hieroglyphica*, plate 62

This print as a whole is filled with the fictitious imaginings of the afterlife. By contrast, this group shows the real Heaven. The True Church (L), devoid of political aspirations, ascends towards God, interested only in the spiritual, Heavenly Jerusalem (M). The Truly Faithful (N) has his Heaven on Earth: knowing that Christ has redeemed him on the Cross, in life he blissfully rejoices in God's mercy. He is shown as a laughing man, who takes whatever life brings him as God's gift and does not fear death.

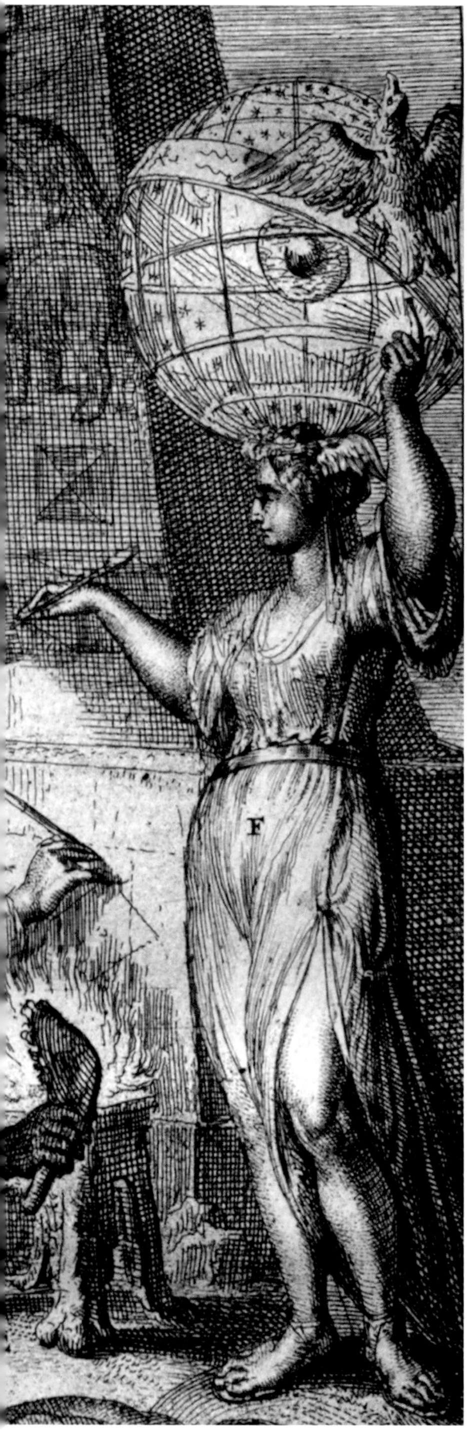

Schilderboeck. By including the Damietta ship, De Hooghe shows he was prepared to contribute to the civic patriotism of his new city of residence.[42]

The main thrust of the *Hieroglyphica*, however, is a critique of the religion of De Hooghe's day. He introduces his subject in the title of the book as 'the progressive decline and corruption of religion through the ages, and its recent reformation continuing until the present day'. At first sight the book provides a chronicle of the history of religions. We see the many gods and demons of the pantheon shared by the civilisations of antiquity, each with their attributes, followed by plates illustrating the rise of Christianity and its relations to the religions of the world. On closer inspection, however, the book is anything but antiquarian. Rather than a history, the *Hieroglyphica* offers a _genealogy_ of religion.[43]

De Hooghe argues that religion originated in the science of astronomy as it was first developed in the ancient Middle East. The astronomers of those days reached high levels of sophistication, which enabled them to determine the best times for sowing, planting and harvesting. They were the first natural scientists. The common people, however, ascribed their knowledge, gained from careful observation and computation, to a presumed access to divine mysteries. In this way, the astronomers, unwilling to disabuse the people of their misconceptions, acquired enormous prestige and a priestly status. Princes deftly exploited the reverence of the common people for the scientist-priests. Between them, princes and scientists produced a cult of the gods, personifications of the forces of nature and deified rulers and heroes. By the grace of these invented gods they won from the common people obedience to their rule and their taxation regimes (fig. 4).

The thesis that all religions are descended from the mythography of the ancients, based on antique astronomy and natural philosophy, runs as a continuous argument through the entire book. Everywhere, the power of princes and the wealth of noble and priestly elites have been based on the local form of religion, however varied the world's religions may seem on the surface. De Hooghe exploits the knowledge – accumulated in the antiquarian scholarship, merchant ventures and voyages of exploration of the Golden Age and made accessible in an extensive travel literature – to present a kaleidoscopic picture of ancient and contemporary religious beliefs and customs. He states that the nature of the various peoples and the degree of religiosity found among them are dependent on the climate zones. Moreover, religions are not static entities, but develop over time, as the cultures in which they are embedded develop. But underneath this dazzling variety in outward appearance, and their variability over time, all religions essentially share this same root in 'applied science'. Christianity is no exception to this rule: it is simply the religion best suited to the nature of Europeans.

The first half of the *Hieroglyphica* presents ancient conceptions about gods, the cosmos, the world, mankind and the human soul. These plates often show the starry heavens, armillary spheres, the signs of the zodiac, and other symbols for the heavenly bodies. From time to time De Hooghe complements his presentation of ancient myths with more recent *faits divers* on exotic peoples and their religious customs. Throughout his work, he compares these mythical and historical data with the biblical narrative. He gradually steers his readers from the pagan world of antiquity to contemporary views on world history, as constructed by humanist scholars on the basis of biblical histories and the chronologies extracted from the Bible. Implicitly, the Bible and Christianity are presented as the products of human projection, as, by their nature, all religions are. De Hooghe boldly states that the most clear-sighted individuals have always been aware of this: the mysteries of religion are 'a great nothing'. But by exposing this 'bowl of Roses' – a euphemism for a used chamber-pot with its malodourous contents – one risked bloody persecution by priests and princes.[44]

The projection theory underpinning the entire argument of the book is further shored up by repeated references to the discussions on philosophy and discoveries in the natural sciences. De Hooghe proves himself well aware of these contemporary discussions and how they transformed the traditional discourse on the history of salvation and undermined belief in the factual truth of Scripture. When presenting the teachings of various religions on the structure of the cosmos, he uninhibitedly ridicules belief in comets and other phenomena as the harbingers of disaster.[45] He scorns the rejection by theologians of the Copernican system. He presents the old Ptolemaic system as a heathen invention, as much as the model of the cosmos from Siam that he had seen in the collection of the father of Jan Swammerdam (1637-1680) (fig. 1).[46] In this context he repeatedly refers to the hypotheses about sentient life on the moon and other celestial bodies as so much poetic licence, meant to taunt overzealous bigotry.[47]

In ironical asides about the reliability of biblical history he draws attention to the inconsistencies and outright contradictions in the biblical narrative. These had been discussed in learned treatises by generations of biblical scholars, but had only recently been presented to a more general audience by radical philosophers like Thomas Hobbes (1588-1679) and Spinoza – and by a rebellious theologian like Balthasar Bekker.[48] In his treatment of prophecy De Hooghe sneers at the 'theologia emblematica' popular among theologians of his days: the art of finding allegorical references to profound religious wisdom in 'ordinary' words and names in the Bible – for instance, in the rather tawdry and technical descriptions of the building of Solomon's Temple.[49] And he shares the sarcasm of Bekker about the popular beliefs in demons and devils in a covert reference to the latter's controversial book *De betoverde weereld*.[50]

The chapters on the ancient gods soon become rather repetitive. De Hooghe seems to have spun his classical mythology out to such an extent with an eye towards the overall composition of the book. In the division of subject matter over the 63 chapters, De Hooghe mirrors the history and morally questionable achievements of religion in pagan antiquity, in chapters up to and including 31, against the gradual corruption of religion in the Christian era from chapter 33 to the end. Chapter 32, exactly in the middle of the *Hieroglyphica*, represents the Incarnation of Christ. This dogma is not represented in a traditional nativity scene, however. For De

4
De Hooghe, *Rakende de naam en eerste gang van de Hieroglyphica of Beeld-spraak-Konst in het gemeen* 1690-1708 (detail: right margin)
etching, *Hieroglyphica* plate 2
Lady Philosophy, her head winged to denote quickness of mind. She researches the heavens (on her head) and the Earth (on which she stands). She records her findings with a sponge-capped stylus, always prepared to amend unsatisfactory formulations. She points to the eagle that flies vertically upwards, the symbol of speculative theology. Fully informed about the natural causes of comets and other frightening phenomena, she none the less interprets them for the common people as signs of divine wrath, to keep them in obedience. This may be deceit, but it is politically expedient.

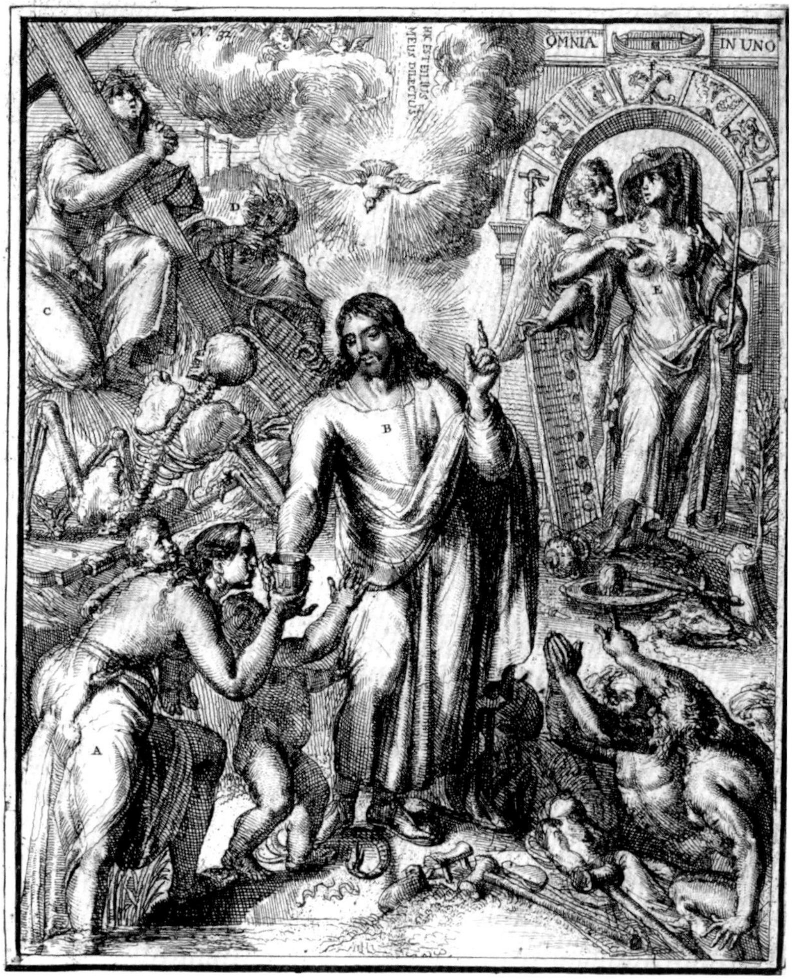

5
De Hooghe, *Van Christus in 't vlees*, 1690-1708
etching, *Hieroglyphica*, plate 32
De Hooghe's True Church, depicted in three stages: in the lower left corner, the simple soul following Christ (A). She stands with one foot in the river Jordan (baptism), and she accepts from Christ's hand the cup of the Last Supper. Above her, the Church that accepts the Cross, overcomes the strictures of the ceremonial Law, and against which Death and the Gates of Hell do not prevail (C). Finally, the Church entering the gate of heaven, surrounded by references to the biblical prophecies (E). The cornucopia she carries denotes their fulfillment.

Hooghe the embodiment of Christ is the Christian Church, represented here in a triad of female figures: as a humble, half-naked mother surrounded by small children, stretching out her arms to Christ; as a woman worshipfully clasping the Cross; and as the accomplished Christian soul, enlightened by the Spirit to see through the outward forms and ceremonies. In this triad De Hooghe constructs an ideal Christianity: a church of true believers, followers of Christ, for whom power, religious authority and theological learning are irrelevant (fig. 5).[51] As the Incarnation is the heart of the Christian Scriptures, so this 32nd chapter is the heart of the *Hieroglyphica*, strategically placed in the centre of its composition – at the price of repetitiousness in the chapters on the classical pantheon.

The factual history of Christianity, however, presents us – according to De Hooghe – with the opposite of this spiritualist ideal. The Early Church was rent by theological conflict, scorning simple piety. The central message of the Gospel that the law had been fulfilled with the coming of Christ could not prevent the imposition of new laws and religious obligations by precisians who pretended to know better than the truth revealed in the Bible.[52] This imperiousness of self-proclaimed ecclesiastical authorities harmed the Church more than the persecutions of pagan princes and led to persecution among Christians. The Peace of God's Church can be attained only when the Church is content with the central tenets of the Christian faith and leaves individual Christians completely free in their study and interpretations of the sacred mysteries.[53]

Prelates and theologians are rarely presented in a favourable light in the *Hieroglyphica*. By accepting their authority and obeying their commands, the common faithful fall prey to slavish dependence, existential doubt, anguish of conscience or theological pedantry.[54] De Hooghe displays a boundless confidence in the natural religiosity of the common people. This is graphically represented in the figure of Balaam's Ass.[55] This animal plays a curious role in the biblical story of the Old Testament prophet Balaam. Seduced by a large sum of money, the prophet had consented to curse the Israelites so that the King of Moab could defeat them in battle. When God sent an angel to intercept Balaam on his way to the battlefield and prevent this evil plan, it was the humble ass that saw the angel and stopped in its tracks, whereas the prophet himself was blind to the heavenly messenger (fig. 6).[56] De Hooghe uses Balaam's Ass as the emblematic representation of the simple faithful who have, as if by instinct, a deeper insight into the nature of religion than kings and prophets, prelates and theologians. In this sense Balaam's Ass figures in two rather central places in the *Hieroglyphica*, in the chapters on the decline of the Church and on the dawn of reformation.[57] Innate religiosity is not typical for Christian lay believers only: in a plate on the nature of Islam, the more exotic dromedary takes the place of the ass in representing the piety of the Muslim faithful – sober, modest, trustworthy and industrious – again as opposed to the cruelty and deceitfulness of their political and religious leaders.[58]

The divisions of Christianity into rival Churches strengthened the

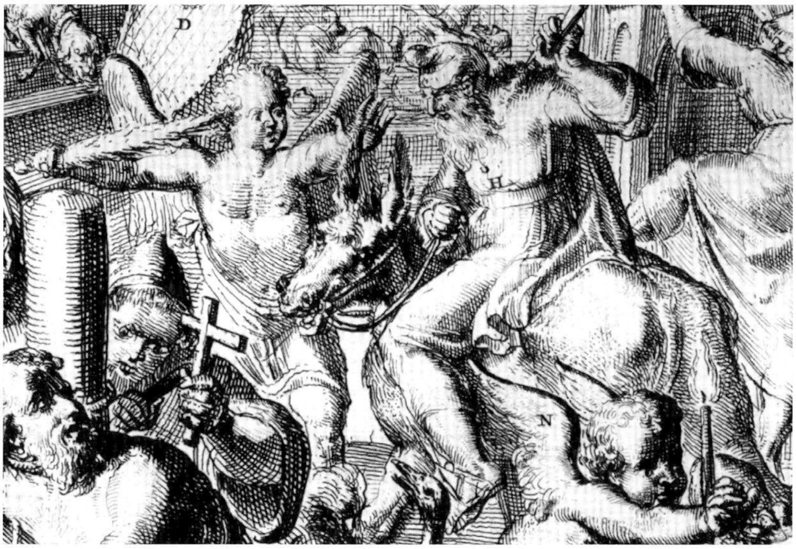

6

De Hooghe, *Van de aankomende Hervorming*, 1690-1708 (detail: centre)
etching, *Hieroglyphica*, plate 56
Balaam's Ass refuses to heed the blows of its unseeing rider as it sees the Angel blocking its path. This image is applied in the text to the simple people of the high valleys of Savoy, Bohemia, the Alps and Pyrenees (Waldenses, Hussites, Albigensians), persecuted for their faith, but eventually redeemed as torchbearers of reformation.

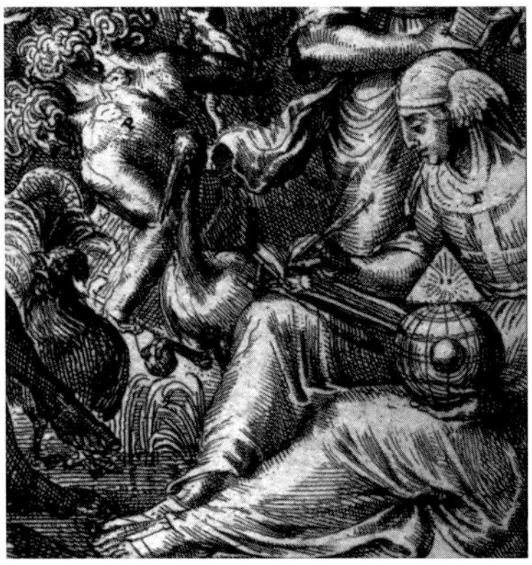

7
De Hooghe, *Van den Oorsprong en Voortgang der Ordens in de Roomsche Kerk*, 1690-1708 (detail: lower right corner)
etching, *Hieroglyphica*, plate 41
Praise for the intellectual labour of the medieval monks, who studied both nature (world orb) and its Creator (triangle with the All-Seeing Eye), teaching and edifying the faithful with their writings.

8
De Hooghe, *Van de Gereformeerde Godsdienst*, 1690-1708
etching, *Hieroglyphica*, plate 61
De Hooghe shows Reformed Religion as a form of True Religion. To the left of the central figure are two sisters, representing the schools of Voetius and Cocceius. During De Hooghe's lifetime these two schools had been embroiled in bitter controversies. Here they stand for the devout life and speculative theology respectively, as two complementary sides of Reformed Religion. Note the two figures in the lower left corner: the Synod of Dordrecht (F) protected by the States of Holland.

power of secular and religious authorities. Persecution of heretics – from Albigensians (presented as precursors of the Huguenots) to Muslims – served to increase their power and wealth. The Crusades produced the Military Orders as the storm troopers of the Catholic Church. Although De Hooghe rejects monastic rules as a straightjacket that inhibits natural religiosity and leads to hypocrisy and other vices, he praises the monastic virtues of voluntary poverty, obedience, scrutinizing of the conscience, chastity, discipline and industry (fig. 7).[59] These he considers not as specifically monastic, but rather as common Christian – or even human – virtues. The only Catholic saint to receive a separate treatment in the *Hieroglyphica* is St. Christopher, characterised by De Hooghe as a mirror of Christian virtue. The legendary St. Christopher was converted from a life of vice and purified his soul by carrying the Christ Child, heavy with the sins of fallen humanity, over the water. For De Hooghe this makes him an example of true piety, and the opposite of the corrupted Balaam.[60]

The pretentions of the medieval Church, supported by the power of princes and the authority of Councils and canon law, were undermined by 'The Dawn of Reformation'.[61] For De Hooghe this phrase does not indicate a specific historical period, but rather a conglomerate of medieval heretical groups. By the end of the seventeenth century these had been co-opted into the genealogy of Protestantism because of their rejection of ecclesiastical institutions, rituals or specific points of doctrine (fig. 6). De Hooghe here follows the views of Gottfried Arnold, with which he was well acquainted, as he designed and etched the illustrations for the Dutch translation of his *Unpartheyische Kirchen- und Ketzer-historie*.[62] For De Hooghe, reform of the Church had not been completed with the establishment of Protestantism. He highly praises the efforts of Martin Luther (1483-1546) and Frederick III, Elector of Saxony (1463-1525), for their contributions to reformation, but the Churches that originated in their wake created opportunities for new forms of religious fraud, power play and the contentiousness of theological schools. De Hooghe devotes a plate and chapter specifically to forms of 'misguided reformation', which disrupt Christian commonwealths through hypocrisy, blasphemy, schism and wars of religion.[63] The Protestant Churches were therefore in need of continual reformation. He presents his ideal for Dutch Reformed Protestantism as a (provisional) end station of this process. According to this ideal, the Church should teach simple imitation of Christ, allowing for a variety of devotional styles, under the tutelage of political power.

Remarkable in this respect is De Hooghe's treatment of the Synod of Dordrecht and the States of Holland: together they represent for him the highest authority in the Dutch Reformed Church (fig. 8). The opinion that the Church was subject to the sovereign power of the State and was bound to maintain the doctrines and church ordinances established by political authority was not uncommon in the Dutch Republic,[64] but against the background of the larger argument unfolded in the *Hieroglyphica* we may confidently see this as a reference to the political philosophies of Hobbes and Spinoza. When the authority of the Church is thus firmly bridled by the State, it can focus upon its true mission: the cultivation of piety. De Hooghe represents true piety in the guise of two

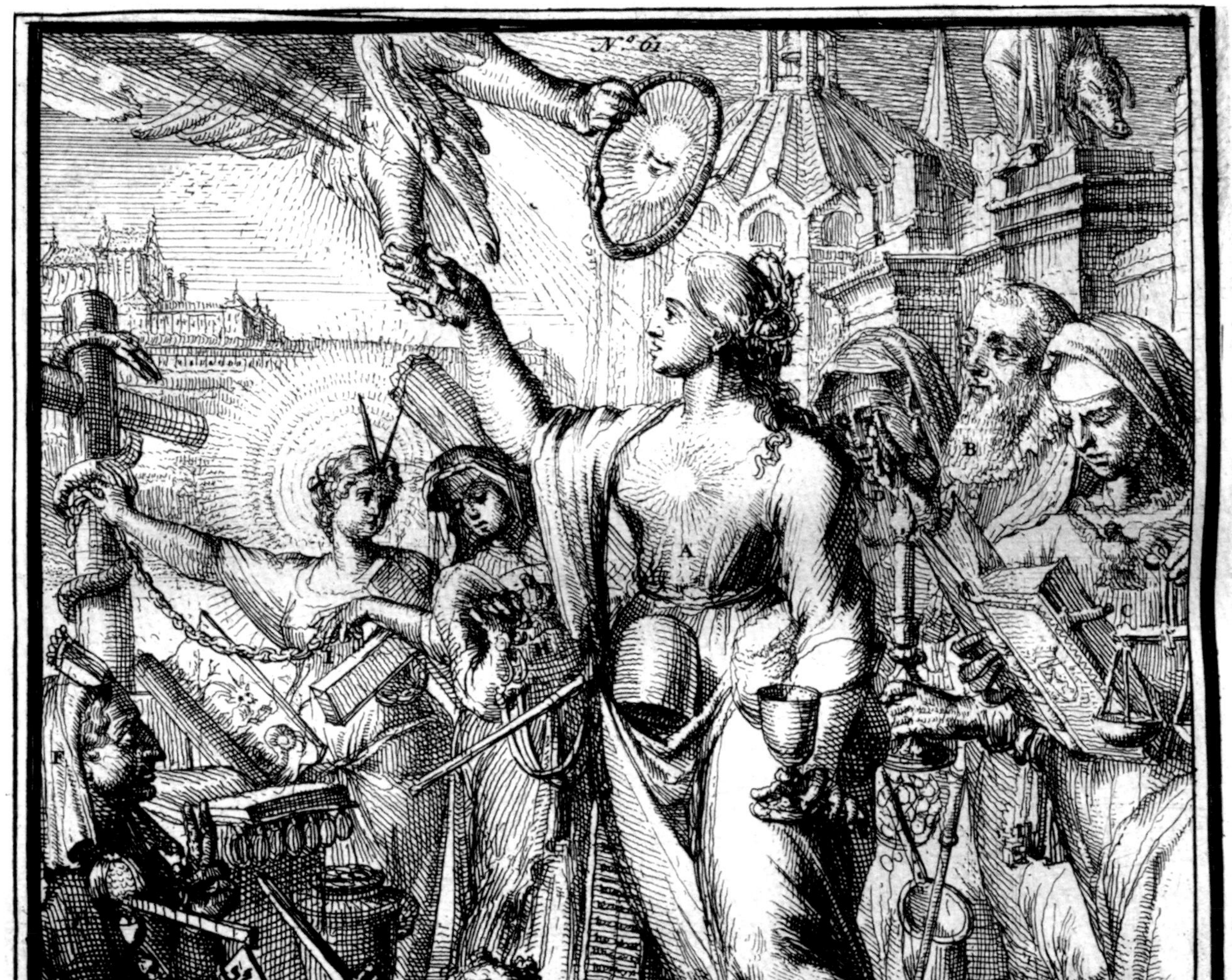
N° 61

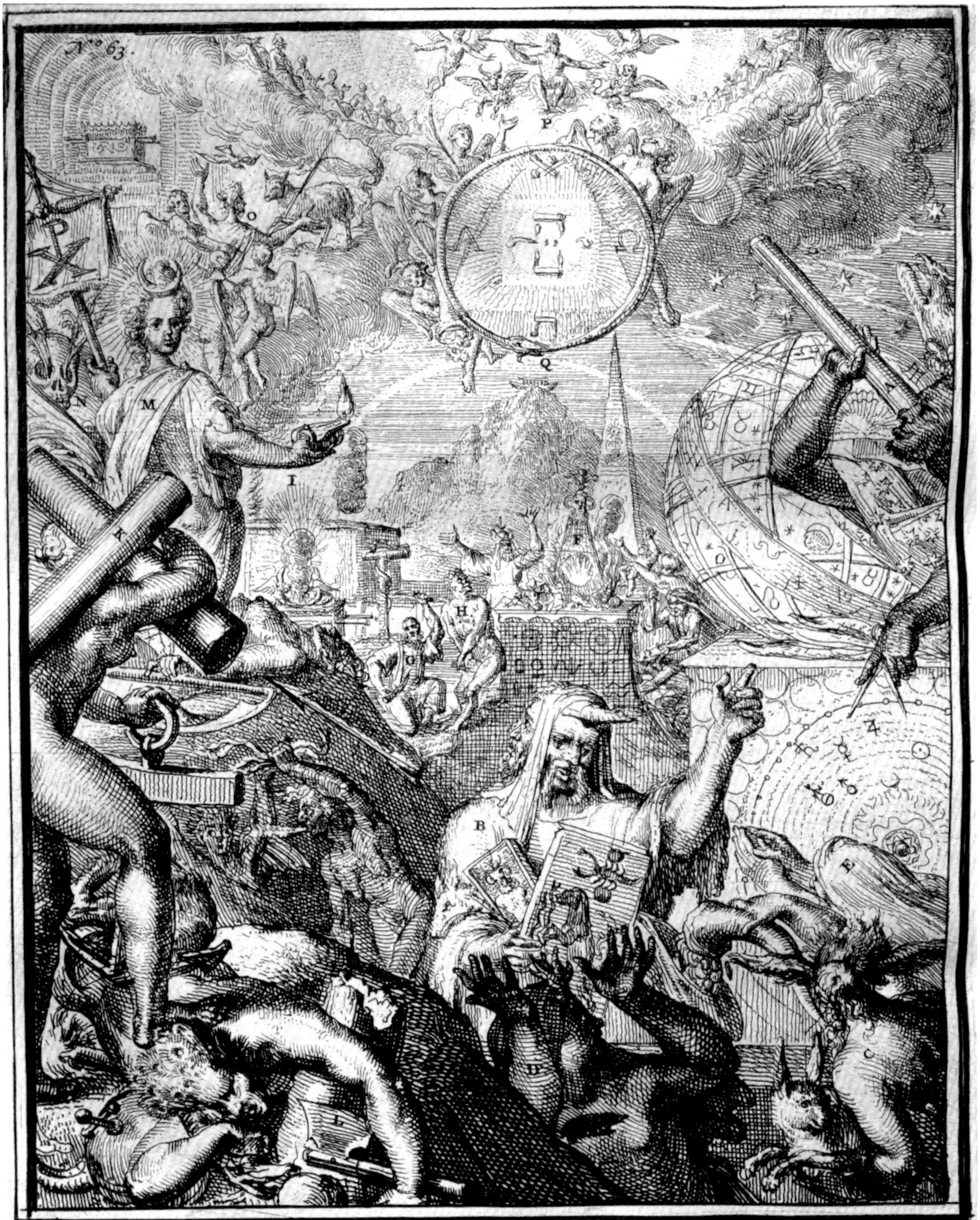
N° 63.
P
O
Q
N
M
I
K
F
H
G
B
E
D
L

sisters, practical and theoretical theology, the embodiments of the schools of Voetius and Johannes Cocceius (1603-1669). As sisters they are equal, and they should compete – not for hegemony over the Church or for the upper hand in a casuistry that perverts the preaching of the Gospel, but for excellence in piety only. The central message of the 32nd chapter on the Christian Church as the Incarnation of Christ on earth is recapitulated here and applied to the recent theological conflicts in the Dutch Reformed Church.[65]

In the final chapter De Hooghe summarises the entire genealogy of religion. Again he starts with its origin in ancient astronomy, visualised in the anachronistic image of a Chaldean magus with an enormous telescope. The magus leans upon a model of the system of the world, identified as the mysterious 'machine' built by the Alkmaar mathematician Cornelis Drebbel (1572-1633) and presented to James II of England (1633-1701) – a model of the world that, according to De Hooghe, was so accurate that it even predicted the weather.[66] From there he restates his general argument: that the one and only legitimate aim of religion is to enable people to find the way to attain the salvation of their souls. The hallmark of every civilised society is that it is religious and offers its members the freedom to perfect their souls and prepare them for ascending to God. The Chaldean magus, a scholarly researcher and seeker of truth before his kind succumbed to the lures of priestcraft, has his telescope trained on Christ in Heaven (fig. 9).[67] The penultimate chapter 62 demonstrates that the salvation De Hooghe has in mind is not the afterlife of conventional Christian theology. In this chapter De Hooghe gives a comparative description of ancient and exotic conceptions of the destination of the human soul after death, arguing that they are all equally the product of projection. The True Church is the imagined community of all those who bear their cross in the imitation of Christ and who joyously strive towards Heaven. And the God ruling this Kingdom is 'the Beginning without beginning and the End without end'. This use of the traditional alpha-and-omega metaphor for God reflects in the case of De Hooghe the Spinozist worldview of God as all-encompassing and unending Nature.[68]

The central argument of De Hooghe's *Hieroglyphica* was entirely too radical to be published during De Hooghe's lifetime. The plates must have been puzzling even to a contemporary intellectual audience, but the vernacular textual legend left little room for doubt about their maker's intentions. Unlike the republican political philosophy of *Spiegel van Staat*, the enlightened critique of religion of the *Hieroglyphica* remained unpublished for decades. The context of its eventual publication and its readership are as yet unexplored.[69] However, it may not be a coincidence that the work finally appeared in 1735, just a few years before the completion of the multivolume *Cérémonies et coutumes religieuses de tous les peuples du monde* by Bernard Picart (1673-1733) and Jean Frederic Bernard (1683-1744). Both works put the universal claims of Christian Churches in perspective and advocated tolerance.[70] Of the two books, De Hooghe's *Hieroglyphica* is the more outspoken and offers a more fundamental critique of formal religion. It reflects the various intellectual strands that went into the radical phase of the early Enlightenment and, more specifically, the impact of the cosmological debates on theology.

9
De Hooghe, *Toepassing*, 1690-1708
etching, *Hieroglyphica*, plate 63
Recapitulation of the over-all argument of the book. Note the Chaldean, symbol for the 'invention' of religion, with his telescope trained on Christ in Glory. The orb behind him represents the mysterious 'machine' devised by Drebbel.

Science and religion

With hindsight, the suspicions of De Hooghe's orthodoxy were certainly correct. In the *Hieroglyphica* he shows his true colours. Playing upon the declared principles of Protestantism – rejection of ecclesiastical hierarchy and scholastic theology and a reading of the Bible unfettered by tradition – he argues for a highly unorthodox religion of humanity. The recent reformation, to which the title of the book refers, is not the emergence of Protestantism but rather of enlightened Christianity, based on freedom of inquiry and freedom of conscience. Individuals imitate Christ according to their own lights, within the constraints of outward conformity to the established Churches and obedience to political authority – as did De Hooghe himself.[71]

Like his artistic style, De Hooghe's views on True Religion seem to have formed early in his life. They inform all his artwork for books on religion. His earliest commissions betray familiarity, if not affinity, with Counter-Reformation Catholic devotional styles. It has been argued that early modern Catholicism, with its 'cult of introspection', was more 'modern' than the various forms of Protestantism.[72] De Hooghe valued this type of spirituality, localised in the individual conscience. Faith based on doctrine alone had nefarious side-effects ranging from cruel persecution to superstitious fear. Freedom of personal choice did not, however, invalidate the State's demands for outward conformity, which was simply a necessary form of social discipline. The Church should be subject to State regulation as well, to preclude priestcraft. Religion was to be used for religious purposes only: to make people pious and virtuous. What is implicit in the title pages of the works De Hooghe illustrated becomes explicit in the text of the *Hieroglyphica*.

De Hooghe was an exponent of the radical Enlightenment of the second half of the seventeenth century. Enlightenment thought, grounded in new epistemology and the recent discoveries in natural philosophy, questioned early modern religious establishments, theology and political theory. The lively debates in this field – not least in cosmology but also in theology and the beginnings of comparative religion – left their traces in the *Hieroglyphica*. De Hooghe's allegories show how even Reformed Protestantism was deformed by the elementary forms of priestcraft inherited from its long genealogy in the ancient Middle East, and which it had in common with the world's religions. Emphatically, De Hooghe advocates a freedom of inquiry that takes seriously even the most far-reaching consequences of the scientific and scholarly revolutions of his time, also when these run counter to the Biblical worldview. True Faith can be based only on a free moral judgement of the believer. The history of religions teaches that gods and demons are the product of human projection and that the gods worshipped by man are either deified men or the powers of nature. In his radical deconstruction of Christianity, De Hooghe reveals himself a full-blooded Spinozist.[73] Like Spinoza, he was not an atheist in the modern sense. His work is infused by an enlightened spirituality. Measured by the religious codes of his own time, he was every inch the libertine his

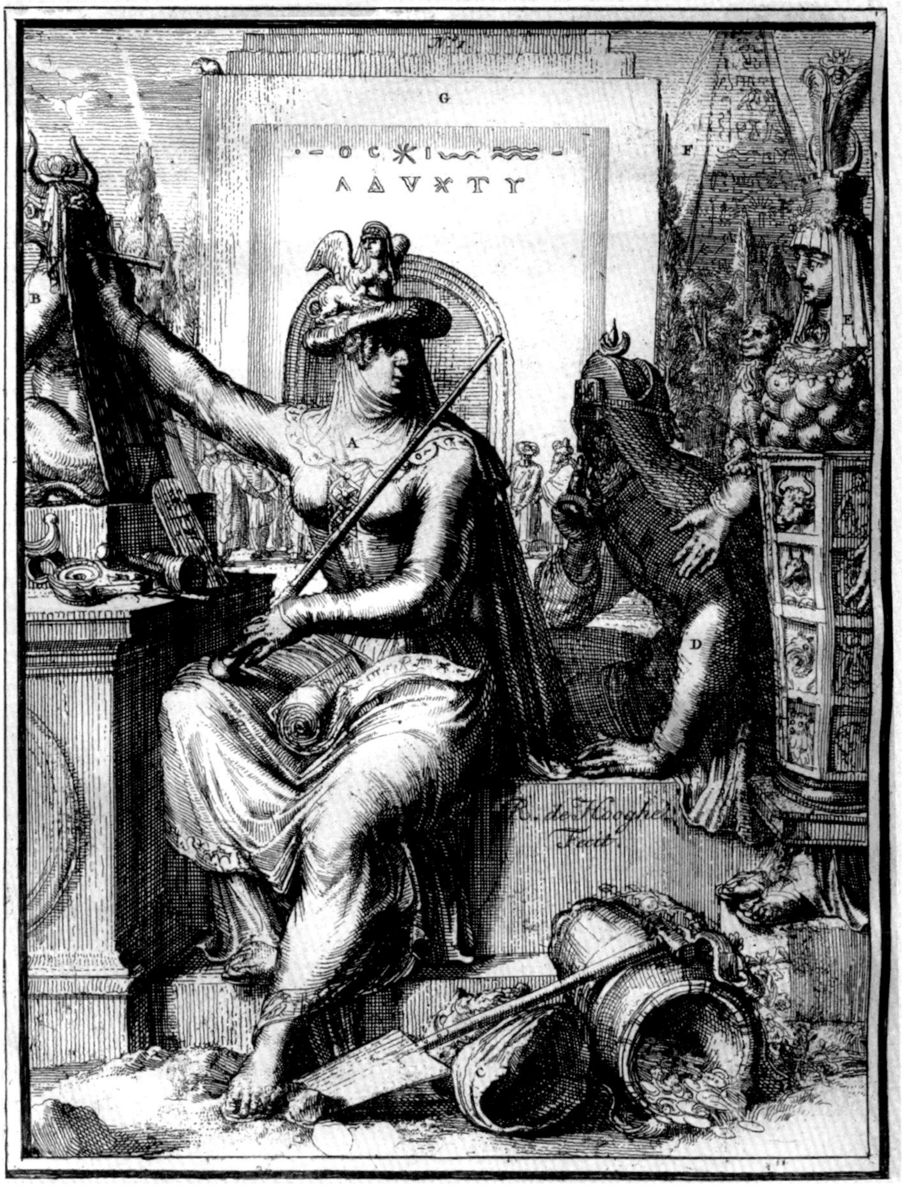

10
De Hooghe, *Uytlegging van de Prent- en Merkbeelden der Egyptenaren. Wat dezelve zyn; haar oorsprong en voortgang,* 1690-1708
etching, *Hieroglyphica*, plate 1
The central figure is the maker of hieroglyphs; the dark figure behind her represents a priest of Harpocrates.

enemies claimed he was, but in a much more interesting way than any pornographer could ever be.[74]

It is not without reason that the *Hieroglyphica* was published only in 1735, long after the author died. Characteristically, the first plate shows the allegorical figure of the inventress of emblems – she who perceives the true nature of things and knows how to depict them allegorically – accompanied by the figure of a priest of Harpocrates: the Egyptian god of secrecy, the one who knows how to interpret emblems but keeps his knowledge to himself, depicted with his finger on his closed mouth (fig. 10). Both are, of course, representations of De Hooghe himself.

Notes

1 De Hooghe 1706, cf. Israel 2006, via the entry 'De Hooghe' in the index.
2 *Hieroglyphica, or emblems of the ancient peoples (...) containing an exhaustive essay on the progressive decline and corruption of religion through the ages, and its recent reformation continuing until the present day.*
3 Iversen 1993. See also the article by Thijs Weststeijn in this volume.
4 De Hooghe 1735, 24-25.
5 Klever 1992; Oudaen 1690; cf. Anonymous 1690; De Hooghe 1690.
6 On the shady reputation of promotions at Harderwijk, see Frijhoff 2000, 26-29.
7 De Hooghe 1735, 126.
8 Elias 1903-1905, vol 1, 453.
9 Andriesse 1993, 119-147.
10 Vermij 2002.
11 Huygens 1689; Seidengart 1981; Dekker 1986.
12 For 'moon-stories' as a satirical genre, see Christianson 1976; Cressy 2006.
13 Bekker 1691-1693, vol 1, 134-135. On the status of mathematics, cf. Van Bunge 2001, 60-61.
14 Fix 1999; Spaans 2007.
15 On this bafflement, also in the world of the microscopically small, see Ruestow 1996.
16 Form, background and intention of *Cosmotheoros* are currently the subject of a master's thesis by Joas van der Schoot at Utrecht University.
17 Van Bunge 2001; Cools *et al.* 2006; Van Bunge 1996; Vermij 2002.
18 Champion 1992; Barnett1999.
19 Weekhout 1998.
20 Cf. Israel, 2001, 275-327.
21 Vermij 2002, 188-237.
22 Van Rooden 2000.
23 Van der Haar 1956; De Haas 2008; Leemans 2008. Leemans misjudges De Hooghe's familiarity with theological debate, and mistakes conformity for orthodoxy.
24 Frijhoff & Spies 1999, 359.
25 Leeflang 2008, 127, 131.
26 Van Hoogstraten 1668. On Van Hoogstraten, see Zuidema 1930.
27 Sucquet 1620, plates 21 (p. 451) and 29 (p. 569); Rosweyde 1631. See airborne cross-bearers in Sucquet, pl. 29 and the title page to the New Testament in De Hooghe 1703. For another Catholic devotional project, see Coppens 1995.
28 Van Dale 1687, 142.
29 Arnold 1701-1729.
30 Cf. Palingh 1659; Bekker 1691-1693.
31 Van Dale 1687, next to p. 81.
32 De Hooghe 1735, plate 39, figure F.
33 Verkruijsse 2008.
34 Although he rejected the idea of a universal language, De Hooghe obviously shared the interest of his contemporaries in pictography; see the article by Thijs Weststeijn in this volume.
35 De Hooghe cites both authors by name in his address to the reader.
36 Van der Coelen1999, 264-265.
37 De Haas 2008.
38 Porteman & Smits-Veldt 2008, 166-167.
39 Spaans 1989, 128-130; Frijhoff 2006.
40 De Hooghe 1735, 340-341, and plate 46.
41 Ibid., 447-448; Wielema 2004, 103-132.
42 Cf. De Haas 2008, 17.
43 Wilson 1974, 314-321, loosely characterises the book as 'an Encyclopedia of World Religions', which entirely misses the point.
44 De Hooghe 1735, 403.
45 Ibid., 27-28.
46 Ibid., 122-123, 126, and plate 12. See also the contribution by Eric Jorink earlier in this volume.
47 De Hooghe 1735, 123, 139.
48 Ibid., 165. For the history of a critical study of the Bible in the Republic, see Van Bunge 2001, 161-162, and Jorink 2010, 88-105.
49 De Hooghe 1735, 186-188. On emblematic theology, see Van Asselt 1997, and on the interest in the building of Salomon's Temple, Van Asselt 1996.
50 De Hooghe 1735, 237.
51 Ibid., 255-257.
52 Ibid., 261.
53 Ibid., 269-270.
54 All four thematised in De Hooghe 1735, plate 39, the figures A, B, C and M.
55 De Hooghe 1735, 291, 405.
56 Numbers 22-24.
57 De Hooghe 1735, 291, 405.
58 Ibid., 338. Plate 46 has the dromedary on the monument in the upper right corner, half hidden by the 'Asian' warrior. See also the contribution by Dániel Margócsy in the present volume.
59 De Hooghe 1735, 303-309, 351-355.
60 Ibid., 387-388.
61 The title of chapter 56 in De Hooghe 1735, cf. 405-407.
62 Arnold 1701-1729.
63 Plate and chapter 60, De Hooghe 1735, 427-433.
64 Schilling 1993.
65 De Hooghe 1735, 435-441 (plate 61).
66 Ibid., 449; cf. Vermij 2002, 92-97.
67 De Hooghe 1735, 454, cf. 119, 185.
68 Ibid., 443-448 (plate 62).
69 *Hieroglyphica* is the subject of the PhD project of Trudelien van 't Hof at Utrecht University.
70 Hunt *et al.* 2010. It is not unlikely that the *Hieroglyphica*, even if as yet unpublished, influenced the work of Picart and Bernard, but no research has been done in this direction as yet.
71 Van Rooden 2000.
72 Nalle 2004, 323.
73 Cf. Van Rooden 1984. Israel 2006, 245 characterises De Hooghe as a disciple of Spinoza for his republicanism. Israel ignores *Hieroglyphica* or De Hooghe's views on religion.
74 Although no outright pornographic work is known from his hand, he did have a taste for the genre. In chapter 27 of his *Hieroglyphica* he writes that he refrains from elaborating on more salacious aspects of Venus, as these often cause offence, but that, given time, he will do so in forthcoming work; De Hooghe 1735, 229.

Bibliography

Andriesse 1993
C.D. Andriesse, *Titan kan niet slapen. Een biografie van Christiaan Huygens,* Amsterdam & Antwerp 1993.

Anonymous 1690
Anonymous, *Het eerste Antwoort tegen Romeyn de Hooge (...)* [1690] (Knuttel 13548A).

Arnold 1701-1729
G. Arnold, *Historie der Kerken en Ketteren van den beginnen des Nieuwen Testaments tot aan het Jaar onses Heeren 1688*, Amsterdam & Groningen 1701-1729.

Barnett 1999
S.J. Barnett, *Idol temples and crafty priests. The origins of Enlightenment anticlericalism*, London & New York 1999.

Bekker 1691-1693
B. Bekker, *De betoverde weereld, zynde een grondig ondersoek van 't gemeen gevoelen aangaande de Geesten, deselver Aart en Vermogen, Bewind en Bedryf; als ook 't gene de Menschen door derselver kraght en gemeenschap doen*, Amsterdam 1691-1693.

Champion 1992
J.A.I. Champion, *The pillars of priestcraft shaken. The Church of England and its enemies, 1660-1730*, Cambridge 1992.

Christianson 1976
G.E. Christianson, 'Kepler's Somnium. Science fiction and the Renaissance scientist', *Science Fiction Studies* 3 (1976), 79-91.

Cools *et al.* 2006
H. Cools, M. Keblusek & B. Noldus (eds.), *Your humble servant. Agents in early modern Europe*, Hilversum 2006.

Coppens 1995
C. Coppens, *Een Ars Moriendi met etsen van Romeyn de Hooghe. Verhaal van een boekillustratie*, Brussels 1995.

Cressy 2006
D. Cressy, 'Early modern space travel and the English man in the moon', *American Historical Review* 111 (2006), 961-982.

De Haas 2008
A. de Haas, 'Commissaris van zijne majesteit en mikpunt van faamrovende paskwillen. Een biografische schets', in: Van Nierop *et al.* 2008, 12-31.

De Hooghe 1690
[R. de Hooghe], *Nieuw oproer op Parnassus. Zijnde een verhaal van de valse Getuygen (...) teegen den Heer en Meester Romein de Hooge (...)*, [1690] (Knuttel 13547).

De Hooghe 1703
R. de Hooghe, *Alle de Voornaamste historien des Ouden en Nieuwen Testaments, verbeeld in uytsteekende Konst-Platen. Met omstandige verklaring der stoffen, en seer beknopte Punt-Digten, van den Eerw. Godsgel. Heer Henricus Vos. Waar by ook gevoegt zyn Nieuwe Kaarten, tot opheldring der zaaken nodig*, Amsterdam 1703.

De Hooghe 1706
R. de Hooghe, *Spiegel van Staat des Vereenigde Nederlands*, Amsterdam 1706.

De Hooghe 1735
R. de Hooghe, *Hieroglyphica of merkbeelden der oude volkeren, namentlyk Egyptenaren, Chaldeeuwen, Feniciers, Joden, Grieken, Romeynen, enz. Nevens een omstandig Bericht van het Verval en voortkruypende Verbastering der Godsdiensten door verscheyde eeuwen; en eyndelyk de Hervorming, tot op deze tyden toe vervolgt.* Overzien en Beschaaft door Arn. Henr. Westerhovius, Amsterdam 1735.

Dekker 1986
E. Dekker, 'Sterrekunde in de zeventiende eeuw', *De zeventiende eeuw* 2 (1986), 84-100.

Elias 1903-1905
J.E. Elias, *De Vroedschap van Amsterdam 1578-1795*, Haarlem 1903-1905.

Fix 1999
A. Fix, *Fallen angels. Balthasar Bekker, spirit belief, and confessionalism in the seventeenth-century Dutch Republic*, Dordrecht 1999.

Frijhoff 2000
W.T.M. Frijhoff, 'De betekenis van de Harderwijkse universiteit in internationaal perspectief', in: *Het Gelders Athene. Bijdragen tot de geschiedenis van de Gelderse universiteit in Harderwijk (1648-1811)*, Hilversum 2000, 15-36.

Frijhoff 2006
W.T.M. Frijhoff, 'Damiette appropriée. La mémoire de croisade, instrument de concorde civique (Haarlem, XVIe-XVIIIe siècle)', *Revue du Nord* 88 (2006), no. 364, 7-42.

Frijhoff & Spies 1999
W.T.M. Frijhoff & M. Spies, *1650. Bevochten eendracht*, Den Haag 1999.

Hunt *et al.* 2010
L. Hunt, M.C. Jacob & W. Mijnhardt, *The book that changed Europe. Picart and Bernard's 'Religious ceremonies of the world'*, Cambridge, MA & London 2010.

Huygens 1689
C. Huygens, *Cosmotheoros. De wereldbeschouwer*, (P. Rabus, transl. (1754); H.A.M. Snelders, ed.), Utrecht 1989.

Israel 2001
J.I. Israel, *Radical Enlightenment. Philosophy and the making of modernity, 1650-1750*, Oxford 2001.

Israel 2006
J.I. Israel, *Enlightenment contested. Philosophy, modernity, and the emancipation of man 1670-1752*, Oxford 2006.

Iversen 1993
E. Iversen, *The myth of Egypt and its hieroglyphs in European tradition*, Princeton 1993.

Jorink 2010
E. Jorink, *Reading the book of nature in the Dutch Golden Age, 1575-1715*, Leiden & Boston 2010.

Klever 1992
W. Klever, 'Inleiding', in: F., *Vrije politijke stellingen*, Amsterdam 1992, 13-119.

Leeflang 2008
H. Leeflang, 'Waarheid, vlugheid en inventie. Ontwerp en uitvoering van de etsen', in: Van Nierop *et al.* 2008, 126-145.

Leemans 2008
I. Leemans, 'De viceroy van de hel. Radicaal libertinisme', in: Van Nierop *et al.* 2008, 32-47.

Nalle 2004
S.T. Nalle, 'Self-correction and social change in the Spanish Counter-Reformation', in: J.D. Tracy & M. Ragnow (eds.), *Religion and the early modern state. Views from China, Russia, and the West*, Cambridge 2004, 302-323.

Oudaen 1690
J. Oudaen Fransz., *Op de regtsvordering van de Heer Mr. Adriaan Bakker, Hooft-Officier der Stad Haarlem, tegen de persoon van Romeijn de Hooge, Wijlen Plaetsnijder tot Amsterdam. Tot voorstand des Gerechts, en vordering der Eere Gods*, [1690] (Knuttel 13544).

Palingh 1659
A. Palingh, *'t Afgerukt mom-aansight der tooverye, daar in het bedrogh der gewaande toverye naakt ontdekt en met gezonde redenen en exempelen dezer eeuwe aangewezen wort*, Amsterdam 1659.

Porteman & Smits-Veldt 2008
K. Porteman & M.B. Smits-Veldt, *Een nieuw vaderland voor de Muzen. Geschiedenis van de Nederlandse literatuur 1560-1700*, Amsterdam 2008.

Rosweyde 1631
H. Rosweidius, *De Naervolginghe Christi van Thomas a Kempis (...). Mitsgaders sijn leven*, Antwerp 1631.

Ruestow 1996
E.G. Ruestow, *The microscope in the Dutch Republic. The shaping of discovery*, Cambridge 1996.

Schilling 1993
H. Schilling, 'Afkeer van domineesheerschappij. Ein neuzeitlicher Typus des Antiklerikalismus', in: P.A. Dykema & H.A. Oberman (eds.), *Anticlericalism in late medieval and early modern Europe*, Leiden 1993, 655-668.

Seidengart 1981
J. Seidengart, 'Les théories cosmologiques de Christian Huygens', in: *Huygens et la France. Table ronde du Centre national de la recherche scientifique*, Paris 1981, 209-222.

Spaans 1989
J. Spaans, *Haarlem na de Reformatie. Stedelijke cultuur en kerkelijk leven 1577-1620*, The Hague 1989.

Spaans 2007
J. Spaans, 'Censorship, satire and religion. A seventeenth-century satirical print on the condemnation of Balthasar Bekker's "The Enchanted World"', in: J.A.C. van Ophem & C.H.E. Verhaar (eds.), *On the mysteries of research. Essays in various fields of Humaniora*, Leeuwarden 2007, 273-287.

Sucquet 1620
A. Sucquet, *Via vitae aeternae*, Antwerp 1620.

Van Asselt 1996
W.J. van Asselt, 'Ultimum tempus nobis imminet. Eschatologische structuren van de theologie van Johannes Coccejus's, *Nederlands Archief voor Kerkgeschiedenis / Dutch Review of Church History* 76 (1996), 189-226.

Van Asselt 1997
W.J. van Asselt, 'De neus van de bruid. De profetische en zinnebeeldige godgeleerdheid van Henricus Groenewegen en Johannes D'Outrein', in: F.M.G. Broeyer & E.M.V.M. Honeé (eds.), *Profetie en godsspraak in de geschiedenis van het christendom. Studies over de historische ontwikkeling van een opvallend verschijnsel*, Zoetermeer 1997, 163-184.

Van Bunge 1996
W. van Bunge, 'Eric Walten (1663-1697). An early Enlightenment radical in the Dutch Republic', in: W. van Bunge & W. Klever (eds.), *Disguised and overt Spinozism around 1700*, Leiden 1996, 41-54.

Van Bunge 2001
W. van Bunge, *From Stevin to Spinoza. An essay on philosophy in the seventeenth-century Republic*, Leiden 2001.

Van Dale 1683, 1687
A. van Dale, *De oraculis ethnicorum dissertationes duae*, Amsterdam 1683 (=*Verhandeling van de oude Orakelen der Heydenen*, Amsterdam 1687).

Van der Coelen 1999
P. van der Coelen, 'Emblemata sacra? Biblical picture books and emblem literature', in: J. Manning, K. Porteman & M. Van Vaeck (eds.), *The emblem tradition and the Low Countries*, Turnhout 1999, 261-278.

Van der Haar 1956
C. van der Haar, 'Romeyn de Hooghe en de pamflettenstrijd van de jaren 1689 en 1690, *Tijdschrift voor Geschiedenis* 69 (1956), 155-171.

Van Hoogstraten 1668
F. van Hoogstraten, *Het Voorhof der Ziele. Behangen met leerzaeme Prenten en Zinnebeelden*, Rotterdam 1668.

Van Nierop *et al.* 2008
H.F.K. van Nierop et al. (eds.) *Romeyn de Hooghe. De verbeelding van de late Gouden Eeuw*, Zwolle & Amsterdam 2008

Van Rooden 1984
P.T. van Rooden, 'Spinoza's bijbeluitleg', *Studia Rosenthaliana* 18 (1984), 120-133.

Van Rooden 2000
P.T. van Rooden, 'Vroomheid, macht, verlichting', *De Achttiende Eeuw* 32 (2000), 57-75.

Verkruijsse 2008
P. Verkruijsse, 'Door Romeijn de Hooghe geschreven en geïllustreerde boeken', in: Van Nierop *et al.* 2008, 258-290.

Vermij 2002
R. Vermij, *The Calvinist Copernicans. The reception of the new astronomy in the Dutch Republic, 1575-1750*, Amsterdam 2002.

Weekhout 1998
I. Weekhout, *Boekencensuur in de Noordelijke Nederlanden. Een verkennend onderzoek naar de vrijheid van drukpers in de zeventiende eeuw*, The Hague 1998.

Wielema 2004
M. Wielema, *The march of the libertines. Spinozists and the Dutch Reformed Church (1660-1750)*, Hilversum 2004.

Wilson 1974
W.H. Wilson, *The art of Romeyn de Hooghe. An atlas of European late Baroque culture*, Cambridge, Mass. 1974 (unpub. diss. Cambridge, MA).

Zuidema 1930
R. Zuidema, 'Hoogstraten, Frans of Francois van', in: P.C. Molhuysen & P.J. Blok (eds.), *Nieuw Nederlands biografisch woordenboek*, Leiden 1911-1937, vol. 8, 833-834.

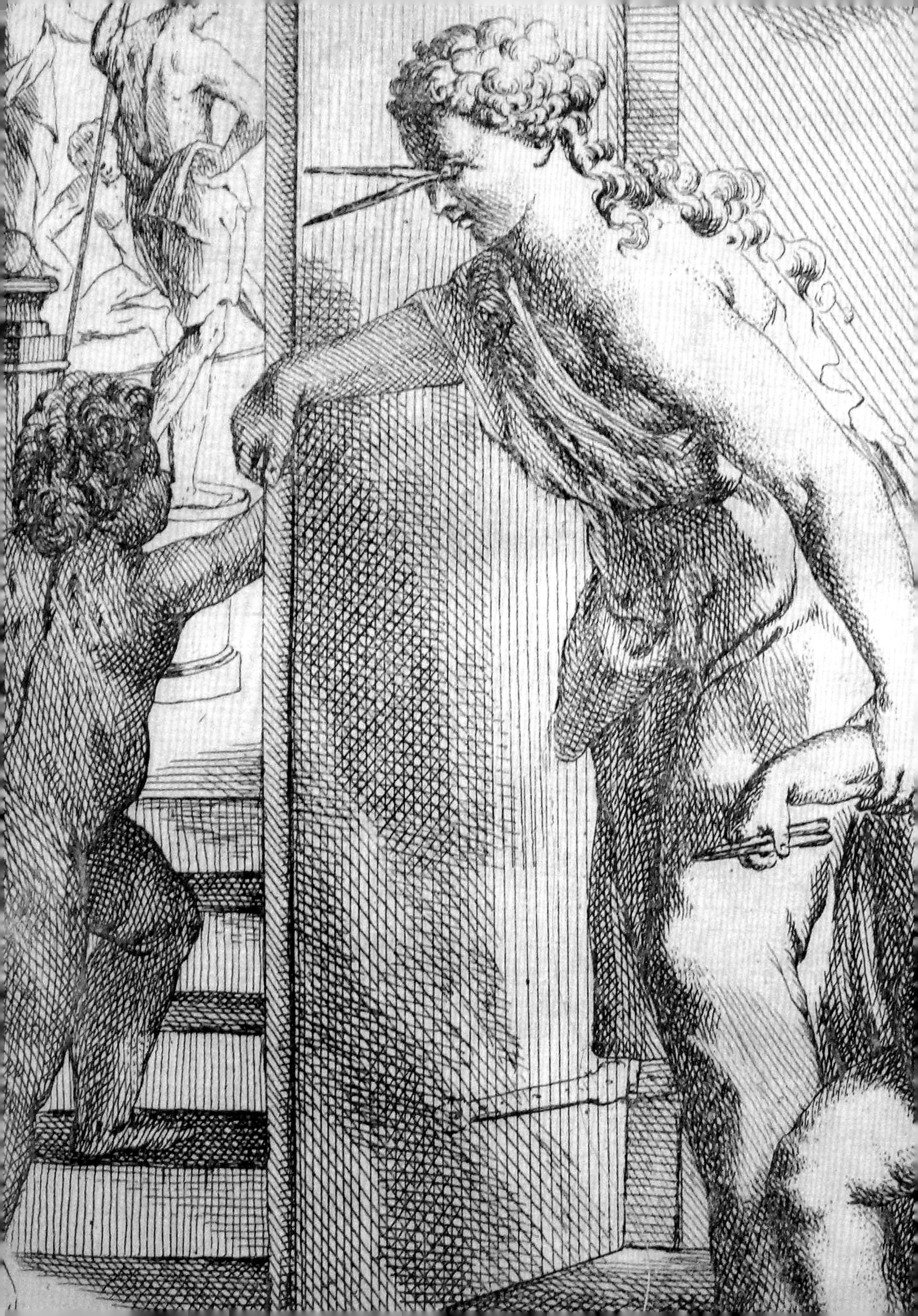

'Will the eye be the sole judge?'

'Science' and 'art' in the optical inquiries of Lambert ten Kate and Hendrik van Limborch around 1710

Fokko Jan Dijksterhuis

Between 1707 and 1713, the polymath Lambert ten Kate (1674-1731) from Amsterdam and the painter Hendrik van Limborch (1681-1759) from The Hague engaged in a project to investigate the nature and properties of light and colors. Their main concern was color mixing, trying to discover the optimal way to make what they called *mezzatints*: gradations of bright and dark in colors. The project involved numerous complex issues, such as the effect and quality of specific pigments, techniques for mixing and applying paints, ways to obtain and maintain standardized outcomes, and – far from trivial – finding words and means to communicate ideas and results. In the end, the project bore little fruit and never left the private papers of Ten Kate and Van Limborch. Still, its historical significance stretches much farther than the mere anecdotal. Ten Kate and Van Limborch inquired into the nature and properties of colors in a quite original way that not only is epistemically interesting but also draws our attention to a line of development that has gone largely unnoticed by historians of optics.

The color-mixing project is at the intersection of 'art' and 'science', two categories that are evidently problem-ridden. When applied to early modern culture, a distinction between 'art' and 'science' is more likely to hamper than help historical understanding. The case of Ten Kate and Van Limborch underscores the fundamental entanglement of artistic and scientific pursuits in the early modern period. This pertains in particular to the domain of optics, where artists and philosophers, as well as doctors and artisans, amassed a diversity of insights into the nature, properties and handling of light. This article shows how early modern optics can be discussed in a historiographically new way while avoiding a priori categorization and treating various forms of knowing on equal terms.[1] If 'science' is provisionally understood as 'inquiry into the nature of things' and 'art' as 'imaging the world', this article considers the ways of knowing that come into being when artists do science. It addresses the question of how to characterize Ten Kate and Van Limborch's inquiry into the nature and properties of light and colors – and 'artist science' in general – without subordinating it a priori to some set category of 'science'.

Detail fig. 3
Pierre Bodart (after drawings by Hoet), ***The 'Compass in the eye'*****, 1605**
The artist of the eighteenth century should be learned in mathematics and philosophy.

The main characters

Ten Kate and Van Limborch were of kindred spirit in several respects. They shared a preference for Italian artists – Raphael in particular – and appreciated classicist tendencies in art. Both valued intellectual approaches to art, aimed at formal qualities of proportion, order and content. Van Limborch endorsed Gerard de Lairesse's (1640-1711) dictum that a painter ought to be learned in philosophy, mathematics, history and art. Ten Kate was exactly such a learned man and a renowned connoisseur with a large collection of drawings and statues. The two were directly connected through their social and religious backgrounds. Both were of distinguished families: Ten Kate came from a wealthy family of Amsterdam merchants; Van Limborch was the son of a prominent lawyer in The Hague. Ten Kate belonged to the liberal Mennonite congregation in Amsterdam that had close relations to the Remonstrant congregation to which Van Limborch's family was connected. An uncle of Van Limborch was a renowned professor at the Remonstrant Seminary in Amsterdam, as was a close acquaintance of Ten Kate, the theologian and publicist Jean Leclerc (1657-1736). Ten Kate and Van Limborch began a longstanding correspondence after they met in the former's Amsterdam home in 1706. Initially they discussed human proportions, trying to find a method to determine the ideal measures, for which they conducted – among other things – series of truly precise measurements of classic statues. In the summer of 1707, they started inquiring into the properties of colors.

Lambert ten Kate was a true polymath, making original contributions to a wide range of often original fields of inquiry. He originally participated in his family's trading company, but around 1705 he left the business and devoted his time to his intellectual interests. He is best known for his work in linguistics and as a key figure in the circle of early Dutch 'Newtonians'.[2] His writings on harmonics, phonetics, optics and esthetics were made public only partially.[3] Ten Kate exemplifies the new type of dilettante collector that developed around 1700 and that systematically composed and organized their collection on the basis of esthetical and philosophical considerations.[4] *Naturalia* and *artificialia* naturally sat together: besides drawings, paintings and statues, his collection included a range of *naturalia* and instruments.[5] The range of his interests has resulted in a rather fragmentary historical picture, because historians of art, philosophy, linguistics and physics have each cut out parts of Ten Kate's *oeuvre* that are relevant to them.

Hendrik van Limborch took his apprenticeship as painter in The Hague and Rotterdam with Jan de Baen (1633-1702) and Adriaen van der Werff (1659-1722), respectively. He was admitted to the guild of The Hague in 1706. He never married, and he worked from his ancestral home. He painted in the style of his teachers, focusing on historical and allegorical scenes. His clientele consisted of a small circle of elite acquaintances, and upon his death much of his production was part of his own collection of prints and drawings.[6] Marina Aarts has argued that his painting activities were first of all aimed at inquiring into the proper rules of proportions and chromatics. In this sense he developed Van der Werff's classicist approach a step further towards art as academic study. Aarts also shows that the artistic studies of Van Limborch can be seen as a critical reading of De Lairesse's *Groot Schilderboek* (1707), developing its accounts of light, color and proportion further and, together with Ten Kate, systematically inquiring into them. De Lairesse, with Van der Werff the main protagonist of classicism in the Netherlands, published his theories of art and painting in the early days of 1707. Only two months later, in one of their first letters, Van Limborch wrote Ten Kate: 'I have been surprised that Lairesse treats these matters so foolishly in his book'.[7]

In the background of Ten Kate and Van Limborch's exchanges was a third person: the painter and engraver Jacob Christoph Le Blon (1667-1741).[8] Le Blon came from Frankfurt and married and settled in Amsterdam in 1705. He collaborated in Ten Kate's artistic pursuits, and his ideas and opinions pervade the correspondence. At the time of the color-mixing project, Le Blon was developing his ideas and techniques on painting with three *couleurs primitives*. Around 1710 he invented the first system of color printing by means of red, yellow and blue mezzotint plates.[9] Around 1717, Le Blon moved to London where he acquired a patent and a privilege for color printing in 1719. Two years later a 'Picture Office' was established, with Le Blon as technical manager. A prospectus written by Le Blon included a list of prints that had been published, including an anatomical print that was to be first in a series of twelve.[10] Before long the affair foundered because of commercial and organizational troubles. Soon after, Le Blon published *Coloritto* (1725), a brief discourse on his system of painting with *couleurs primitives* that included a few color prints.[11] Around 1735 Le Blon moved to Paris where he acquired another privilege and set up a new business. Le Blon's ideas and techniques were well received; in Amsterdam, London and Paris he maintained good relationships with savants and financiers. Besides *Coloritto* he did not publish on his ideas and techniques, but, in London as well as Paris, informative and favorable reports were published in journals like *Philosophical Transactions* and *Mémoires de l'Académie*.[12] Although Le Blon's color prints were of high quality and he was good at mobilizing savants and financiers, his commercial enterprises were less successful, and his Paris successors did not manage to maintain the quality of his printing.

Le Blon may have been instrumental in introducing the theme of color mixing to Ten Kate and Van Limborch, as well as in the particularities of their inquiry. The precise interaction cannot be determined because the contributions of Le Blon are documented only indirectly in Ten Kate's letters. We may assume a fair degree of symbiosis in the project because the themes and questions of Le Blon, Ten Kate and Van Limborch ran fairly parallel. We should note, however, important differences. Le Blon's main interest was *coloritto*, the mixing of (primary) colors into hues, for which he developed materials and techniques. Ten Kate and Van Limborch were mainly interested in the gradation of given colors and the determination of rational rules. In both cases, however, the principal focus was on the incarnadine, the color of flesh. For Ten Kate and Van Limborch, the main issue was to find out how a given color of a particular body – old, young, dead or healthy – could be gradated properly.

The background of the project

The project on color mixing started after a visit by Ten Kate to Van Limborch in The Hague early in August 1707.[13] The basic question was quite simple: how much white and black, respectively, should be mixed with a color to acquire a natural gradation of bright and dark that maintains the hue of the color.[14] The goal of the project was to develop a rational method of painting shades of various colors. The project was recorded in the correspondence of Ten Kate and Van Limborch that has recently been published in a magnificent edition by Hessel Miedema.[15]

Ten Kate and Van Limborch's main interest was in the painting of nudes. This is no coincidence: the color of flesh was a notorious painterly challenge, and the quality of the *Inkarnat* was generally regarded as the touchstone of an artist's mastery. The sixteenth and seventeenth centuries had witnessed a tremendous development in the realistic depiction of flesh in oils, most notably by Rubens and Rembrandt. *Lijfverf*, as it was referred to, required the careful observation of the color of a particular body; the expert choice, preparation and mixing of pigments; and the masterly application of paint and brushing techniques.[16] In books on painting, much attention was given to the color of nudes. Ann-Sophie Lehmann points out that such accounts were relatively practical in nature, largely consisting of recipes for materials and techniques. It is telling, she says, that in his chapter on the color of nudes, De Lairesse, who advocated a formal and intellectual approach to art theory, offered a 'curious mix of detailed practical instruction and general theoretical reflection'.[17] Ten Kate and Van Limborch wanted to surpass the common recipes for materials and techniques and develop a rational system of color mixing. They investigated the way colors mix in a systematic way, intending to discover the exact, standardized proportions for combining paints. In so doing, they shifted the perspective from paints to colors. While experimenting with painted colors and gradations, they wanted to know *how colors mix* rather than how paints mix.

The seventeenth century had witnessed notable developments in the natural philosophical understanding of colors as well. First of all, René Descartes (1596-1650) had discarded the classical distinction between real and apparent colors – that is, the colors of bodies and the colors appearing in rainbows and fountains.[18] In *Les météores*, one of the essays of *Discours de la méthode* (1637), he not only developed a mathematical theory of the rainbow to calculate the angles at which the colors are seen, but also gave an account of how light is modified by the successive refractions and reflections in a raindrop to create the various colors of the rainbow. Subsequently, he argued that any color is a result of such an interaction between light and matter. Descartes did not discard the classical notion that colors are mixtures of light and dark, but already in the early seventeenth century scholars had begun to adopt the artist's distinction between color and light and the notion of primary colors.[19] Isaac Newton's (1642-1727) 'New theory about white light and colors' is generally regarded as a watershed in the optical theory of colors. His basic idea, first published in an article in 1672, was that colors are primary and determined by the degree to which they are refracted, and that white light is a mixture of colored rays. He elaborated it into a comprehensive account of colors produced in prisms, thin films and thick layers of refracting mediums in *Opticks* (1704). In *Opticks* Newton also discussed color mixing, including a circle to determine the relative proportions of primary colors in a mixture, but this was entirely based on transferring his doctrine of colored rays to pigments and employing seven primary colors.[20]

Such philosophical issues were in the background of Ten Kate and Van Limborch's project. Ten Kate in particular was acquainted with contemporary natural philosophical ideas regarding light and colors. Both regarded knowledge of mathematics and philosophy imperative to the artist, but, as we shall see, current theories of light and color were far from directive to their investigations. Besides, they did not investigate the disembodied colors of the philosophers; they studied the nature and properties of colors *in* paint. In this way they transformed the issue of *Lijfverf* by abstracting from the painter's know-how on the one hand and keeping philosophical concepts material on the other.

Line patterns and mezzatints

On 17 August 1707 Ten Kate reported on his first experiments with mixing colors.[21] He had performed these together with Le Blon. They started out by determining the relative 'mixing power' of black and white. To this end they painted narrow lines of alternately black and white. When this pattern is observed at a distance, the distinction between the lines disappears, and a gray is perceived. Then they mixed white and black to produce the same gray. It turned out that black was three times as powerful as white: three measures of white with one measure of black produced the required gray. To facilitate the further investigation of mixing colors, they adjusted the black they used. They added as much 'over burnt ivory black' to weaken the black so that it became equally powerful as their white: equal measures of white and this 'harmonious black' now produced the required gray.[22] Having established their black and their white, they went on to investigate the 'power' of colors relative to black and white. They concluded that black is four times as powerful as a color, and colors are two to three times as powerful as white. In order, therefore, to maintain the hue of a color in shadows, colors ought to be mixed with white and black respectively in measures in a proportion of one to eight (yellow) or one to twelve (red and blue).

> So that then, (...) a flesh tint made out of two or three of the colors will have to be mixed in 11 or 12 measures of tint to the harmonious black, against 1 measure of tint to the white; if one desires that the dark represents the same art of color as the light, which was the aim of the color mixing.[23]

Ten Kate added that the 'accidentals', such as a reddish bloom or bluish temples, can be achieved in the glazing – the finishing with transparent layers of paint.

The experiment with line patterns is quite remarkable for its time and seems to have been an original idea of Ten Kate and Le Blon. The similarities to Le Blon's engraving and his efforts to develop a color printing technique are interesting. Le Blon was skilled in mezzotint printing and used mezzotints in the primary colors to create color prints. The basis of a mezzotint is a metal printing plate that is roughened, producing a structure of minute grooves. The lines that are subsequently printed produce varying tones of gray; mezzotints are renowned for their luxurious quality of tones. This suggests that Le Blon's expertise was in some way instrumental in the original setup of the experiment. The letter does not explain what pigments they used for the primary colors. Le Blon himself would never explain what exact pigments he used, though it was crucial to his color printing in which he used mezzotint plates for yellow, red and blue layers.[24] In his development of color printing he concluded that transparent colors were needed, as opposed to the covering colors used in painting.[25]

The word Ten Kate and Van Limborch used for gradations of bright and dark is a bit confusing: *mezzatints*. The term 'mezzatint' or 'mezetint' was adopted in the sixteenth century from Italian writers on painting and had been used throughout the seventeenth century to denote gradations of colors.[26] It has nothing to do with the mezzotint printing technique. Because Ten Kate and Van Limborch used the term quite consistently, I will also use 'mezzatint'.

After Ten Kate and Le Blon's first experiments, Van Limborch visited them and discussed the results. On 13 October he sent Ten Kate a letter, in reply to his account of the experiments.[27] Van Limborch explained that when he applied the proportions they had found to mix a range of mezzatints, the colors became far too full.[28] Ten Kate then further investigated the proportions of the intermediate gradations, employing a combination of numerical analysis and color tests.[29] He calculated the measure of color in a mixture of white and black to which color was added according to the proportions determined initially, and tested its power by comparing it with a line pattern of gray and color. He explained to Van Limborch that the arithmetic progression they had implicitly used for the proportions of mixing color to successive shades of gray was too simple. He proposed a new, geometric progression and calculated improved proportions.[30] A comparison of both series explained why the original mezzatints had become too full.

These first experiments, calculations and tests were the beginning of an exchange that would continue for over five years. The inquiry into an optimal method of color mixing was not easy and was restrained by Ten Kate and Van Limborch's specific ideas about the desired outcome. Unsatisfactory results of proposed mixing proportions could, for example, be corrected by painting techniques. They discussed such options, such as weakening a color by adding some blue.[31] Using painting techniques to produce good mezzatints was, however, not the solution they were looking for:

> Among other things I have often thought of making the mezzatints seem bluish without applying blue color, by brushing thinly the true flesh color, but I cannot get that quite right or the rule is not generally good.[32]

The whole point of the project was to develop a rational method for mixing gradations of bright and dark that was based on systematic inquiry of the properties of *colors* rather than clever ways of preparing and mixing paints. This ambition gave rise to detailed discussions of the proper series to be developed and specific ways to mix paints, apply mezzatints, and so forth. Various optical effects complicated the seemingly straightforward issue of developing a proper gradation of colors. Van Limborch explained that the color of an object was not only determined by the amount of direct light, but also affected by the light reflected from colored objects in the vicinity and the weakening effects of distance.[33] On 20 March 1709, Ten Kate and Van Limborch met at the home of Le Blon to investigate the latter phenomenon. They placed three sheets of paper at regular distances and then tested how much the colors were softened.[34] Optical effects such as these further complicated the color-mixing project.

In order to solve the problems of the original gradation of colors, Van Limborch proposed to calculate a measure of gray to correct the excess of color for each mezzatint.[35] This, however, reversed the problem: instead of determining the measure of color for each gradation between bright and dark, the amount of gray was required to give each mezzatint the right hue. To this end a double series of gradations had to be determined. Ten Kate calculated several series and they exchanged several test samples without finding a definite solution.[36] In fact, the issue was complicated because the samples they sent each other made clear that variations of the pigments used greatly affected the power of the colors. Eventually Van Limborch made a precise inventory of the relative powers of colors produced by different pigments.[37] In late January or early February 1712, they convened in Amsterdam, together with Le Blon, to discuss and test the ideas. Ten Kate wrote a long letter to Van Limborch recording the proceedings and main conclusions.[38]

By this time their opinions had begun to diverge. Ten Kate did not really think that Van Limborch's approach of determining separate gradations for mezzatints and grays was feasible. Despite the negative conclusion of Ten Kate's report, Van Limborch made several proposals for a 'mixing triangle', combining numerical series for gradations of bright and dark and of mezzatints.[39] He sent Ten Kate his final numerical triangle in a letter of 17 December 1712 (fig. 1).[40] Ten Kate responded that he did not quite understand it and found it unfeasible – evidently he did not really take it seriously. So their discussion of color mixing came to an end and the project petered out. They continued their correspondence, though, discussing other themes of painting and art. Although Le Blon shared their interest in color mixing and took an active part in the discussions and experiments, he does not seem to have been very much taken by their ideal of finding rational rules. He considered his system of

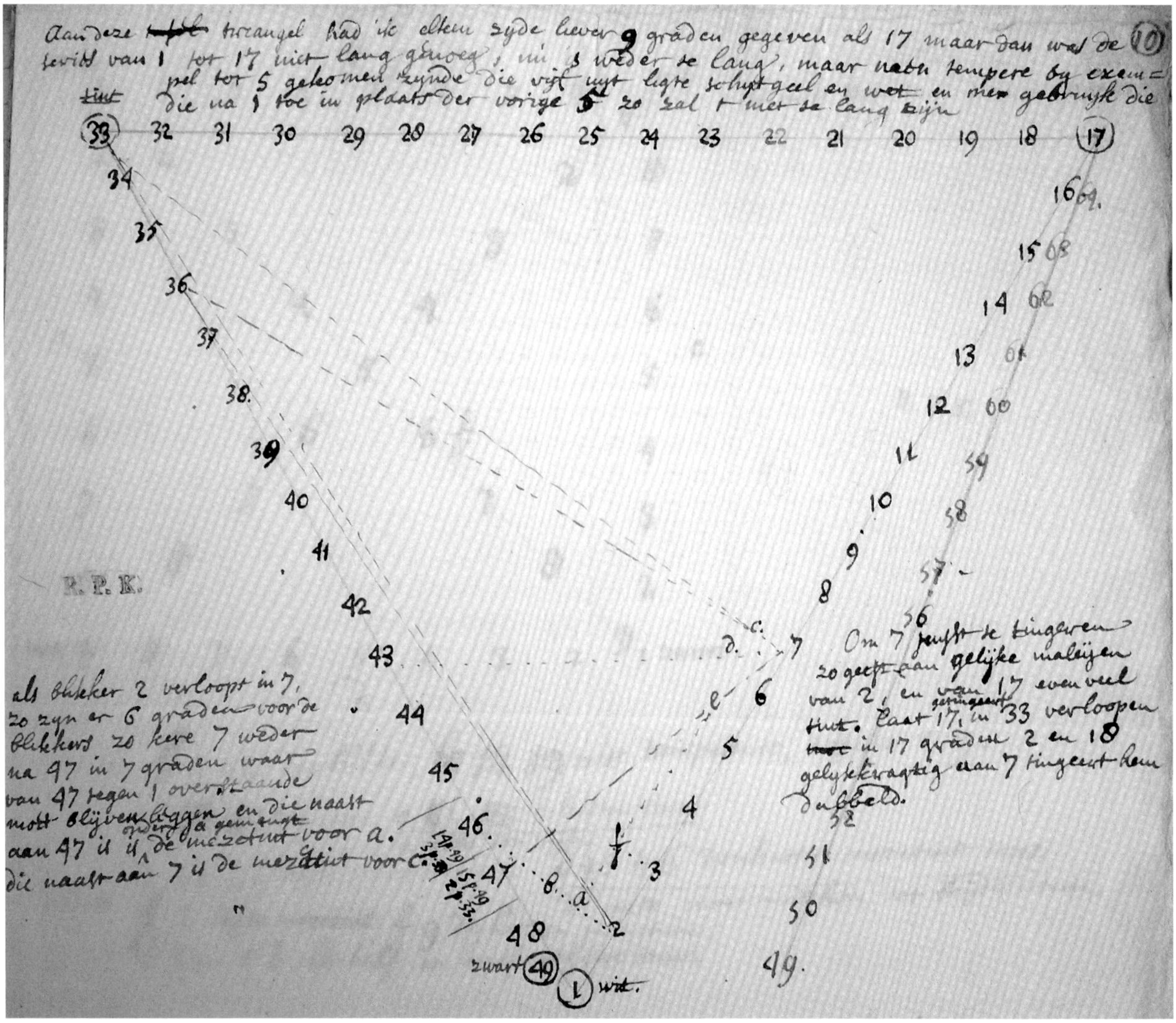

I

Draft of the 'mixing triangle' Van Limborch sent Ten Kate, 1712

His manuscript notes contain numerous sketches and calculations to make a scheme for the double mixing series.

Amsterdam Rijksprentenkabinet, RP-D-00-1062 sign 310 K (folder 8).

primary colors a rational foundation of his art, but in the closing lines of *Coloritto* he made clear that determining the proportions in hues, making pigments and paints '(...) must be learn'd only by long *Mechanical* Practice'.[41]

Expertises and exchanges

The color-mixing project was complicated not only by the nature and properties of light, color, paints and pigments, but also by differences of expertise and preference between the two men. They had specific esthetic ideas and marked opinions about painting techniques. Despite the fact that Ten Kate was an amateur, they discussed these on equal terms. At the same time, Van Limborch openly acknowledged Ten Kate's superiority in matters of mathematics and philosophy. He was '(...) in the numbering not so far (...)' that he could employ a means of calculating proportions.[42]

At one point he encountered a problem with one of Ten Kate's series. He noticed irregularities when he calculated the increase in the amount of color to be added in the mix.[43] He added a test to show that the measures of color did not increase 'rationally'.[44] Ten Kate then responded to explain that the series was not arithmetic – in which case the increase would indeed be constant – but geometric and gave numerical examples of proportional increase (fig. 2).[45] Van Limborch's understanding of 'regular' and 'rational' thus turned out to be too restricted. Ten Kate did not, for that matter, explain how to actually mix increasing measures of paint. Later on in the correspondence, Ten Kate would try to correct Van Limborch on the physics of light and colors, but in this domain he did not find an equally docile pupil. In order to defend his position regarding mezzatints, Van Limborch appealed to the teachings of the *Physici*. He argued that the color of an object is void of any daylight and that perceived colors are produced when light is reflected at the superficies of an object and are thus a mixture of color and light.[46] Ten Kate replied with a range of arguments about the physics of reflecting surfaces as well as the difference between mixed and hatched paint, but in this case Van Limborch was not convinced.[47]

Discussing colors in writing created difficulties in its own right. The correspondence displays a to-and-fro of carefully phrased considerations and meticulously recorded experiments that nevertheless did not always clear up all misunderstandings. In an early stage of the project on color mixing, Van Limborch wrote Ten Kate:

> And we will never understand the other well as long as there is not a foundation about which we can reason. This foundation could be a paper on which was painted a spot of your harmonious black and the 7 grades of darkness that you pose in the same letter. Because otherwise I cannot examine whether my test and yours were made in the same way.[48]

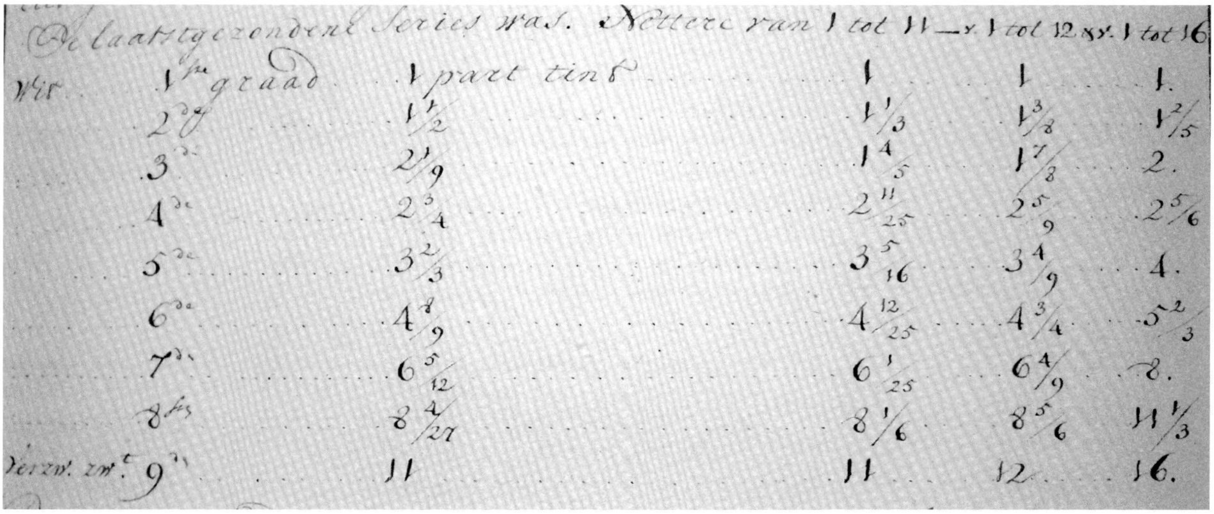

		1 tot 11	1 tot 12	1 tot 16
1ste graad	1 part tint	1	1	1.
2de	1 1/2	1 1/3	1 3/8	1 2/5
3de	2 1/9	1 4/5	1 7/8	2.
4de	2 3/4	2 11/25	2 5/9	2 5/6
5de	3 2/3	3 5/16	3 4/9	4.
6de	4 8/9	4 12/25	4 3/4	5 2/3
7de	6 5/12	6 1/25	6 4/9	8.
8ste	8 4/27	8 1/6	8 5/6	11 1/3
9de	11	11	12	16.

2

Table by Ten Kate explaining the proper calculations of geometric series, 1708

From the transcripts by Cornelis Ploos van Amstel, Amsterdam, Library University of Amsterdam, 63 U.B. I C. 23, p. 127.

Such an exchange of proofs could reveal problems that could not be described in letters. Still, this did not resolve all difficulties, and from time to time they would pay each other a visit '(...) to keep your honor no longer with comments on your reasonings, that are much easier resolved by verbal conversation and direct experiments than with the pen (...)'.[49] Convening made it possible, for example, to compare the way each mixed paints on their palette.[50] These meetings did not, as we saw, resolve all differences of opinion, and neither did they solve all problems of understanding and judging colors and their effects. They understood that the properties of colors were not innate to objects but depended on the judgment of an experienced observer. In order to evaluate the results of their methods, Ten Kate emphasized, these first should be applied quite rigorously to let the eye accustom itself to the gradations. To judge colors, the eye ought to be trained and civilized.[51] This learning of the eye was a mixture of epistemic concerns of determining the optical strength of colors, and esthetic convictions of assessing works of art that, I will argue, was at the heart of the project.

Knowing materials and knowing eyes

The color-mixing project of Ten Kate and Van Limborch sprang from a mixture of artistic interests in the judgment of paintings and the development of pictorial techniques – including Le Blon's color-print project in the background – and philosophical interests in empirical investigation and theoretical interpretation. The methods of investigation they developed were likewise a mixture of embodied inquiries of techniques of color making and painting on the one hand, and experiments and reflections on the nature and properties of light and color on the other. Unlike Le Blon, and many other painters in their accounts of color mixing, who investigated the properties of material colors in order to improve painterly techniques, Ten Kate and Van Limborch investigated the properties of *colors* in paints. Without dissociating their inquiry from material colors of paints and pigments and turning to the 'philosophical' colors of rainbows and prisms, they investigated the properties and effects of colors proper. The project was thus a symbiosis of 'artistic' and 'scientific' perspectives.

The project of Ten Kate and Van Limborch is historically interesting because optical inquiry of this kind seems to have been then generally uncommon. What makes it even more significant, however, is their original and inventive approach to the subject matter. They employed systematic and disciplined means of investigating colors and their effects. The goal of the project was to develop a rational and quantitative method of color mixing. With this theme it would not have been surprising if they had turned to number juggling or to analysis on the basis of some other theory of colors. Such a line was quite common in early modern science: in music theory, for example, much of the inquiry consisted of analysis of the numerical proportions of consonants.[52] Yet, Ten Kate and Van Limborch did not take this line of approach. The foundation of their inquiry consisted of experiments, combined with mathematical analysis that was again submitted to experimental test.

They realized that experiments with colors and their effects involved considerable epistemic problems. Color perceptions were not readily exchangeable, and objective definitions and measures of colors and tones did not exist. In addition, the eye is unreliable and inexperienced in judging colors:

> How will one know whether the power of the yellower nude is about the same as of the white one? Will the eye be the sole judge? It can be addicted to habit and it is fickle in the verdict, first one thing then another. So that this part does not impart difficulty but benefit to my said method.[53]

In order to counter such problems, Ten Kate and Van Limborch employed ingenious experimental means to investigate colors and their effects. To determine the relative power of colors they compared line patterns to mixtures. The effect of colors was established on the basis of *perceptions*. In other words, the nature and properties of colors were considered to be determined by visual experience rather than some measuring device or rational rule. The perception of colors – by a learned eye – was the foundation of Ten Kate's theory of colors, and for this they developed an original mode of inquiry.

This perceptual approach in optical inquiry is historically remarkable. First of all, inquiry into the nature and properties of light and colors is rarely found in the study of painters. Secondly, this particular approach to optical inquiry is not expected at this time, the early eighteenth century. It brings to mind the work of Pierre Bouguer (1698-1758) on light intensity, *Essai d'optique sur la gradation de la lumière* (1729). Bouguer experimentally determined the quantity of light loss through transparent mediums in order to assess the effect of the atmosphere in astronomical observations. To determine the intensity of light, he compared light sources observed at various distances and through varying mediums, using the eye as a null-indicator.[54] In the history of optics, Bouguer's work is usually presented as a singular achievement: firstly, because he was the first to study light intensity in a systematic, empirical way and with unambiguous results; secondly, because visual perception formed the basis of his experiments. Yet, in the light of the project of Ten Kate and Van Limborch, Bouguer turns out to be less original. Although the project was rather obscure, hidden in their private quarters and private correspondence, and although it bore less definite fruit, Ten Kate and Van Limborch did develop methods that are similar to Bouguer's and did so at an earlier date. How should we read this?

Histories of optics

In order to assess Ten Kate and Van Limborch's project historically, we need to step back and consider some features of the history of optics as it is usually written. The history of optics is a modest part of the history of science and is still much informed by present-day interests concerning theoretical and experimental physics. Consequently, the narrative is

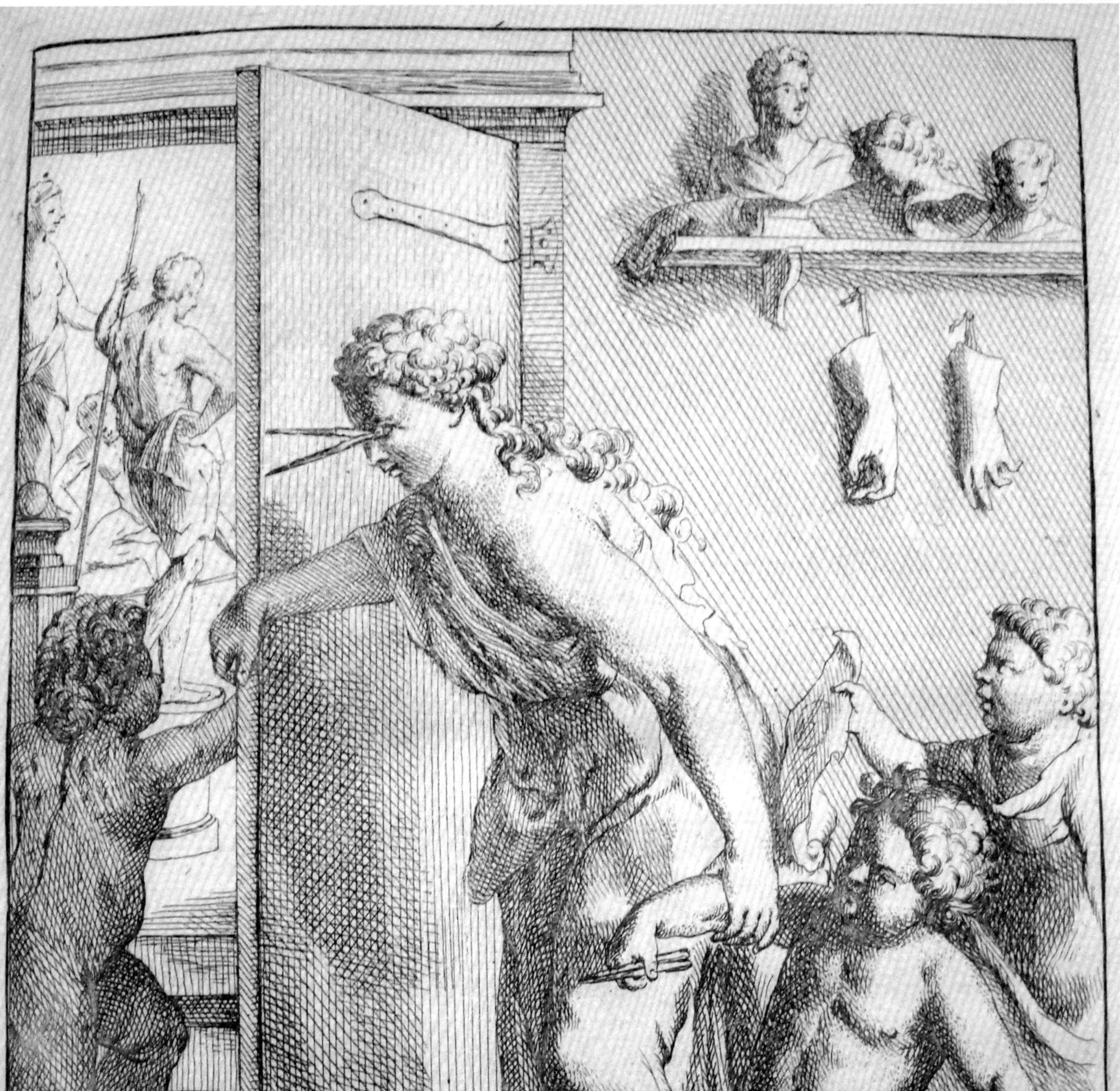
P. Bodart fec:

dominated by the development of theories of light and optical instruments and the precursors of modern conceptions. Themes that fall outside this scope and advocates of antiquated ideas are largely absent. The topic of light intensity, for example, is scarcely discussed in historical literature.[55] In this case, the fact that investigations were often pursued by 'non-scientists' – from a present-day perspective – has made these virtually invisible. This combination of predilections explains why a fascinating set of manuscript documents has not attracted the attention of historians of science. Ten Kate and Van Limborch are not recognized as natural philosophers, and the domain of art is not considered directly relevant to the history of optics.[56]

The project of Ten Kate and Van Limborch is not, however, merely interesting in an anecdotal sense. Its historical significance has to do with a more fundamental issue in the history of optics. Their perceptual approach to optics does not fit the received narrative of early modern optics. This narrative begins with the groundbreaking work of Johannes Kepler (1571-1630). In *Ad vitellionem paralipomena* (1604) he developed a new foundation for optics by introducing a new understanding of image formation. Images, he explained, are not formed by single rays but by the focusing of bundles of rays from points of the object into points of the image. This focusing can be brought about in various ways: pinholes, mirrors and lenses. The eye is a combination of a pinhole and a lens, and images are 'painted' on the retina.[57] In his account of vision, Kepler then drew an explicit boundary between optics and philosophy: optics explains how images are 'painted' on the retina, but the cognitive question of what happens with the image beyond the retina and how an image is perceived is referred to philosophers. Optics thus became a science of light propagation that excluded issues of cognition and perception. However, these issues had been foundational for optics until Kepler. The medieval science of *perspectiva* was essentially a science of *sight* rather than *light*.[58] As Peter Bexte has argued, the optics of Kepler, Descartes and their followers was actually an optics of the blind, replacing perception with mathematical instruments like compasses and prisms.[59] Gerard Hoet's *Ontslote deure der tekenkunst* (Unlocked door of drawing, 1713) offers a wonderful illustration of this idea, with his engraving of the 'compass in the eye' (fig. 3). In the history of early modern optics, Kepler is generally considered to have created a revolution that brought this tradition to a close. The new optics that was further developed by Descartes, Chistiaan Huygens (1629-1695), Newton, and so forth, was essentially about theories of light, the primary concern of modern physical optics.[60]

Historians of optics have largely confined themselves to this history of theories of light. In this history the interest of early modern students of optics in issues of vision is largely unnoticed, apparently because these are considered to belong – echoing Kepler – to the domain of philosophers like Berkeley. Still, vision was a central topic in the most important eighteenth-century textbook on optics, Robert Smith's *Compleat system of opticks* (1738).[61] In the received narrative, issues of perception returned to optics only in the nineteenth century with the work of Hermann von Helmholtz (1821-1894) on physiological optics and the optical trickery of

3
Pierre Bodart (after drawings by Hoet), The 'Compass in the eye', 1605
engraving, Gerard Hoet, *Ontslote deure der tekenkunst* (from the French edition, *Les principaux fondements du dessein,* Leiden 1723).

stereoscopes, the Kinematoscope, and so on, in which the visual system acquired an active role in image formation.[62] In this grand narrative Bouguer is indeed the odd one out. The use of the eye as an instrument does not seem to fit the gradual removal of the senses from scientific observation in the seventeenth and eighteenth centuries.[63] But Bouguer is saved in the narrative because his inquiry was into the properties of light, and he used the eye only as a null-indicator, for lack of other instrumental means to determine intensity of light. And he is saved because he was a 'scientist': an established *académicien* contributing to the mathematical sciences.

Closer inspection reveals, however, that Bouguer, Ten Kate and Van Limborch were not the only exceptions in early modern optics. Outside the received canon of optics, quite a few examples can be found of a perceptual approach to the subject. Isaac Vossius (1618-1689) based his theory of light on the perception of refracted images.[64] He was criticized by his contemporaries – and even ridiculed by his acquaintance Huygens – but he found his approach supported by the work of Willebrord Snel (1580-1626). Snel is famous for the discovery of the sine law of refraction, and a brilliant reconstruction by Klaus Hentschel shows that he employed traditional perspectivist means to arrive at this result. Snel constructed the *refractaria*, the locus of images of a line seen through a refracting medium, and derived the relationship between angles of incidence and refraction.[65] There were more inquirers who approached optics in terms of perception. Besides a savant like William Molyneux (1656-1698), there were medical inquirers like Vopiscus Plempius (1601-1671) and Petrus Camper (1722-1789) and students of art like Samuel van Hoogstraten (1627-1678).[66] The optics of Frans Hemsterhuis (1721-1790) was essentially a *perspectiva*, as were Johann Heinrich Lambert's (1728-1777) optical inquiries.[67] Historians of early modern optics seem to have, however, a rather blind eye for writers such as these, who are not readily perceived as 'scientists'.[68]

An early modern line of 'perceptual optics' may well provide a historical background to Goethe's infamous color theory. Goethe's optics plays a crucial role in the transformation of nineteenth-century optics, but his particular conception of science has made it difficult to assess his role.[69] Historians of optics are mostly ill at ease with the *Farbenlehre*. On the one hand Goethe is acknowledged as a precursor of the physiological turn in nineteenth-century optics; on the other hand, his anti-Newtonianism is seen to make him a non-scientific outsider. This has not only been a problem for historians; Helmholtz himself took great pains to come to terms with his indebtedness to Goethe. In 1853 he gave a lecture on Goethe's scientific works, racking his brains over the question of how someone who had proved to be a terrific empiricist and a great thinker in botany and morphology could have erred so tremendously in optics.[70] Helmholtz explained that Goethe's poetic approach to nature was fruitful in the descriptive sciences but failed in the explanatory sciences because he confused reality and appearance. Twenty years later, Helmholtz added a postscript to the lecture when he realized that this was precisely the crux of Goethe's science: the phenomena *are* the reality, and this is where the laws of nature reside.[71] This was after Helmholtz's *Wahrheitsgewissheits-*

verlust of the 1870s, when he came to the conclusion that mechanistic explanations were not the ultimate key to reality.[72] In his epistemic conceptions, Goethe regarded Hemsterhuis as his example, being a philosopher of immediate sensation and aesthetic experience.[73] In his optics too, Goethe directly built on Hemsterhuis, who had interacted with Camper's study of the physiology of vision and was directly linked to Ten Kate.[74] Cornelis Ploos van Amstel (1726-1798) collected the papers of Ten Kate and translated tracts of Hemsterhuis. Working further backwards may reveal links between earlier students of optics and thus shed new light upon the historical place of the *Farbenlehre*.

For now we can conclude that Ten Kate and Van Limborch were not so much an exception, but rather, they represent an approach to optics and a type of inquirer that largely has escaped the attention of historians. Without claiming the existence of a specific tradition of 'perceptual optics', I would say that the break brought about by Kepler was not as complete as is often suggested. Throughout the seventeenth and eighteenth centuries a range of savants approached optics in terms of the perception of light, color and images. A historical investigation of early modern perceptual approaches to optics may help relieve our uneasiness with the *Farbenlehre*. It would not be a bad idea to start with Goethe's own history of optics in which someone like Vossius figures prominently. Goethe also discusses Le Blon, but unfortunately he drew on the critical account of his Paris successor Jacques-Fabien Gautier d'Agoty (1716-1785), who turned Le Blon into a Newtonian.[75] Anticipating the results of a more systematic investigation, a common feature of such a 'perceptual optics' can be discerned: an artistic and esthetic context of optical inquiry. The question then is, akin to Helmholtz's problem, how to assess this mode of inquiry from a scientific perspective. To explore this question, I will return to Ten Kate.

A science of connaisseurs

An early modern 'perceptual optics' is not a derivative optics, somehow subordinate to the optics of Kepler, Descartes and Newton. Rather, it is a separate line in optics where students largely had their own agenda and style of inquiry. A case in point is the color-mixing project of Ten Kate and Van Limborch. The topic and timing of their project suggest indebtedness on their part to Newton's *Opticks* of 1704. Miedema, the editor of the correspondence, immediately refers to Newton when Ten Kate writes about prisms and the harmonic division of the spectrum.[76] However, Newton is conspicuously absent in the correspondence, and the editor's references are not substantiated by direct references. Ten Kate knew about the *Opticks* at least by 1706 and may have known the contents as well, making it all the more remarkable that he did not refer to it in his research on color mixing. In 1716 he performed an experiment with prismatic colors that directly referred to the *Opticks*. Closer examination reveals that he not so much followed his alleged master as that he corrected Newton, namely on the harmonic division of the spectrum.[77] This in turn is remarkable because Ten Kate had a prominent role in the

'Newton hype' of the mid-1710s in the Low Countries. Apparently he regarded Newton as a guide in the ideology of the new philosophy but not in the understanding of light and colors.

Evidently the *Opticks* was not the ubiquitous gauge of eighteenth-century optics, as many historians presume. This is not that strange, though, for Newton was concerned with issues other than those that concerned men like Ten Kate. His agenda was primarily natural philosophy – with brief excursions to instrumentation. The theory of different refrangibility and primacy of colors addressed positions of Aristotle, Descartes, and the like. On the topics that interested people like Ten Kate, Newton had little new to say. The idea that colors are not modifications of white light may have been news for natural philosophers, but for artists and connoisseurs it was rather stale.[78] Le Blon was explicit about the difference. In the preliminaries of *Coloritto*, he stated:

> I am only speaking of *Material* Colours, or those used by *Painters*; for a *Mixture* of *all* the primitive *impalpable* Colours, that cannot be felt, will not produce *Black*, but the very Contrary, *White*; as the Great Sir Isaac Newton has demonstrated in his Opticks.[79]

Le Blon accepted Newton, but he was far from a 'Newtonian' in his concepts and theories. Like Ten Kate, he proceeded independently of Newton's optical theories. The same goes for the methodological model Newton's science is said to have provided for men like Ten Kate. The Dutch 'Newtonians' did not need Newton to become experimental philosophers. They had developed their methods well before Newton hit the scene, building upon the work of savants like Robert Boyle (1627-1691), Jacques Rohault (1618-1672) and Burchardus de Volder (1643-1709). The 'Newtonianism' that developed from the 1710s was a metaphysical program, and 'Newton' primarily served rhetorical purposes to distinguish their experimental philosophy from the unfounded speculations of Cartesians and the detested atheism of Spinozists.[80]

Ten Kate's collaboration with Van Limborch and Le Blon fitted a research agenda that spanned a large part of his seemingly desperate pursuits in physics, linguistics, art and theology. Miedema has made an impressive effort to paint a coherent picture of his ideas and activities, judged on their own merits.[81] Ten Kate was searching for harmony, in terms of regularity, beauty and piety. This was not, however, the classical Pythagorean harmony and its Renaissance renewal. First of all, Ten Kate combined the study of the classics with empirical and mathematical investigations of spoken languages, of statues and drawings, and of light and colors. Secondly, harmony for Ten Kate was not so much in the world – *in* Nature – as in the *senses*, in our perception of the world. Van Veen has shown that this epistemic, esthetic and ethical conviction can be traced back to Ten Kate's particular background of well-to-do liberal Mennonites in Amsterdam circa 1700 who sought devotion in the experience of the ordinary.[82] This approach gave rise to a series of remarkably original inquiries: linguistic patterns, phonetics, precision measurements of human proportions, the mixing properties of colors, and so on. Ten Kate's 'perceptual optics' thus was an offshoot of his particular style of inquiry and philosophical agenda.

It seems difficult to assess Ten Kate's investigations – and artistic inquiry in general – on their merits as science. Artistic pursuits are often categorized in terms of artisanal knowledge, but this rarely does justice to the scope of artistic inquiry. In the color-mixing project, the development of pictorial techniques was integrated into an inquiry after the nature and properties of colors. Such problems of assessment flow from a specific conception of science that has dominated the history of science for a long time. In this conception, 'science' is considered as a single overarching entity of knowledge production that is identified with analytical modes of reasoning aimed at theory development. This conception of science as a homogeneous and essentially intellectual practice has been criticized by pointing out the heterogeneity of ways of knowing as well as the practical and material inventiveness that is at the heart of natural inquiry.[83] Within this framework, the inquisitiveness of men like Ten Kate can be assessed on its own merits without subordinating it to the analytical science of a Newton that dominates our conception of science (as Helmholtz eventually realized too). Characteristic to it is a particular empiricism focused on the variety and 'whatness' of objects and materials and aimed at esthetic evaluation that is best characterized by the Dutch word *kennen*.[84] It is a science of 'connoisseurs' that is typical of the collector that became prominent in early modern natural history and can also be linked to the rise to prominence of chemistry around 1700.[85]

Such a reorientation on the history of early modern science also opens opportunities to connect the writing of the history of science with the writing of the history of art. If in the history of science the focus on theory development has given way to recognition of practical and material pursuits, the same can be said for the Panofskian focus on iconology and style development in history of art.[86] As regards pigment, paint and color mixing, Monika Wagner has argued that this focus results in a triple dematerialization in which pigments disappear in oils, paints in images, and colors in ideas.[87] Miedema draws attention to a pre-Panofskian account of eighteenth-century art, in which the knowledge of materials and practices are much more emphasized.[88] In this account the role of amateur art collectors is crucial to the development of esthetics and connoisseurship, thus linking up rather nicely with the role of the connoisseur in early modern science. The esthetic category of 'connoisseur' of the history of art thus has its epistemic counterpart in the history of science. This should not surprise us at all: early modern connoisseurs considered the natural and the artful equal sources of knowledge and developed styles of inquiry of their own.[89] In the color-mixing project we see two sides of the same medal: for Ten Kate art was the key to nature; for Van Limborch nature was the key to art. Together with Le Blon they pursued a 'connoisseur science' that produced an inventive and original instance of 'perceptual optics'.

Notes

1 I am indebted to the inspiring works of Svetlana Alpers, Pamela Smith, Sven Dupré, Eileen Reeves and Horst Bredekamp.

2 Ten Kate 1710; Ten Kate 1716; Ten Kate 1723; Ten Cate 1987; Noordegraaf & Van der Wal 2001; Vermij 2003. For a critical discussion of his 'Newtonianism' see Dijksterhuis 2011.

3 Ten Kate wrote a manuscript treatise on phonetics in 1699, parts of which appeared in Ten Kate 1723, his major publication on linguistics. The manuscript is discussed in Van der Hoeven 1896. For optics, Ten Kate 1757; for esthetics, Ten Kate 1732.

4 Van Gelder 1970, 144-157.

5 Van Gelder 1970; Zuidervaart 1999, 450.

6 Aarts 2003, 48.

7 Miedema 2006, vol. I, 32: 'Ik heb my verwonderd dat Lares, in zyn boek, deeze materie zo onnozel behandelt.'

8 An extensive biography with a critical discussion of available sources is Lilien 1985.

9 Zacharias Conrad von Uffenbach visited Amsterdam in 1711 and gives the first account of Le Blon's color printing. Lilien 1985, 22-23.

10 Lilien 1985, 29-36: 'One anatomical preparation after the system of Dr. Cockburn, from the object prepared and injected by St. André, painted by Le Blon, height 10 ¾ inches by 8. Price 5 shillings.'

11 Lilien 1985, 94-97 gives a convincing argument and source evidence for the dating.

12 In 1732 he published a translation of Ten Kate's account of esthetics, *Le beau ideal*. More details on Le Blon's printing technique are found in Mortimer 1731 and Gaultier de Montdorge 1756.

13 Miedema 2006, vol. I, 86.

14 'If one desires that the dark represents the same art of color as the light'. Miedema 2006, vol. I, 89: 'indien men begeert dat het donker eenderleij natuer van koleur representere als het Licht'.

15 Miedema 2006. The manuscripts are in the Rijksprentenkabinet, Amsterdam (originals from the estate of Van Limborch; RP-D-00-1062, 301 K) and the Library of the University of Amsterdam (18th-century transcripts by Cornelis Ploos van Amstel; Bibliotheek Universiteit van Amsterdam 63 U.B. I C. 21.-24). Miedema discusses the manuscripts and their provenance in detail; his edition includes all of the remaining letters as well as relevant manuscript notes and writings, together with a searching commentary.

16 Lehmann 2008, 99.

17 De Lairesse 1712, 35-38. This edition is identical to the first publication of 1707. Lehmann 2008, 87-88.

18 Descartes 1637; Descartes 1656; Gaukroger 1995.

19 Scholars such as Franciscus Aguilonius (1567-1617), a close acquaintance of Rubens, and Marin Cureau de la Chambre (1594-1669). An insightful overview is provided by Gage 1993, 153-176. See also the contribution of Karin Leonhard to this volume, who draws attention to the role of mineralists.

20 Newton 1704, 110-117 (book I, part II, prop. V, theorem IV, exp. XV; and prop. VI, problem II).

21 Miedema 2006, vol. I, 89.

22 Miedema 2006, vol. II, 85 explains that ivory black is made by charring waste of ivory in an airtight receptacle. If, however, air is admitted to the receptacle, a white powder is produced. This is the 'over burnt ivory black' they used to weaken their black.

23 Miedema 2006, vol. I, 89: 'zulks dat dan, onder gelijke quantiteijt Wit & verzwakt Zwart met oly vermengt zijnde, een tint, uijt 2 a 3 van de Coleuren bestaande, zal moeten vermengt worden 11 a 12 delen tint bij het verzwakte zwart, tegens 1 deel tint onder 't wit; indien men begeert dat het donker eenderleij natuer van koleur representere als het Licht het welk het doelwit van de koleurmenging is.'

24 Lilien 1985, 114-127.

25 Ten Kate reports on this insight to Van Limborch on 3 February 1710 and 1 January 1712. Miedema 2006, vol. I, 162, 203.

26 Miedema 2006, vol. II, 85-86.

27 Van Limborch sent a note with his letter, in which he briefly indicated the manner of mixing paints. Miedema 2006, vol. I, 98-99.

28 Ibid., 91-92.

29 Ibid., 95-97.

30 Ibid., 95-96.

31 Ibid., 102-103.

32 Ibid., 91: 'Onder ander heb ik nog veelmaal gedagt om zonder blaauwe koleur aan te leggen de half koleuren blaauw agtig te doen schijnen door 't dun overhalen van de ware vleeskoleur maar daar mede kan ik niet wel regt komen of den regel is niet generaal goet'.

33 Miedema 2006, vol. I, 114-115. This too, according to Aarts 2003, was a direct criticism of De Lairesse.

34 Miedema 2006, vol. I, 150-151.

35 Ibid., 113-114; Miedema 2006, vol. II, 103-104.

36 Miedema 2006, vol. II, 151.

37 Miedema 2006, vol. I, 196-197.

38 Ibid., 210-215.

39 The collection of manuscripts at the Rijksprentenkabinet in Amsterdam (RP-D-OO-1062, 310 K) contains many scraps and notes on which he made calculations and tabulations, edited in Miedema 2006, vol. I, 255-308.

40 Ibid., 223.

41 Le Blon 1725, appendix, 6.

42 Miedema 2006, vol. I, 120: 'Indien ik by uw Ed.' was verbeeld ik mij dat ik wel zou kunnen beduyden een middel om het in weynig tijts uijt te rekenen maar zelf ben ik in 't cijfferen zo ver niet dat ik het middel kan te werk stellen (...)'.

43 Ibid., 140: 'Dezen opgang schynt mij niet egaal te springen (...)'.

44 Ibid., 141: 'Dat nu de parten tint niet rationeel opklimmen kan verder uijt het proefje A blyken'.

45 Ibid., 143.

46 Ibid., 205.

47 Ibid., 21.

48 Ibid., 103-104: 'En nooit zullen wy den anderen wel kunnen verstaan zo lang 'er niet een fondament is daar wij over kunnen raisonneren. Dit fondament kon zyn een papiertje daar op geschildert was een plekje van uw harmonieus zwart en de 7 graden van donkerheit die uw E poseert in des selfs brief. Om dat ik anders niet net kan onderzoeken of myn proeve en de uwe wel op ene wyze werden genomen.'

49 Ibid., 135. 'En om uw E: niet verder op te houden met aanmerkingen op uwe raisonnementen, welke door mondeling gesprek en dadelijke proeven merkelijk gemakkelijker zijn op te lossen, dan met de penne, te meer dewijl we dezelve zonder hinder in uw oogmerk, mijns agtens, kunnen voorbij gaan (...)'.

50 Ibid., 210.

51 Ibid., 122-123.

52 Cohen 1984.

53 Miedema 2006, vol. I, 128: 'Hoe zal men weten of de kragt van 't geelder

naakt omtrent dezelfde is, als die van 't blanke? Zal het ooge de eenigste Scheidsman wezen? dit kan aan gewoonte verslaaft zijn, & is omtrent zulke differentien wispeltuerig in 't vonnissen, dan dus dan zoo: zoo dat dit stuk geen zwarigheijd maar voordeel aan mijne gemelde methode toebrengt'.

54 Ariotti & Marcolongo 1976; Knowles Middleton 1960.

55 Dijksterhuis 2007.

56 Historical studies exist on the intersection of optics and art in history; see Gage 1993, and Kemp 1990, who focuses on perspective and *disegno*.

57 The idea that the retina is the sensitive organ and that images are not created in the lens was a recent development in ocular anatomy on which Kepler built. He also employed Dürer's techniques of image analysis and consciously used artistic concepts like *pictura*; see Straker 1976, and Alpers 1983.

58 Smith 1983; Smith 1981. Note that Smith's interpretation is nuanced in Vanagt 2010.

59 Bexte 1999.

60 This conception is usually more or less implicit, but very explicit in Gal & Chen-Morris 2010.

61 Dijksterhuis 2008.

62 Crary 1999.

63 Kuykendall *et al.* 1995.

64 Dijksterhuis 2012.

65 Hentschel 2001.

66 Weststeijn 2008; Vanagt 2010.

67 Petry 1985.

68 Le Blon is conspicuously absent, as are most of the 'artistic' investigators of optics. Shapiro 1993; Cantor 1983; Boyer 1959; Sepper 1988; Hall 1993; Wilde 1838. In his work on the early history of the telescope, Sven Dupré has drawn our attention to the decisive role of 'practical' knowledge of optical instruments that has also gone largely unnoticed in the history of early modern optics; Dupré 2005.

69 Crary 1999.

70 Helmholtz 1875.

71 Helmholtz 1875, 54 ('Nachschrift, [geschrieben 1875]'): 'Goethe's Versuch, seine Anschauungen an dem Beispiel der Farbenlehre praktisch durchzuführen, können wir freilich nicht als gelungen betrachten, aber das Gewicht, was er selbst auf diese Richtung seiner Arbeiten legte, wird verständlich'.

72 Schiemann 1997.

73 Petry 1985.

74 Petry 1985, 219.

75 Goethe 2003, vol. 4, 230-238. Gautier did so to create a distinct profile for himself; the *Coloritto* and other sources easily show that matters are much more nuanced. Goethe did not use these sources and presented Le Blon as one of the victims of Newton's eighteenth-century dominance.

76 Miedema 2006, vol. II, 136, 145.

77 Dijksterhuis 2011.

78 Shapiro 1994.

79 Le Blon 1725, 6. His French on p. 7: 'Je ne parle ici que des *Couleurs Materielles*, c'est à dire, des Couleurs dont se servent les Peintres; car le mélange de toutes les Couleurs primitives impalpables, ne produit pas le *Noir*; mais precisément le contraire, c'est à dire, le *Blanc*; comme l'a démontré l'incomparable Mons. le Chevalier Newton dans son Optique'.

80 Vermij 2003.

81 Miedema 2006, vol. II, 39-42.

82 Van Veen 1995, 68-72.

83 Heterogeneity of science is conceptualized in Kuhn 1977, Pickstone 2000, Pickstone 2007, and Crombie 1995. These studies do not, however, include the domain of art. See also Smith 2004, Turnbull 2003, Klein 2005, and Roberts *et al.* 2007.

84 Cook 2007, 13-20; Pickstone 2000; Klein & Spary 2009.

85 Cook 2007, combined with Klein & Spary 2009.

86 Van de Wetering 1997; Alpers 1983, to name a few.

87 Wagner 2001, 18-22.

88 Miedema 2006, vol. II, 23-24.

89 Bredekamp 2000.

Bibliography

Aarts 2003
M. Aarts, 'Achter de tronie van Van Limborch', *Tableau* 25 (2003) no. 6, 45-49.

Alpers 1983
S. Alpers, *The art of describing: Dutch art in the seventeenth century*, Chicago 1983.

Ariotti & Marcolongo 1976
P.E. Ariotti & F.J. Marcolongo, 'The law of illumination before Bouguer (1729). Statement, restatements and demonstration', *Annals of Science* 33 (1976) 331-340.

Bexte 1999
P. Bexte, *Blinde Seher. Wahrnehmung von Wahrnehmung in der Kunst des 17. Jahrhunderts*, Dresden 1999.

Boyer 1959
C.B. Boyer, *The rainbow. From myth to mathematics*, Princeton 1959.

Bredekamp 2000
H. Bredekamp, *Antikensehnsucht und Maschinenglauben. Die Geschichte der Kunstkammer und die Zukunft der Kunstgeschichte*, Berlin 2000.

Cantor 1983
G.N. Cantor, *Optics after Newton. Theories of light in Britain and Ireland, 1704-1840*, Manchester 1983.

Cohen 1984
H.F. Cohen. *Quantifying music. The science of music at the first stage of the scientific revolution, 1580–1650*, Dordrecht 1984.

Cook 2007
H.J. Cook, *Matters of exchange. Commerce, medicine, and science in the Dutch Golden Age*, New Haven 2007.

Crary 1999
J. Crary, *Techniques of the observer. On vision and modernity in the nineteenth century*, Cambridge 1999.

Crombie 1995
A.C. Crombie, *Styles of scientific thinking in the European tradition. The history of argument and explanation especially in the mathematical and biomedical sciences and arts*, London 1995.

De Lairesse 1712
G. de Lairesse, *Groot Schilderboek*, Amsterdam 1712.

Descartes 1637
R. Descartes, *Discours de la méthode pour bien conduire sa raison, et chercher la vérité dans les sciences. Plus la dioptrique, les meteores et la geometrie, qui sont des essais de cete methode*, Leiden 1637.

Descartes 1656
R. Descartes, *Principia philosophiae. Nunc demum hac editione diligenter recognita, & mendis expurgate*, Leiden 1656.

Dijksterhuis 2007
F.J. Dijksterhuis, 'Clair & distinct. Seventeenth-century conceptualizations of the quality of images', in: W. Lefèvre, *Inside the camera obscura. Optics and art under the spell of the projected image*, Berlin 2007, 105-115.

Dijksterhuis 2008
F.J. Dijksterhuis, 'Reading up on the *Opticks*. Refashioning Newton's theories of light and colors in eighteenth-century textbooks', *Perspectives on Science* 16 (2008), 309-327.

Dijksterhuis 2011
F.J. Dijksterhuis, 'Low Country opticks', in: E. Jorink & A. Maas (eds.), *Newton & the Netherlands*, Leiden 2011 (forthcoming).

Dijksterhuis 2012
F.J. Dijksterhuis, 'A view from the mountaintop. The development of Isaac Vossius' optics, 1658-1666', in: E. Jorink & D. van Miert (eds.), *The challenger. Isaac Vossius (1618-1689) and the European world of learning*, Leiden 2012 (forthcoming).

Dupré 2001
S. Dupré, 'De Optica van Galileo Galilei. Interactie tussen kunst en wetenschap', *Verhandelingen van de Koninklijke Vlaamse Academie van België voor Wetenschappen en Kunsten*, Nieuwe Reeks, Nummer 5, Brussels 2001.

Dupré 2005
S. Dupré, 'Ausonio's mirrors and Galileo's lenses. The telescope and sixteenth-century practical optical knowledge', *Galilaeana. Journal of Galilean Studies* 2 (2005), 145-180.

Gage 1993
J. Gage, *Colour and culture. Practice and meaning from antiquity to abstraction*, London 1993.

Gal & Chen-Morris 2010
O. Gal & R. Chen-Morris, 'Baroque optics and the disappearance of the observer. From Kepler's optics to Descartes' doubt', *Journal of the History of Ideas* 71 (2010), 191-217.

Gaukroger 1995
S. Gaukroger, *Descartes. An intellectual biography*, Oxford 1995.

Gaultier de Montdorge 1756
A. Gaultier de Montdorge, *L'Art d'imprimer les tableaux. Traité d'après les ecrits, les opérations & les instructions verbales de J.C. Le Blon*, Paris 1756.

Goethe 2003
J.W. Goethe, *Farbenlehre*, Herausgegeben von Gertrud und Gerhard Ott, Stuttgart, 1979, 2003, 5 vols.

Hall 1993
A.R. Hall, *All was light. An introduction to Newton's Opticks*, Oxford 1993.

Helmholtz 1875
H. Helmholtz, 'Über Goethe's naturwissenschaftliche Arbeiten', *Allgemeine Monatsschrift für Wissenschaft und Literatur* (1853), 383-98; republished in: H. Helmholtz, *Populäre wissenschaftliche Vorträge*, vol. 1, Braunschweig 1875, 30-54.

Hentschel 2001
K. Hentschel, 'Das Brechungsgesetz in der Fassung von Snellius. Rekonstruktion seines Entdeckungspfades und eine Übersetzung seines lateinischen Manuskriptes sowie ergänzender Dokumente', *Archive for History of Exact Sciences* 55 (2001), 297-344.

Jansen 1998
G.M.C. Jansen, 'De *Notitie der dagelijxe schilderoeffening* van Henrik van Limborch (1681-1759)', *Bulletin van het Rijksmuseum* 46 (1998), 26-67.

Kemp 1990
M. Kemp, *The science of art. Optical themes in Western art from Brunelleschi to Seurat*, New Haven 1990.

Klein 2005
U. Klein, 'Technoscience avant la lettre', *Perspectives on science* 13 (2005), 226-266.

Klein & Spary 2009
U. Klein and E.C. Spary (eds.), *Materials and expertise in early modern Europe. Between market and laboratory*, Chicago 2009.

Knowles Middleton 1960
W.E. Knowles Middleton, 'Bouguer, Lambert, and the theory of horizontal visibility', *Isis* 51 (1960), 145-149.

Kuhn 1977
T.S. Kuhn, 'Mathematical versus experimental traditions in the development of physical science', in: T.S. Kuhn, *The essential tension. Selected studies in scientific tradition and change*, Chicago 1977, 31-65.

Kuykendall *et al.* 1995
L.D. Kuykendall, W.J. Hunter, M.M. Zein & L.L. Roberts, 'The death of the sensuous chemist. The 'new' chemistry and the transformation of sensuous technology', *Studies in History and Philosophy of Science Part A* 26 (1995), 503-529.

Le Blon 1725
J.C. Le Blon, *Coloritto, or the harmony of colouring in painting. Reduced to mechanical practice, under easy precepts, and infallible rules*, London [1725].

Lehmann 2008
A.-S. Lehmann, 'Fleshing out the body. The 'colours of the naked' in workshop practice and art theory, 1400-1600', in: A.-S. Lehmann &

H. Roodenburg (eds.), *Body and embodiment in Netherlandish art*, Zwolle 2008 (= *Nederlands Kunsthistorisch Jaarboek* 59), 86-110.

Lilien 1985
O.M. Lilien, *Jacob Christoph Le Blon 1667-1741. Inventor of three- and four-colour printing*, Stuttgart 1985.

Miedema 2006
H. Miedema, *Denkbeeldig schoon. Lambert ten Kates opvattingen over beeldende kunst*, Leiden 2006, 2 vols.

Mortimer 1731
C. Mortimer, 'An account of Mr. James Christopher Le Blon's principles of printing, in imitation of painting, and of weaving tapestry, in the same manner as brocades', *Philosophical Transactions* 419 (1731), 101-109.

Newton 1704
I. Newton, *Opticks, or, a treatise of the reflexions, refractions, inflexions and colours of light*, London 1704.

Noordegraaf & Van der Wal 2001
J. Noordegraaf & M. van der Wal, 'Lambert ten Kate (1674-1731) and linguistics', introduction to L. ten Kate Harmensz., *Aenleiding tot de kennisse van het verhevene deel der Nederduitse sprake*, Alphen aan den Rijn 2001, 2-32.

Petry 1985
M.J. Petry, 'Hemsterhuis on mathematics and optics', in: J.D. North & J.J. Roche (eds.), *The light of nature. Essays in the history and philosophy of science presented to A.C. Crombie*, Dordrecht 1985, 209-234.

Pickstone 2000
J.V. Pickstone, *Ways of knowing. A new history of science, technology and medicine*, Manchester 2000.

Pickstone 2007
J.V. Pickstone, 'Working knowledges before and after circa 1800. Practices and disciplines in the history of science, technology and medicine', *Isis* 98 (2007), 489-516.

Reeves 1997
E.A. Reeves, *Painting the heavens. Art and science in the age of Galileo*, Princeton 1997.

Reeves 2008
E.A. Reeves, *Galileo's glassworks. The telescope and the mirror*, Cambridge 2008.

Roberts *et al.* 2007
L.L. Roberts, S. Schaffer & P. Dear (eds.), *The mindful hand. Inquiry and invention from the late Renaissance to early industrialisation*, Amsterdam 2007.

Roodenburg 2010
H. Roodenburg, 'Over *habitus* en de codes van de *honnêteté*. De wereld van de achttiende-eeuwse adel verkend', *De Achttiende Eeuw* 42 (2010) no.1, 120-140.

Schiemann 1997
G. Schiemann, *Wahrheitsgewissheitsverlust. Hermann von Helmholtz' Mechanismus im Anbruch der Moderne. Eine Studie zum Übergang von klassischer zu moderner Naturphilosophie*, Darmstadt 1997.

Sepper 1988
D.L. Sepper, *Goethe contra Newton. Polemics and the project for a new science of color*, Cambridge 1988.

Shapiro 1993
A.E. Shapiro, *Fits, passions, and paroxysms. Physics, method, and chemistry and Newton's theories of colored bodies and fits of easy reflection*, Cambridge 1993.

Shapiro 1994
A.E. Shapiro, 'Artists' colors and Newton's colors', *Isis* 85 (1994), 600-630.

Smith 1981
A.M. Smith, 'Getting the big picture in perspectivist optics', *Isis* 72 (1981), 568-589.

Smith 1983
A.M. Smith, 'Introduction', *Witelonis perspectivae liber quintus. Book V of Witelo's Perspectiva* (A.M Smith, ed. & transl.) (Studia Copernicana XXIII), Wroclaw 1983.

Smith 2004
P.H. Smith, *The body of the artisan. Art and experience in the scientific revolution*, Chicago 2004.

Straker 1976
S. Straker, 'The eye made "Other". Dürer, Kepler, and the mechanization of light and vision', in: A. Knafla, M.S. Staum & T.H.E. Travers (eds.), *Science, technology, and culture in historical perspective* (University of Calgary Studies in History, no. 1), Calgary 1976, 7-25.

Ten Cate 1987
C.L. ten Cate, *Lambert ten Kate Hermansz. (1674-1731). Taalgeleerde en konstminnaar*, Utrecht 1987.

Ten Kate 1710
L. ten Kate, *Gemeenschap tussen de Gottische Spraeke en de Nederduytsche*, Amsterdam 1710.

Ten Kate 1716
L. ten Kate, *Den schepper en zyn bestier te kennen in Zyne schepselen, volgens het licht der reden en wiskonst*, Amsterdam 1716.

Ten Kate 1723
L. ten Kate, *Aenleiding tot de kennisse van het verhevene deel der Nederduitsche sprake*, Amsterdam 1723.

Ten Kate 1732
L. ten Kate, *The beau ideal, by the late ingenious and learned Hollander Lambert Hermanson ten Kate, translated from the original French by James Christopher Le Blon, author of the Coloritto*, London 1732.

Ten Kate 1757
L. ten Kate, 'Proef-ondervinding over de scheyding der coleuren, bevonden, door een prisma, in de volgende orde der muzyk-toonen in navolging eener proef-ondervinding in *Newtons Gezigtkunde*.

Eertyds waargenomen, en nu uit de nalatenschap van *Lamb. ten Katen, Hz.* medegedeeld door den voorgem. *Joh: Nettis.*', *Verhandelingen uitgegeeven door de Hollandse Maatschappy der Weetenschappen, te Haarlem.* Derde Deel (1757), 17-30.

Turnbull 2003
D. Turnbull, *Masons, tricksters and cartographers. Comparative studies in the sociology of scientific and indigenous knowledge*, London 2003.

Van de Wetering 1997
E. van de Wetering, *Rembrandt. The painter at work,* Amsterdam 1997.

Van der Hoeven 1896
A. van der Hoeven, *Lambert ten Kate (De 'Gemeenschap tussen de Gottische Spraeke en de Nederduytsche' en zijne Onuitgegeven geschriften over Klankkunde en Versbouw.)*, The Hague 1896.

Van Gelder 1970
J.G. van Gelder, 'Lambert ten Kate als kunstverzamelaar', *Nederlands Kunsthistorisch Jaarboek* 21 (1970), 139-186.

Van Veen 1995
H.T. van Veen, 'Devotie en esthetiek bij Lambert ten Kate', *Doopsgezinde Bijdragen* 21 (1995), 63-96.

Vanagt 2010
K. Vanagt, *De emancipatie van het oo. V.F. Plempius (1601-1671) en de vroegmoderne medische denkbeelden over het zien*, Enschede 2010.

Vermij 2003
R. Vermij, 'The formation of the Newtonian philosophy. The case of the Amsterdam mathematical amateurs', *The British Journal for the History of Science* 36 (2003), 183-200.

Wagner 2001
M. Wagner, *Das Material der Kunst. Eine andere Geschichte der Moderne*, Munich 2001.

Weststeijn 2008
T. Weststeijn, *The visible world. Samuel van Hoogstraten's art theory and the legitimation of painting in the Dutch Golden Age*, Amsterdam 2008.

Wilde 1838
E. Wilde, *Geschichte der Optik. Vom Ursprunge dieser Wissenschaft bis auf die gegenwärtige Zeit*, Berlin 1838-1843.

Zuidervaart 1999
H.J. Zuidervaart, *Van 'Konstgenoten' en hemelse fenomenen. Nederlandse sterrenkunde in de achttiende eeuw*, Rotterdam 1999.

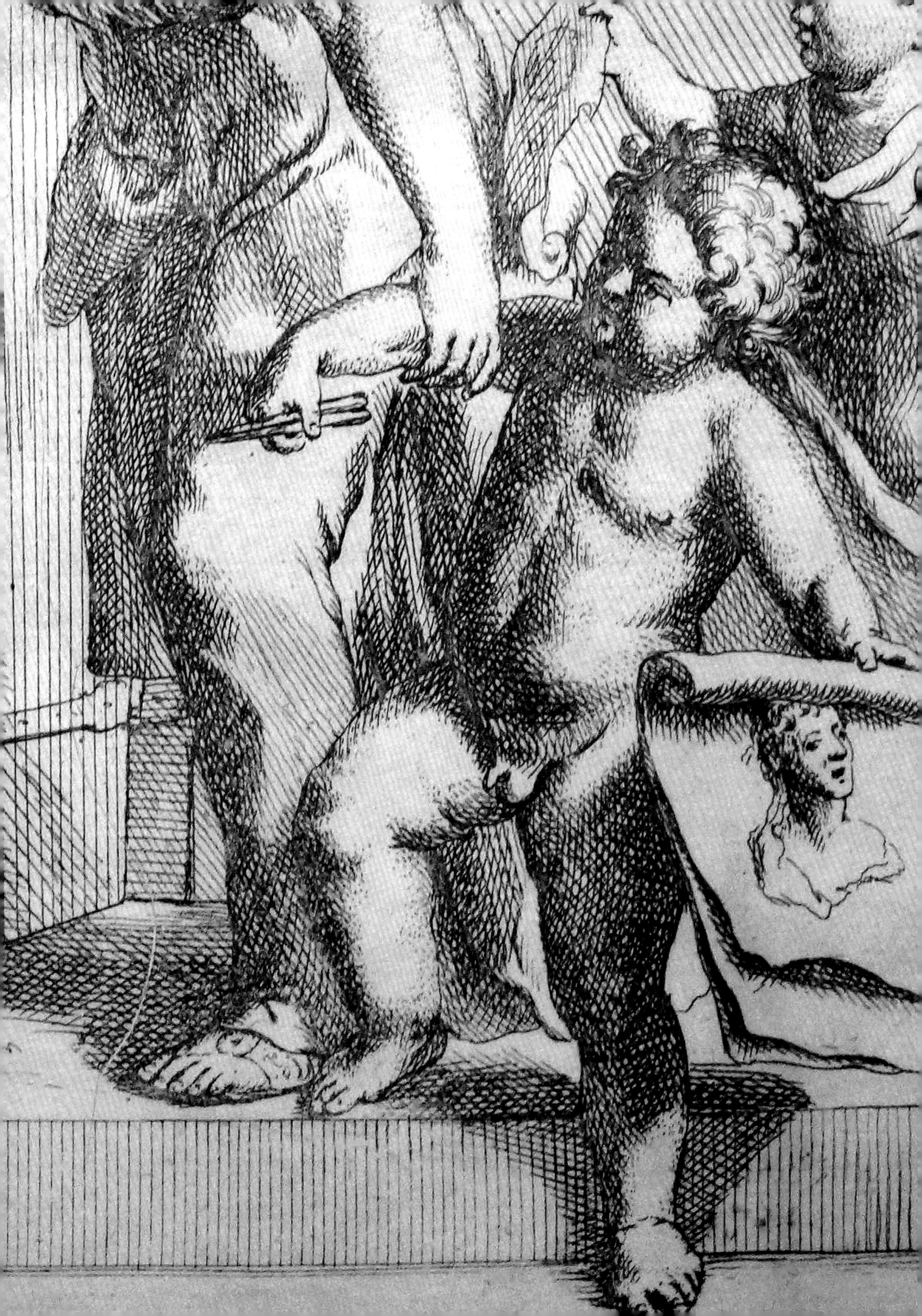

HET LAND,

IN BRIEVEN.

MET PLAATEN.

DERDE DRUK.

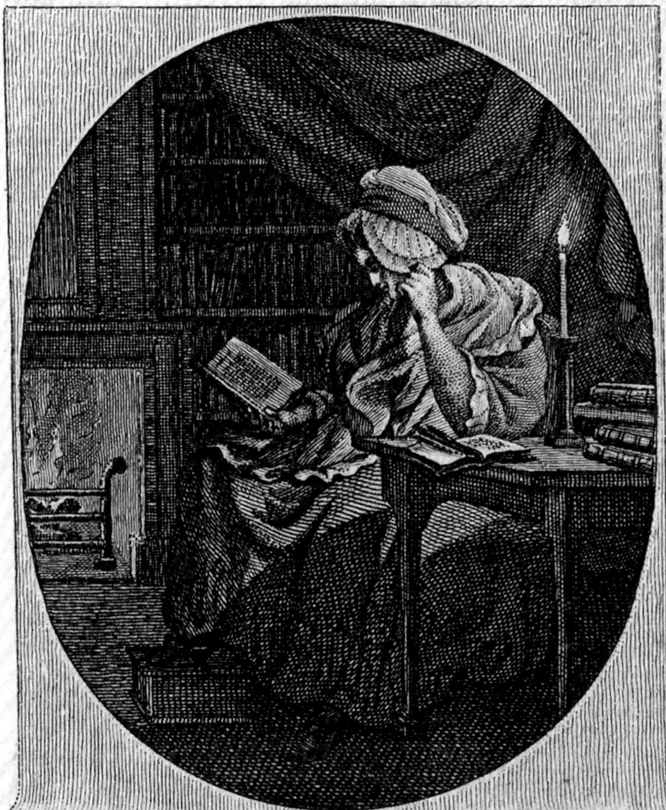

Reinr. Vinkeles, inv. del. et sculp.

TE AMSTERDAM, bij
JOHANNES ALLART,
MDCCLXXXVIII.

Staging nature
Observation, imagination and experience in E.M. Post's *Het land, in brieven* (1788)

Bart Ramakers

Introduction

One entry into the canon of the eighteenth-century Dutch epistolary novel, aside from much more widely known examples by the writing duo Elizabeth Wolff-Bekker (1738-1804) and Aagje Deken (1741-1804), is *Het land, in brieven* by Elisabeth Maria Post (1755-1812) (fig. 1). The novel appeared in 1788 and saw a second and third edition that same year, followed by a fourth in 1792.[1] *Het land, in brieven* – a title perhaps best translated as *The countryside, in letters* – portrays life in the countryside. We can read the word 'land' in the title as meaning 'nature' or 'landscape', which forms the main subject of the 55 letters that the novel comprises.[2] Post published only one other epistolary novel and several volumes of poetry – all in all, a small body of work. The two novels are highly autobiographical, in that the events, situations and characters described in them can be traced back to the author's life and the people close to her. For the reading of *Het land, in brieven* it is also relevant that the landscape described in the book is very similar to that surrounding the village of Amerongen in the province of Utrecht, where Post spent part of her life. It is characterised by a combination of flat river landscape surrounded by wooded hills. Equally important is that the author is introduced in the preface as someone whose intellect and education predisposes her to describe and accurately represent nature.

This article concerns itself with the observation, imagination and experience of nature in *Het land, in brieven*. It attempts to show that the author was driven by a desire to understand the inner workings of nature and that she did so along the lines of what was considered to be scientific at the time. She used faculties that were increasingly being used in the approach to nature in the second half of the eighteenth century: perceptual observation, literary imagination and emotional experience. The characters in her novel do not only trust the objective, empirical results of contemporary research (to which we find references here and there in the novel) but, in their approach to nature, also trust in subjective insights that result from their imagination and their feelings. That approach presupposes a certain view on empiricism – namely, 'that knowledge grew not from sensory experience alone, but from a combination of sensation and sentiment'. This was a mode of natural science described as 'sentimental empiricism'.[3]

1
R. Vinkeles, *The author reading*
engraving, Elisabeth Maria Post, *Het land, in brieven* (Amsterdam, 3rd ed., 1788), University of Groningen Library, KW B 1544, title page (photo: Dirk Fennema), quotation: 'O die stille abundstunden, köstlich, herrlich sint sie mir', opening line from a poem by Johann Kaspar Lavater (1741-1801).

Art as an instrument to arrive at knowledge and to express it – that is what this contribution aims to make clear and where the connection to this volume of articles lies. We may think of art primarily as visual art, but literature, through the use of imagination – the word itself is significant here – was capable of evoking and representing images too. Imagination played a role in both arts in the eighteenth century. Where the artistic representation of nature was concerned, the correspondences between the two disciplines were found to be stronger than usual. In their representations of landscape, the characters in Post's novel, as they themselves point out, operate in a very visual, one could even say painterly, way. This includes the repeated descriptions of landscapes as scenic or picturesque. Additionally, where they describe observed nature as a grand spectacle that is performed in words, a space is created for the imagination and the experience of the sublime – that category of experience that started to rise in both literature and painting during the second half of the eighteenth century and that permitted the aesthetical observer of nature to come to the core, the essence, of nature. For Post, that core or essence was metaphysical and religious, and connected to the divine, with God as starting and ending point for all (knowledge of) nature.

Approach and plan

This contribution leans heavily on results of the study of the artistic treatment of nature, especially landscape, in the second half of the eighteenth century in the German lands. In that study, a connection is made between the artistic treatment of nature – in both literature and painting – and the empirical research into it. As far as literature is concerned, there are frequent references to some influential German authors, whose names occur several times in *Het land, in brieven* and who are even quoted on several occasions.[4] (By contrast, English authors are almost completely absent, French authors even entirely so.)[5] This prominence corresponds with a growing influence accorded to German literature (in the broadest sense of the word) in the Netherlands in the second half of the eighteenth century. Thus, the obvious step is to explicate the work of Post through attention to the artistic treatment of nature in the German lands. To be clear, the German authors that this contribution refers to may have influenced her directly – this is the case at least for the authors that she herself names or quotes – or indirectly, through other, possibly also Dutch, publications in which their ideas found expression. At any rate, a study of the course of that influence on Post, and the influence of these authors in general on the Republic, falls outside this article's scope.

The representation of nature in this period cannot be disconnected from its precise observation. Stronger still, much in Post's novel indicates that her representation of nature emerged from – and was impossible without – an intense sensory engagement with it. It is not simply that the characters observe nature meticulously, they also let it transport them, in two senses of the word: they allow it to carry them to other places in thought, both inside and outside of perceptible reality, and they let themselves be emotionally affected. Their observations lead to imagination, to the epistolary description of *Nature* – the one with a capital N – but also to experiences, to feelings and ideas, that touch their own, inner *nature* – the one with a lower case n. Thus, imagination in this novel is seen not only as the voicing on paper of observation and experience, which is to say literary representation, but also as inner observation and experience themselves, which precede that voicing and are used to attempt to reach the essence, the very core of nature, both outside and inside man. In that case we mean imagination as a mental category.

The 'interior' of nature was accessible through the inner self of the knowing subject, of the observers of nature themselves. The empirical approach to nature with the help of eye and reason – eventually aided by microscope and telescope and supported through knowledge of the natural sciences – was supplemented with the approach through emotion (the heart or soul). Both made use of imagination in order to fathom nature, the first primarily through (re)productive imagination, the second through creative imagination. In the scholarly literature the latter has been associated chiefly with Romantic authors, but it is already apparent in Post. The creative imagination enabled her to attain knowledge of nature in the double sense described above. She did not work *ex nihilo*, but built on the other stages of imagination, rooted in precise observation. The subjective experience of nature was the extension of the objective approach to it. In other words, the subjective experience was seen as a kind of objectivity. Subject and object coincided. In this way literature, like the visual arts, was able to become an instrument for the reflection on nature.

In the following section, the importance of the genre of *Het land, in brieven* – the epistolary novel – and its most important motif, female friendship, are discussed. Both had a stimulating effect on the perception, imagination and experience of nature and landscape in the second half of the eighteenth century. After a short overview of the developments in the scientific approach to nature in the eighteenth century, the predisposition of Post and the characters of her novel towards that observation, imagination and experience is explored, and in what passages they exhibit this knowledge of nature is examined. After these sections explications follow dedicated to imagination in contemporary thinking on literature, landscape art and science respectively. These deliver insights with which the observation, imagination and finally the experience of nature and landscape in Post's novel may be analysed. We end on a discussion of what Post's experience of nature culminates in: the sublime.

Epistolary novel and friendship

The letters in Post's novel are exchanged primarily between two friends: Emilia and Eufrozyne.[6] The other correspondents are also female. A novel entirely devoted to life in nature, written by a woman, in the form of letters written by women, is unique on a European scale. In the early novel, nature was often accorded little attention, which has been attributed to the fact that the most important audience of the genre consisted of women of the (upper) middle class, who had limited travel experience.[7] These women had seen very little nature. The novel allowed them the freedom to manifest themselves in literature, not bound as it was to the rules that governed the so-called higher genres with classical roots, which were moreover the domain of male authors.

That women felt themselves especially drawn to the epistolary novel, both in reading and in writing, is also no coincidence. The eighteenth century is known as the century of the letter. Women played an important role in this epistolary culture, certainly where it concerned the sensitive letter.[8] Additionally, the epistolary novel enjoyed the status of 'exhibiting life in its true state'.[9] Through the letter – and specifically the 'familiar letter', exchanged between private persons – the epistolary novel was able to grow into 'the specific medium of communication for European sensibility and the language of emotion'.[10] It was the genre *par excellence* of self-expression.[11] One could let one's heart speak and give direct expression to observations and experience.[12] The characterisation of the epistolary novel as 'written, as it were, to the moment' even speaks to a view of the genre as a kind of empiricism.[13]

Thus, the epistolary novel lends itself exceptionally well to the double approach to nature that is central here. It was reinforced through the role that friendship plays in it. Post's novel describes the friendship between Emilia and Eufrozyne over the course of one year or, better yet, over one cycle of the seasons, starting with letter I, dated December 3 of an unspecified year (only the century is consistently given: '17–'), and ending with letter LV, dated November 10 of the following year. In the eighteenth century, the letter was seen as the most important medium for the expression of friendship. In German, this type of letter was even separately named: the 'freundschaftliche Brief'.[14] Post herself uses this term.[15] Friendship is indeed 'the central motif' of the novel,[16] but certainly not its central theme. That, as previously noted, is landscape or nature. Most of the letters are dedicated to the observation and experience of it, as well as the thoughts and feelings that result. Friendship, however, did create the circumstances, especially the intimate familiarity, to express all those observations and experiences, both orally and in the form of letters. True friendship presupposed the ability to observe and experience nature in a sensitive way, 'sympathetisch', or 'empfindsam', as it would be expressed in German.

The development of the friendship between Emilia and Eufrozyne, the growing closeness of the two women, is dependent upon and expressed through the increasing ability of Eufrozyne to observe and experience the land, landscape and nature in the same sensitive way that Emilia does. Emilia is predisposed to do so through the circumstances of her life. Not only does she live on the land, but she also lives there alone – in a house called Carefree (Dutch: Zorgenvrij) (fig. 2). In his influential work *Über die Einsamkeit* (1783-1785), Johann Georg Zimmermann (1728-1795) – whose name, incidentally, is not mentioned by Post– wrote that life in seclusion creates the ideal conditions for a studious and writerly

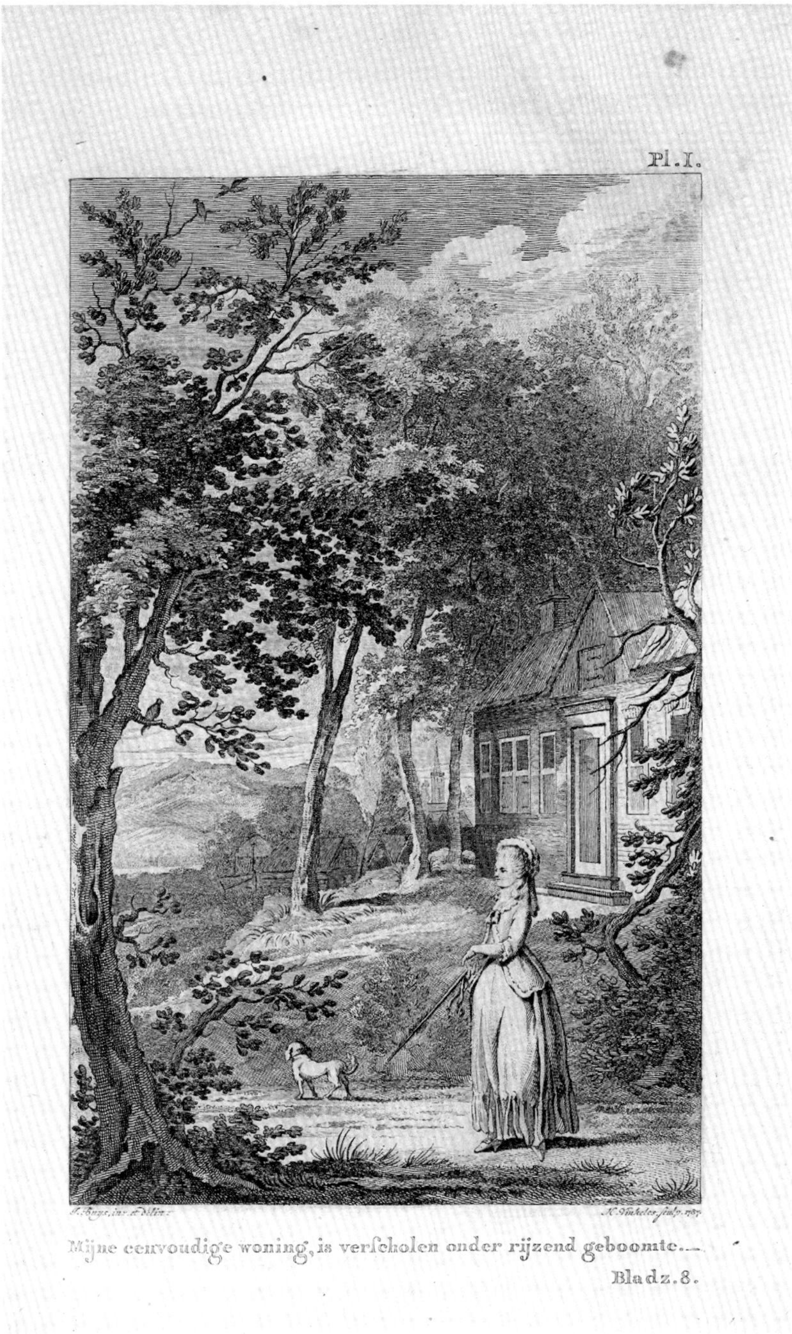

2

H. Vinkeles after J. Buys, *Emilia before her house*

engraving, Elisabeth Maria Post, *Het land, in brieven* (Amsterdam, 3rd ed., 1788) University of Groningen Library, KW B 1544, Pl. I (photo: Dirk Fennema), quotation: 'My simple dwelling is hidden beneath rising trees' (p. 8).

existence.[17] Eufrozyne, by contrast, lives with her mother in the city and leads a busy life there – busy in terms of noise, but also in terms of the many social obligations, through which urban life differs in a negative sense from rural life in pastoral literature.

Eufrozyne is, however, aware of the superiority of life on the land and life in nature. Descriptions of such life in Emilia's letters feed Eufrozyne's desire to come to similar observations and experiences and to subsequently describe it. (That she possesses the ability to do so – has both the capacity for observation and the capacity for imagination – is evident from the letters in which she, in turn, describes life in the city.) This reinforces the two women's feelings of friendship that, owing to the similarities between them just mentioned, match those for nature. Therefore they also feed Eufrozyne's desire to join Emilia on the land. From letter XXIX onwards, their closeness through friendship is also a physical fact, as they are together. This occurs just over halfway through the 55 letters and slightly less than halfway through the text. Their stay together is described in ten letters, all written by Eufrozyne and addressed to another friend, Elize – obviously not to Emilia, with whom she is after all currently staying.

One passage illustrates very well the intensifying effect of their friendship on their experience of nature. This concerns words spoken by Emilia to Eufrozyne, which she cites in her first letter to Elize. On a favourite spot in nature, Emilia states (fig. 3):

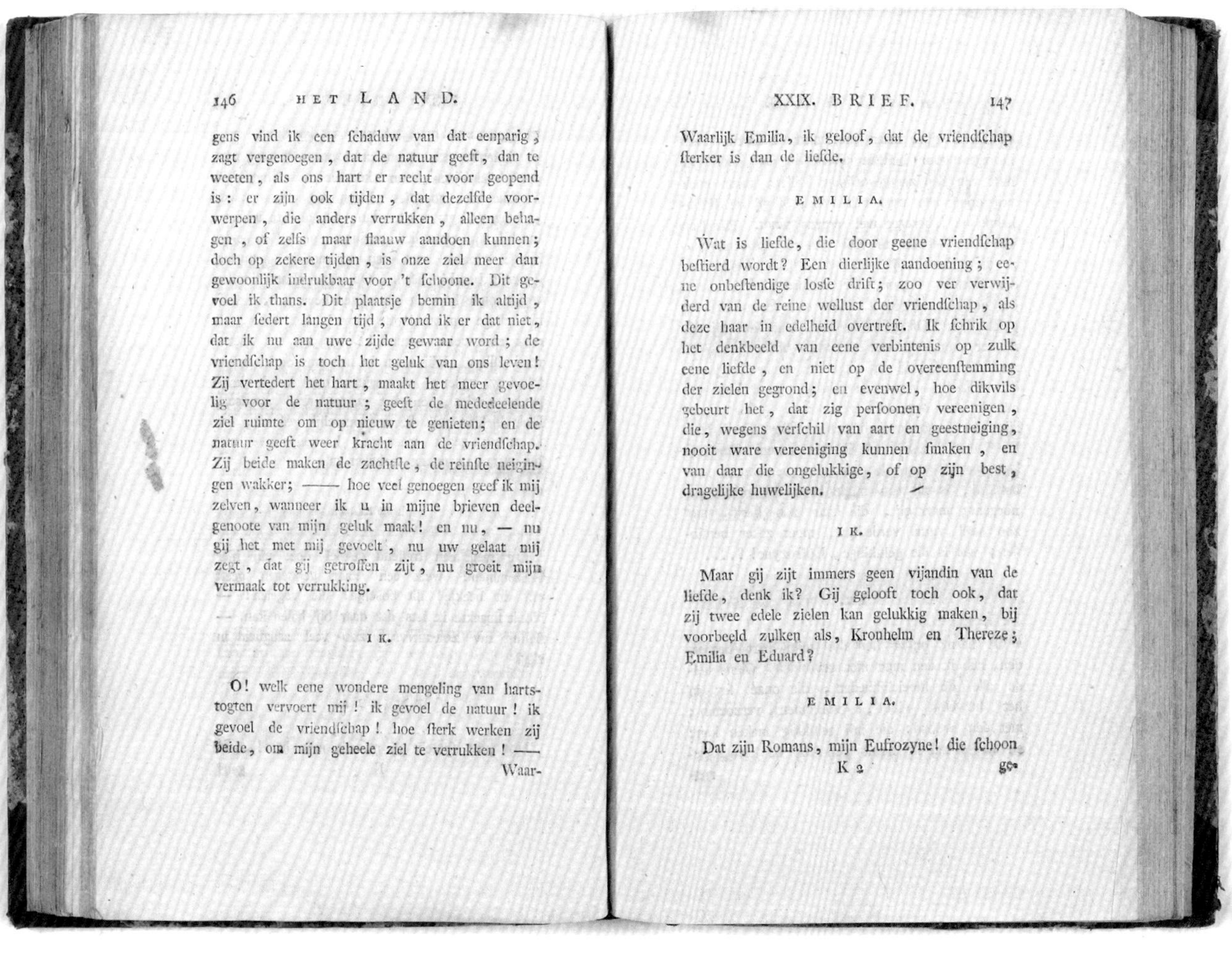

146 HET LAND.

gens vind ik een ſchaduw van dat eenparig, zagt vergenoegen, dat de natuur geeft, dan te weeten, als ons hart er recht voor geopend is: er zijn ook tijden, dat dezelfde voorwerpen, die anders verrukken, alleen behagen, of zelfs maar flaauw aandoen kunnen; doch op zekere tijden, is onze ziel meer dan gewoonlijk indrukbaar voor 't ſchoone. Dit gevoel ik thans. Dit plaatsje bemin ik altijd, maar ſedert langen tijd, vond ik er dat niet, dat ik nu aan uwe zijde gewaar word; de vriendſchap is toch het geluk van ons leven! Zij vertedert het hart, maakt het meer gevoelig voor de natuur; geeft de mededeelende ziel ruimte om op nieuw te genieten; en de natuur geeft weer kracht aan de vriendſchap. Zij beide maken de zachtſte, de reinſte neigingen wakker; —— hoe veel genoegen geef ik mij zelven, wanneer ik u in mijne brieven deelgenoote van mijn geluk maak! en nu, — nu gij het met mij gevoelt, nu uw gelaat mij zegt, dat gij getroffen zijt, nu groeit mijn vermaak tot verrukking.

IK.

O! welk eene wondere mengeling van hartstogten vervoert mij! ik gevoel de natuur! ik gevoel de vriendſchap! hoe ſterk werken zij beide, om mijn geheele ziel te verrukken! —— Waar-

XXIX. BRIEF. 147

Waarlijk Emilia, ik geloof, dat de vriendſchap ſterker is dan de liefde.

EMILIA.

Wat is liefde, die door geene vriendſchap beſtierd wordt? Een dierlijke aandoening; eene onbeſtendige loſſe drift; zoo ver verwijderd van de reine wellust der vriendſchap, als deze haar in edelheid overtreft. Ik ſchrik op het denkbeeld van eene verbintenis op zulk eene liefde, en niet op de overeenſtemming der zielen gegrond; en evenwel, hoe dikwils gebeurt het, dat zig perſoonen vereenigen, die, wegens verſchil van aart en geestneiging, nooit ware vereeniging kunnen ſmaken, en van daar die ongelukkige, of op zijn best, dragelijke huwelijken.

IK.

Maar gij zijt immers geen vijandin van de liefde, denk ik? Gij gelooft toch ook, dat zij twee edele zielen kan gelukkig maken, bij voorbeeld zulken als, Kronhelm en Thereze; Emilia en Eduard?

EMILIA.

Dat zijn Romans, mijn Eufrozyne! die ſchoon ge-

K 2

3

Mise-en-page of letter XXIX

Elisabeth Maria Post, *Het land, in brieven* (Amsterdam, 3rd ed. 1788)

University of Groningen Library, KW B 1544, pp. 146-147 (photo: Dirk Fennema).

> I have always loved this place, but for a long time I did not find here what I am now aware of, at your side. Friendship is the greatest happiness of our lives! It softens the heart, attunes it to nature, gives the passionate soul space to enjoy again. And nature in turn lends strength to friendship. They both awaken the most tender, the purest impulses. How much joy do I give myself when I share my happiness with you in my letters! And now, now that you feel this with me, now that your face tells me that you are touched, now my pleasure grows to delight.[18]

Friendship, nature and harmony

The ten letters from Eufrozyne to Elize take up less than a fifth of the total number of letters, but they constitute almost a third of the total text. Thus, they are much longer on average. They are also more varied. Eufrozyne reports not only on her own observations and experiences, but also on those of Emilia and those shared by her and Emilia. In this way she describes Emilia's current life, but also her past – namely, her love for the now-deceased Melidor, whose grave in the forest they visit during a hike at night (fig. 4). Their joint experiences and observations regard walks in nature, like the one just mentioned. This section contains not only many more quotations from eighteenth-century poetry – Dutch as well as translated German – than the rest of the novel, but also more monologues (like the one quoted above) and dialogues. The intensifying function of friendship on the experience of nature features prominently here. The observations, experiences and resulting considerations become stronger and more profound as they are exchanged and combined in description, exclamation, citation and in questions and answers.

Friendship rose to prominence during the second half of the eighteenth century.[19] It was seen as the ideal form of human interaction, superior even to love.[20] The literary cult of friendship was dominated by men,[21] but nature offered, as is evident in Post's novel, opportunities for the development of female friendships too.[22] Nature was, after all, far off the beaten path of public life so dominated by men.[23] In her first long letter to Elize, at the start of her stay in the countryside, Eufrozyne writes:

> What, I wonder, is the reason that our sex, which is after all described as the most attuned to the purer, more gentle and more lively of passions, should be less suited to the constancy of a friendship that is not based on fickle caprices, but on wise choice and harmony of soul, and that is governed by the aim of perfecting the pleasure of one another's happiness? (...) No, Eufrozyne! True friendship between noble, alike souls, even when placed in female bodies, cannot fail to endure.[24]

Pl. III.

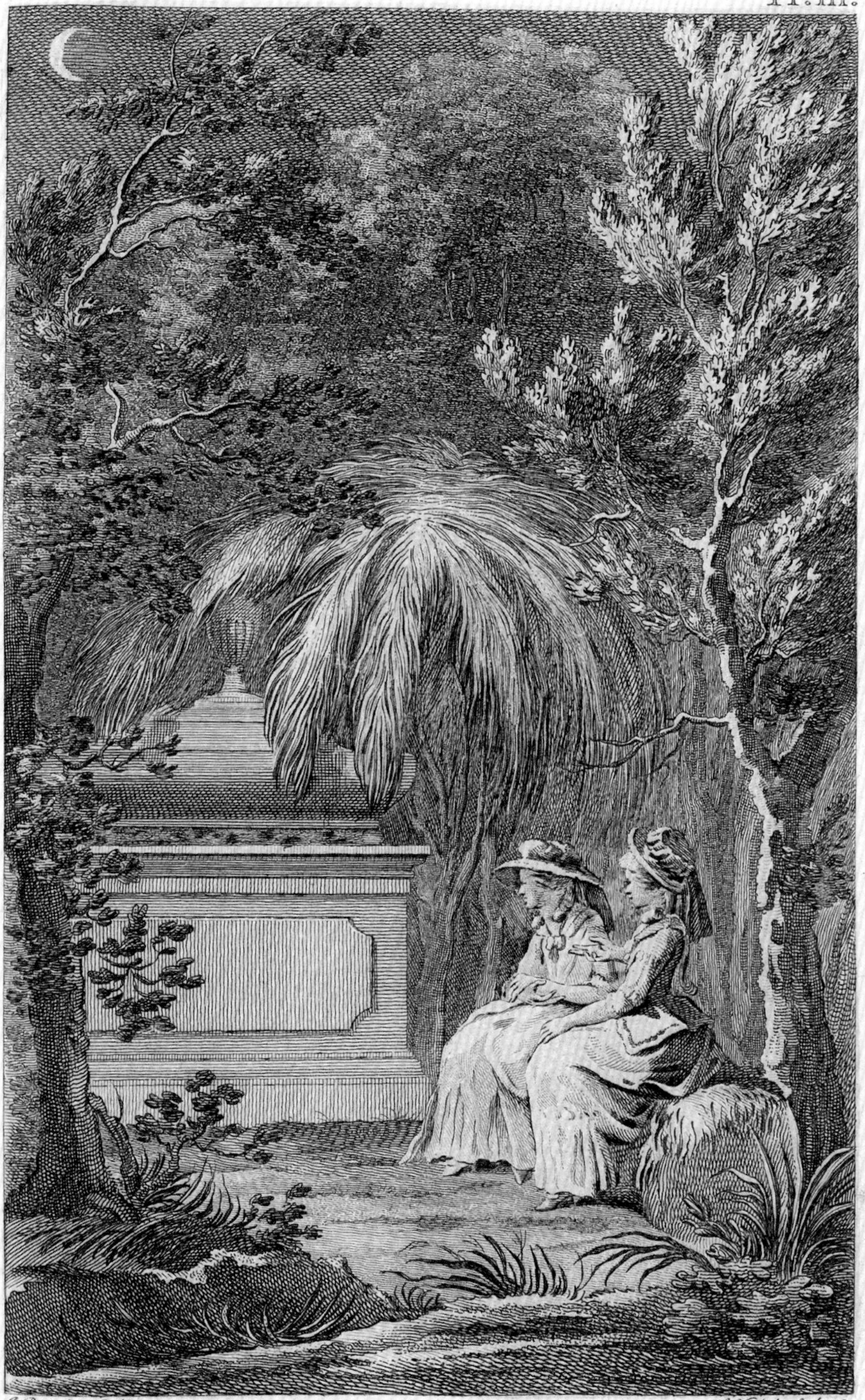

J. Buys. inv. et delin. H. Vinkeles. sculp. 1787.

—Het schijnt een grafplaats, waar gij mij brengt. Bladz. 172.

4
H. Vinkeles after J. Buys, *Emilia and Eufrozyne at Melidor's tomb*
engraving, Elisabeth Maria Post, *Het land, in brieven* (Amsterdam, 3rd ed., 1788) University of Groningen Library, KW B 1544, Pl. III (photo: Dirk Fennema), quotation: 'It seems a burial site where you bring me' (p. 172).

Additionally, a strong experience of nature matches the tendency in this period to internalise and emotionalise friendship.[25] It was also intensified, as friendship was seen as the highest expression of nature. Because nature was seen as divine, love of nature brought man to the highest rapture, a love identified with love for fellow man. In *Über die Einsamkeit*, Zimmermann wrote that in nature, man could experience nothing but friendship, simplicity and innocence. It was in nature that the 'natürliche und angeborene Vereinigungstrieb' arose, which found its greatest fulfilment in the 'Trieb zu hauslicher Geselligkeit und vertrautem Umgange'.[26] So, while Zimmermann did praise solitude, he also pointed out the importance of social contact in order to inspire new images and new ideas through mental and physical stimulation.[27] According to him, the aesthetical experience of nature was a natural, religious and moral fact: it matched the love of the Creator and attempted to share that love with fellow humans.[28] The experience of nature and the experience of friendship were unified through the realisation that Creation had emerged through love.

The aspect of friendship also plays a role in the *Theorie der Gartenkunst* (1779-1785) by Christian Cay Lorenz Hirschfeld (1742-1792), an author who, unlike Zimmerman, is mentioned by Post and who most certainly influenced her.[29] In that work, Hirschfeld emphasised that the pleasure provided by nature is amplified through its shared experience.[30] Friendship (just like love) was comparable to nature. Both represented a universal natural sympathy or harmony.[31] Whoever experienced nature, experienced friendship, and vice versa. In *Das Landleben* (1767) Hirschfeld describes the intensity of the joint experience of nature in the following way:[32]

> (...) keine Vereinigung van Freunden kommt mir würdiger vor, als in der Absicht, die Schönheiten der Natur gemeinschaftlich zu betrachten, und in ihnen ihren unsichtbaren Urheber zu bewundern. Ja, die sanfte Heiterkeit des Gemüthes, und das frohe Gefühl der Freundschaft erregen sich wechselsweise (...).[33]

In *Rhapsodie, oder Zusätze zu den Briefen über die Empfindungen* (1761), Moses Mendelssohn (1729-1786) had written that man can witness supreme perfection in nature.[34] Experiencing perfection – thus experiencing nature – heightened the desire to express this perfection and to increase this experience. That desire was as strong as the survival instinct of man. In that way, taking pleasure in the beautiful, and thus also in the beauty of nature, led man to friendship, which in turn was reinforced through the shared experience of it and the private – intimate – conversation on the experience of nature.[35] Friendship, says Emilia, softens the heart and attunes it to nature.[36] Two entwined trees, she writes, express her friendship with Eufrozyne.[37] Such friendship, experienced in nature, was seen as a balm for the wounds that society – and primarily life in the city – inflicted.[38] It was seen as so exceptional that it became almost religious and became a kind of happiness that was experienced as heaven on earth.[39] After Eufrozyne's stay with Emilia, the latter again connects nature, friendship and harmony in a letter to her:

> The brook, which flows with an even murmur here next to me and makes little waterfalls out of the roots of trees, was always pleasant, but is now foremost in full harmony with nature and my heart. However, this is a harmony that I can feel, but cannot describe. Over there is a tree – I can see its top – in whose bark I once carved the name of Eufrozyne. I feel and say loudly: O nature! O silence! O friendship! How bountiful are you! With your threefold pleasures I can live happily.[40]

It is essential here to note that this harmony could be felt in the heart, where the subjective experience of nature took place. The similarity between nature and friendship, the parallel course of the two, is evident from the season in which Eufrozyne visits Emilia: summer.[41] While nature in winter symbolises the lingering hope for a reunion with the absent friend, nature in spring represents the growing desire to bring closer this being together, which in turn is realised in summer, after which nature in autumn first represents a melancholic lacking, then a fear of loss, and finally the loss of the friendship itself. The correspondences between nature and friendship are thus an indication of those between nature and life. The cycle of seasons mirrors in this way that of all existence, including that of man. In Post's novel, Eufrozyne falls ill and dies halfway through autumn. Emilia is informed of her passing by Sofia, another friend of Eufrozyne, who is thus presented as a new (pen) friend. This ends one cycle and introduces a new one in the coming winter. This cyclical thinking also has a metaphysical aspect: Emilia believes that her friend Eufrozyne's death, like that of her beloved Melidor, points to a new life, the eternal life, where she will be reunited with both after her own passing.

Developments

Scholars have thus far placed *Het land, in brieven* only in a physico-theological context, the philosophical approach that assumes that God reveals himself to man in nature.[42] Fitting into this is the remark by Emilia in letter IX that she ventures 'outside' in order to 'admire, on the advice of the honourable MARTINET, in every well-shaped [snow]flake the eternal Creator'. This 'MARTINET' is the reverend Johannes Florentius Martinet (1729-1795), whose 'advice' Emilia must have derived from his immensely popular *Katechismus der Natuur* (*Catechism of Nature*) (1777-1779),[43] which helped spread the philosophy of physico-theology widely throughout the Netherlands.[44] Regardless of the degree to which this approach to nature also led to moralising and theologising and so contained a strong teleological nature,[45] it rested on an inductive, empirical approach to reality. Nature was reduced as a whole, but even more so in its parts and down to the smallest organisms, to divine providence. This referred to God as an artist. Hence terms such as 'kunstig' (artful) and 'kunstvermogen' (artistic ability) are often used in the novel in relation to Creation (especially of animal creatures).[46]

In the tradition of physico-theology, the description of nature was predominantly static. Similar to the natural sciences, which aimed for order and classification (especially in botany), nature prose and poetry could predominantly be categorised as a collection of images of nature.[47] The imagination was limited to the representation of individual organisms (the 'small') and delineated scenes of nature. The emphasis was on the material, sensory perception of nature as it had been created by God. However, in the sensitive approach to nature, which developed in the second half of the eighteenth century alongside and in conjunction with physico-theology, the empathic approach to nature led to a much more dynamic description of it, a description aimed much more at the entirety of nature that the onlooker was exposed to. In this context, scholarly texts often use the term 'landscape', in the sense of an aesthetically experienced 'nature as a whole'. Nature prose and poetry increasingly started to represent landscapes of mood, which had a much more subjective character than the descriptions of nature in the tradition of physico-theology. The main concern was no longer the exact description of nature, no matter how aesthetical the experience of it, but rather how nature moved its observer. Nature or landscape started to mirror the soul. To that end, it was transformed through the use of imagination.

The connection with contemporary science lay in the dynamic character of this description of nature. Insofar as literature did not just represent separate moods but also reflected the change or movement of those moods, it can be compared to what Romantic science considered as the creative power of nature itself. Nature was no longer seen as 'being', as in physico-theology, but as 'becoming', and was accorded a status comparable to that of God himself (*Deus sive Natura*).[48] Despite the fact that this approach did not come to fruition until the rise of the Romantics and is, for that reason, always associated with them, it already manifested itself in the second half of the eighteenth century, albeit without being taken to extremes by removing God from the image of nature.

In this context, Peter Hanns Reill speaks of a vitalistic approach to nature.[49] This approach was characterised by attention to changes in nature (instead of classification) in order to see the workings of the interior, formative powers in organisms; it was a way of looking at nature whereby every detail was seen in relation to the whole (of the organism itself and its surroundings), and observation was linked to comparison (analogical reasoning).[50] Influenced by the rise of chemistry, organisms were moreover seen increasingly in terms of combinations of elements, of attraction.[51] This resulted in a greater attention to dynamics, to movement in nature and to the interaction between nature and humankind. Man himself, as a part of nature, was capable of attaining a profound understanding of nature through self-reflection and, vice versa, was able to gain insight into himself through the study of nature. An important part was played in this process by his imagination.

Scientific insights

From the start of the novel, Post defines her alter ego Emilia as an intellectually established woman, a savante, who displays a scholarly attitude. In letter IV she describes her day-to-day activities to Eufrozyne. Concerning her religious duties, Emilia is brief: she does not wish to write about them. That remark reinforces the impression given by the novel as a whole: that religion is primarily experienced in and through nature. The actual writing of letters and the subject to which she dedicates most of her letters, the description of nature, seem only a small part of her activities. At any rate she mentions her nature descriptions only at the end of a long list of activities and even then as a by-product of her country walks. She will, as she promises Eufrozyne, describe one of those – the start of a long line of letters that are almost all dedicated to observations of nature.

The essence and representation of these should be seen against the backdrop of her other activities. Emilia sings, plays the organ and the flute and draws 'beautiful landscapes'.[52] However, she spends most of her time in her study (fig. 1).[53] It is there that she dedicates herself to the higher purposes for which the female soul, too, has been created.[54] (Higher than housekeeping chores; on these she spends no more time than is strictly necessary.) Her bookcases contain, aside from religious literature, reading that may be summarised under the categories of 'History' (Historie), 'Natural Science' (Natuurkunde), 'Literature' (Literatuur) – all three written with a capital letter.[55] The latter includes poetry, drama and novels, and she mentions the names of several protagonists of (epistolary) novels: 'Clarisse', 'Grandison', 'Sofia' and 'Burgheim'.[56]

Natural science, then, was a kind of daily activity alongside artistic endeavours. That situation was familiar and entirely normal for women in the second half of the eighteenth century. Microscopes and telescopes could be found in many middle-class households and fulfilled a social role.[57] Easily obtainable textbooks gave access to the state of affairs in the natural sciences,[58] sometimes written by men, like Martinet's *Katechismus der Natuur*,[59] but also by women.[60] Translations of popular textbooks from abroad also appeared.[61]

Thus, it appears that when Emilia writes about all those artfully crafted snowflakes, she is envisioning the ice crystals that make up those flakes. Their hexagonal, symmetrical structure was one of those miraculous phenomena that early modern science, especially that of physico-theological origin, liked to point out.[62] Moreover, Emilia tells Eufrozyne that she has a microscope in her home: 'The microscope also furnishes me with agreeable activities'.[63] Later, Eufrozyne writes to Elize about how she and Emilia

> [saw] some insects through the microscope, the structure of the wings of butterflies and more wonders of nature, which the naked eye cannot discern. And [it is] truly happy that our senses are so crude or these creatures so small, otherwise we, surrounded by such horrifying and heavily armed animals, would live in constant fear, and offer our bodies as a resting place to none of them.[64]

It is revealed that Emilia also owns a telescope when she writes to Eufrozyne about how she has used it as a spyglass in order to see the city in which her friend lives.[65] Earlier she had asked Eufrozyne whether she could also see that the sky was of a bluer colour because of the 'thinning' ('t verfijnen) of the atmosphere as a result of the winter cold – a reference to meteorological insights.[66] Knowledge of atmospheric light phenomena is also shown in the reference to northern lights and the general lack of daylight in the polar regions.[67] Coincidentally, she herself at one point observes the aurora borealis from her window.[68] And on two separate occasions she refers to the speed of light.[69]

On butterflies, she remarks that the caterpillar '[weaves] itself a little cell, wherein it spends some time, as in deep sleep, until, gifted with wings, it emerges as a beautiful butterfly'.[70] It was Johannes Swammerdam (1637-1680) who discovered that caterpillars do not die, but rather, through a mysterious process or metamorphosis, the caterpillar itself transforms into a butterfly; it already contained all the rudimentary parts of the butterfly.[71] The fact that Emilia writes that the caterpillar sleeps rather than dies indicates that she is aware of this by then generally accepted insight. Post's knowledge of human physiology shows in a remark by Eufrozyne in a letter to Elize: 'The drawing in of fresh outside air, the regular circulation of body fluids, caused by continuous motion, makes me healthier than I ever was.'[72] We can conclude from this that Post was aware of the theory of the circulatory system and of the role breathing played in it.

Interest in and knowledge of insights in the field of astronomy show from phrases by Eufrozyne regarding the 'dazzling infinity' of the universe,[73] that 'endless space', 'that untold multitude of fixed stars [which] are even so many suns, orbited by other, to us invisible globes'.[74] That these globes continuously completed their orbit with mathematical precision was universally known, but Eufrozyne's speculation that 'on other planets, too, moral creatures of another nature have sinned, who shall share in the salvation delivered by the Saviour of our world'[75] is an indication that Post not only adhered to a heliocentric world view, but also took into account the possible existence of intelligent life in outer space.[76] And even if that did not exist, writes Eufrozyne, 'shall we [then] judge God's wisdom by the standards of our limited intellect?'.[77] In other words, God's knowledge, his power of creation, exceeds human understanding. Much more than scepticism regarding man's intellectual capacities, however, the sentence conveys the conviction that one could and was allowed to cross the limits of what could be visually perceived and penetrate into the world of the thinkable and speculative. With that, imagination as a scientific method arises alongside observation.

Imagination and literature

Judging by Gert-Jan Johannes's study of the literary thinking on imagination in the Netherlands, it seems out of the question to attribute a creative character to Post's imagination, nor that it would have played a role in an approach to nature. Eighteenth-century theoreticians described imagination primarily as a dual capacity: reproductive, which is to say the capacity to recall images of a previously experienced reality; and productive, which is to say the capacity to modify those images of the past.[78] The latter was also referred to as 'poetic power' (dichtvermogen) or 'poetic imaginative power' (dichtende verbeeldingskracht).[79] The imagination operated 'painterly' (schilderend), which accounts for the term 'painterly imagination' (schilderende verbeelding).[80] In the artistic process there was a fluid transition from the imitative to the more creative aspect of poetry.[81]

Whoever possessed a rich imagination – the hallmark of the true poet – was literally able to identify, to become one, with nature. The poet animated life, as it were, with his own thoughts and emotions.[82] The recalling and recombining of images – reproduction and production – allowed man to realise the miraculous essence of Creation, and it affected him: it brought him into ecstasy and rapture.[83] Here lies the basis of the aesthetical approach to nature. The poetic imagination seems in the description of some theoreticians to have been truly creative and to have inspired poets to create an organically nurtured whole, which could not simply be reduced to the sum of its parts.[84] However, on the other hand several references indicate an assumption of the literary work of art as a construct rather than an organism. In the end, writing a literary work of art did not involve creative imagination.[85] Thus, the latter also had no role in understanding nature.[86]

It seems the poet was being given something only to have it taken away again almost immediately. Sometimes poets themselves seemed to be complicit, such as Hieronymus van Alphen (1746-1803) – cited by Post in her novel. Van Alphen distinguishes between the representation of 'actually present nature', of 'imagined, fictitious nature' and of 'a creation wholly one's own'. In the first case, the poet was, according to Van Alphen, 'a follower of nature', in the second case 'a follower of pure nature' and in the third case 'a Creator'.[87] The term 'imitate' (navolgen) led to painterly imagery, where it was befitting to describe the products of the imagination as 'lively' and to call them 'paintings'.[88] In those cases, nature was indeed imitated.[89] However, where Van Alphen describes the third form, it is, as shown in his phrasing, abstracted from the concept of imitation:

> Poems in which the poet represents only his own emotions and sensations are no copies or imitations of the unfettered ways of nature, but they are representations of nature herself (...) when he produces his own feelings, his own emotions, it is not imitation, but we hear nature herself.[90]

This kind of poetry is, according to Van Alphen, a matter of softheartedness or sensibility, not of prowess of imagination. Yet, the poet must represent this interior in order to be able to imagine 'experiences and afflictions'. The representation of his inner self, 'of nature herself', as Van Alphen puts it, thus remains a kind of imitation. But is the poet's imagination then devoid of all creativity? Characteristic of landscape literature was that inner and outer nature were connected to one another. It was construed from both external observation and the resulting internal experience of nature. As that nature was God's creation, the description of its beauty and all of the resulting feelings, moods and sensations that coincided with the realisation of that divine origin made the poet a second creator, who did not just imitate, but also created.

Imagination and landscape

In his study of German landscape painting in 1770-1840, Timothy F. Mitchell employs a distinction between the terms 'Einbildungskraft' and 'Dichtungskraft', which he derives from the *Allgemeine Theorie der schönen Künste* (1771-1774) by Johann Georg Sulzer (1720-1779): 'In general *Einbildungskraft* denotes the power of the mind to build a network of associations from raw experiences, while *Dichtungskraft* stands for the special ability of the creative imagination to construct whole, unique entities with a coherent life of their own'.[91] Although Sulzer, unlike the later Romantics, saw nature as lifeless,[92] he did regard it as artistically valuable, because of its capacity to incite feelings, moods and sensations, which in turn would lead to ideas through association.[93] 'Empfindsame' authors such as Friedrich Gottlieb Klopstock (1724-1803) – also named by Post – did not distance themselves from nature, but surrendered to it completely, especially to movement in nature.[94] This represented a clear move away from the primacy of reason and logic in the approach to nature that had dominated the description of landscape in the first half of the eighteenth century.[95]

This 'shift' did not happen at the expense of the importance attached to observation – on the contrary. An important next step in the theory of landscape painting was made by Johann Heinrich Merck (1741-1791). In an article entitled 'Über die Landschaftsmalerei' (1777) he advised artists to devote themselves to 'long and assiduous observation of the wide range of natural forms in their different conditions, seasons, and times of day'.[96] Here we may speak of a correlation between 'aesthetic theory and the wider revolution taking place in philosophy and the natural sciences'.[97] The value of landscape art did not reside in rules and plans formulated a priori, but in the precision with which the observed reality was represented: 'As logic was now inductive, the finer the experiences, the sharper and more detailed the observation, the keener and more penetrating would be the resulting knowledge'.[98] This, however, did not lead to attention to detail coming at the cost of attention to the whole. The landscape artist and poet Salomon Gessner (1730-1788) had warned of this in his 'Briefe über die Landschaftsmalerei' (1770-1772; translated into Dutch in 1786).[99] Coincidentally, Post mentions Gessner in her novel, because of his *Idyllen*, a collection of very popular pastoral tales.[100]

A parallel development is that artists became convinced that the imagination could work only with the materials, experiences and impressions of nature itself. In the representation of nature, people made more and more use of their own local surroundings[101] – something we also see in Post. The close connection here between landscape painting and landscape literature shows in the ideas of Karl Ludwig Fernow (1763-1808), who maintained a distinction between three kinds of landscape art: 'natural, aesthetic, and poetic'.[102] This seems to correspond with the trinity of reproduction, production and creation of Van Alphen. The first rested on the representation of what was seen, the second on the representation within a considered composition and the third on the expression of an idea inspired by nature. That last category can be linked to Sulzer's 'Dichtungskraft'.[103]

While, according to Sulzer, the knowing subject forgot himself for the sake of the object of his knowledge, the experiencing subject became more and more aware of himself through the object of his experiences. Thus, in the aesthetical experience, the experience of beauty, the object of the experience remained present in the sensation.[104] It provided an empirical confirmation of the speculative assumption (through metaphysics) that man possessed a destiny that transcended the material world. Nature aided man in experiencing that destiny.[105] This often concerned emotional experiences. Those could not be attributed solely to man's sensory capacity, but resulted from his imagination as well.[106] The experience of a realistic landscape, the sensation of it, the observation or even the study of it, released not only pleasant (emotional or otherwise) feelings, but also memories of past experiences and observations.[107] Imagination even led to experiences outside of empirically observable reality.[108] That Post, for example, was able to imagine that the beauty of the earth and the sky could be admired not just by mortal creatures, that there had to be extraterrestrial life at all, might have been the result of logical reasoning, but it was the imagination that allowed authors to arrive at a conception of the nature of that life or, as is the case with Post, at sensations that accompanied this insight.[109]

Imagination and natural science

Imagination thus became a tool for gathering knowledge. Regardless of how much physico-theology continued to play a part in the background, there was a clear shift in the approach to nature. Where physico-theology pointed to God as a unifying force, that function was assumed in landscape art by the experiencing subject.[110] All experience of nature became dependent on that subject's sensibilities.[111] In that sensory experience, nature came alive, and the only true nature was living nature.[112] That way, imagination began to play a part in natural science. Not only artists but also scientists were expected to make use of creative imagination in their dealings with nature, finding in it a source of comparable creativity.[113] Nature's unity could, in the end, only be fathomed by the artist using his sensitivity, through landscape. After all, not just nature presented itself as landscape – the soul did too. That made it possible to describe the one using imagery of the other.[114]

The point is that nature in landscape literature at one point stopped being simply nature and was seen as an expression of something higher, of the divine.[115] While natural science attempted to understand nature as an object or product, art attempted to understand it as producing, as a process that was repeated through aesthetical production. Natural science occupied itself with nature as creation (*natura naturata*), art with creative nature (*natura naturans*).[116] In Romantic views, literature even became a medium that was capable of representing formal changes in nature.[117] In that process reason and feeling did not have to exclude each other. Through its careful study nature was not deprived of its mystique. In fact, through its study man was able to experience the enchantment of its limitless existence.[118] The view of nature as endlessly creative, as though it were itself an artist (*Deus sive Natura*),[119] was not limited to just the Romantics. Before them, sentimental poets, as Friedrich Schiller (1759-1805) called them, had already felt an attraction to nature, because they had come to view the *nature* of the artist – the one with a lower case n – as part of *Nature* – the one with a capital N.[120] Thus, nature went from object to subject.[121]

The rise of the aesthetical approach to nature has been called one of the most important developments in the cultural history of the eighteenth century, but it has received very little scholarly attention.[122] It was not universal or general, but specific to a moment, place and point of view (observation) and coloured by sensations, feelings and reflections of the aesthetical subject.[123] This did not concern a 'being' but rather a 'becoming' nature.[124] The subject (man) experienced 'die unerschöpfliche und wandelbare Mannigfaltigkeit ihrer Erscheinungen'.[125] This aesthetical experience gained existential meaning within the aesthetics of feeling in the second half of the eighteenth century: what man experienced and enjoyed in and of nature was himself, his own existence. The sentimental experience of nature became the opposite of its rational observation.

What we may conclude from the preceding exploration is that as far as the representation of nature is concerned, the highest form of imagination, the form that was called creative or poetic, did not operate *ex nihilo*. It arose from nature itself, which the artist – the painter or poet – could only himself be inspired by if he ventured personally into that nature, experienced and studied it himself independently and accurately, and allowed himself to be brought to feelings and thoughts of a higher order by it. This presupposes that landscape art in its highest form – the creative or poetic – was not separate from observed nature and did not move away from it but, on the contrary, flowed from that nature and needed it. Stronger still, without first carefully observing nature, one could not allow it to inspire oneself and bring oneself to insights into its essence, origin and meaning.

Post and observation

That Post works along these lines is evident first of all from the descriptions of the flora and fauna, especially the world of insects, which primarily Emilia studies during her walks. These can be located within the physico-theological context of the novel. They demonstrate, as noted, the capacity for careful observation, direct and independent, as well as the capacity for careful description, also direct and independent.

That contemporaries saw *Het land, in brieven* as a product of this double capacity, or at least saw that the novel was presented as such, shows from the preface by the Reverend Ahasverus van den Berg (1715-1769), written in the form of a letter addressed to the poet Margriet van Essen-van Haeften (1751-1793). In it, he introduces Post as someone who, in terms of mind and intellectual development, is predisposed towards the description of nature. She is blessed with a very strong memory, a very quick understanding, a very lively intelligence and a very delicate, pure taste. She has spent the greater part of her life in the countryside and has gathered the requisite knowledge of it in its silence.[126] This has made her into an astute observer of nature.[127] Van den Berg is adamant in emphasising the originality of Post's work, its subjectivity in a literal, positive way. This is because she works quite differently from writers who derive the beauty of nature from the writings of others and who present them with only small changes or additions as new and original, just as painters who construct their scenes of nature from the works of other painters.[128] Then, by contrast, there is Post:

> She possesses that valour of spirit that, transcending slavish imitation, *observes and invents by herself.* Her *accurate, observing eye* wanders everywhere, and everywhere finds inexhaustible sources of beauty, where hundreds of others could find nothing.[129] (my emphasis)

According to Van den Berg, the beauty that she discovers in nature is not a physical, visible, aesthetical beauty, but a metaphysical, invisible, religious beauty:

> She draws our attention to objects that seemed to us barely worthy of mention, and in her hands these become amazing artworks by the wise and powerful hand that creates and sustains all.[130]

In her descriptions, Post brings the divine origin of nature to the surface, transforms, as it were, nature to a work of art and reveals nature's maker, the greatest of all artists, God himself. This, however, does not diminish the importance of her own role: she is just as much an artist as God is, as we may interpret the words of Van den Berg. This verbal art – indeed, a verbal art, as it arises in prose – is the result of accurate observation, as shown in the quotation, which is then reported 'in the normal, civilised language of everyday life'.[131] She prefers 'natural and simple descriptions of striking matters and truths'.[132] In this she distinguishes herself from sentimental authors, whom the Reverend blames for a 'cowardly and ridiculous (...) vogue of writing',[133] filled with exclamation points, dashes and tender, clear-sounding words.

The reference here is to the contemporary debate on sentimentalism, in which were given warnings concerning the excessive sensitivity and the exaggerated emotional expression of authors.[134] Van den Berg clears Post beforehand of any suspicion of that kind. This does not mean that he claims sentiment or feeling play no role in her novel. On the contrary. He describes its effects thus: Post 'pleases and stirs'.[135] Moreover, the transformation 'in her hands' of 'objects that seemed to us barely worthy of mention' into 'amazing artworks by the wise and powerful hand that creates and sustains all' indicates the role of the imagination in the representation of nature in relation to God. According to the most important Dutch theoretician of sentimentalism, Rhijnvis Feith (1752-1823), whose work is quoted by Post, authors ought to have phrased their feelings in a style that appealed to the heart and the imagination.[136] The sentimental author's job was to return the reader to nature.[137] Typical of Dutch sentimentalism is furthermore its religious orientation.[138] Additionally, several of the foreign authors quoted in *Het land, in brieven* were seen as representatives of an acceptable form of sentimentalism.[139] We may conclude from the words of Van den Berg that Post not only possessed the capacity for careful observation and description, but was also capable of connecting insights to the observations of nature regarding its essence, origin and meaning. This happened by allowing nature to influence one's feelings and presupposed the activities of the imagination.

Post and imagination

Imagination is frequently mentioned in *Het land, in brieven*. This concerns the imaginations of both Emilia and Eufrozyne, and in the double capacity of sender and receiver. The previously mentioned characters from novels form the basis for the first statement – by Emilia – on imagination. Why should she condemn novels, she wonders. After all, she loves Christ's parables, which are just as much 'paintings' and 'which owed their existence only to his divine, inventive ingenuity'.[140] While the word 'paintings' may indicate the workings of the (re)productive imagination, the 'divine, inventive ingenuity' points to creative or creating imagination.

When speaking of the imagination, and especially the direct representation of landscape, it made sense to use terms from painting. Literary landscape art, just as visual landscape art, arranged the landscape through perspective, not in the spatial sense but temporally, through the chronological description of sections of the perceived landscape.[141] The parallels with landscape painting show from Post's use of language. There we encounter various conjugations of the verb 'imagine'. The friends furthermore describe their activities as a kind of 'painting', 'portrayal' or 'depiction'.[142] In contemporary poetry, the ability of a poet to paint with words was seen as a sign of true artistry.[143] The desire to find painterly, picturesque landscapes gripped both painters and poets in the second half of the eighteenth century.[144] Since poets painted with words, the verb 'describe' appears often. Given the primary use of seeing in experience, the 'eye' is also frequently mentioned, and the declension of the verb 'see' is frequently used.[145]

Emilia stimulates Eufrozyne's imagination: she supplies it, she writes.[146] In that vein, she describes her walks in nature in such a way that Eufrozyne can follow them in thought.[147] Conversely, Eufrozyne sets Emilia's imagination in motion through the description of the unhappy end of an avaricious old woman.[148] Eufrozyne herself was not present, but the imagination has given her a vivid picture of it, stimulated by the image of the funeral procession for the deceased that passed through her street at night.[149] In this way, the two friends word their experiences and observations so effectively through the use of imagination that the other, using that same imagination, can easily conjure up that same image. Imagination thus works both productively and reproductively, for both the receiver as well as the sender.

Reading, too, could stimulate the imagination. That Post was influenced in this regard by Klopstock seems evident. Reading his epic *Der Messias* (1748), Eufrozyne's thoughts turn to Christ's resurrection:

> When, at nightfall, I sit down with KLOPSTOCK in my hand and start pondering, my imagination sometimes takes me to His [Christ's] grave. Sitting at that cavern, I hear dark cypresses rustle above it. I see my forlorn loved ones wander aimlessly; I hear, it seems, the night's silence broken by the crack of dawn, by the hollow creaking of quaking rock. I see the radiant angel open the cavern, and Jesus, who died, arises living, in a triumphant manner and with a divine joy on his visage.[150]

In most of the cases, it is the perception of nature, of flora and fauna but especially of total nature in the form of landscape, that fires up the imagination. Eufrozyne writes that the descriptions of nature (the Dutch word 'natuurbeschrijvingen' [nature descriptions] appears literally) that flow from Emilia's pen are doubly beautiful.[151] Emilia writes Eufrozyne about how, on a winter morning, she opens her window and watches a snowy landscape:

> Through this sight, my imagination led me to the eternally snowy Alps. I imagined the dangers, which unprepared travellers often find there, the icy chunks of snow that often come down with a thunderous clamour and may unexpectedly kill cattle and men that cannot be saved through timely escape. The beauty, the scale, the overwhelming sense of nature in that miraculous place had the utmost effect on my enraptured soul, and for a moment [it seemed] as though I myself was witness to it and yet in matter of fact I was at CAREFREE (…).[152]

A landscape that is personally witnessed thus evokes the imagining of another landscape, one she once saw for herself or one she is familiar with through description or image, and brings her soul to rapture and thus effects a sentimental experience – in this case a sublime experience (more on which in the final section).

While this passage concerns itself with the landscape, the next passage, from a letter from Eufrozyne to Emilia, containing an account of a sea journey, moves imagination to the invisible life below the water's surface:

> And my thoughts went below this surface, to her chasms and gorges, I imagined countless armies of useful or as yet unknown fish, so different in shape, size and nature as their Creator is rich in wisdom

> and power, I imagined the mutually arranged households of so many different creatures, from the icy sea monsters to the smallest shellfish, their different ways of reproducing and span of life, and how a wakeful Providence that supports all [creatures] keeps [them] safe and makes them happy according to their different needs (...).[153]

According to Eufrozyne, Emilia possesses 'poetic feeling', which moves her to very sensitive descriptions[154] – too sensitive sometimes, in Eufrozyne's view, who sometimes has to laugh at them. Yet this sensitivity seems to her to be very desirable. According to Emilia, their imagination reaches further than the reproductive and productive stages and ascends beyond the (representation of) experienced nature. It seems truly creative:

> Indeed, we cannot make a great distinction between the joys of the eyes and those of the imagination. A vivid imagination, with which we are both gifted, is an excellent treasure for the sensitive soul. It furnishes us from a single joy not less than threefold pleasure. And sometimes, because it works freely and unifies everything that nature divides, it is greater than sensory pleasure.[155]

Imagination, then, can do more than sensory perception. It offers the chance for people with a 'sensitive soul' not just to reconstitute diverse elements in nature, but to combine them into an organic whole that cannot simply be reduced to the sum of its parts. The creative elements must be sought in this 'threefold pleasure'. Imagination leads not just to representation of (a combination of elements from) landscape, but also to the expression of feelings and thoughts that are evoked by that landscape. These are pre-eminently unique and personal, flowing from within the author. Eufrozyne seems to indicate the expression of such feelings when she brings up the 'poetic imagination' (dichterlijke verbeelding) in a letter to Elize – the only time that term is mentioned. She is describing 'a group of eminent trees':

> They all seemed, in a pensive form, to listen to the gurgling of the spring, and answered the polyphonic song of the water with a peaceful rustling. The poetic imagination would find a sanctuary here.[156]

Following which, that imagination immediately goes to work, leading Eufrozyne's thoughts to such things as Jesus' stay in the desert, making her interpret the sound of the spring as a paean to God and, finally, to say: 'I am all feeling'.[157]

Post and experience

There are a great number of passages in *Het land, in brieven* where the imagination is set in motion and an expression of feelings arises, through experiences during a walk in nature in the form of landscape. Most letters are constructed following this pattern. One quotation can illustrate this:

> Not far from CAREFREE there is a forest, on the slope of a mountain, where the thick branches create a pleasant night and the paths are all covered with leafy canopies. (...) the turtledove cooed, the woodpecker whistled, the cuckoo called, a hundred voices came together with harmonious tones. Here, a small brook with a barely heard murmur wanders under the arches of overhanging bushes, and it loses itself in a coarse, lonely valley, where I could find no footprints. I saw various, almost invisible streams trickle from a small sandy plain. These are the source springs from which the little brook emerges. Here I sat in the soft moss, shielded from the hot air, hidden away under a bush. I felt caught in that caressing shiver, that holy awe, that occupies these lonely hiding places. For more than an hour, I was completely given over to my musings.[158]

The use of 'pensive' (above) and 'musings' indicates a farther-reaching or more profound kind of thinking that is put into motion through creative imagination. The empathic way of experiencing nature leads to 'feeling' and 'feelings', a 'mixture of experiences'.[159] In the continuation of fragments such as the preceding two, the character arrives at a mostly sentimental verbalisation of a single 'image' or of multiple 'images' regarding death and loss, but also regarding the nature, origin and meaning of Creation. Often these lead to God or Christ. It is in such moments, when the letter writers are 'all feeling' or are 'entirely swept away' in their 'musings', that subject and object come together in their experience of nature. At such moments, nature becomes a reflection of the heart or attains complete harmony with the heart.[160] The soul then nearly crumples under the weight of feeling.[161] Nature makes the soul 'susceptible to all tender and noble afflictions' – so much so that Eufrozyne exclaims in a letter to Elize, 'O nature! Along how many ladders do you lead our spirit upwards'.[162] At such moments, Eufrozyne has transcended experience: 'my most accurate description would make all this beauty disappear: this is only for feeling', she writes.[163] And when Emilia becomes depressed after parting with Eufrozyne, her friend calls upon her to 'open her heart' to 'the influence of nature and religion'.[164] As such, the heart functions, instead of reason, as a place of knowledge and becomes an instrument of metaphysical reflection.

In the imagining of landscape and the representation of the feeling it evokes, Post seems to owe much to Hirschfeld's *Theorie der Gartenkunst*. In the eighteenth-century description of the landscape, there was a great deal of intertextual reference.[165] Hirschfeld extensively analysed the effects of natural beauty as it appeared to man – the walker – in the form of landscape. He propagated an approach to landscape that took the many impressions that it gave rise to, owing to its varied nature as an intensifying pleasure, as a 'gedoppeltes Vergnügen'.[166] Post seems to make the same point where, in the passage quoted earlier, she speaks of 'the threefold pleasure' from 'a single joy.'[167] Hirschfeld studied systematically how natural beauty affected mankind.[168] The nature of landscape influenced not just the imagination (Einbildungskraft) but also the capacity for feeling (Empfindungsvermögen).[169] The latter is evident from descriptions of the effects that different kinds of landscapes had on the feelings of the walker.[170]

Man became so entwined with nature that he could not escape the influence it had on his mind.[171]

Hirschfeld's influence is clear primarily in the way in which Post orders and expresses her observation and experience of nature. But the landscape she describes was absolutely not identical to the landscape garden created by Hirschfeld. What is typical about *Het land, in brieven* is, after all, that the experience of nature by Emilia and Eufrozyne has an autobiographical basis. As noted, that experience is inspired by the surroundings of Amerongen, where Post spent part of her life: near the Rhine and the farmland, heaths and forests of the Utrecht Hill Ridge. The novel alludes to it in several places.[172] After all, Van den Berg does emphasise that *Het land, in brieven* is founded on personal experience. The attraction of the book, especially for readers who read the work as a nature diary and took it on their own walks,[173] must have been the fact that it contained authentic descriptions of nature. Readers tried to reproduce the effects on their own feelings. Thus, where that authenticity was concerned, Van den Berg was not the only guide. Apparently, the novel itself was already able to present itself as a kind of empiricism – an empiricism that encompassed described nature as well as evoked feelings. Thoughts, ideas and emotions, after all, could appear authentic only when the experiences on which they were based were authentic too.

In closing: Post and the sublime

In Post's writing, 'scene' or 'scenes' is one of the many words used, though the most often used, to describe what imagination leads to. It is used in 'homely scenes' in a pastoral setting (fig. 5),[174] for descriptions of memories of the past and, most notably, in combination with the adjective 'great', to denote described landscapes or natural phenomena in a summarising way.[175] These scenes serve a purpose. In her first letter to Elize, Eufrozyne speaks of 'scenes imparting wisdom', in which Emilia is capable of linking up 'lessons'.[176] Post connects the terms 'grand' and 'grandeur' in a majority of cases to the origin of Creation and to the Creator himself, God. Two quotations illustrate this, the first from a letter by Emilia:

> The bleak, dreary winter may be a sad, but also grand scene of God's omnipotence, and of this concept: 'Nature is not death, but peace. She prepares herself to shine even more brilliantly in the future'.[177]

And during her stay by the sea (fig. 6), Eufrozyne writes:

> The sea seemed to me a scene of the immense grandeur and majesty of the Creator; I felt it with humility, wonder and worship.[178]

Earlier, it was established that the aesthetical experience provided empirical confirmation of the speculative assumption that man possessed a destiny beyond the material world and that nature helped him reach that destiny. In the passages just quoted, especially the last one, something

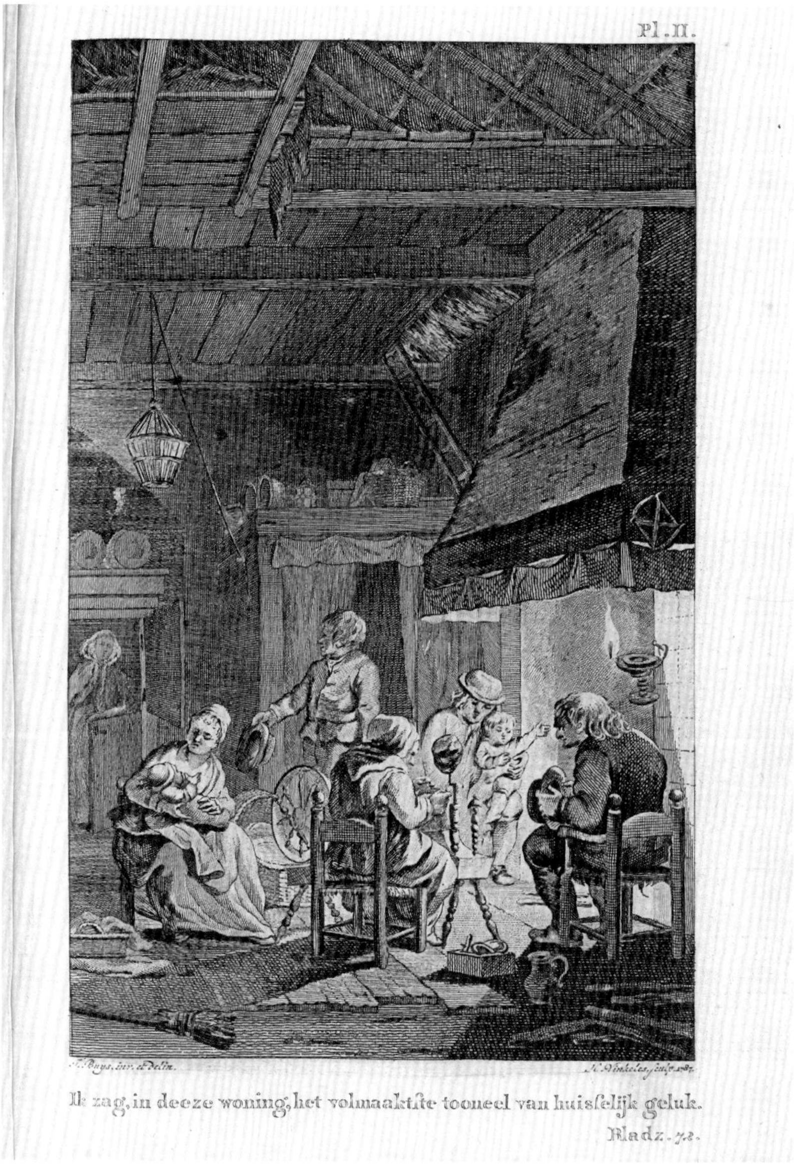

5
H. Vinkeles after J. Buys, *Emilia enters a farmer's house*
engraving, Elisabeth Maria Post, *Het land, in brieven* (Amsterdam, 3rd ed., 1788) University of Groningen Library, KW B 1544, Pl. II (photo: Dirk Fennema), quotation: 'I witnessed, in this house, the most perfect scene of family happiness' (p. 72).

of that destiny is already alluded to. The clearest passages, however, are those in which the most important category of experience in the contemporary aesthetics of feeling is referred to: that of the sublime – 'het verhevene' in Dutch, 'das Erhabene' in German.[179] The only quotation in the previous section already contained an example of it: the description of mood or the impression of nature as a 'caressing shiver', which hardly presupposes a 'reasonable approach' to nature.[180] Paradoxical phrases such as 'caressing shiver' fit the mixed experiences that nature was able to evoke through its isolation, desolation, savageness, height, depth, darkness, sound and movement.[181] Other examples that appear in the novel are 'pleasant gloom', 'pleasant melancholy' and 'sombre pleasure'.[182] (In

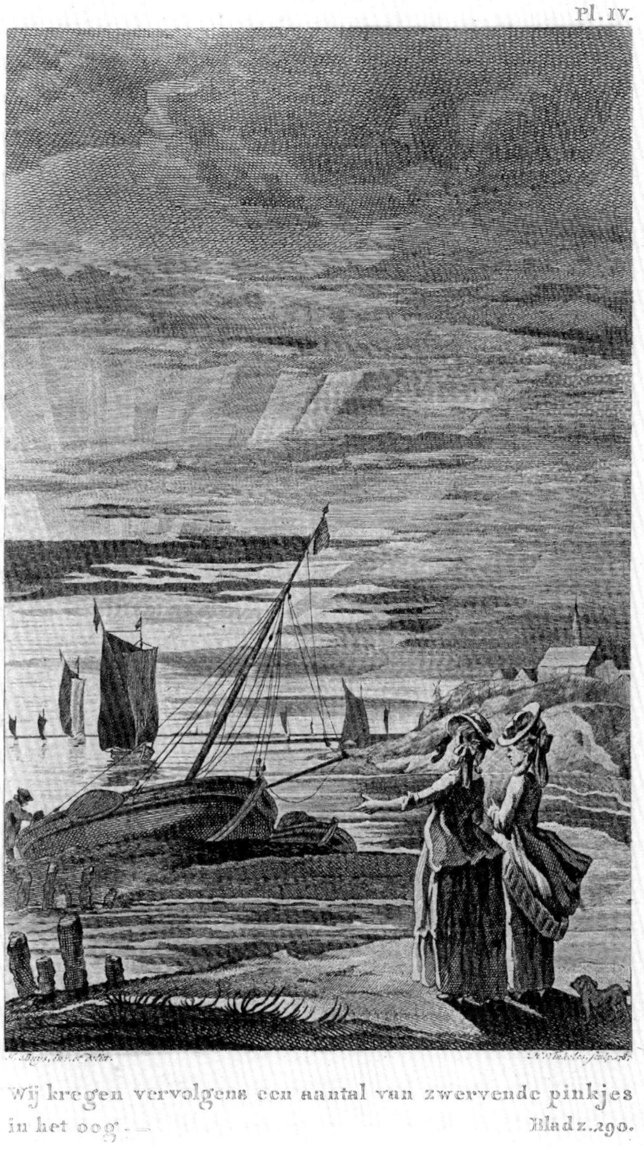

6
H. Vinkeles after J. Buys, *Eufrozyne and Emilia at the seaside*
engraving, Elisabeth Maria Post, *Het land, in brieven* (Amsterdam, 3rd ed., 1788) University of Groningen Library, KW B 1544, Pl. IV (photo: Dirk Fennema), 'Then our eye was caught by a number of gliding flatboats' (p. 290).

comparable German texts, there is mention of 'sanfte Melancholie' or 'süße' or 'sanfte Schwermuth'.)[183] The double sensation that the sublime evoked coincided with the special place of man within Creation. Nature might have an intimidating effect on its observers because of its overwhelming beauty (hence the shiver, gloom or melancholy), but the observer was simultaneously aware of his privileged position within Creation and his ability to control nature through knowledge (hence 'pleasant', 'caress' or 'pleasure').

In the preceding section, a passage was quoted wherein Emilia describes how imagination reminds her of the 'beauty, the scale, the overwhelming sense of nature' in the Alps. That landscapes with a sublime effect were to be found only in Germany, and not in the Netherlands, was generally accepted.[184] Thus Eufrozyne writes to Elize:

> Just now I was reading the pleasant HIRSCHFELD. How splendidly does he depict this life on the land. I can already feel its beauty, but cannot yet understand it. In order to paint like this, one does not just need the brush of HIRSCHFELD; one also has to live in his country. Here nature is beautiful, but the unification of the grand and the beautiful, of order and savageness, happens there, in rapture. But am I to put down the pen because I cannot draw like that? No, friend! You ask me about my pleasure and my daily life. And so I convey it to you (...).[185]

Again we see the terms 'grand', 'beautiful' and 'rapture'. Various passages show, however, that Dutch landscapes and seascapes were both also able to inspire lofty feelings – a sunrise (or sunset), for example:

> Never in my life did I see something so beautiful! Never did I feel such elevated pleasure. – I am beside myself! – As long as I breathe, I shall think of this morning! (...) A grander and more fitting sight cannot be found in nature. You are touched for a reason. At the same time, the setting sun possesses something that most of the time engenders an even more pleasant sensation in my mind.[186]

And, of course, thunder, the most impressive natural phenomenon:[187]

> A terrible thunderstorm awoke me; my darkened bedchamber was awesomely lit with incessant lightning bolts. I stood, trembling, went from one room to the next; all the while the terror of the thunderstorm and the fierce downpour prevented me from going into the open air. I felt God's grandeur, but not with childlike reverence. Fear clouded my mind and every rumbling thunderclap seemed to me to be a herald of God's wrath.[188]

The experience of the grandeur of both Creation and the Creator led to the realisation of change, as was shown especially in the first quotation in this section. For Post, that change still more or less coincided with the cyclical movement of the seasons and with God's influence on nature. And yet, there are signs – be they careful – of a historical understanding of nature, of a sense of being a part of a long history of nature. In a letter to Eufrozyne, Emilia writes:

> Nature shapes my heart in her image, and even as she dies, she pleases me. And why should she not? She teaches me wisdom, she reminds me of the changing of all things! My imagination brings startling apparitions before my eye. Here I see ruined cities and destroyed lands, there mountains sinking into the abyss with a dreadful rumbling, while from the boiling ocean islands rise again. I see rigid cliffs [be] ground down by time and the surface of the earth [be] subject to constant change. They tell me that the world is passing! [189]

In this passage, Emilia arrives at an image of changes that are not entirely cyclical, doing so through observation of nature and aided by imagination. It is predominantly the sublime that was able to evoke such images or experiences of intrinsic change in nature.[190] Sublime experiences were the peak of the experiential approach to nature that, as noted, based itself on a more empirical approach. Both approaches made use of the observer's imagination in order that he be able to first think or feel such experiences and to subsequently put them into words. At the moment the observer lost himself entirely in nature, became all feeling, he merged with the object of his observation, imagination and experience, and inner and outer nature coincided. That is the case here too. It was a moment in which Creation was fathomed to its deepest core. It was observing, imagining and experiencing man in whom and by whom this knowledge was given shape. Through the unification of inner and outer nature, he could discover the essence of Creation, the force that guided its development – God and change.

Thus, we, together with Post, find ourselves in the laboratory of Romanticism and of the Romantic approach to nature, in a period where art was given a place next to natural science as a means of observation, imagination and experience, where the demarcation between both disciplines could not be clearly drawn and one could engage in both to one's heart's content – and where women did so as well, driven by friendship and using both sentiment and reason.

Notes

1 Paasman 1974, 111.

2 For biographical and bibliographical information on the author, see Paasman 1974.

3 Riskin 2002, 4. As a side-note, the sentimental elements of Post's novel have been previously highlighted. See Brandt Corstius 1955, 12-13, 63-64, 106-108. He refers to the work of Post as 'pre-romantic' partly for that reason. On this term, see Van den Berg 1999 (who rejects its usage) and Brodey 2005.

4 See the notes in Post 1987, 234-238. The influence of the mentioned (and other contemporary) authors is also discussed by Brandt Corstius, especially in chapter 6 (Brandt Corstius 1955, 72-99).

5 See also Brandt Corstius 1955, 18-19.

6 In Emilia the female pendant of Émile has been recognised, the protagonist of *Émile, ou de l'éducation* (1762) by Jean-Jacques Rousseau (1712-1778). Eufrozyne is named after Joy, one of the three Graces (Paasman 1987, 223). Post could also have derived the name Emilia from the *Geschichte der Fräuleins von Sternheim* (1771) by Sophie von La Roche (1730-1807), in which a character named Emilia serves as one of the addressees (Becker-Cantarino 1990, 99).

7 Wolf 1999, 630-631.

8 Runge 1997, 14; Anderegg 2001, 20.

9 Wolf 1999, 624.

10 Sauder 1989, 165; Birklund Andersen 2004, 162. On the epistolary novel in general, see Bray 2003.

11 Koepke 1990, 266.

12 Anderegg 2001, 15; Stiening 2005, 10.

13 Stiening 2005, 8.

14 Meyer-Krentler 1991, 16-17.

15 Post 1788, XL, 252.

16 Paasman 1987, 223.

17 Wagner-Egelhaaf 2000, 269, 273.

18 Post 1788, XXIX, 145-146: 'Dit plaatsje bemin ik altijd, maar sedert langen tijd, vond ik er dat niet, dat ik nu aan uwe zijde gewaar word; de vriendschap is toch het geluk van ons leven! Zij vertedert het hart, maakt het meer gevoelig voor de natuur; geeft de mededeelende ziel ruimte om op nieuw te genieten; en de natuur geeft weer kracht aan de vriendschap. Zij beide maken de zachtste, de reinste neigingen wakker; – hoe veel genoegen geef ik mij zelven, wanneer ik u in mijne brieven deelgenoote van mijn geluk maak! en nu, – nu gij het met mij gevoelt, nu uw gelaat mij zegt, dat gij getroffen zijt, nu groeit mijn vermaak tot verrukking'.

19 Becker-Cantarino 1990, 94; Meyer-Krentler 1991, 15-16.

20 Kord 1996, 232. Post seems to share this opinion. On this, see Everard 1994, 45-46.

21 Becker-Cantarino 1991, 55-56; Meyer-Krentler 1991, 19.

22 Everard 1994; M. Meijer 1984, 167, 169, 171.

23 Meyer-Krentler 1991, 21; Roulston 1991, 216.

24 Post 1788, XXIX, 150: 'Wat, bid ik, zou de reden zijn, dat onze sex, die men zelfs beschrijft als meestal gevoeliger voor 't schoone, zagter van aart, en levendiger van hartstogten, ongeschikter zijn zou tot standvastigheid in eene vriendschap, die niet op grillige invallen, maar op eene verstandige keuze, en harmonie der zielen gegrond is, en die bestierd wordt door eene poging, om het wederzijds geluk in elkaârs genot te volmaken? (...) Neen, Eufrozyne! ware vriendschap tusschen edele overeenstemmende zielen, schoon in vrouwelijke lichamen geplaatst, kan niet missen bestendig te zijn'.

25 Becker-Cantarino 1991, 55; Kehn 1991, 167.

26 Kehn 1991, 174.
27 Wagner-Egelhaaf 2000, 268.
28 Kehn 1991, 175.
29 Van Hamel 1915-1916; Brandt Corstius 1955, 17-18.
30 Kehn 1995, 19. On the importance of experience, see also Barck 1995, and Ebbinghaus 1997, 109.
31 Breidbach 2006, 212.
32 Kehn 1991, 176.
33 Cited ibid., 177.
34 Ibid., 172.
35 Ibid., 177.
36 Post 1788, XXIX, 146.
37 Ibid.
38 Kehn 1991, 181.
39 Ibid., 182.
40 Post 1788, XLIV, 269: 'Het beekje dat hier naast mij, met een effenbaar gefluister, henen vloeit en kleine watervallen van boomwortels maakt, ruischt altijd schoon, maar nu vooral geheel harmonisch met de natuur en mijn hart; maar eene harmonie, die ik wel gevoelen, doch niet beschrijven kan. – Ginds ver staat de boom, hier zie ik zijn top, op wiens bast ik eens den naam van Eufrozyne sneed. – Ik gevoel, en zeg het luide: ô natuur! ô stilte! ô vriendschap! hoe groot zijn uwe goederen! bij uw drievoudig genot kan ik gelukkig leven'. Additional places where friendship and harmony are mentioned are XXIX, 153, and XXXV, 213.
41 Letter XXIX is dated June 14, letter XXXVIII, August 5.
42 Bots 1971. Also see Paasman 1987, 219; Groh & Groh 1989, 67-71; Van Zonneveld 1997, 97; Jürgens 2004, 70, note 5; Reill 2005, 57-64. On the background of this movement in the Republic, see also Vermij 1991. An overview of practitioners of natural science in the Republic is given by Hooykaas 1981.
43 Bots 1971, 67-73; Baggerman 2006.
44 Van Zonneveld 1997, 97-98.
45 Paasman 1987, 291.
46 Post 1788, VI, 30; XI, 63; XXXII, 191; XXXVII, 237.
47 A recent study that describes this approach for the Republic is Van de Roemer 2005.
48 Weber 1989a, 120. On the more philosophical basis of the concept of nature, see Spaemann 1967; Mainzer 1989; Mittelstrass 1989; Söring 1999.
49 Reill 2005. For the Republic, see also Knoeff 2004.
50 Reill 2005, 54-55. Carl Linnaeus (1707-1778) is seen as representing a mechanistic natural philosophy and strived for a universal ordering of nature on the basis of deduction. Georges-Louis Leclerc, Comte de Buffon (1707-1788), with his *Histoire naturelle* (1749), is seen as the founder of the vitalistic school (Reill 2005, 27; see also Weber 1989b, 113-114). Physico-theology was founded on preformationism, vitalism on epigenesis. In his *Handbuch der Naturgeschichte* (1778), Johann Friedrich Blumenbach (1752-1840) took up the latter idea and spoke of 'Bildungstrieb' (formative drive) (Reill 2005, 166; Snelders 1970, 196). See also, on Romantic science, Snelders 1994.
51 Reill 2005, 76-77.
52 Post 1788, IV, 19: 'schoone landgezigten'.
53 M. Meijer 1984, 167-168.
54 Post 1788, IV, 16.
55 Ibid., IV, 17.
56 Ibid., IV, 18. Referred to here are *The history of Clarisse Harlowe* (1747-1748); *The history of Sir Charles Grandison* (1753-1754) by Samuel Richardson (1689-1761); *Sophiens Reise von Memel nach Sachsen* (1769-1773) by Johann Timotheurs Hermes (1738-1821); *Geschichte Karls von Burgheims und Emiliens von Rosenau* (1778-1779) by Johann Martin Miller (1750-1814).
57 Saine 1976, 71; Sturkenboom 2004, 45, 54; Baar-de Weerd, 181-187.
58 Sturkenboom 2004, 58.
59 Ibid., 63.
60 Ibid., 69.
61 Zuidervaart 2010.
62 Van Raak 1996.
63 Post 1788, IV, 19: 'Het mikroskoop geeft mij ook eene aangename bezigheid'.
64 Ibid., XXXV, 212: 'door 't mikroskoop eenige insekten [bezagen], stof van de vleugelen der kapellen; en meer wonderen der natuur, die het ongewapend oog niet ontdekken kan: en waarlijk gelukkig, dat onze zintuigen zo grof, of, deze diertjes zo klein zijn, anders zouden wij, omringd van zulke ijzige en sterkgewapende dieren, in eenen gestadigen angst leven, en aan geen hunner ons lichaam tot een rustplaats vergunnen'.
65 Ibid., XXVI, 134. On the development of the telescope, see Ishihara 2005, 52-55. Telescope and microscope symbolised the ability of man to penetrate Creation (Saine 1976, 61).
66 Post 1788, III, 10.
67 Ibid., VI, 30.
68 Ibid., XIII, 77.
69 Ibid., XXXI, 178; XXXVI, 223.
70 Ibid., XXXI, 178: 'zig een celletje [weeft], waarin het eenigen tijd, als in eenen ongevoeligen slaap doorbrengt, – tot dat het, met vleugelen voorzien, als een schoone kapel te voorschijn komt'.
71 Jorink 2006, 229-247. See also his contribution to this volume.
72 Post 1788, XXXV, 211: 'De inademing der frisse buitenlucht, de ordenlijke omloop der levenssappen, door eene gedurige beweging veroorzaakt, maakt mij gezonder, dan ik immer was'. The same thought is expressed in IV, 19.
73 Ibid, XXVI, 135: 'bedwelmende oneindigheid'.
74 Ibid., XXXVI, 220: 'die ongetelde menigte van vaste sterren [die] zo vele zonnen zijn, om welke weder andere, voor ons onzigtbare wereltbollen wentelen'.
75 Ibid., XXXVI, 221: 'ook op andere klooten, zedelijke wezens van eenen andere aart gezondigd hebben, die ook deelen zullen, in het heil, door den Verlosser van onze wereld aangebragt'.
76 Astronomy was very popular in the second half of the eighteenth century (Ishihara 2005, 47-48). Observing the stars led to bliss and the experience of the sublime (ibid., 55-57). Regarding the vastness of the universe, Johann Wolfgang von Goethe (1749-1832) spoke of the 'Mehrheit der Welten'. He expressly accounted for the existence of extraterrestrial life (ibid., 62-64). See also Brandt Corstius 1955, 77-78. On the popular practice of astronomy in the Republic, see Zuidervaart 1999, 69-82, 337-398.
77 Post 1788, XXXVI, 221: 'zullen wij Gods wijsheid naar onze naauwbeperkte begrippen omtrent het mogelijke afmeten?'.
78 Johannes 1992, 47.
79 Ibid., 48.
80 Ibid., 57.
81 Ibid., 58.
82 Ibid., 59.
83 Ibid., 99.
84 Ibid., 102, note 55. In this context, Johannes cites Cornelis van Engelen, *Verhandeling* (1783), 295, who in turn cites Shaftesbury and describes the poet as a secondary creator, as a Prometheus under Jupiter. See also Johannes 1992, 130-131.
85 Ibid., 102-103.
86 The imagination in its reproductive form could be a link between senses and

mind, and an aid to reason (ibid., 80-81). The same seemed to be valid for poetic imagination. It was described as the interpreter of the truth, not only of material reality but also of divine truth, not directly, but through revelation in the Bible and in nature (ibid., 95). But nature's range was immediately restricted. The poet did not have direct insight into higher, divine truth (ibid., 96).

87 Ibid.,126.
88 Ibid.,122-123.
89 Ibid., 125.
90 'Digtstukken, waar in de digter louter zijne eigene gewaarwordingen en aandoeningen voorstelt, zijn geen nabootsingen of navolgingen op een vrijen trant, van de natuur, maar zij zijn, voorstellingen van de natuur zelf (...) wanneer hij zijne eigen gemoedsbewegingen, zijne eigene gewaarwordingen voorstelt, dan is er geen nabootsing, maar hooren wij de natuur zelf' (cited from Johannes 1992, 132). According to Johannes, Van Alphen maintains a view of the imagination as a kind of imitation, that is to say, not as creative 'in the meaning that some Romantics attribute to the term' (ibid., 131, 133-134; 133, note 142). On Van Alphen, see also Oosterholt 1998, 21-24.
91 Mitchell 1993, 13. See also Engell 1981, 104-107; Kehn 1995, 20; Ottmann 2002, 374-375.
92 Mitchell 1993, 13.
93 Ibid., 14; Ottmann 2002, 364-365.
94 Frenzel 1984, 193. On movement in nature and verbal dynamics in the description of landscapes, see Langen 1948-1949, and Ebbinghaus 1997, 95-96, 122-123. On Klopstock and interiorisation, see Martin 1995. Also see, on nature and landscape in German literature, Dethloff 1995; Duesberg 1996; Sheffy 1998; Brüggeman 2009.
95 Mitchell 1993, 14. On the more observing and classifying character of early eighteenth-century descriptions of nature, see Frenzel 1984, 191, Saine 1976, 64-65, and Kehn 1995, 5.
96 Cited from Mitchell 1993, 17, and note 18.
97 Ibid., 30.
98 Ibid., 32. See also Kehn 1995, 10.
99 Streng 1998, 271, note 56.
100 Post 1788, XXIV, 127. Emilia mentions 'der Wunsch'. See also, on Gessner, F. Schneider 2004, 75-80.
101 Mitchell 1993, 35.
102 Ibid., 60.
103 Ibid., 61.
104 Kehn 1995, 11.
105 Ibid., 13.
106 Ibid., 14.
107 Ibid., 15.
108 Ibid.,18.
109 Ibid., 21; Holmes 2008, 91-94.
110 Wanning 2005, 45. Also see, on Romantic science and sensation, Jackson 2008.
111 Ibid., 47.
112 Ibid., 51.
113 Richards 2002, 12. According to Johannes, Dutch theoreticians saw only the creative kind of imagination as characteristic of the poetic genius, and also of the scientific genius (Johannes 1992, 60, 61, note 113). In this context, he refers among other things to the *Theorie der schönen Künste und Wissenschaften* (1767, reprinted 1774; Dutch translation 1780) by Friedrich Justus Riedel (1742-1785), and *Über die Einbildungskraft* (1778; Dutch translation 1782) by Leonhard Meister (1741-1811). Coincidentally, these theoreticians held that the creative imagination got in the way of an empirical, objective representation of reality. In the Classical view it had to lead to a heightened or beautified nature, the *belle nature* (Johannes 1992, 62; Rusam 1992, 16-19). The ever-present danger there was exaggeration and zeal.
114 Wanning 2005, 48-49, 51.
115 Ibid., 72.
116 Wellbery 1984, 224-225; Köchy 2002, 6.
117 Wanning 2005, 93, especially note 218.
118 Köchy 2002, 11.
119 Richards 2002, 111.
120 Ibid., 431. On Schiller and nature, see Braungart 2005.
121 Weber 1989b, 127. Joachim Ritter speaks of the 'Entzweiungsstruktur der modernen Gesellschaft' (Ritter 1963, 21; Zimmermann 1982, 122; Groh & Groh 1989, 57-58; Rusam 1992, 15, note 32; Zelle 2001, 147; Pouvreau 2002, 42; Wanning 2005, 91). Through the physical and technical approach to nature, man had supposedly become estranged from it. He therefore attempted to restore unity with nature through an aesthetical approach. This led to 'sentimentale Dichtung', to use Schiller's words. The experience of nature was accompanied by a feeling of melancholy due to the distance from it (Wells 1966, 498; H. Schneider 1980, 290-291). The development of an aesthetical understanding of nature can be seen as part of the reaction to the mechanisation of the world (Zimmermann 1982, 129-130). In Romantic discourse, the emotional, empathic approach to nature was used to come to a better understanding of nature (Wanning 2005, 30-33). It was complementary to the empirical approach (Richards 2002, 12). In early eighteenth-century nature description, the aesthetical description and experience by a feeling subject was much less prominent (H. Schneider 1980, 299-300). Also see H. Schneider 1981; H. Schneider 2005; Pennone 2008.
122 Zimmermann 1982, 130.
123 Pouvreau 2002, 42.
124 Weber 1989b, 120.
125 Zimmermann 1982, 131.
126 Post 1788, ii.
127 Ibid., iii.
128 Ibid., iv.
129 Ibid., vi: 'Zij bezit die koenheid van geest, die, boven slaafsche navolging verheven, zelf opmerkt en uitvindt. Haar naauwkeurig waarnemend oog zwerft alomme in het ronde, en overal vindt zij onuitputlijke rijkdommen van schoonheden, waar honderd anderen niets vinden'. See also Brandt Corstius 1955, 63.
130 Post 1788, vi: 'Zij houdt onze opmerking stil, bij voorwerpen, die ons naauwlijks noemenswaardig schenen, en ze worden, onder hare handen, verbazende kunststukken van de wijze en magtige hand, die alles voordbrengt, en in wezen houdt'.
131 Ibid.: 'in de gewoone beschaafde taal van het gemeene leven'.
132 Ibid.: 'natuurlijke en eenvoudige beschrijvingen van treffende zaken en waarheden'.
133 Ibid.: 'laffe en belachlijke (...) schrijftrant'.
134 A. Meijer 1998, 124-125. On Post, see also ibid., 107-108, and Brandt Corstius 1955, 53-55.
135 Post 1788, vi: 'behaagt en roert'. See also Van Elferen 2006, 16-19.
136 A. Meijer 1998, 81.
137 Ibid., 82.
138 Ibid., 112-113.
139 Ibid., 95-96, 111.
140 Post 1788, IV, 18: 'schilderingen, die hun bestaan alleen in zijn Goddelijk, vindingrijk vernuft hadden'.
141 Pouvreau 2002, 41; Wolf 1999, 633.
142 If Eufrozyne possessed the same

'painterly pen' as Emilia, she would 'paint scenes' (toonelen afmalen) of city life (Post 1788, VII, 32). Furthermore, she does not see herself capable of making 'poetic paintings' (poëtische schilderstukken) (ibid., XXIX, 144). What the women picture in their letters can also be called an 'image' (beeld) and more often still a 'sight' (gezicht), 'scene' (tafereel), 'painting' (schilderij) or 'depiction' (schildering).

143 Oosterholt 1998, 55.

144 See Andrews 1989.

145 Of the other senses, hearing comes second (through conjugations of the verb 'hear' (horen) and smell (or scent) third. In landscape art, the ear (hearing) and nose (smelling) were emphasised alongside the eye (vision) – the first two in descriptions of nighttime walks (Ebbinghaus 1997, 104). Impressions of various senses reinforced sensations and feelings (ibid., 105).

146 Post 1788, XII, 113.

147 Ibid., XII, 68.

148 Ibid., IX, 49-52.

149 Ibid., X, 54.

150 Ibid., XXXI, 181-182: 'Als ik hier, tegen het vallen van den avond, met KLOPSTOCK in de hand, nederzit; en zelf aan 't peinzen geraak, voert mij mijne verbeelding wel eens naar zijn graf. Bij dat rotshol neergezeten, hoor ik boven het zelve de donkere Cijpressen ruischen. Ik zie mijne verlatene lievelingen hopeloos omdolen; ik hoor, dunkt mij, in de morgenschemering de nachtstilte afgebroken, door het hol gekraak der bevende rotzen. Ik zie den luistervollen hemeling het hol openen; en, de gestorvene Jezus treedt levendig te voorschijn, met eene zegevierende houding, en eene goddelijke vreugde op zijn gelaat. Met zijne verrijzing zie ik alle ellende van mij weggevaagd, en mijne zaligheid, in zekerheid gesteld: – dan, aanbid ik hem vol gevoel, en noem God vertrouwlijk mijnen Vader'.

151 Ibid., XII, 70.

152 Ibid., XI, 61-62: 'Mijne verbeelding voerde mij op dit gezicht naar de eeuwig besneeuwde Alpen henen; ik verbeeldde mij de gevaren, die onbedagte reizigers daar dikwijls ontmoeten; de ijselijke sneeuwbrokken, die, met een donderend geraas dikwils nederploffen, en vee en menschen, zoo geen tijdige vlugt hen redden kan, onverwacht doen omkomen. – Het schoone, het grootsche, het ontzettende der natuur, in dien verwonderlijken oord, had de sterkste werking op mijn verrukte ziel; even of ik er zelf getuige van was, en nogtans welbezien, bevond ik mij op ZORGENVRIJ (...)'.

153 Ibid., XLVIII, 289: 'En drongen mijne gedachten verder dan dit oppervlak door, tot hare kolken en afgronden; verbeeldde ik mij daar de ontelbare heiren van nuttige, of nog onbekende visschen – zoo verschillend in vorming, in grootte en aart, als hun Schepper rijk in wijsheid en magt is; verbeeldde ik mij de onderlinge geregelde huishoudingen van zoo verschillende wezens, van de ijselijke zeegedrogten af tot de kleinste schulpdiertjes toe, hunne verschillende wijze van voortplanting, en duur, en hoe eene wakende Voorzienigheid die allen onderhoudt, beveiligt, en naar hunne verschillende vatbaarheid gelukkig maakt (...)'.

154 Ibid., XII, 68.

155 Ibid., XXII, 113: 'In de daad, er is zulk een groot onderscheid niet tusschen de vermaken der oogen en der verbeelding. Eene levendige verbeeldingskragt, met welke wij beide beschonken zijn, is voor eene gevoelige ziel, een uitmuntende schat. Zij geeft ons het driedubbel genot van een enkel genoegen; en somtijds, wijl zij vrij werkt, en alles bij elkander voegt, dat de natuur verdeelt, heeft zij iets vooruit boven het zintuigelijk genoegen. Nu, om tot de zaak te komen, moet ik uwe verbeelding met voorraad voorzien'.

156 Ibid., XXXIV, 207: '(...) eene verzameling van eerwaardige boomen. (...) Zij schenen allen in eene peinzende gedaante, te luisteren naar het klateren der bron, en beantwoordden het veeltoonig waterlied met een vreedzaam gesuis. De dichterlijke verbeelding zou hier een heiligdom vinden. (...) Ik ben geheel gevoel'.

157 In the representation of those experiences incited by nature, the words 'affliction' (aandoening), 'feeling' (gevoel) – this word also appears as a verb, 'feel' (voelen) – and 'awareness' (gewaarwording) are used. They are often 'touched' (geroerd), 'charmed' (bekoord) – something can, moreover, 'charm' (bekoren) – 'pleased' (vergenoegd), 'cheerful' (vrolijk) or 'overwhelming' (verstom(men)d). Once, a 'soul stirring scene' (zielroerend toneel) is mentioned (ibid., VIII, 46). The nature of those feelings is repeatedly described with the verb 'enjoy' (genieten) or 'elate' (verrukken). Frequently, one of the two writers is 'elated' (verrukt) or witnesses something that is 'elating' or 'rapturous' (verrukkend) or an 'elation' or 'rapture' (verrukking).

158 Post 1788, XXV: 128-129: 'Niet ver van ZORGENVRY ligt een bosch, op 't hangen van den berg, waar de dichte takken eenen aangenamen nacht vormen, en de wandellanen als met loofgewelven overdekken. (...) De tortel kirde; de spegt floot; de koekkoek schreeuwde; honderd stemmen vermengden zig met harmonische toonen; hier dwaalt een smal beekje met een naauwlijks hoorbaar gefluister, onder het gewelf van de overhangende struiken, en het verliest zig in een ruuw, eenzaam dal, daar ik geen voetspoor vinden kon. Uit een kleine zandige vlakte zag ik verscheiden, bijna onzigtbare straaltjes, zijpelen: dit zijn de bronadertjes waar uit het beekje voordspruit. Hier zat ik in het zagte mosch, beschermd tegen de heete lugt, onder een struik weggedooken: ik gevoelde mij door die streelende huivering, dien heiligen eerbied, die deze eenzame schuilplaatsen bewonen, bevangen, meer dan een uur zat ik hier geheel weggevoerd in gepeinzen'.

159 Ibid., XV, 88.

160 Ibid., XVI, 91; XXXV, 213.

161 Ibid., XXIII, 121.

162 Ibid., XXXIV, 208: 'O natuur! Langs hoe vele ladders leidt gij onzen geest naar boven'.

163 Ibid., XXXIII, 194: 'mijne naauwkeurigste tekening zou al het schoon doen verdwijnen: dit is alleen voor 't gevoel'.

164 Ibid., XL, 253: 'zet uw hart weer open voor den invloed van de natuur en den Godsdienst'; also see XXIX, 146.

165 Wanning 2005, 67.

166 Kehn 1995, 6.

167 Post 1788, XXII, 113: 'het driedubbel genot van een enkel genoegen'.

168 Kehn 1995, 6. The experience of nature and a landscape was founded on a combination of mutually reinforcing aesthetical forces, respective of the experiences that it give rise to (Empfindungen) (ibid., 8-9; Ebbinghaus 1997, 116).

169 Kehn 1995, 11; Ebbinghaus 1997, 108. Also see, on the landscape garden, Gaier 1989; Trotha 2004; Baggerman & Dekker 2004, 161.

170 Kehn 1995, 12; Ottmann 2002, 352-355; Ebbinghaus 1997, 111.

171 Ebbinghaus 1997, 115.

172 Paasman 1974, 62-64, 71-72; Paasman 1987, 221. On autobiography and natural science, see Kuhn 2009.

173 Paasman 1987, 218.

174 Post 1788, XII, 72; XIV, 81; XXXVII, 232.

175 Post 1788, III, 9, 11.

176 Ibid., XXIX, 144: 'de wijsheidleerende tooneelen (...) wier lessen door Emilia worden aangedrongen'.

177 Ibid., III, 9: 'De gure, de woeste winter, is wel een treurig, maar toch een grootsch toneel van Gods alvermogen: en dit denkbeeld; 'De Natuur is niet dood – maar rust; zij bereidt zig, om eerlang met grooter luister te schitteren'. See also XXVI, 135.

178 Ibid., XLVIII, 289: 'De zee was voor mij een tooneel van de ontzachverwekkende grootheid en majesteit des Scheppers; ik gevoelde die met ootmoed, met verbaasdheid en aanbidding'. See, for similar descriptions, XLVIII, 287 and 290.

179 Zimmermann 1982, 132; Hall et al. 1988, 112; Groh & Groh 1989, 79-82; Rusam 1992, 14.

180 In contrast to the assumptions in Johannes 1997, 8.

181 Mitchell 1993, 13.

182 Post 1788, III, 16: 'aangename droefgeestigheid'; III, 16; XI, 65: 'somber vermaak'; XLIX, 295: 'aangename weemoedigheid'.

183 Ebbinghaus 1997, 111, 117; Reill 2005, 27-28.

184 Madelein & Pieters 2008, 11-13. On the sublime in the Low Countries, see Madelein 2007.

185 Post 1788, XXXIV, 125: 'Zoo even las ik in den bevalligen HIRSCHFELD; hoe verrukkend maalt hij dit Landleven af! Ik gevoel er al de schoonheid van, maar kan hem niet navolgen. Om zoo te schilderen, moet men niet slegts HIRSCHFELDS penseel hebben, maar ook in zijn land wonen. Hier is de natuur schoon, maar daar vereenigt zig het grootste en het schoone, de order en woestheid, op eene verrukkende wijs. Doch zal ik nu de pen nederleggen om dat ik zoo niet tekenen kan? – Neen, vriendin! gij vraagt mij naar mijn genoegen, naar mijne levenswijs. Die deel ik u dan mede (...)'

186 Ibid., XXXIII, 197-198: 'Nooit in mijn leven zag ik iets zoo schoon! Nooit voelde ik zulk een verheven vermaak. – Ik ben buiten mij zelven! – Zo lang ik adem, zal ik aan dezen morgen denken! (...) Grootscher en treffender gezigt levert de natuur niet op; met reden zijt gij zo getroffen; evenwel heeft eene ondergaande Zon iets dat mijnen geest meestijds noch eene aangenamer gewaarwording verwekt'.

187 Wanning 2005, 60.

188 Post 1788, XLII, 260: 'Een vreeslijk onweder deed mij verschrikt ontwaken; mijn donker slaapvertrek wierd ontzachlijk verlicht door onophoudelijke bliksemvlammen. Ik stond bevend op, begaf mij van de eene kamer in de andere; terwijl de vreeslijkheid van het onweder, en de geduchte stortregen mij het verblijf in de open lucht belette. – Ik gevoelde Gods grootheid, maar niet met kinderlijken eerbied, de vrees bedwelmde mijne reden, en elke dreunende donderslag scheen mij eene aankondiging van Gods toorn te zijn'. See also XLII, 262.

189 Ibid., LI, 309: 'De natuur vormt mijn hart naar hare gedaante, ook stervende behaagt zij mij; – en zou zij niet? zij leert mij wijsheid, zij doet mij aan de verandering van alles denken! Mijne verbeelding voert ontstellende verschijnsels voor mijn oog. – Hier zie ik vernielde steden, en verwoeste landstreken, daar bergen met een vervaarlijk gedruis in den afgrond verzinken; terwijl uit den kokenden oçeaan weder eilanden oprijzen: ik zie ongemerkt de harde rotzen door den tijd vergruisd, en de oppervlakte der aarde aan gestadige veranderingen onderhevig, die mij zeggen, dat de wereld voorbij gaat!'.

190 Weber 1989b, 117, 121.

Bibliography

Anderegg 2001
J. Anderegg, *Schreibe mir oft! Zum Medium Brief zwischen 1750 und 1830* (with a contribution by E.A. Kunz), Göttingen 2001.

Andrews 1989
M. Andrews, *The search for the picturesque. Landscape aesthetics and tourism in Britain, 1760-1900*, Aldershot 1989.

Baar-de Weerd 2009
C. Baar-de Weerd, *Uw sekse en de onze. Vrouwen en genootschappen in Nederland en in ons omringende landen (1750-ca. 1810)*, Hilversum 2009.

Baggerman 2006
A. Baggerman, 'Children's walks in the Book of Nature. The reception of J.F. Martinet's *Katechismus der Natuur* around 1800', in: K. van Berkel & A. Vanderjagt (eds.), *The Book of Nature in early modern and modern history*, Leuven, etc. 2006, 141-154.

Baggerman & Dekker 2004
A. Baggerman & R. Dekker, 'Het spook van Sion. Veranderende visies op jeugd en natuur in de late achttiende eeuw', *De Achttiende Eeuw* 36 (2004), 143-156.

Barck 1995
K. Barck, '"Umwandlung des Ohrs zum Auge". Teleskopisches Sehen und ästhetische Beschreibung bei Alexander von Humboldt', in: B.J. Dotzler & E. Müller (eds.), *Wahrnehmung und Geschichte. Markierungen zur Aisthesis materialis*, Berlin 1995, 27-42.

Becker-Cantarino 1990
B. Becker-Cantarino, 'Freundschaftsutopie. Die Fiktionen der Sophie La Roche', in: H. Gallas & M. Heuser (eds.), *Untersuchungen zum Roman von Frauen um 1800*, Tübingen 1990, 92-113.

Becker-Cantarino 1991
B. Becker-Cantarino, 'Zur Theorie der literarischen Freundschaft im 18. Jahrhundert am Beispiel der Sophie La Roche', in: Mauser & Becker-Cantarino 1991, 47-74.

Birklund Andersen 2004
O. Birklund Andersen, 'The letter-writing culture of the eighteenth century and the novel. The familiar letter and the epistolary novel, with special reference to Richardson's *Pamela*', in: K.-M. Simonsen, M. Ping Huang & M. Rosendahl Thomsen (eds.), *Reinventions of the novel. Histories and aesthetics of a protean genre*, Amsterdam & New York 2004, 161-175.

Bots 1971
J. Bots, *Tussen Descartes en Darwin. Geloof en natuurwetenschap in de 18e eeuw in Nederland*, Assen 1972.

Brandt Corstius 1955
J.C. Brandt Corstius, *Idylle en realiteit. Het werk van Elisabeth Maria Post in verband met de ontwikkeling van de Europese literatuur in de tweede helft van de achttiende eeuw*, Amsterdam 1955.

Braungart 2005
G. Braungart (ed.), *Schillers Natur. Leben, Denken und literarisches Schaffen*, Hamburg 2005.

Bray 2003
J. Bray, *The epistolary novel. Representations of consciousness*, London & New York 2003.

Breidbach 2006
O. Breidbach, 'Freundschaft und all das Andere – Zu Freundschaft, Sympathie und Liebe in Goethes Naturlehre', in: K. Manger & U. Pott (eds.), *Rituale der Freundschaft*, Heidelberg 2006, 207-220.

Brodey 2005
S.B. Brodey, 'On pre-Romanticism or sensibility. Defining ambivalences', in: M. Ferber (ed.), *A companion to European Romanticism*, Malden, MA & Oxford 2005, 10-28.

Brüggemann 2009
H. Brüggemann, 'Religiöse Bild-Strategien der Romantik. Die ästhetische Landschaft als Andachtsraum und Denkraum', in: *Romantik und Moderne. Moden des Zeitalters und buntscheckige Schreibart; Aufsätze*, Würzburg 2009, 17-64.

Dethloff 1995
U. Dethloff (ed.), *Literarische Landschaft. Naturauffassung und Naturbeschreibung zwischen 1750 und 1830*, St. Ingbert 1995.

Duesberg 1996
P. Duesberg, *Idylle und Freiheit. Ein Entwicklungsmodell der frühromantischen Landschaft in der Wechselwirkung von äußerer und innerer Natur*, Frankfurt a. M., etc. 1996.

Ebbinghaus 1997
H. Ebbinghaus, *Der Landschaftsgarten. Natur und Phantasie in der deutschen Literatur des 18. und frühen 19. Jahrhunderts*, Frankfurt a. M., etc. 1997.

Engell 1981
G. Engell, *The creative imagination. Enlightenment to Romanticism*, Cambridge, MA & London 1981.

Everard 1994
M. Everard, *Ziel en zinnen. Over liefde en lust tussen vrouwen in de tweede helft van de achttiende eeuw*, Groningen 1994.

Frenzel 1984
E. Frenzel, 'Stufen der deutschen Naturlyrik von Brockes bis Eichendorff. Erkenntnis und poetische Erfassung der außermenschlichen Umwelt', in: Th. Wolpers (ed.), *Motive und Themen romantischer Naturdichtung [...]*, Göttingen 1984, 190-200.

Gaier 1989
U. Gaier, 'Garten als inszenierte Natur im 18. Jahrhundert', in: Weber 1989a, 133-158.

Groh & Groh 1989
R. Groh & D. Groh, 'Von den schrecklichen zu den erhabenen Bergen. Zur Entstehung ästhetischer Naturerfahrung', in: Weber 1989a, 53-95.

Hall *et al.* 1988
C. Hall, P. Perreten, J. Shinehouse & D. Visser, 'Paradox in paradise. Nature and art in the eighteenth-century landscape garden', in: D.C. Mell Jr., T.E.D. Braun & L.M. Palmer (eds.), *Man, God, and nature in the Enlightenment*, East Lansing, MI 1988, 107-119.

Holmes 2008
R. Holmes, *The age of wonder. How the romantic generation discovered the beauty and terror of science*, London 2008.

Hooykaas 1981
R. Hooykaas, 'De natuurwetenschap in "de eeuw der genootschappen"', in: H.A.M. Snelders & K. van Berkel (eds.), *Natuurwetenschappen van Renaissance tot Darwin. Thema's uit de wetenschapsgeschiedenis*, Den Haag 1981, 131-167.

Ishihara 2005
A. Ishihara, *Goethes Buch der Natur. Ein Beispiel der Rezeption naturwissenschaftlicher Erkenntnisse und Methoden in der Literatur seiner Zeit*, Würzburg 2005.

Jackson 2008
N. Jackson, *Science and sensation in British romantic poetry*, Cambridge 2008.

Johannes 1992
G.J. Johannes, *Geduchte verbeeldingskracht. Een onderzoek naar het literaire denken over de verbeelding – van Van Alphen tot Verwey*, Amsterdam 1992 (diss. University of Amsterdam).

Johannes 1997
G.J. Johannes, *De lof der aalbessen. Over (Noord-)Nederlandse literatuurtheorie, literatuur en de consequenties van kleinschaligheid 1770-1830*, Den Haag 1997.

Jorink 2006
E. Jorink, *Het Boeck der Nature. Nederlandse geleerden en de wonderen van Gods Schepping 1575-1715*, Leiden 2006.

Jürgens 2004
H. Jürgens, 'Van Gods akker tot spiegel der natuur. Veranderende percepties van de Indiase natuur in berichten van Duitse zendelingen (1750-1810)', *De Achttiende Eeuw* 36 (2004), 67-91.

Kehn 1991
W. Kehn, '"Die Schönheiten der Natur gemeinschaftlich betrachten". Zum Zusammenhang von Freundschaft, ästhetischer Naturerfahrung und "Gartenrevolution"', in: Mauser & Becker-Cantarino 1991, 167-193.

Kehn 1995
W. Kehn, 'Ästhetische Landschaftserfahrung und Landschaftsgestaltung in der Spätaufklärung. Der Beitrag von Christian Cay Lorenz Hirschfelds Gartentheorie', in: H. Wunderlich (ed.), *'Landschaft' und Landschaften im achtzehnten Jahrhundert. Tagung der Deutschen Gesellschaft für die Erforschung des 18. Jahrhunderts, Herzog-August-Bibliothek Wolfenbüttel, 20. bis 23. November 1991*, Heidelberg 1995, 1-23.

Knoeff 2004
R. Knoeff, 'Het begrip "Natuur" in Boerhaaves oraties', *De Achttiende Eeuw* 36 (2004), 112-122.

Köchy 2002
K. Köchy, 'Das ganze der Natur – Alexander von Humboldt und das romantische Forschungsprogramm', *Humboldt im Netz* 3, 5 (2002), 5-16.

Koepke 1990
W. Koepke, 'Epistolary fiction and its impact on readers. Reality and illusion', in: F. Burwick & W. Pape (eds.), *Aesthetic illusion. Theoretical and historical approaches*, Berlin & New York 1990, 263-274.

Kord 1996
S.T. Kord, 'Eternal love or sentimental discourse? Gender dissonance and women's passionate "friendships"', in: A.A. Kuzniar (ed.), *Outing Goethe & his age*, Stanford, CA 1996, 228-249.

Kuhn 2009
B. Kuhn, *Autobiography and natural science in the age of Romanticism. Rousseau, Goethe, Thoreau*, Aldershot & Burlington, VT 2009.

Langen 1948-1949
A. Langen., 'Verbale Dynamik der dichterischen Landschaftsschilderung', in: *Zeitschrift für deutsche Philologie* 70 (1948-1949), 249-318.

Madelein 2007
C. Madelein, 'Kant in de fout? Het verhevene in de Nederlanden', *De Achttiende Eeuw* 39 (2007), 77-93.

Madelein & Pieters 2008
C. Madelein & J. Pieters (eds.), *Bilderdijk, Kinker & Van Hemert. Als van hooger bestemming en aart*, Groningen 2008.

Mainzer 1989
K. Mainzer, 'Von der Naturphilosophie zur Naturwissenschaft. Zum neuzeitlichen Wandel des Naturbegriffs', in: Weber 1989a, 11-31.

Martin 1995
D. Martin, 'Klopstocks *Messias* und die Verinnerlichung der deutschen Epik im 18. Jahrhundert', in: K. Hilliard & K. Kohl (eds.), *Klopstock an der Grenze der Epochen*, mit Klopstock-Bibliographie 1972-1992, Berlin & New York 1995, 97-116.

Mauser & Becker-Cantarino 1991
W. Mauser & B. Becker-Cantarino (eds.), *Frauenfreundschaft – Männerfreundschaft. Literarische Diskurse im 18. Jahrhundert*, Tübingen 1991.

A. Meijer 1998
A. Meijer, *The pure language of the heart. Sentimentalism in the Netherlands 1775-1800*, Amsterdam & Atlanta, GA 1998

M. Meijer 1984
M. Meijer, 'Vrome en geleerde hartsvriendinnen in de achttiende eeuw in Nederland', in: M. Duyves, G. Hekma & P. Koelemij (eds.), *Onder mannen, onder vrouwen. Studies van homosociale emancipatie*, Amsterdam 1984, 167-181.

Meyer-Krentler 1991
E. Meyer-Krentler, 'Freundschaft im 18. Jahrhundert. Zur Einführung in die Forschungsdiskussion', in: Mauser & Becker-Cantarino 1991, 1-22.

Mitchell 1993
T.F. Mitchell, *Art and science in German landscape painting 1770-1840*, Oxford 1993.

Mittelstrass 1989
J. Mittelstrass, 'Der idealistische Naturbegriff', in: Weber 1989a, 159-175.

Oosterholt 1998
J.T.W. Oosterholt, *De ware dichter. De vaderlandse poëticale discussie in de periode 1775-1825*, Amsterdam 1998 (diss. University of Amsterdam).

Ottmann 2002
D. Ottmann, 'Gebändigte Natur. Garten und Wildnis in Goethes *Wahlverwandschaften* und Eichendorffs *Ahnung und Gegenwart*', in: W. Hinderer (ed.), *Goethe und das Zeitalter der Romantik*, Würzburg 2002, 345-395.

Paasman 1974
A.N. Paasman, *Elisabeth Maria Post (1755-1812)*, Amsterdam 1974.

Paasman 1987
B. Paasman, 'Nawoord', in: Post 1987, 218-233.

Pennone 2008
F. Pennone, *Literarische Landschaftsbilder*, Freiburg 2008.

Post 1788
E.M. Post, *Het land, in brieven* (3rd ed.), Amsterdam 1788.

Post 1987
E.M. Post, *Het land, in brieven* (B. Paasman, ed.), Amsterdam 1987.

Pouvreau 2002
R. Pouvreau, *Schöpferische Weltbetrachtung. Zum Verhältnis von Einbildung und Erkenntnis in Texten der deutschen Romantik*, Amsterdam & New York 2002.

Reill 2005
P.H. Reill, *Vitalizing nature in the Enlightenment*, Berkeley & Los Angeles 2005.

Richards 2002
R.J. Richards, *The romantic conception of life. Science and philosophy in the age of Goethe*, Chicago & London 2002.

Riskin 2002
J. Riskin, *Science in the age of sensibility. The sentimental empiricists of the French Enlightenment*, Chicago 2002.

Ritter 1963
J. Ritter, *Landschaft. Zur Funktion des Ästhetischen in der modernen Gesellschaft*, Münster 1963.

Roulston 1998-1999
C. Roulston, 'Separating the inseparables. Female friendship and its discontents in eighteenth-century France', *Eighteenth-Century Studies* 32 (1998-1999), 215-231.

Runge 1997
A. Runge, *Literarische Praxis von Frauen um 1800*, Hildesheim, etc. 1997.

Rusam 1992
A.M. Rusam, *Literarische Landschaft. Naturbeschreibung zwischen Aufklärung und Moderne*, Wilhelmsfeld 1992.

Saine 1976
T.P. Saine, 'Natural science and the ideology of nature in the German Enlightenment', *Lessing Yearbook* 8 (1976), 61-88.

Sauder 1989
G. Sauder, 'Subjektivität und Empfindsamkeit im Roman', in: W. Hinck (ed.), *Sturm und Drang*, Frankfurt a. M. 1989, 163-174.

F. Schneider 2004
F. Schneider, *Im Brennpunkt der Schrift. Die Topographie der deutschen Idylle in Texten des 18. Jahrhunderts*, Würzburg 2004.

H. Schneider 1980
H.J. Schneider, 'Naturerfahrung und Idylle in der deutschen Aufklärung', in: P. Pütz (ed.), *Erforschung der deutschen Aufklärung*, Königstein 1980, 289-315.

H. Schneider 1981
H.J. Schneider, 'Erinnerte Natur. Einleitende Bemerkungen zur literarischen Geschichte der deutschen Landschaft', in: *Deutsche Landschaften*, Frankfurt a. M. 1981, 7-18.

H. Schneider 2005
H.J. Schneider, 'Selbstbescherung. Zur Phänomenologie des Landschaftsblicks in der Empfindsamkeit', in: K. Garber (ed.), *Das Projekt Empfindsamkeit und der Ursprung der Moderne. Richard Alewyns Sentimentalismusforschungen und ihr epochaler Kontext*, München 2005, 129-138.

Sheffy 1998
R. Sheffy, 'How do textual models relate to reality? The formation and status of nature and landscape description in late eighteenth-century German prose-fiction', in: B.F. Scholz (ed.), *Mimesis. Studien zur literarischen Repräsentation / Studies on literary representation*, Tübingen & Basel 1998, 245-257.

Snelders 1970
H.A.M. Snelders, 'Romanticism and Naturphilosophie and the inorganic natural sciences 1797-1840. An introductory survey', *Studies in Romanticism* 9 (1970), 193-215.

Snelders 1994
H.A.M. Snelders, *Wetenschap en intuïtie. Het Duitse romantisch-speculatief natuuronderzoek rond 1800*, Baarn 1994.

Söring 1999
J. Söring, *Rousseauismus. Naturevangelium und Literatur*, Frankfurt a. M., etc. 1999.

Spaemann 1967
R. Spaemann, 'Genetisches zum Naturbegriff des 18. Jahrhunderts', *Archiv für Begriffsgeschichte* 11 (1967), 59-74.

Stiening 2005
G. Stiening, *Epistolare Subjektivität. Das Erzählsystem in Friedrich Hölderlins Briefroman 'Hyperion oder der Eremit in Griechenland'*, Tübingen 2005.

Streng 1998
T. Streng, 'De waardering van het landschap in Nederland in de eerste helft van de negentiende eeuw', in: R.L. Falkenburg *et al.* (eds.), *Nature and Landscape in Netherlandish Art 1500-1850*, Zwolle 1998 (=*Netherlands Yearbook for History of Art* 48 [1997]), 259-317.

Sturkenboom 2004
D. Sturkenboom, *De elektrieke kus. Over vrouwen, fysica en vriendschap in de 18de en 19de eeuw*, Amsterdam & Antwerpen 2004.

Trotha 2004
H. Trotha. 'Der Landschaftsgarten des 18. Jahrhunderts als literarisches Phänomen', *Hölderlin-Jahrbuch* 33 (2004), 13-34.

Van den Berg 1999
W. van den Berg, 'De preromantiekconceptie in de Nederlandse literatuurgeschiedenis', in: *Een bedachtzame beeldenstorm. Beschouwingen over de letterkunde van de achttiende en negentiende eeuw*, Amsterdam 1999, 13-39.

Van de Roemer 2005
G.M. van de Roemer, *De geschikte natuur. Theorieën over natuur en kunst in de verzameling van zeldzaamheden van Simon Schijnvoet (1652-1727)*, Amsterdam 2005 (diss. Amsterdam, UvA).

Van Elferen 2006
I. van Elferen, '"Gij zult op mijn zerk wenen…". Sentimentele, metafysische en performatieve tranen in Elisabeth Maria Posts *Het land*', *De Achttiende Eeuw* 38 (2006), 3-19.

Van Hamel 1915-1916
A.G. van Hamel, 'E.M. Post en Hirschfeld', *Tijdschrift voor Nederlandse Taal- en Letterkunde* 34 (1915-1916), 184-210.

Van Raak 1996
R. van Raak, 'De sneeuwtheologie van Jan Engelsman. Een poging tot Newtoniaanse wijsbegeerte', *Geschiedenis van de wijsbegeerte in Nederland* 7 (1996), 99-116.

Van Zonneveld 1997
P. van Zonneveld, 'Groots, woest of bekoorlijk? Het romantische landschap en de Nederlandse literatuur (1750-1850)', in: W. Loos, R.-J. te Rijdt & M. van Heteren (eds.), *Langs velden en wegen. De*

verbeelding van het landschap in de 18de en 19de eeuw, Blaricum & Amsterdam 1997, 95-106.

Vermij 1991
R.H. Vermij, *Secularisering en natuurwetenschap in de zeventiende en achttiende eeuw. Bernard Nieuwentijt*, Amsterdam 1991.

Wagner-Egelhaaf 2000
M. Wagner-Egelhaaf, 'Unheilbare Phantasie und heillose Vernunft. Johann Georg Zimmermann, *Über die Einsamkeit* (1784/85)', in: A. Assmann & J. Assmann (eds.), *Einsamkeit. Archäologie der literarischen Kommunikation VI*, Munich 2000, 265-279.

Wanning 2005
B. Wanning, *Die Fiktionalität der Natur. Studien zum Naturbegriff in Erzähltexten der Romantik und des Realismus*, Berlin 2005.

Weber 1989a
H.-D. Weber (ed.), *Vom Wandel des neuzeitlichen Naturbegriffs*, Konstanz 1989.

Weber 1989b
H.-D., Weber, 'Die Verzeitlichung der Natur im 18. Jahrhundert', in: Weber 1989a, 97-131.

Wellbery 1984
D.E. Wellbery, 'Die Grenzen des Idyllischen bei Goethe', in: W. Barner, E. Lämmert & N. Oellers (eds.), *Unser Commercium. Goethes und Schillers Literaturpolitik*, Stuttgart 1984, 221-240.

Wells 1966
G. Wells, 'Schiller's view of nature in *Über naïve und sentimentalische Dichtung*', *The Journal of English and Germanic Philology* 65 (1966), 491-510.

Wolf 1999
W. Wolf, 'Nature, landscapes, and the early English novel', in: S. Totosy de Zepetnek, M.V. Dimic & I. Sywensky (eds.), *Comparative literature now. Theories and practice / La litterature comparée a l'heure actuelle. Theories et realisations*, Paris 1999, 625-636.

Zelle 2001
C. Zelle, 'Die Geburt der Natur aus dem Geiste der Rhetorik. Zur Schematisierung von Natur und Genie bei Dennis und Goethe', in: M. Scheffel (ed.), *Erschriebene Natur. Internationale Perspektiven auf Texte des. 18. Jahrhunderts*, Frankfurt a. M., etc. 2001, 145-167.

Zimmermann 1982
J. Zimmermann, 'Zur Geschichte des ästhetischen Naturbegriffs', in: *Das Naturbild des Menschen*, München 1982, 118-154.

Zuidervaart 1999
H.J. Zuidervaart, *Van 'konstgenoten' en hemelse fenomenen. Nederlandse sterrenkunde in de achttiende eeuw*, Rotterdam 1999.

Zuidervaart 2010
H.J. Zuidervaart, 'Science for the public. The translation of popular texts on experimental philosophy into the Dutch language in mid-eighteenth century', in: S. Stockhorst (ed.), *Cultural transfer through translation. The circulation of enlightened thought in Europe by means of translation*, Amsterdam & New York 2010, 231-262.